7th International Conference on Gravitation and Cosmology 2011

Vishwa Mimansa: An Interpretive Exposition of the Universe

AA002941

Journal of Physics: Conference Series Volume 484

Goa, India
14-19 December 2011

Editors:

B. S. Sathyaprakash
Tejinder P. Singh

ISBN: 978-1-63266-705-2
ISSN: 1742-6588

Printed from e-media with permission by:

Curran Associates, Inc.
57 Morehouse Lane
Red Hook, NY 12571

CURRAN ASSOCIATES INC. proceedings.com

Some format issues inherent in the e-media version may also appear in this print version.

Printed by Curran Associates, Inc. (2014)

Institute of Physics
Dirac House, Temple Back
Bristol BS1 6BE UK

Phone: 44 1 17 929 7481
Fax: 44 1 17 920 0979

techtracking@iop.org

Additional copies of this publication are available from:

Curran Associates, Inc.
57 Morehouse Lane
Red Hook, NY 12571 USA
Phone: 845-758-0400
Fax: 845-758-2634
Email: curran@proceedings.com
Web: www.proceedings.com

TABLE OF CONTENTS

PREFACE

CONFERENCE PROGRAMME

PAPERS

KEYNOTE ADDRESS

PLENARY LECTURE

MINI-SESSION: 1 GRAVITY AS AN EMERGENT PHENOMENON

Coordinator: T. Padmanabhan

MINI-SESSION: 2 DARK ENERGY

Coordinator: Varun Sahni

MINI-SESSION: 3 ASTRONOMY WITH A GLOBAL NETWORK OF GRAVITATIONAL WAVE DETECTORS

Coordinator: B.S. Sathyaprakash

WORKSHOP: 1 CLASSICAL GENERAL RELATIVITY AND GRAVITATIONAL WAVES

Coordinators: Sukanta Bose, Sanjay Jhingan

WORKSHOP: 2 COSMOLOGY

Coordinators: E. Komatsu, L. Sriramkumar

WORKSHOP: 3 QUANTUM GRAVITY AND EARLY UNIVERSE

Coordinators: J. David, S. Mukohyama

7th International Conference on
Gravitation and Cosmology
14-19 December 2011, Goa, India
Gμν
ä

Journal of Physics: Conference Series **484** (2014) 011002 doi:10.1088/1742-6596/484/1/011002

VISHWA MIMANSA

An interpretative exposition of the Universe

Proceedings of the
7th International Conference on Gravitation and Cosmology (ICGC)
14th–19th December, 2011 GOA, India

(The conference component of the ICTS Programme:
Frontiers of Cosmology and Gravitation
1st–23rd December, 2011)

The Silver Jubilee Meeting of the ICGC series organized jointly by

International Centre for Theoretical Sciences, TIFR
&
Indian Association for General Relativity and Gravitation (IAGRG)

Publication sponsor:
Sir Dorabji Tata Trust and the Allied Trusts

Editors:
B S Sathyaprakash and Tejinder P Singh

Assistant Editor:
V Chellathurai

Conference Co-sponsors:
Inter-University Centre for Astronomy and Astrophysics, Pune
Association of Friends of Astronomy, Goa
Centre for AstroParticle Physics, SINP, Kolkata
Department of Science, Technology and Environment, Goa
Foundational Questions Institute, USA
Harish-Chandra Research Institute, Allahabad
Infosys Science Foundation, Bangalore
Institute of Mathematical Sciences, Chennai

Preface

The International Conferences on Gravitation and Cosmology (ICGCs) are a four yearly conference series organized in India, under the auspices of the Indian Association for General Relativity and Gravitation (IAGRG). The first ever ICGC was organized in December 1987 in Goa, a pretty coastal state located in western India along the Arabian Sea. The 7th Silver Jubilee ICGC returned to its original venue, and was held during 14th–19th December, 2011. A unique feature of ICGC-2011 was that it was organized as the conference component of the programme, "Frontiers of Cosmology and Gravitation" (FOCG, 1st–23rd December, 2011). FOCG was a programme of the newly formed International Centre for Theoretical Sciences (ICTS) of TIFR. Apart from ICGC-2011, it consisted of a School on Cosmology and Gravitational Waves (1st–11th December, held at IUCAA, Pune), and a workshop on Gravitational Wave Astronomy (20th–23rd December, again held at IUCAA, Pune). The occasion was also utilized to organize the ICTS Public Lectures at TIFR, Mumbai, on 13th December, 2011 by Dick Bond, John Ellis, Jim Peebles, and Kip Thorne. Detailed information about these various activities is available at the conference website http://www.tifr.res.in/ icgc2011 and in the conference brochure.

About 250 scientists attended the ICGC-2011, and about 200 presentations were made. These included keynote addresses, plenary lectures, and workshop presentations (oral and poster). In addition, there were three dedicated mini-sessions on Gravity as an Emergent Phenomenon, Dark Energy, and Gravitational Wave Astronomy. The Journal of Physics Conference Series (JPCS) has kindly agreed to publish the proceedings of ICGC-2011, after subjecting the submitted papers to a peer-review process. The publication of these proceedings was made possible by a generous grant from the Sir Dorabji Tata Trust and Allied Trusts, to whom we are grateful.

We express our sincere thanks to the Assistant Editor V Chellathurai, who handled the painstaking task of proofreading and copyediting the accepted papers. We also thank the SOC, and the Workshop Chairs for their help in putting together an excellent scientific programme, the LOC for the wonderful organization, and the referees of papers for their help in improving the quality of the papers. Finally, we would like to thank all the authors who have made the publication of these proceedings possible, through their contribution.

B S Sathyaprakash and Tejinder P Singh

Scientific Organizing Committee

- Pijushpani Bhattacharjee (SINP, India)
- J Richard Bond (CITA, Canada)
- Manuela Campanelli (RIT, USA)
- Debajyoti Choudhury (Delhi University, India)
- Richard Ellis (Caltech, USA)
- Gary Gibbons (DAMTP, UK)
- Pankaj Joshi (TIFR, India)
- Romesh Kaul (IMSc , India)
- Subhabrata Majumdar (TIFR, India)
- Don Marolf (UCSB, USA)
- Ramesh Narayan (CfA, Harvard University, USA)
- Lyman Page (Princeton University, USA)
- Misao Sasaki (YITP, Japan)
- B S Sathyaprakash (Cardiff University, UK)
- Ashoke Sen (HRI, India)
- Tarun Souradeep (IUCAA, India) [Chair]
- P S Sreekumar (ISRO, India)
- Alexei Starobinsky (Landau Institute, Russia)
- Sumati Surya (RRI, India)
- C S Unnikrishnan (TIFR, India)
- Spenta Wadia (TIFR, India)

Local Organzing Committee

- Ghanashyam Date
- A Gopakumar
- Subhabrata Majumdar
- D Narasimha
- T P Singh (Chair)

Supporting Staff

- V Chellathurai [Scientific Programmes Coordinator]
- Ashok Deshpande [Accounts Officer]
- Margaret D'Souza [Secretary]
- Shobha Jagtap [Secretary]
- Nishikant Kadam [Secretary]
- Vijay Kadam [Assistant]

Journal of Physics: Conference Series **484** (2014) 011003 doi:10.1088/1742-6596/484/1/011003

List of Participants

Name	**Affiliation**
Eric Adelberger	University of Washington, USA
Rana Adhikari	Caltech, USA
Aniket Agrawal	IIT, Delhi, India
Suhail Ahmad	Jamia Millia Islamia, India
Moumita Aich	IUCAA, India
P Ajith	Caltech, USA
Ujjaini Alam	Los Alamos National Laboratory, USA
Amna Ali	Jamia Millia Islamia, India
Bruce Allen	Max Planck Institute, Germany
Gareth Amery	University of KwaZulu-Natal, South Africa
Rizwan ul Haq Ansari	IUCAA, India
Frederico Arroja	Ewha Woman's University, South Korea
Abhay Ashtekar	Penn State University, USA
Kumar Atmjeet	University of Delhi, India
Abhishek Atreya	Institute of Physics, India
Luca Baiotti	Osaka University, Japan
Y Bala Murali Krishna	Indian Science Writers Association, India
Debades Bandyopadhyay	SINP, India
Fernando Barbero	IEM, Spain
Sukratu Barve	CMS, India
Rupal Basak	TIFR, India
Soumen Basak	University of Paris, France
Rudranil Basu	S N Bose Centre, India
Yumnam Bembem Devi	Manipur University, India
Pijushpani Bhattacharjee	SINP, India
Srijit Bhattacharjee	SINP, India
Dipankar Bhattacharya	IUCAA, India
Pinaki Bhattacharya	Jadavpur University, India
Swastik Bhattacharya	HRI, India
J Richard Bond	CITA, Canada
Sukanta Bose	Washington State University, USA
Francois Bouchet	IAP, France
Marco Bruni	University of Portsmouth, UK
Luke Butcher	Kavli Inst. of Cosmology, Cambridge, UK
Rong-Gen Cai	Institute of Theoretical Physics, China
Marie-Noelle Celerier	Meudon, France
Sukanya Chakrabarti	Florida Atlantic University, USA
Manoneeta Chakraborty	TIFR, India
Shuvendu Chakraborty	Seacom Engineering College, India
N Chandrachani Devi	Jamia Millia Islamia, India
Tzu-Ching Chang	CITA, Canada
V Chellathurai	IUCAA, India
Pravabati Chinganbam	IIA, India
S M Chitre	Centre for Basic Sciences, India
Eugenio Coccia	University of Rome, Italy

Marina Cortes	LBNL, USA
Margaret D'Souza	TIFR, India
Carlos Filipe Da S Costa	INPE, Brazil
Naresh Dadhich	IUCAA, India
Mihalis Dafermos	DAMTP, UK
Santanu Das	IUCAA, India
Sudeep Das	UC Berkeley, USA
Sudipta Das	Visva Bharati, India
Suratna Das	TIFR, India
Patrick Das Gupta	University of Delhi, India
Ghanashyam Date	IMSc, India
Justin David	IISc, India
Adam Day	IOP Publishing, UK
Josue De Santiago S	UNAM, Mexico
Ujjal Debnath	BESU, India
Walter Del Pozzo	Nikhef, Netherlands
Ashok Deshpande	TIFR, India
Umananda Dev Goswami	Dibrugarh University, India
Sanjeev Dhurandhar	IUCAA, India
Amol Dighe	TIFR, India
Suresh Doravari	Caltech, USA
Olivier P Dore	JPL, USA
Gustavo Dotti	University Nacional de Cordoba, Argentina
John Ellis	King's College, UK
Stephen Fairhurst	Cardiff University, UK
Brendan Foster	Foundational Questions Institute, USA
Anne Franzen	University of Utrecht, Netherlands
Joshua Frieman	Fermilab, USA
Ryuichi Fujita	RRI, India
Koyel Ganguly	St Xavier's College, Kolkata, India
S H Ghadiali	TIFR, India
Abhik Ghosh	IIT Kharagpur, India
Shaon Ghosh	Washington State University, USA
Suman Ghosh	IISER, Thiruvananthapuram, India
Sushant Ghosh	Jamia Millia Islamia, India
Tuhin Ghosh	IUCAA, India
Jinn-Ouk Gong	CERN, Switzerland
Gabriela Gonzalez	Lousiana State University, USA
A Gopakumar	TIFR, India
Gaurav Goswami	IUCAA, India
Rituparno Goswami	University of Cape Town, South Africa
Gabriel Govender	University of KwaZulu-Natal, South Africa
Julien Grain	University of Paris, France
Samuel Gralla	University of Maryland, USA
Pierre Gratia	University of Chicago, USA
Pathik Guha	Anand Bazar Patrika, India
Sarbari Guha	St Xavier's College, Kolkata, India
Tapomoy Guha Sarkar	HRI, India
Anuradha Gupta	TIFR, India

Gaveshna Gupta	Jamia Millia Islamia, India
Rajalakshmi Gurumurthy	TIFR, India
Yuan Ha	Temple University, USA
Sudan Hansraj	University of KwaZulu-Natal, South Africa
Tomohiro Harada	Rikkyo University, Japan
Dhiraj Kumar Hazra	HRI, India
Suraj Hegde	Bengaluru, India
Steven Hergt	Friedrich-Schiller University, Germany
Gilbert Holder	McGill University, USA
Gary Horowitz	UC Santa Barbara, USA
Ngangbam Ibohal	Manipur University, India
Ningombam Ibotombi Singh	Manipur University, India
Stephane Ilic	Universite Paris-Sud, France
Bala Iyer	RRI, India
Shobha Jagtap	TIFR, India
Deepak Jain	University of Delhi, India
Rajeev Kumar Jain	University of Geneva, Switzerland
Sanjay Jhingan	Jamia Millia Islamia, India
Charles Jose	IUCAA, India
Nidhi Joshi Pant	Jamia Millia Islamia, India
Pankaj S Joshi	TIFR, India
Narayan K	CMI, India
Arun K G	CMI, India
Rakesh Kabir	University of Delhi, India
Nishikant K. Kadam	TIFR, India
Vijay Kadam	TIFR, India
Laishram Kapil	Manipur University, India
Manoj Kaplinghat	UC Irvine, USA
Sayan Kar	IIT Kharagpur, India
Mansi Kasliwal	Caltech, USA
Laxman Katkar	Shivaji University, India
David Keitel	AEI, Germany
Ajit Kembhavi	IUCAA, India
Robert Kirshner	Harvard University, USA
Sergey Klimenko	University of Florida, USA
Sanved Kolekar	IUCAA, India
Eiichiro Komatsu	University of Texas at Austin, USA
Atanu Kumar	SINP, India
Devaky Kunneriath	Astronomical Institute, Czech Republic
Fabien Lacasa	University of Paris-Sud, France
Alok Laddha	Penn State University, USA
Luis Lehner	Perimeter Institute, Canada
Tjonnie Li	Nikhef, Netherlands
Yen-Ting Lin	ASIAA, Taiwan
Eric Linder	UC Berkeley, USA
Kinjalk Lochan	TIFR, India
Sunil Maharaj	University of Kwazulu-Netal, South Africa
Debasish Majumdar	SINP, India
Subhabrata Majumdar	TIFR, India

Suman Majumdar	IIT Kharagpur, India
Daniele Malafarina	TIFR, India
Gautam Mandal	TIFR, India
Felipe Marin	Swinburne University of Tech., Australia
Nairwita Mazumder	Jadavpur University, India
Shiraz Minwalla	TIFR, India
Chandra Kant Mishra	RRI, India
Priti Mishra	TIFR, India
Sanjit Mitra	IUCAA, India
Sourav Mitra	HRI, India
Subhendra Mohanty	PRL, India
Anupreeta More	KICP Chicago, USA
Surhud More	KICP Chicago, USA
Amitabha Mukherjee	University of Delhi, India
Arunava Mukherjee	TIFR, India
Shinji Mukohyama	IPMU, Japan
Marcello Musso	Abdus Salam ICTP, Italy
Sharvari Nadkarni-Ghosh	S N Bose Centre, India
Yasusada Nambu	Nagoya University, Japan
Roshina Nandra	Cambridge, UK
Gaurav Narain	IMSc, India
D Narasimha	TIFR, India
Jayant Narlikar	IUCAA, India
Priyamvada Natarajan	Yale University, USA
Akhilesh Nautiyal	HRI, India
William Nelson	Penn State University, USA
Samaya Nissanke	Caltech, USA
Ernesto Nungesser	AEI, Germany
Anne Marie Nzioki	University of Cape Town, South Africa
Seiju Ohashi	Kyoto University, Japan
T Padmanabhan	IUCAA, India
Isha Pahwa	University of Delhi, India
Archana Pai	IISER Thiruvananthapuram, India
N Panchapakesan	Delhi, India
Maria Allessandra Papa	Max Planck Institute, Germany
Aseem Paranjape	Abdus Salam ICTP, Italy
Rajesh Parwani	National University of Singapore
Kishor Patil	B D College of Engineering, India
Mandar Patil	TIFR, India
P James E Peebles	Princeton University, USA
Valeria Pettorino	SISSA, Italy
Dmitri Pogosyan	University of Alberta, Canada
Anirudh Pradhan	Hindu P G College, India
Jayanti Prasad	IUCAA, India
A R Prasanna	Ahmedabad, India
Reinhard Prix	AEI, Germany
Kangujam Priyokumar Singh	Manipur University, India
Surversperi Suryakumari Rajah	Durban University of Tech., South Africa
Raghavan Rangarajan	PRL, India

Anais Rassat	EPFL, Switzerland
Satyanarayan Ray P M	University of Massachusetts, USA
Sourya Ray	CECS, Chile
Subharthi Ray	University of KwaZulu-Natal, South Africa
Snehal Rebello	Hindustan Times, India
David Reitze	Caltech, USA
Pablo A Rosado	AEI, Germany
Graziano Rossi	KIAS, Korea
Aditya Rotti	IUCAA, India
Carlo Rovelli	Centre de Physique Theorique, France
Dibakar Roychowdhury	S N Bose Centre, India
Anirban Saha	West Bengal State University, India
Varun Sahni	IUCAA, India
Pradyumn Kumar Sahoo	BITS-Pilani, Hyderabad, India
Satyabrata Sahu	TIFR, India
Hiromi Saida	Daido University, Japan
Mairi Sakellariadou	King's College, UK
Prakash Sarkar	IUCAA, India
B S Sathyaprakash	Cardiff University, UK
Seema Satin	IMSc, India
P S Saumia	Institute of Physics, India
Bernard Schutz	AEI, Germany
Edward Seidel	NSF, USA
Anjan Ananda Sen	Jamia Millia Islamia, India
Anand Sengupta	University of Delhi, India
Arman Shafieloo	Ewha Woman's University, Korea
S Shankaranarayanan	IISER Thiruvananthapuram, India
Pankaj Sharan	Jamia Millia Islamia, India
Ranjan Sharma	P D Women's College, India
Masaru Shibata	YITP, Japan
Chandra Prakash Singh	Delhi Technological University, India
Suprit Singh	IUCAA, India
Tejinder P Singh	TIFR, India
Navin Sivanandam	Cape Town, South Africa
Tristan Smith	UC Berkeley, USA
Rafael Sorkin	Perimeter Institute, Canada
Tarun Souradeep	IUCAA, India
L Sriramkumar	IIT Chennai, India
Angom Sumati Devi	Manipur University, India
Sanasam Surendra Singh	Manipur University, India
Sumati Surya	RRI, India
Jacek Tafel	Institute of Theoretical Physics, Poland
Manuel Tessmer	Friedrich-Schiller University, Germany
Itzadah Thongkool	HRI, India
Kip Thorne	Caltech, USA
Rakesh Tibrewala	IMSc, India
Ramesh Tikekar	Pune, India
Tommaso Treu	UC Santa Barbara, USA
Pranjal Trivedi	University of Delhi, India

Sandip Trivedi	TIFR, India
C S Unnikrishnan	TIFR, India
Sanil Unnikrishnan	IUCAA, India
Mauri Valtonen	Tuorla Observatory, Finland
Chris Van Den Broeck	Nikhef, Netherlands
Madhavan Varadarajan	RRI, India
Nijo Varghese	Cochin University of Science and Tech., India
Cenalo Vaz	University of Cincinnati, USA
Filippo Vernizzi	Institut de PhysiqueTheorique, France
Francesca Vidotto	Centre de Physique Theorique, France
Alexander Vikman	CERN, Switzerland
C V Vishveshwara	Bengaluru, India
Ram Gopal Vishwakarma	Zacatecas, Mexico
Salvatore Vitale	Nikhef, Netherlands
Spenta Wadia	TIFR, India
Kameshwar C Wali	Syracuse University, USA
David Wands	University of Portsmouth, UK
Neils Warburton	University of Southampton, UK
John T Whelan	Rochester Institute of Technology, USA
Stanley Whitcomb	Caltech, USA
Moslem Zarei	Isfahan University of Technology, Iran

The ICGC Meetings: 1987-2011

When India became an independent nation in 1947, the presence of Jawaharlal Nehru on the scene, as the Prime Minister was a fortunate circumstance. For, Nehru was a visionary, who believed that the future of the new nation should be guided by science and technology (S & T), and so with like-minded scientists like, Shanti Swarup Bhatnagar, Homi Bhabha, Meghnad Saha, and Daulatsingh Kothari in leading positions, Nehru set about creating an extensive infrastructure for S & T.

The initiatives taken in and around 1947 began to yield dividends in a couple of decades, and by the late 60s and early 70s, major S & T organizations and associated trained human resources were making significant contributions to the country's development, both material and intellectual. Amongst these may be included, the creation of various scientific societies devoted to specific subjects. These came in response to the urge of scientists to meet as a group for discussions in their fields of research.

So, it was in 1969, a number of research workers in the field of general relativity and gravitation (GRG), who had gathered for a workshop in Ahmedabad, resolved to form a society for GRG. The initiative was taken by P C Vaidya, who pointed out the advantages of forming such a recognized group, mentioning that the international community of workers in GRG had already formed such a society. The Indian Association for General Relativity and Gravitation (IAGRG), thus, came up in 1969 with the veteran relativist V V Narlikar, as its first President. Some three years later, the astronomical community responding to a similar urge, formed the Astronomical Society of India (ASI), under the presidency of Vainu Bappu, the Founder Director of the Indian Institute of Astrophysics, Bangalore.

Since the days of discovery of quasars in 1963, there has been an alliance between gravitation theorists and astrophysicists, resulting in the Texas Symposia, which are held once every two years. The International Society of GRG, referred above, also has a world congress (numbered GR-n) once in four years. The ASI and IAGRG have, naturally, many members in common, and over the years, these organizations have grown. Their national meets take place approximately once in two years. It is customary to invite overseas scientists to these meets, if they happen to be visiting India.

In the early 80s, the scientific community in India felt more ambitious, and debated the possibility of hosting a truly international meet in India. Thus, in 1985 the astronomers successfully hosted the General Assembly of the International Astronomical Union. The GRG members, likewise, were pressing for hosting a meeting in the GR-n series. Since the human resources and institutional base of relativists were not as wide and strong as of the astronomers, they were advised to proceed in stages. "First successfully hold a miniature version of GR-n, and once your credibility and capability are established, bid for the big one." This was the advice of some international veterans. That is how the ICGC-series of conferences came about.

The first ICGC

As the name (International Conference on Gravitation and Cosmology) implies, the conference sought to combine the interests of the Texas Symposia as well as of GR-n. The meetings were planned to be once in four years starting in mid-December 1987. (This was the blank year for the Texas Symposium.) To launch them in an impressive manner, the opening meeting was planned to be in the famous tourist resort state of Goa. As IAGRG lacked the infrastructure to organize such a conference, the main organizational responsibility had to be taken up by one of the research institutes; and the biggest amongst these was the Tata Institute of Fundamental Research (TIFR), Mumbai. The TIFR, therefore, was approached, and agreed. As the 'most interested person' from TIFR, I was put in charge of the enterprise. Of course, I had the backing of my senior colleague Kumar (S.M.) Chitre and the younger one, Ajit Kembhavi, and the organizational infrastructure of TIFR. In particular, the then administrator dealing with scientific meetings, Parelkar enthusiastically joined me to set up the local boundary conditions in Goa.

To get our priorities right, we first picked upon Hotel Mandovi in Panjim (Capital of Goa) as the nodal hotel for catering, accommodation, and arrangements of the academic sessions. We were expecting around 100 participants with 50% from abroad. Kamat, the Manager of Hotel Mandovi, promised all help including beds in nearby hotels to make up any shortfall. Regarding the quality of food, for which Hotel Mandovi was justly famous, its in-charge, Satish Prabhu took care of all our worries. Delegates may still recall the various Goan delicacies featured on the conference menus. Parelkar negotiated the basic deal with the hotels to extract the maximum benefits for the conference. But we were still rather vulnerable in lack of the local conditions.

An unexpected helping hand came to make up for our lack of local knowledge. The amateur astronomer organization, called "Friends of Astronomy in Goa", led by their venerable President, Percival Noronha offered to run errands for us and play the role of local volunteer. Noronha had been associated with the state government of Goa, since the days when it was a Portuguese colony. His 'inside knowledge' came in useful, especially when he advised us to request the Chief Minister, Pratapsingh Rane to entertain the conference participants in the state catamaran. Somewhat hesitatingly, I wrote a request letter, and to my delightful surprise, the CM agreed. Indeed, the boat-ride followed by the CM's riverside banquet was the entertainment high point of the conference. The food, Goan drinks, and musical entertainment were well appreciated by the participants, and an uninitiated person might be forgiven for thinking that ICGC stood for "International Conference on Goan Conviviality".

However, it was the academic menu of ICGC-1987 that took precedence over anything else. The Scientific Organizing Committee, consisting of Vishveshwara (Vishu, if your tongue is not up to pronouncing the full name) as the Chairman, and Ashtekar, Barrow, Bertotti, Brill, Carr, Hartle, Hu, Iyer, MacCallum, Narlikar, Novikov, Penrose, and Sciama reflected a nice international mix as well as a mix of active branches of Gravitation and Cosmology.

Later meetings

The original intention of meeting every four years in some part of India was honoured by the community, and accordingly, the ICGC met in Ahmedabad (1991), Pune (1995), Kharagpur (2000), Cochin (2004), and Pune (2007) before reconvening in Goa in 2011. These meetings were more modestly planned, and instead of trying to cover many fields, each meeting was limited to some specialized topics. Thus, the Kharagpur meeting had highlighted gravitational waves, while the Pune (1995) meeting had emphasized the Raychaudhuri equation and the singularity theorems. These conferences did help to put India firmly on the map of GR and Cosmology, to the extent that the GR-15 was held successfully in Pune with Naresh Dadhich from IUCAA playing the main coordinator's role.

It has been a pleasure to see young blood being attracted to these areas of research. Two major developments are on the horizon at the time of writing this piece. The LIGO is planning to set up an extra detector of gravitational waves in India with local help, both in hardware and software. Likewise, steps are being taken for India to join the TMT mega-telescope project as an international partner. These heartening developments help us to bear the loss of old doyens like, V V Narlikar, P C Vaidya, and A K Raychaudhuri.

Though 'small is beautiful' is a truism, the decision was taken to return to Goa with a wider and richer academic menu to recall and celebrate the first ICGC. This is how we all find ourselves in Goa and its pleasant surroundings. As they say: "Your problem is simplified if you choose the right boundary conditions!"

J V Narlikar
Inter-University Centre for Astronomy and Astrophysics (IUCAA)
Post Bag 4, Pune University Campus, Ganeshkhind, Pune 411007, India
E-mail: jvn@iucaa.ernet.in

From jets to cosmos to cosmic censorship

P S Joshi
Tata Institute of Fundamental Research, Homi Bhabha Road, Colaba, Mumbai 400005, India
E-mail: psj@tifr.res.in

1. Introduction

At the outset, I should like to acknowledge that part of the title above, which tries to capture the main flavour of this meeting, and has been borrowed from one of the plenary talks at the conference. When we set out to make the programme for the conference, we thought of beginning with observations on the Universe, but then we certainly wanted to go further and address deeper questions, which were at the very foundations of our inquiry, and understanding on the nature and structure of the Universe. I believe, we succeeded to a good extent, and it is all here for you in the form of these Conference Proceedings, which have been aptly titled as `Vishwa Mimansa', which could be possibly translated as 'Analysis of the Universe'!

It is my great pleasure and privilege to welcome you all to the ICGC-2011 meeting at Goa. The International Conference on Gravitation and Cosmology (ICGC) series of meetings are being organized by the Indian Association for General Relativity and Gravitation (IAGRG), and the first such meeting was planned and conducted in Goa in 1987, with subsequent meetings taking place at a duration of about four years at various locations in India. So, it was thought appropriate to return to Goa to celebrate the 25 years of the ICGC meetings. The recollections from that first meeting have been recorded elsewhere here in these Proceedings.

The research and teaching on gravitation and cosmology was initiated quite early in India, by V V Narlikar at the Banares Hindu University, and by N R Sen in Kolkata in the 1930s. In course of time, this activity grew and gained momentum, and in early 1969, at the felicitation held for the 60 years of V V Narlikar at a conference in Ahmedabad, P C Vaidya proposed the formation of the IAGRG society, with V V Narlikar being the first President. This was how IAGRG was born, and currently the association has about 350 members, both from within India and abroad.

2. The IAGRG

The main goal and activities of the IAGRG have been to connect the Indian researchers in the field of gravitation and cosmology to the latest developments in this area of knowledge, and to the latest international research streams and avenues [1]. Towards this aim, the organization holds the national IAGRG meetings once in about every two years at various locations in India, and also it organizes the international level ICGC meetings taking place every four years, which have been apparently quite popular and useful. Furthermore, specialized meetings, workshops and schools are organized when possible, and when certain interesting developments in the field warrant the same.

An important activity of the IAGRG has been the organization of the Vaidya-Raychaudhuri lectures [2]. This lecture series was established by the association to celebrate and honour the seminal scientific contributions that P C Vaidya and A K Raychaudhuri have made in the fields of gravitation and cosmology, in the form of what are well-known today as the Vaidya metric, and the Raychaudhuri equation. These lectures have been generally held at the time of the biannual national conferences of IAGRG until recently. However, with the pace of rapid developments in the field, and with the need to reach out to a larger student body and the university faculty, these are now being held annually, once at the time of conference, and the next at a suitable college or university, where a wide body of students, faculty and general public also may be able to participate.

All these activities are supposed to enhance and add quality to the research as well as teaching of gravity and cosmology within the country, while aiming at an increasing international participation.

3. The gravity research scenario in India

It may be relevant here to summarize the current scene on gravitation and cosmology research in India, and the main interests and focus areas at the research frontier that various groups are pursuing. During the earlier decades, the Indian relativists had pursued mainly exact solutions to Einstein's equations, and certain related applications, and the scenario has changed quite considerably in past two or three decades, with many new directions emerging in both theoretical and observational or experimental activities. I mention below some of these activities briefly: My aim here is, of course, to be indicative and not to be exhaustive on all that is happening, and the idea being just to provide a few indications on main streams of research.

Cosmology has been a field pursued in India from early decades of the twentieth century, at both mathematical and theoretical levels. As is well-known, the Raychaudhuri equation in cosmology, found and published in 1955, has played an important role in the theoretical developments in the field, leading us to a much deeper understanding on the structure of spacetime, and especially on the occurrence of spacetime singularities in both gravitational collapse of massive stars and cosmology within the framework of the general theory of relativity. Over the years, the Indian researchers reported many useful solutions to the Einstein's equations, which have been useful in relativistic astrophysics as well as cosmology. More recently, with the observations getting much more precise and accurate, there is a considerable interest developed in India in various research groups to examine various topics in physical cosmology, such as the cosmic microwave background research, structure formation in the universe and related simulations, theoretical connections to the Planck and other missions and related issues. At the mathematical and theoretical levels, there have been progress on the causal and topological structures of the universe, and understanding the structure of spacetime singularities in cosmology.

Another major area that has drawn attention in India is the gravitational waves research. This has been connected with the data analysis aspects and techniques mainly, and the related developments, but also there has been much attention in recent years to get actively involved in the gravity waves detection efforts happening at the international level. Such a possibility should provide an active thrust to both theoretical as well as observational and experimental activities in gravity physics related areas in India.

Relativistic astrophysics and theoretical gravitation are other important areas, where much activity has taken place in India in recent decades. In particular, the issue of final fate of a massive star collapsing continually on exhausting its internal nuclear fuel at the end of its life cycle, and the related issue of the cosmic censorship hypothesis have been investigated extensively. These are problems fundamental to the theoretical developments in black hole physics and also its modern astrophysical applications today. The new results obtained over a past couple of decades show that under physically reasonable conditions, both black holes and visible singularities result as collapse end states, the later being super-ultra dense regions in the universe, where important quantum gravity effects may be taking place.

The other major areas being pursued include quantum gravity research. In this case, there is no unique pathway available as yet, and various approaches including the string theory, loop quantum gravity, and other approaches are being pursued.

While exciting new results have been obtained in certain areas of research mentioned above, in some other arenas a beginning is being made. I think, the key point here would be to emphasize that all these fields of research, even if they may look disconnected at times, are deeply inter-connected at another level. It is clear that a much deeper understanding of gravity and the spacetime physics is necessary to make any sustainable and path breaking progress in any of these. The tendencies seen at times to see and consider various fields in a very isolated manner have to be avoided, and that is where broad based meetings such as ICGC could be of great help.

4. Universities versus research institutions

If the research is to prosper and progress well, an important issue that has to be addressed within the Indian context is the dichotomy observed between the so called 'Institutes', where focus is essentially on research only, and the 'Universities', where both teaching and research are supposed to take place simultaneously. In the past decades, especially after the independence, most of the research resources and probably talent also have focused in the Institutes mainly, whereas the Universities have been neglected to a great extent.

An effort is being made now to recover the balance and that needs to be intensified. One possible way out of this dilemma would be to directly encourage the individuals, in terms of funds as well as facilities, who have shown already certain research strengths in terms of good research results and publications, which are or have been taken note of. If such persons are offered a reward or help in terms of research grants directly, with a minimum of bureaucracy involved, that could be a help, and would encourage more such cases, building up a good research pool in due course of time. It will be the most beneficial, of course, if all new research institutes of future will be located within one or the other University campuses. Such a geographical vicinity will provide a great opportunity to bright young students to possibly interact and learn directly with active researchers in various fields.

5. Our future vision and goals

Emerging from the strength of a strong tradition as well as background, the Indian researcher in gravitation and cosmology today will clearly like to make the best of the international opportunities available presently. Within the Indian context, talent has not been a problem any time. The key issue has been nurturing and projecting the same, so that the talent develops and flourishes to make an international impact. With very many organizations and agencies in India working today towards this goal, it is all the more possible that such a vision would become a reality.

A key concern voiced many times is that the science endeavour today has become too much of a big project driven phenomenon. That young individual researcher with bright new ideas finds much less help and support unless they are part of a big project science team. Do we really need big science and management, or is science actually about key intellectual breakthroughs on our understanding of the Universe?

Perhaps both are needed in the prevailing environment today, and I personally think that while big science is frequently and certainly needed for major international projects to come through, not necessarily all have to fall in a project mode.

It is an arduous task, however, to chart out a trajectory and path, which balances the requirements of both the big science as well as individual creativity. That is where probably the real advantage of a community such as the one gathered here lies. The true benefit of such an association in my view would be, we try together so that the stream of scientific creativity continues, develops, and is always nourished.

I am sure the wonderful academic menu just ahead of us will no doubt help us achieve these targets and clarify our vision and the goals ahead. I am very happy to initiate the ICGC-2011, a very special meeting that celebrates completing twenty five years of the ICGC series.

I would like to take this opportunity to warmly thank the colleagues, who have put in a great effort to make this meeting a success. In particular, my warm thanks go to the Scientific Organizing Committee and the Local Organizing Committee. Without their hard work and keen efforts, the meeting in this format would not have been possible. Our special thanks go to the International Center for Theoretical Sciences (ICTS) of TIFR for their keen interest in this meeting, and for the great encouragement and support they have given us through out, without which this meeting would not have been possible.

References

[1] For further details on IAGRG and its activities, please see, http://www.imsc.res.in/~iagrg/
[2] For a complete list of the Vaidya-Raychaudhuri lectures given so far, please see the above website .

ICGC-2011: PROGRAMME

Wednesday, December 14, 2011

Time		
1700 – 1900	**REGISTRATION**	
1800 – 1900	***Refreshments***	
1500 – 1900	Poster Session W1	Classical General Relativity and Gravitational Waves
1800 – 1900	Poster Viewing	

Thursday, December 15, 2011

Time	Speaker	Title
0930 – 0945	**PANKAJ S JOSHI** President, IAGRG	Inaugural Address
0945 – 1045	[Chairperson: Tejinder Singh] **KIP THORNE**	Keynote Address: **Black Hole Research: A New Golden Age**
1045 – 1115	***Coffee Break***	
1115 – 1215	[Chairperson: Tarun Souradeep] **P JAMES E PEEBLES**	Keynote Address: **The Natural Science of Cosmology**
1215 – 1400	***Lunch Break***	
1300 – 1400	Poster Viewing	
1400 – 1530	**Parallel Sessions**	**Hall A.** Mini-Session: 1. Gravity as an Emergent Phenomenon **[T PADMANABHAN]** **Hall B.** W1: Gravitational Waves **Hall C.** W2: Cosmology
1530 – 1600	***Coffee Break***	
1600 – 1730	**Parallel Sessions**	**Hall A.** Mini-Session: 1. Gravity as an Emergent Phenomenon **Hall B.** W1: Gravitational Waves **Hall C.** W2: Cosmology
1730 – 1800	***Break***	
1800 – 1815	**SPENTA R WADIA**	The International Centre for Theoretical Sciences – An Introduction
1815 – 1930	[Chairperson: Ajit Kembhavi] **JAYANT NARLIKAR** **C V VISHVESHWARA**	**The ICGCs: A Retrospective** ICGC-87: As I See Today ICGCs – A Cartoon Chronicle
0930 – 1930	Poster Session W1	Classical General Relativity and Gravitational Waves
1930 – 2130	***Welcome Reception***	Holiday Inn Lawns

Journal of Physics: Conference Series **484** (2014) 011006 doi:10.1088/1742-6596/484/1/011006

Friday, December 16, 2011

Time	Speaker	Title
0900 – 2000	Poster Session W2	Cosmology
0930 – 1100	[Chairperson: S M Chitre] **ROBERT KIRSHNER** **MIHALIS DAFERMOS**	Plenary Lectures **Exploding Stars and the Accelerating Universe** **The Black Hole Stability Problem**
1100 – 1130	***Coffee Break***	
1130 – 1300	[Chairperson: Naresh Dadhich] **ERIC ADELBERGER** **DIPANKAR BHATTACHARYA**	Plenary Lectures **Laboratory Tests of Gravity from Micron to Galactic Scales** **Observational Evidence of Relativity in and Around Compact Stars**
1300 – 1500	***Lunch Break***	
1400 – 1500	Poster Viewing	
1500 – 1630	**Parallel Sessions**	**Hall A.** Mini-Session: 2. Dark Energy **[VARUN SAHNI]** **Hall B.** W1: Classical General Relativity **Hall C.** W3: Early Universe
1630 – 1700	***Coffee Break***	
1700 – 1830	**Parallel Sessions**	**Hall A.** Mini-Session: 2. Dark Energy **Hall B.** W1: Classical General Relativity **Hall C.** W3: Early Universe
1830 – 1900	***Break***	
1900 – 2000	[Chairperson: Ghanashyam Date] **ABHAY ASHTEKAR**	Evening Lecture **The Big Bang and the Quantum**
2000 – 2200	***Conference Banquet***	Holiday Inn Seaside Lawns

Saturday, December 17, 2011

Time	Speaker	Title
0900 – 1430	Poster Session W2	Cosmology
0930 – 1100	[Chairperson: Sanjeev Dhurandhar] **BERNARD SCHUTZ** **J RICHARD BOND**	Plenary Lectures **Gravitational Waves: Astronomy for the 21st Century** **Probing the Cosmic Theory of Early and Late Universe Physics**
1100 – 1130	***Coffee Break***	
1130 – 1300	[Chairperson: A R Prasanna] **LUIS LEHNER**	Plenary Lectures **Membranes and Black Holes: From Jets to**

	SHIRAZ MINWALLA	**Cosmic Censorship** **Fluid Dynamics from Gravity**
1300 – 1430	***Lunch Break***	
1330 – 1430	Poster Viewing	
1430 – 2100	Conference Excursion	Cuncolim Village, and Dinner at Sunset Beach

Sunday, December 18, 2011

Time	Speaker	Title
0900 – 2000	Poster Session W3	Quantum Gravity and Early Universe
0930 – 1100	[Chairperson: Bala Iyer] **STAN WHITCOMB** **FRANCOIS BOUCHET**	Plenary Lectures **Laser Interferometer Gravitational Wave Detectors: Advancing Toward a Global Network** **Cosmic Microwave Background Anisotropies: Status and Prospects**
1100 – 1130	***GROUP PHOTO, Coffee Break***	
1130 – 1300	[Chairperson: Sumati Surya] **RAFAEL SORKIN** **MASARU SHIBATA**	Plenary Lectures **Geometry from Order and Number: Causal Sets** **Coalescence of Binary Neutron Stars and Black Hole-Neutron Star Binaries**
1300 – 1500	***Lunch Break***	
1400 – 1500	Poster Viewing	
1500 – 1630	**Parallel Sessions**	**Hall A.** Mini-Session: 3. Astronomy with a Global Network of Gravitational Wave Detectors **[B S SATHYAPRAKASH]** **Hall B.** W3: Quantum Gravity **Hall C.** W2: Cosmology
1630 – 1700	***Coffee Break***	
1700 – 1830	**Parallel Sessions**	**Hall A.** Mini-Session: 3. Astronomy with a Global Network of Gravitational Wave Detectors **Hall B.** W3: Quantum Gravity **Hall C.** W2: Cosmology

Monday, December 19, 2011

Time	Speaker	Title
0900 – 1500	Poster Session W3	Quantum Gravity and Early Universe
0930 – 1100	[Chairperson: N Panchapakesan]	Plenary Lectures

	PRIYAMVADA NATARAJAN **DAVID WANDS**	**The Formation of the First Black Holes in the Universe** **Primordial Non-Gaussianity from Inflation**
1100 – 1130	***Coffee Break***	
1130 – 1300	[Chairperson: Amol Dighe] **GARY HOROWITZ** **JOHN ELLIS**	Plenary Lectures **Using General Relativity to Study Superconductivity** **Particle Physics and Cosmology in Light of the LHC**
1300 – 1330	Announcement of Poster Prizes, Closing Remarks	
1330 – 1500	***Lunch Break***	
1400 – 1500	Poster Viewing	
1500	End of Conference	

MINI-SESSIONS

1. Gravity as an Emergent Phenomenon

Coordinator: T Padmanabhan

Recent research suggests that gravity could be an emergent phenomenon like elasticity or fluid mechanics. Such an idea is fairly old – possibly dating back to Sakharov – and different researchers have tried to interpret, and implement this idea in different ways. This mini-session will examine the evidence for such a paradigm, some of the approaches, their current status and open issues. It will consist of a few short presentations with some time made available for open discussion.

Schedule of Talks

Thursday, December 15, 2011

Venue: Hall A

Time	Speakers
1400 – 1430	Rong-Gen Cai
1430 – 1500	Shiraz Minwalla
1500 – 1530	T Padmanabhan

1530 – 1600	***Coffee Break***
1600 – 1630	Kip Thorne
1630 – 1700	Rafael Sorkin
1700 – 1730	Group Discussion

2. Dark Energy

Coordinator: Varun Sahni

The Dark Energy session will focus on observational and theoretical aspects of an accelerating cosmology. It will cover future surveys, including DES, non-Einsteinian approaches to cosmic acceleration, and the affect of inhomogeneities on cosmological inferences of dark energy. It will also include an open discussion on various aspects of the accelerating universe and dark energy.

Schedule of Talks

Friday, December 16, 2011

Time: 1500 – 1830

Venue: Hall A

Speakers
Varun Sahni
Joshua Frieman
Eric Linder
Coffee Break
Marie-Noelle Celerier
Open Discussion

3. Astronomy with a Global Network of Gravitational Wave Detectors

Coordinator: B S Sathyaprakash

Gravitational wave detectors have quadrupolar antenna patterns with a wide sky coverage. Even a single detector would be sensitive to a third of the sky, but their ability to resolve a transient source is very poor. A global network of three or more detectors is essential to fully reconstruct the incident radiation and extract the best possible science. Since the current network of two detectors (LIGO) in the US, two in Europe (Virgo and GEO600), and Japan (LCGT) are roughly coplanar and don't fully exploit the global baseline that could, in principle, significantly improve the measurement of the source's parameters. The mini-session will discuss the advantages of extending the global network, in particular, of building a detector in India or Australia, with regard to improvements in the angular resolution of the network, enhanced accuracy in the estimation of the source's distance, polarization of the waves, and the inclination of the binary with respect to the line of sight.

Schedule of Talks

Sunday, December 18, 2011

Venue: Hall A

Time	Speakers
1500 – 1530	Chris Van Den Broeck
1530 – 1600	Stephen Fairhurst
1600 – 1630	Sergey Klimenko
1630 – 1700	***Coffee Break***
1700 – 1830	Round Table Discussion

W1: Classical General Relativity and Gravitational Waves

Topic: Gravitational Waves

Thursday, December 15, 2011

Venue: Hall B

Time	Speaker	Title
Topic: Gravitational Waves		
1400 – 1412	Bruce Allen	The Einstein@Home search for new neutron stars
1412 – 1424	Maria Alessandra Papa	Deep searches for continuous gravitational wave signals
1424 – 1436	John Whelan	Directed searches for continuous gravitational waves
1436 – 1448	C S Unnikrishnan	The design and characteristics of the TIFR gravitational wave prototype detector
1448 – 1500	Luca Baiotti	Gravitational waves from simulations of binary neutron stars
1500 – 1512	Luke Butcher	Localizing the energy and momentum of linear gravity
1512 – 1524	Chandra Kant Mishra	Nonspinning inspiralling compact binaries in quasi circular orbits: 2.5 PN linear momentum loss and associated recoil
1524 – 1536	Mansi Kasliwal	The transient electromagnetic sky
1536 – 1600	***Coffee Break***	
1600 – 1612	Samaya Nissanke*	Localizing compact binary inspirals on the sky using ground-based gravitational wave interferometers
1612 – 1624	Steven Hergt	On the comparison of results regarding the post-Newtonian approximate treatment of the dynamics of extended spinning compact binaries
1624 – 1636	P Ajith	Addressing the spin question in gravitational-wave searchers: Spin distribution of compact binary populations
1636 – 1648	Satyanarayan Ray Pitambar Mohapatra	Detectability of binary black hole merger signals in ground based gravitational wave detectors
1648 – 1700	Sanjit Mitra	Probing a stochastic gravitational wave background using a network of laser interferometric detectors
1700 – 1712	Arun K G	Possible bounds on vector modes of generic metric theories of gravity using gravitational wave observations
1712 – 1724	Tjonnie Li	Coalescing binary neutron stars and black holes as laboratories for testing general relativity

* This talk will be given by Mansi Kasliwal

W1: Classical General Relativity and Gravitational Waves

Topic: Classical General Relativity

Friday, December 16, 2011

Venue: Hall B

Time	Speaker	Title
Topic: Classical General Relativity		
1500 – 1515	Tomohiro Harada	High-velocity collision of particles around a rapidly rotating black hole
1515 – 1530	Mauri Valtonen	Black hole binary OJ287 as a testing platform for general relativity
1530 – 1545	Samuel Gralla	Why do spinning black hole binaries bob and kick?
1545 – 1600	Gustavo Dotti	Instabilities in super extreme Kerr spacetime and black hole interiors
1600 – 1615	Mandar Patil	Naked singularities as particle accelerators
1615 – 1630	Jacek Tafel	Static, spherically symmetric black holes with scalar field
1630 – 1700	***Coffee Break***	
1700 – 1715	Sayan Kar	Confinement and focusing of geodesics in warped spacetimes
1715 – 1730	Ernesto Nungesser	Late-time asymptotics of some Bianchi A spacetimes with collisionless matter
1730 – 1745	Sunil Maharaj	New models with heat flow and Lie symmetries
1745 – 1800	Sushant Ghosh	Inhomogeneous dust collapse in 5D Einstein-Gauss-Bonnet gravity
1800 – 1815	Seiju Ohashi	Spherical collapse of inhomogeneous dust cloud in the Lovelock theory
1815 – 1830	Daniele Malafarina	Static configurations from gravitational collapse

W1: Classical General Relativity and Gravitational Waves

POSTERS (35)

December 14 – 15, 2011

No.	AUTHOR	TITLE
01	Gareth Amery	Exact solutions for isometric embeddings of pseudo-Riemannian manifolds
02	Rudranil Basu	Local symmetries of non-expanding horizons
03	Carlos Filipe Da Silva Costa	Spherical gravitational wave detectors: Mario Schenberg and Mini GRAIL
04	Patrick Das Gupta	Gravitational waves in the presence of a dynamical four-form
05	Suresh Doravari	Third harmonic resonant side band extraction and interferometer control at the Caltech 40 m prototype

06	Anne Franzen	The interior of Kerr black holes
07	Ryuichi Fujita	Gravitational waves from extreme mass ratio inspirals to the 14th post-Newtonian order
08	Shaon Ghosh	Searching for short duration gamma ray bursts in ligo with large sky position uncertainties
09	A Gopakumar	Gravitational wave phasing for spinning compact binaries
10	Gabriel Govender	Generalised junction conditions for radiating star
11	Sarbari Guha	Gravity in five-dimensional warped product space-times with time-dependent warp factor
12	Anuradha Gupta	Compact binaries in hyperbolic orbit encounters
13	Rajalakshmi Gurumurthy	Short range force measurement employing prototypes of interferometric gravity wave detectors
14	Sudan Hansraj	Perfect fluid spacetimes conformally related to the Schwarzschild exterior metric
15	Suraj Hegde	Analytical theory for optical black hole analogs
16	Sanjay Jhingan	Depletion of energy from naked singular regions during gravitational collapse
17	Laishram Kapil	Hawking radiation from Reissner-Nordstrom-Vaidya black hole by using Hamilton-Jacobi method
18	David Keitel	An F-statistic based multi-detector veto for detector artifacts in searches for continuous gravitational waves
19	Priti Mishra	Quadrupolar gravitational back-reaction and flat galaxy rotation curves
20	Arunava Mukherjee	Recent observations of extremely strong gravity phenomena and its interesting consequences
21	Archana Pai	Tests of post-Newtonian theory using singular value decomposition in parameter estimation
22	Reinhard Prix	Bayesian detection methods for continuous gravitational waves
23	Sourya Ray	Birkhoff's theorem in higher derivative theories of gravity
24	Pablo Antonio Rosado Gonzalez	Gravitational wave background from binary system
25	Aditya Rotti	A new window into the stochastic gravitational wave background
26	Anand Sengupta	Cosmology with the network of advanced gravitational wave detectors
27	Pankaj Sharan	Variational principle in the extended phase space formalism
28	Ranjan Sharma	Spacetime inhomogeneity and gravitational collapse
29	Manuel Tessmer	Full-analytic time Fourier-domain gravitational waveforms from inspiralling eccentric compact binaries through 2PN
30	Ramesh Tikekar	Spherical fluid collapse with radiation on in-

		homogeneous spacetime background
31	Nijo Varghese	Behaviour of massive scalar field around a black hole in Horava gravity
32	Ram Gopal Vishwakarma	Pressure in the relativistic representation of matter
33	Salvatore Vitale	Effect of detectors calibration errors on parameter estimation of gravitational waves from inspiral compact binaries
34	Niels Warburton	A fast frequency-domain algorithm for gravitational self-force: Eccentric orbits in Schwarzschild spacetime
35	John Whelan	Treatment of calibration uncertainty in multi-baseline cross-correlation searches for gravitational waves

W2: Cosmology

Schedule of Talks

Thursday, December 15, 2011

Venue: Hall C

Time	Speaker	Title
Topic: Dark Matter (30 minutes)		
1400 – 1420	Manoj Kaplinghat	Fermi-LAT constraints on WIMP dark matter
1420 – 1430	Sukanya Chakrabarty	A new probe of the distribution of dark matter in galaxies
Topic: Cluster of Galaxies (50 minutes)		
1430 – 1450	Yen-Ting Lin	A WISE-Chandra view of baryon content evolution in galaxy clusters
1450 – 1510	Aseem Paranjape	The extreme tail of the non-Gaussian mass function
1510 – 1520	Graziano Rossi	Halo shapes in the cosmic web
1520 – 1550	***Coffee break***	
Topic: Large Scale Structure (50 minutes)		
1550 – 1610	Marco Bruni	Relativistic effects and large scale structure: Post-Newtonian dynamics
1610 – 1620	Felipe Marin	High-order clustering of WiggleZ galaxies
1620 – 1630	Surhud More	Galaxy-dark matter connection: A cosmological perspective
1630 – 1640	Charles Jose	Weighing neutrinos using high redshift galaxy luminosity functions
Topic: Dark Energy (50 minutes)		
1640 – 1700	Filippo Vernizzi	Structure formation with clustering quintessence
1700 – 1715	Anjan Ananda Sen	Deviation from ΛCDM: Pressure parametrization

Journal of Physics: Conference Series **484** (2014) 011006 doi:10.1088/1742-6596/484/1/011006

1715 – 1730	Valeria Pettorino	Interacting dark energy: Impact on CMB and structure formation

W2: Cosmology

Schedule of Talks

Sunday, December 18, 2011

Venue: Hall C

Time	Speaker	Title
Topic: Gravitational Lensing (70 minutes)		
1500 – 1520	Sudeep Das	A new view of the cosmic microwave background with ACT
1520 – 1540	Gil Holder	Gravitational lensing of the CMB and the South Pole Telescope
1540 – 1600	Tommaso Treu	Cosmology from gravitational time delays
1600 – 1620	Anupreeta More	The SL2S arcs sample
1620 – 1630	Anne Marie Nzioki	A geometrical approach to strong gravitational lensing in f(R) gravity
1630 – 1700	***Coffee break***	
Topic: Cosmic Microwave Background (90 minutes)		
1700 – 1710	Stephane Ilic	CMB/CIB cross-correlation for ISW detection
1710 – 1720	Marina Cortes	On the prior dependence of constraints on tensor-to-scalar ratio
1720 – 1730	Tristan Smith	Constraints on neutrino interaction using cosmological observations
1730 – 1740	P S Saumia	Probing the anisotropic expansion history of the universe with cosmic microwave background
1740 – 1750	Nidhi Pant	Statistics of statistical anisotropy measures
1750 – 1800	Pravabati Chinganbam	Search for non-Gaussianity in the CMB with geometrical and topological quantities
1800 – 1810	Dmitri Pogosyan	Geometrical statistics of non-Gaussian cosmological fields
1810 – 1820	Fabien Lacasa	Non-Gaussianity of point-sources foreground
1820 – 1830	Pranjal Trivedi	CMB non-Gaussianity sourced by primordial magnetic fields

W2: Cosmology

POSTERS (28)

December 16 – 17, 2011

No.	AUTHOR	TITLE
01	Moumita Aich	Evolution of primordial statistical isotropy violation
02	Ujjaini Alam	Non-parametric reconstruction of dark energy equation of state from diverse datasets
03	Rizwan ul Haq Ansari	Perturbations in dark energy models with evolving speed of sound
04	Soumen Basak	A needlet ILC analysis of WMAP 7-year temperature data
05	Ningombam Chandrachani Devi	Evolution of spherical over density in thawing dark energy models
06	Santanu Das	Leakage of power from dipole to higher multipoles due to non-circular WMAP beam
07	Walter Del Pozzo	Precision measurements of the Hubble constant with gravitational waves
08	Abhik Ghosh	Characterizing the diffuse foregrounds for redshifted 21-cm HI signal: GMRT 150 MHz observation
09	Tuhin Ghosh	Characterisation of CMB foregrounds
10	Rituparno Goswami	On the shear-free perturbations of FLRW universe
11	Tapomoy Guha Sarkar	The cross-correlation of 21-cm signal and the Lyman-alpha forest: A cosmological probe
12	Deepak Jain	Observational cosmology and the cosmic distance duality relation
13	Fabien Lacasa	Non-Gaussianity of point-sources foregrounds
14	Suman Majumdar	Matched filter detection of ionized bubbles in simulated redshifted 21-cm maps of epoch of reionization
15	Sanjit Mitra	Effect of asymmetric beams in CMB experiments: Importance and remedy
16	Sourav Mitra	Joint QSO – CMB constraints on reionization history
17	Sharvari Nadkarni-Ghosh	Extending the domain of validity of Lagrangian perturbation theory
18	Roshina Nandra	The interplay between cosmological expansion and massive objects
19	Anne Marie Nzioki	A geometrical approach to strong gravitational lensing in f(R) gravity
20	Isha Pahwa	Higher dimensional cosmological models - An alternative explanation for late time cosmic acceleration
21	Jayanti Prasad	Cosmological parameter estimation using Particle

		Swarm Optimization
22	Anais Rassat	New statistical methods for 3D analysis of cosmological surveys
23	Prakash Sarkar	The luminosity, colour and morphology dependence of galaxy structures in the Sloan Digital Sky Survey
24	Arman Shafieloo	The crossing statistic: Dealing with unknown errors in the dispersion of type Ia supernovae
25	Ismael Tereno	COSMOS weak-lensing constraints on modified gravity
26	Itzadah Thongkool	How delicate are the f(R) gravity models with disappearing cosmological constant?
27	Pranjal Trivedi	CMB non-Gaussianity sourced by primordial magnetic fields
28	Sanil Unnikrishnan	Dark energy versus f(R) theories of modified gravity

W3: Quantum Gravity and Early Universe

Schedule of Talks

Topic: Early Universe

Friday, December 16, 2011

Venue: Hall C

Time	Speaker	Title
Topic: Early Universe		
1500 – 1510	Marcello Musso	Galilean symmetry in the effective theory of inflation: New shapes of non-Gaussianity
1510 – 1520	Jinn-Ouk Gong	Multi-field inflation: Formulation, effective theory and phenomenology
1520 – 1530	Frederico Arroja	Large and strong scale dependent bispectrum in single field inflation from a sharp feature in the mass
1530 – 1540	Gaurav Goswami	Features in scalar primordial power spectrum and their consequences
1540 – 1550	Suratna Das	Effects of radiation on primordial non-Gaussianity
1550 – 1600	Dhiraj Kumar Hazra	On the discriminating power of f_{nl}
1600 – 1610	Raghavan Rangarajan	Non-Gaussian fluctuations of the inflaton and constancy of correlations of zeta outside the horizon
1610 – 1620	Yasusada Nambu	Classical and quantum correlations of scalar field in the inflationary universe
1620 – 1650	***Coffee Break***	
1650 – 1700	Mairi Sakellariadou	Non-commutative spectral geometry

1700 – 1710	Akhilesh Nautiyal	Warm natural inflation
1710 – 1720	Abhishek Atreya	Baryon inhomogeneities in early universe due to CP violating quark scattering from QCD Z (3) domains
1720 – 1730	Rajeev Kumar Jain	Can slow roll inflation induce relevant helical magnetic fields?
1730 – 1740	Moslem Zarei	Gauge vector fields at the preheating stage
1740 – 1750	Rajesh Parwani	Nonlinear quantum cosmology: Nonperturbative effects
1750 – 1800	Josue De Santiago Sanabria	Generalizing a unified model of dark matter, dark energy, and inflation with a noncanonical kinetic term
1800 – 1810	Amitabha Mukherjee	Scalar field theory with a time-dependent potential and its application to cosmology
1810 – 1820	Dibakar Roychowdhury	Thermodynamics of Hawking-Page phase transition in AdS black holes

W3: Quantum Gravity and Early Universe

Schedule of Talks

Topic: Quantum Gravity

Sunday, December 18, 2011

Venue: Hall B

Time	Speaker	Title
Topic: Quantum Gravity		
1500 – 1515	Carlo Rovelli	Loop quantum gravity
1515 – 1530	Fernando Barbero	Detailed black hole state counting in loop quantum gravity
1530 – 1540	Sumati Surya	Markov chains with causal sets
1540 – 1550	Rakesh Tibrewala	Modified constraint algebra in LQG and spacetime interpretation?
1550 – 1600	Madhavan Varadarajan	Towards a consistent quantum dynamics for canonical loop quantum gravity
1600 – 1610	Francesca Vidotto	Spinfoam cosmology
1610 – 1620	Julien Grain	The primordial universe as a potential probe of loop quantum cosmology
1620 – 1630	William Nelson	Inhomogeneities in loop quantum cosmology

1630 – 1700	***Coffee Break***	
1700 – 1710	Sanved Kolekar	Action for gravity: An emergent perspective
1710 – 1720	Alexander Vikman	Non-canonical scalar field theories with kinetic gravity braiding
1720 – 1730	Cenalo Vaz	Quantum gravitational collapse
1730 – 1740	K Narayan	Cosmological singularities in string theory
1740 – 1750	Srijit Bhattacharjee	Gauge invariant coupling of fields to torsion: A string inspired model
1750 – 1800	Hiromi Saida	Universal property of quantum gravity implied by uniqueness theorem of Bekenstein-Hawking entropy
1800 – 1810	S Shankaranarayanan	Entanglement entropy in all dimensions
1810 – 1820	Anirban Saha	Upper-bound of the time-space noncommutative parameter from gravitational quantum well experiment

W3: Quantum Gravity and Early Universe

POSTERS (14)

December 18 – 19, 2011

No.	AUTHOR	TITLE
01	Suhail Ahmad	Sub-leading corrections to black hole entropy formula in black holes (convergences and divergences)
02	Kumar Atmjeet	Cosmological magnetogenesis
03	Shuvendu Chakraborty	Higher dimensional cosmology with normal scalar field and tachyonic field
04	Ujjal Debnath	Thermodynamics in Vaidya spacetime
05	Archisman Ghosh	Singularity resolution in cosmologies using AdS/CFT
06	Suman Ghosh	Higher-order field derivatives and deviations from entropy-area law
07	Yuan K Ha	An underlying theory for gravity
08	Kinjalk Lochan	Extension of trace dynamics to spacetime
09	Azadeh Maleknejad	Non-Abelian gauge field inflation
10	Gaurav Narain	Analogue of Wilson-Fisher fixed point in quantum gravity
11	Pradyumn Kumar Sahoo	Plane symmetric domain walls and cosmic strings in bimetric theory

12	Chandra Prakash Singh	Viscous FRW model with particle creation in the early universe
13	Navin Sivanandam	Adiabaticity and non-Gaussianity
14	Sanasam Surendra Singh	Particle creation in cosmology with varying speed of light

Peer review statement

All papers published in this volume of *Journal of Physics: Conference Series* have been peer reviewed through processes administered by the proceedings Editors. Reviews were conducted by expert referees to the professional and scientific standards expected of a proceedings journal published by IOP Publishing.

The natural science of cosmology

P J E Peebles
Joseph Henry Laboratories, Princeton University, Princeton NJ 08544, USA
E-mail: PJEP@Princeton.edu

Abstract. The network of cosmological tests is tight enough now to show that the relativistic Big Bang cosmology is a good approximation to what happened as the universe expanded and cooled through light element production and evolved to the present. We have explained the reasons to reach this conclusion, commented on the varieties of philosophies informing searches for a still better cosmology, and offered an example for further study, the curious tendency of some classes of galaxies to behave as island universes.

1. Introduction

Cosmology aims to draw large conclusions from exceedingly limited evidence, but we have to conclude, to our surprise, that the considerations reviewed in Section 2 make a convincing case that the relativistic Friedman-Lemaître expanding world model is a good approximation. Cosmology is a work in progress, of course, and many lines of research are aimed at establishing a still better theory. We have offered in Section 3, thoughts on directions this research has taken and likely will take. One direction is the search for anomalies, where the reconciliation of theory and observation seems particularly difficult. A good anomaly will teach us something, either a better understanding of the standard model or clues to an improved one. In Section 4, we have given an example, for the apparent behaviour of pure disk and elliptical galaxies as island universes.

2. The cosmological tests

The standard ΛCDM cosmology assumes the general theory of relativity. This is an extrapolation of some 14 orders of magnitude in length scale from the precision tests on the scales of the Solar System and smaller. It assumes that 95% of the present mass of the universe is in two hypothetical forms, dark matter and dark energy. And it assumes initial conditions that agree with simple versions of the inflation picture of the very early universe, which is encouraging, but inflation is an idea, not a theory. Substantiating this considerable variety of bold assumptions is a demanding challenge. But the considerations reviewed here, lead us to conclude that the challenge has been met, largely.

Detecting the relativistic curvature of the redshift-magnitude relation [1, 2] is a rightly celebrated contribution to the cosmological tests. But one gets some feeling for the challenge of substantiating cosmology by imagining the situation in an alternative universe, in which the only significant cosmological test is the redshift-magnitude relation. That would include the demonstration of a close to linear relation between redshift and distance at low redshift, which is in line with Einstein's argument for homogeneity. We have expected the demonstration of the departure from linearity at high redshift by the SNe Ia measurements would split the cosmology community in this alternative universe into three camps. Those who already favoured the Steady State cosmology on philosophical grounds would point out that the curvature is in line with the Steady State prediction, and would rightly claim strong support for this cosmology. Among those who had a philosophical preference for the relativistic Friedman-Lemaître cosmology one camp would add Einstein's cosmological constant Λ to their cosmology. The other camp would point out that the indicated value of Λ is absurdly different from any reasonable estimate from well-established quantum physics. They would point out that in the Friedman-Lemaître cosmology, the universe at $z > 1$ was different from now, and SNe Ia then, therefore, had to have been at least somewhat different from present-day ones. They would acknowledge that carefully explored tests have given no indication of evolution of the properties of SNe Ia, but still insist that supernovae cannot be examined in close detail, and there is no way to exclude the possibility that some subtle evolution has escaped the tests. Each of these three camps has a reasonable case.

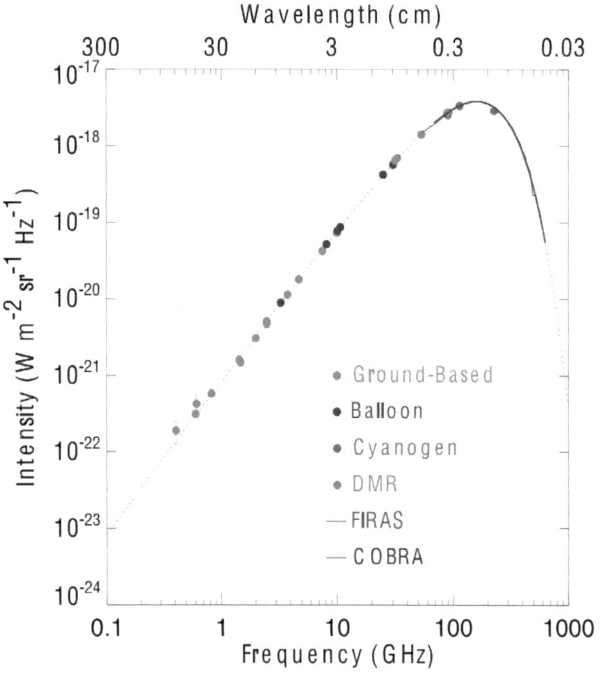

Figure 1. Measurements of the CMB spectrum, compiled by Alan Kogut. The dotted curve is a thermal Planck spectrum.

The point is that establishing cosmology requires a network of tests that check for systematic errors, both in the measurements and in the theory used to interpret the measurements. For example, suppose cosmologists in the alternate universe had, in addition to the redshift-magnitude relation, measurements of the cosmic microwave background radiation (the CMB), which show that the CMB is isotropic to better than a part in 10^5 and that the spectrum is very close to thermal, as shown in Figure 1 (compiled and plotted by Alan Kogut [3], NASA GSFC). The only reasonable way to reconcile this measured thermal spectrum with the classical Steady State cosmology is to postulate that the universe is optically thick to the absorption and emission of microwave radiation at very small redshifts, for otherwise the CMB spectrum would be an unacceptable mix of temperatures set by the distribution of redshifts summed out past unit optical depth. Since people in the alternate universe could measure the CMB, they would have detected discrete sources at CMB wavelengths. If they had adequate optical telescopes, they would see that some were distant radio galaxies, meaning the universe was optically thin. This serious challenge to the classical Steady State cosmology is an example of a contribution to a network of tests which, if dense enough, may establish cosmology.

Returning to our universe, let us note that no natural science is complete, and that certainly includes the established relativistic cosmology. For example, we have a few constraints on what happened in the very early universe. One can save the Steady State philosophy by postulating that our universe behaves in a Quasi-Steady State manner [4], where the continual creation of the classical version is replaced by periodic bursts of creation at densities large enough to be capable of thermalizing the radiation. If the bursts of creation built up an appropriate distribution of matter and radiation at densities greater than that of the Big Bang at light element nucleosynthesis, and were followed by the expansion described by the relativistic model, then QSS would pass all cosmological tests, and merit a place with variants of inflation in models for the very early universe that were viable but need work.

Because the value of Einstein's Λ in the standard Big Bang cosmology seems so unlikely within quantum physics, it is natural and important to consider whether the evidence for detection of Λ results from systematic error in the application of the relativistic cosmology. An example is the proposal that the strongly inhomogeneous distribution of mass on scales less than a few tens of megaparsecs may significantly affect the general rate of expansion of the universe [5]. We believe, Siegel and Fry [6] have first answered this question. One may start with the observation that deflection angles $\delta\theta$, in gravitational lensing are small, $\delta\theta \sim G\delta M/rc^2 \lesssim 10^{-5}$ around a mass concentration δM of radius r. This angular deflection is a measure of the departure from the homogeneous Friedman-Lemaître geometry $|\delta g_{\mu\nu}| \lesssim 10^{-5} g_{\mu\nu}$. (Black holes are exceptions but irrelevant, because at any distance of interest for this discussion they behave as ordinary particles.) One is invited to compute the spacetime geometry to second order in $\delta g_{\mu\nu}$ sourced by the mass auto-correlation function, which is reliably measured on the scales of significantly non-linear mass clustering. The result (which requires a modest extension of the calculation in [6]) is that non-linear mass fluctuations modelled as a gas of particles with mass m and mean number density $\bar{n}$ appear in the Friedman equations as an effective contribution to the mean mass density $\rho_{\rm eff} = m\bar{n}(1+K+W)$, and to the pressure $p_{\rm eff} = m\bar{n}(2K+W)/3$, where $K = \langle v^2\rangle/2$ is the kinetic energy of peculiar motion per unit mass and $W = -\frac{1}{2}Gm\bar{n}a^2 \int d^3x \langle\delta(\mathbf{r})\delta(\mathbf{r}+\mathbf{x})\rangle/\mathbf{x}$ is the gravitational energy per unit mass belonging to the departure from homogeneity. The effective pressure $p_{\rm eff}$ can be negative as for Λ, but it is tiny, five orders of magnitude smaller than the standard estimate of Λ. This is not a way out of Λ or dark energy.

Another proposed way out supposes that Λ is small or zero, and that the mass density and expansion rate vary with distance from us in such a way as to make the angular size distance vary with redshift in the same way as in a homogeneous universe with the standard value of Λ. We would have to be very near the centre of an accurately spherically symmetric universe, for otherwise the Sachs-Wolfe effect would produce an unacceptably large CMB anisotropy. Our favoured position is arguably unlikely, but more important is that this model is challenged by other cosmological tests. Intergalactic plasma scatters CMB photons, meaning the spectrum in Figure 1 is a mix of spectra originating at a range of distances. If the temperature at decoupling were not very close to homogeneous, the CMB spectrum would be a mix of thermal spectra at different temperatures, contrary to the measurements. In the standard cosmology, acoustic waves in the baryons and radiation prior to decoupling produced waves in the CMB anisotropy power spectrum, and in the power spectrum of the large-scale galaxy distribution. Analyses [7, 8] pointing to consistency of the measurements of the CMB and galaxy distributions depend on, and test, many elements of the cosmology, but our point is that measurements of the angular distribution of the radiation probe what happened near the edge of the observable universe, and measurements of the spatial distribution of the galaxies probe what happened much closer to us. The consistency of these measurements within the homogeneous ΛCDM cosmology argues that conditions actually were quite similar across the sky at these two quite different ranges of distance, or else that there was yet another conspiracy.

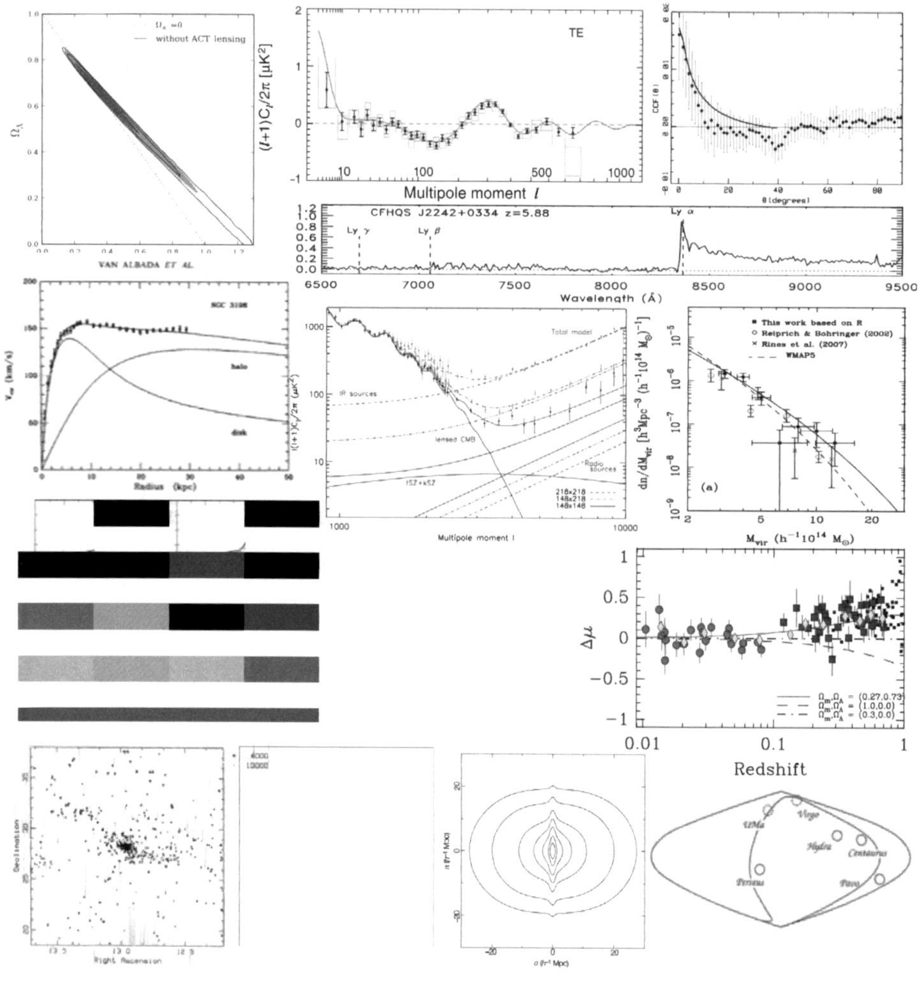

Figure 2. Figures taken from papers on cosmological tests, for the purpose of illustrating the broad variety of phenomena, and means of measurement and analysis in the network of tests.

This discussion could continue at length, but we, instead, refer to Figure 2 for a schematic illustration of the broad variety of phenomena and means of measurement employed in the cosmological tests. The consistency of results from these many and varied approaches convincingly argues against the possibility that we have been seriously misled by systematic errors in the observations or errors in the theory by which the observations are interpreted.

We hope it is understood that we are not arguing that cosmology after baryogenesis is complete: that is, contrary to the experience that natural sciences advance by successive approximations. But we can be sure that if there is a better cosmology, it will predict a universe that looks much like ΛCDM, because the tests look at the universe from many sides now, and find that it looks much like ΛCDM.

3. The future

We have got some sense of the future of this subject by recalling how we got to where we are now. We have seen several paths common to cosmology and other parts of natural science.

3.1. Serendipitous discovery of suggestive phenomena

In relativity physics, one might think of the apparently curious behaviour of the velocity of light. In cosmology, one thinks of the linear redshift-distance relation, which by 1936 Hubble and Humason [9] had tested out to $z = 0.1$. This relation was the centrepiece of most discussions of cosmology until the CMB, in 1965. Serendipity strikes seldom, but let us remember that it happens.

3.2. The philosophical appeal of ideas

Einstein followed this path to wonderfully good effect in his discovery of general theory of relativity, and his proposal that a philosophically satisfactory universe is homogeneous in the large-scale average. He had no justification for homogeneity from phenomenology, and only a few hints to GR: Electromagnetism and the Eötvös experiment. It is striking that Nature on occasion agrees with our ideas of elegance. A contrary example is the fate of the classical Steady State cosmology, though the idea arguably reappears in eternal inflation. We have included as modern contenders cosmic strings, superstring cosmology, the multiverse and the anthropic principle.

3.3. Mathematical incompleteness

We take it that Nature abhors the singularities in general relativity. The community favourite remedy for the singularity in ΛCDM at infinite redshift is a variant of inflation, which needs work.

3.4. Testing ideas

The central example in cosmology is the programme of cosmological tests that has succeeded remarkably well, after some re-negotiation of ideas about Λ. An influential idea driving research now is that there may be a signature of inflation in the polarization of the CMB. A successful detection would have a wonderfully strong effect on the status of inflation. Laboratory searches may detect dark matter particles that behave much like the hypothetical cold dark matter in ΛCDM. If this detected matter along with massive neutrinos saturates the mass in matter in the dark sector, this will close a line of research as far as cosmology is concerned. But resolution of an issue more often raises new ones. May be some component of the dark matter will prove to do something interesting, leading us to new lines of research. The fascinating enigma of Einstein's Λ motivates the study of modifications of the physics of gravity and the dark sector, and the research programmes aimed at detecting or bounding evolution of Λ. Testing variation of Λ is good science, though we must say, we are uneasy about such heavy expenditure of resources on a shot in the dark. Let us support it, but also support wider casts of the net.

3.5. Anomalies

Two celebrated examples are in Lord Kelvin's essay [10], *Nineteenth Century Clouds over the Dynamical Theory of Heat and Light.* One cloud is exemplified by the curious behaviour of mercury vapour. A spark causes the vapour to radiate at sharply defined frequencies seemingly scattered at random across the visible spectrum. The vapour must contain resonators that radiate at definite frequencies when excited. The anomaly is that these radiators ought to be excited when the vapour is heated, yet the heat capacity of mercury vapour is quite close to that of an ideal gas of structureless particles. Lord Kelvin knew physics is well founded: He owed his fortune and peerage to his great contributions to physics, to say nothing of his income from physics-based patents for transatlantic telegraph cables among other things. Our impression of Kelvin's position, put in modern jargon, is that physics is solid and the rates for thermal excitation of internal energies of gasses must be slow enough that heat capacities are measured in systems that have not relaxed to statistical equilibrium. At the time Planck was struggling with another heat capacity problem, that of thermal radiation, another precursor of the perfect storm of new physics to come. Kelvin acknowledged that the other cloud in his essay, the curious behaviour of the velocity of light, is "very dense". He mentioned with approval "the brilliant suggestion made independently by Fitz Gerald and by Lorentz of Leiden that the motion of ether through matter may slightly alter its linear dimensions." That was yet another precursor of new physics, though we imagine Kelvin had in mind a mechanical effect of the ether on the matter, to be dealt with in the standard physics of the day.

We put Milgrom's [11] MOND (Modified Newtonian Dynamics) in this path. His idea that flat rotation curves of late-type galaxies could signify a departure from Newtonian gravity rather than the presence of dark matter seems quite out of line with the cosmological tests discussed in the last section, but MOND gives a remarkably successful prediction of the relation between circular velocity and detected baryonic mass in stars and neutral gas in these galaxies [12]. One has to be impressed that this prediction was observed over a far broader range of masses than was known when Milgrom proposed MOND. We should support continued research on this cloud over cosmology, by a few capable people, for it may teach us something of value. We offer in the next section two other anomalies that, whether real or apparent, also seem likely to be informative.

3.6. Learning to compute

Consider Kelvin's opinion of heat capacities, and the un-enthusiastic community response to MOND. We have a well-founded cosmology, just as Kelvin had a well-founded physics. Kelvin saw the need for better computation of rates of relaxation to statistical equilibrium of internal energies of gasses. The cosmology community by and large is optimistic that better methods of computation will show that the standard cosmology predicts the observed baryon mass-circular velocity correlation, and the curious behaviour of the dwarf galaxy spatial distribution and baryonic mass function [13], and other such issues. This echos the old admonition in physics, "just trust the theory", which often proves to be a useful guide. Trust in the standard cosmology, as in explaining the MOND prediction within ΛCDM, is a good idea too, but for a different reason. The cosmological tests are far less dense than tests of the more well developed branches of natural science, and trust, therefore, more questionable. But by concentrating almost exclusively on the one standard cosmology rather than scattering attention across alternatives, the community has a far better chance of discovering whether within the available evidence there really is something wrong with the chosen standard model. On the other hand, we should bear in mind the outcome of more flexible thinking about Kelvin's clouds. It is good strategy for some to scout more flexible thinking about cosmology, which may aid discovery of problems with given ideas, if there are any, and advance discovery of remedies, if they are needed.

3.7. Adding decimal places

In Lord Kelvin's time, some felt that all that was left to physics was adding decimal places. But this example reminds us why improving accuracy with no other goal in mind is not to be deprecated: it can reveal anomalies. And despite the optimistic new title for our subject, precision cosmology, we still are proud to see measurements that get beyond the first decimal place.

Figure 3. The face-on pure disk galaxy M 101.

4. Island universes

We have offered two examples along the path mentioned in Subsection 3.5, apparent anomalies in the behaviour of pure disk galaxies and elliptical galaxies. We have mentioned these particular cases because both seem to us to be resistant enough to explanation within the ΛCDM cosmology to be very likely to teach us something.

A pure disk galaxy is shown in Figure 3. The main image of this nearby face-on spiral galaxy was made by R Gendler by piecing together Hubble Space Telescope images and ground-based data. The dark lanes in the spiral arms are caused by obscuration by interstellar dust. The inset is a single HST image obtained by John Kormendy *et al* [14]. It shows that the dust lanes run inward to a central star cluster with radius ~ 0.1 per cent of the radius of the luminous part of the galaxy. This galaxy appears to be in a near relaxed state, in which the dust will have settled onto the plane of the disk. The prominent dust lanes, extending very close to the centre of the galaxy indicate the bulk of the stars, also are confined to the plane, and that the structure of this galaxy is largely supported by rotation. Kormendy *et al* have argued that this situation does not seem to be uncommon, because they have found that about half the 20 nearest big galaxies are largely supported by motions in the plane. The striking whirlpool appearance of M 101 seems to require that this galaxy grew by dissipative settling of gas or plasma onto the growing disk prior to significant star formation. If stars had formed in significant amounts in matter before it had settled into the disk it would have populated a stellar halo or classical bulge, which were not prominent components of this or other pure disk galaxies.

Numerical simulations of structure formation in ΛCDM seem to predict that galaxies were assembled by hierarchical mergers of fragments (as illustrated in Figure 3 in [15]). The larger fragments would have been massive and dense enough that pressure support required matter temperatures large enough to have collisionally ionized hydrogen. The plasma would lose energy by thermal bremsstrahlung, causing the fragments to contract until some energy source, presumably star formation, prevented it. But, if too many stars formed in these fragments, it would have produced a distinct stellar component outside the disk in the final product of the hierarchy of mergers. That is quite acceptable for some galaxies, such as our nearest large neighbour M 31, which has the substantial bulge of old stars that one might have expected to have resulted from the merging seen in structure formation simulations. Recent simulations [16, 17] do produce impressive pure disk galaxies, through careful attention to models for the conditions that promote and prevent star formation, but the mystery remains. The global star formation rate density was at its peak at $4 \gtrsim z \gtrsim 2$. How could the progenitor fragments of pure disk galaxies have "known" not to have participated in this generally high global star formation rate? One piece of the matter tumbling together according to the ΛCDM picture of the formation of the pure disk galaxy in Figure 3 "knew" it was going to host the growing disk, and start growing it at redshift well above unity if the age of the disk of the Milky Way [18] was typical of pure disk galaxies, while the rest of the fragments "knew" they had to hold off star formation until they had reached the growing disk. It is a curious situation.

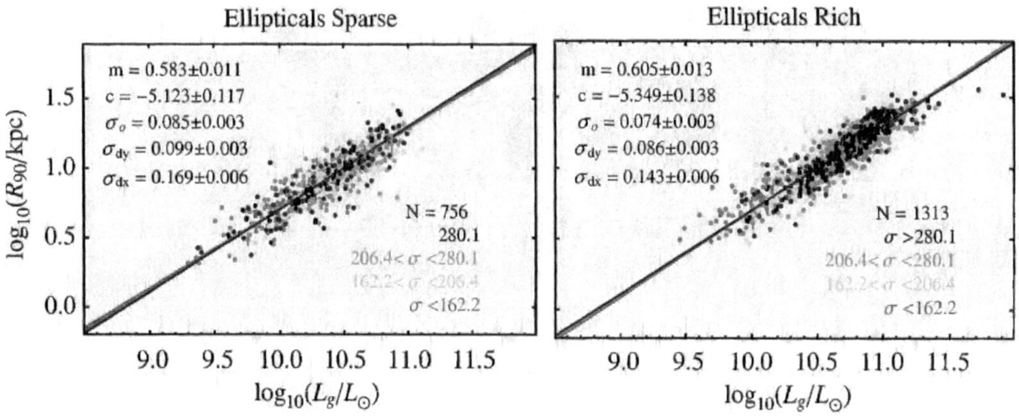

Figure 4. Comparison of relations among elliptical galaxies found in regions of lower and higher ambient densities.

A possibly related issue is why overall properties of another large class of galaxies, ellipticals, are so little sensitive to environment. The phenomenon is illustrated in Figure 4, taken from Nair, van den Bergh and Abraham [19] (and based on the Sloan Digital Sky Survey). The vertical axis is a measure of radius. The authors have taken special care with this parameter, because galaxies do not have edges: they just trail off into intergalactic space. The sample is separated according to ambient density, lower in the left-hand panel, higher in the right-hand panel. One sees several classical regularities. The Kormendy relation is the tendency for ellipticals with larger luminosities L to have larger radii, as indicated by the solid lines. The Faber-Jackson relation is the tendency for ellipticals with larger L to have larger stellar velocity dispersions σ, as one can make out from the colour coding for σ. The figure also illustrates the preference of more luminous ellipticals for denser environments: In the higher ambient density sample in the right-hand panel, the data points more densely populate the scatter plot near luminosity $\sim 10^{11}$ Solar luminosities and extend beyond that to luminosities three times the largest in the lower ambient density sub-sample. This preference is predicted by the ΛCDM structure formation simulations. Also, Nair *et al* [19] have found that radius R_e that contains half the starlight of an elliptical shows more scatter, and more distinctly so in the lower density subset, than the radius R_{90} used in Figure 4 that contains a larger fraction of the starlight. This is an environmental effect, perhaps rearrangement of internal structure by galaxy interactions [19]. Environment certainly matters, but to us the striking and surely significant phenomenon is the overall insensitivity of the elliptical galaxy size-luminosity relation to the present environment.

We have seen a similar phenomenon in the Zhu, Blanton and Moustakas [20] comparison of mean spectra of central parts of ellipticals classified by the central stellar velocity dispersion σ

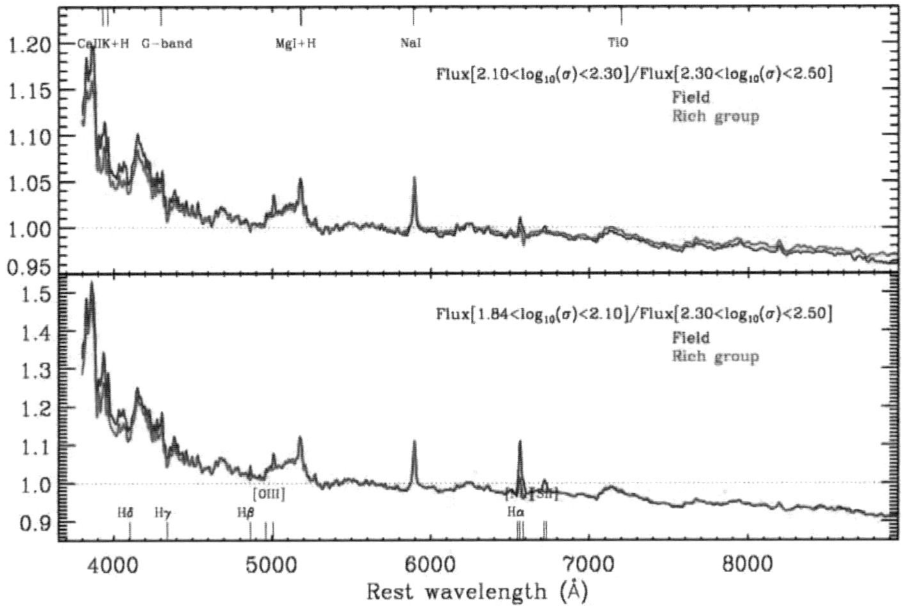

Figure 5. Comparison of mean spectra of elliptical galaxies found in regions of lower and higher ambient densities for three ranges of stellar velocity dispersion.

and ambient density. Figure 5 (kindly provided by Gunagtun Zhu using the results in [20], and based on SDSS) shows ratios of mean spectra in the three bins in σ indicated in the two panels, separately for the higher and lower density sub-samples. The ratios of spectra, with the lower σ divided by the larger σ mean spectrum, increase to the left, toward the blue. This illustrates the classical colour-magnitude relation: More luminous ellipticals, that tend to have larger σ, tend to be redder. The scale is larger in the lower panel that compares the lowest to highest σ bins than in the upper panel that compares the intermediate to highest σ bins: the relation is monotonic across the sample. The ratios for the lower density (field) and higher density (rich group) subsets are plotted in blue and red. The blue spike in the lower panel is a hydrogen recombination line likely from star-forming regions that are more common in ellipticals in lower density regions. The wonderful SDSS stability also reveals in these data a very clear relation between spectrum and ambient density, as shown by the systematic tilt of blue relative to red curves. Figure 11 in [20] shows that at given σ the spectra are bluer at lower density. There are exceptions to the very modest sensitivity to environment illustrated in Figure 5. The brightest galaxy in a cluster, which usually is near the cluster centre defined by the X-ray emission [21], tends to be redder than other early-type cluster members with the same σ [22], or it may host significant star formation [23]. That is, environment certainly matters to ellipticals. But again, to us the central point is that in the general class of ellipticals, the variation of spectrum with environment is wonderfully small compared to the variation with velocity dispersion.

We may say that Nair *et al* have classified ellipticals by a measure of internal conditions, the luminosity L, and a measure of external conditions, the ambient density. They have found that another measure of internal conditions, the radius, is much more sensitive to L than to the ambient density. Zhu *et al* have classified by the measure σ of internal conditions and a measure of external conditions, and found that another measure of internal conditions, the spectrum, is much more sensitive to σ than ambient density. Other examples along the same lines are the relation between colour and luminosity [24], between red galaxy mass-to-light ratio and radius [25, 26], and between galaxy stellar mass and radius [27]. One might read this to mean that elliptical galaxy evolution has an attractor behaviour that pulls ellipticals with given present L to near common spectra, radii, and mass-to-light ratios. This is conceivable, and good science requires us to keep it in mind, but it seems unlikely to the author. The alternative is that the present environment does not much matter to ellipticals either, because they have been formed at high redshift, when there is little variation in environment, and then passively evolved with little interaction with the environment, or else they grow more slowly, but by rearrangement of matter that is already part of the proto-galaxy evolving as a closed system. This straightforward reading of the evidence is not what pure dark matter simulations have led us to expect from ΛCDM.

Other broad classes of galaxies do show signatures of interaction with their environment. The transition of spiral galaxies into more nearly gas-free S0 galaxies seems to be a particularly clear example [28, 29]. Galaxies are observed in the act of merging, and theory and intuition argue that the numerous close galaxies at high redshift will have suffered many more mergers [30]. These observations are in line with the hierarchical assembly of structures by mergers seen in simulations of the standard cosmology, and it is entirely reasonable, therefore, that mergers have figured heavily in discussions of how the galaxies formed. Less reasonable, we think, is the scant attention to the evidence that two broad classes of galaxies, pure disks and ellipticals, have evolved in near isolation from their surroundings, as island universes. This aspect of galaxy evolution also must be telling us something important about the origin of cosmic structure.

5. Concluding remarks

Cosmology has advanced far more than we, or suppose anyone, imagined a half century ago when we started looking into this subject. Perhaps witnessing the many wrong turns taken, and right turns overlooked so long, has conditioned us to be more doubtful about the stability of the present standard ΛCDM cosmology than we have seen in the thinking of many younger colleagues. But we can agree that the advances of the observational and theoretical bases are substantial and of lasting importance. This review leaves us with the impression that there is no shortage of interesting problems for study by the next generation. We have mentioned the island universe puzzle, because it seems to us that the opportunities for research are even broader than generally advertised.

Acknowledgements

The author has benefitted from discussions with Michael Blanton, John Kormendy, Yen-Ting Lin, Piero Madau, Chris McKee, Surhud More, John Moustakas and Gunagtun Ben Zhu.

References

[1] Riess A G *et al* 1998 *Astron. J.* **116** 1009
[2] Perlmutter S *et al* 1999 *Ap. J.* **517** 565
[3] Kogut A J 2011 *Private communication*
[4] Narlikar J V 1997 *J. Ap. & A.* **18** 353
[5] Kolb E W, Matarrese S and Riotto A 2006 *New J. Phys.* **8** 12
[6] Siegel E R and Fry J N 2005 *Ap. J. Lett.* **268** L1
[7] Komatsu E *et al* 2011 *Ap. J. Supp. Ser.* **192** 18
[8] Percival W J *et al* 2001 *MNRAS* **327** 1297
[9] Hubble E 1936 *The Realm of the Nebulae* (New Haven: Yale University Press)
[10] Kelvin (Thomson W) 1901 *Phil. Mag., Sixth Ser.* **2** 1

[11] Milgrom M 1983 *Ap. J.* **270** 365
[12] McGaugh S S 2012 *Astron. J.* **143** 40
[13] Macciò A V, Kang Xi, Fontanot S, Somerville R S, Koposov S and Monaco P 2010 *MNRAS* **402** 1995
[14] Kormendy J, Drory N, Bender R and Cornell ME 2010 *Ap. J.* **723** 54
[15] Peebles P J E and Nusser A 2010 *Nature* **465** 565
[16] Brook C B *et al* 2011 *MNRAS* **415** 1051
[17] Guedes J, Callegari S, Madau P and Mayer L 2011 *Ap. J.* **742** 76
[18] Aumer M and Binney J J 2009 *MNRAS* **397** 1286
[19] Nair P, van den Bergh S and Abraham RG 2011*Ap. J. Lett.* **734** L31
[20] Zhu G, Blanton M R and Moustakas J 2010 *Ap. J.* **722** 491
[21] Lin Y-T and Mohr J J 2004 *Ap. J.* **617** 879
[22] Roche N, Bernardi M and Hyde J 2010 *MNRAS* **407** 1231
[23] O'Dea K P *et al* 2010 *Ap. J.* **719** 1619
[24] Hogg D W *et al* 2004 *Ap. J. Lett.* **601** L29
[25] Bernardi M, Nichol R C, Sheth R K, Miller C J and Brinkmann J 2006 *Astron. J.* **131** 1288
[26] Magoulas C, Colless M, Jones H, Mould J and Springob C 2010 *Highlights Astron.* **15** 84
[27] Maltby D T *et al* 2010 *MNRAS* **402** 282
[28] Just D W, Zaritsky D, Sand D J, Desai V and Rudnick G 2010 *Ap. J.* **711** 192
[29] Kormendy J and Bender R 2010 *Ap. J. Supp. Ser.* **198** 2
[30] López-Sanjuan C *et al* 2012 [arXiv:1202.4674]

Using general relativity to study superconductivity

G T Horowitz

Department of Physics, University of California, Santa Barbara, California 93106 USA

E-mail: gary@physics.ucsb.edu

Abstract. It has recently been shown that in addition to describing black holes, gravitational waves, and other gravitational phenomena, general relativity can also describe aspects of non-gravitational physics including condensed matter. This is a result of a remarkable gauge/gravity duality that has emerged from string theory. The author has explained this surprising development, and illustrated it by showing how general relativity can reproduce aspects of superconductivity. This novel approach may eventually yield new insight into the known high temperature superconductors.

1. Introduction

We may be witnessing the birth of a new field - one that was unimaginable just five years ago. This field is based on surprising connections between gravity and condensed matter physics. In this article, we have focused on the application to superconductivity. In fact, this year marks the 100th anniversary of the discovery of superconductivity, so it seems appropriate to consider an entirely new approach to the subject. The starting point is the following remarkable claim:

In addition to describing gravitational phenomena, general relativity can also describe other fields of physics, including aspects of superconductivity.

We begin by reviewing black hole thermodynamics since that will play an important role in our discussion. In the next Section, we have described the remarkable gauge/gravity duality, which is the main tool that allows us to use general relativity to describe other areas of physics. In Section 4, we have applied this to superconductivity, and in Section 5, we have constructed a gravitational description of a Josephson junction (which consists of two superconductors separated by a weak link). The final Section has some concluding remarks.

2. Black hole thermodynamics

Following the seminal work by Bekenstein and Hawking, it is known that black holes are thermodynamic objects with a temperature related to the surface gravity κ by $T = \hbar\kappa/2\pi$, and entropy $S = A/4\hbar G$, where A is the area of the event horizon. Although, black hole thermodynamics is often studied with asymptotically flat boundary conditions, this is really not suitable for discussing equilibrium configurations. If we try to put a black hole in thermal equilibrium with its Hawking radiation, there is non-zero energy density out to infinity, so the solution is not really asymptotically flat. Instead, it resembles a cosmology and starts to expand or contract.

The situation is much better with anti-de Sitter (AdS) boundary conditions. AdS is the maximally symmetric solution of Einstein's equation with negative cosmological constant. It has constant negative curvature and this negative curvature acts like a confining box, so there are static black holes in thermal equilibrium with their Hawking radiation [1]. The study of black holes is also much richer in AdS. Unlike asymptotically flat black holes which must be spherical, AdS black holes exist with spherical, planar, or hyperbolic horizon geometry.

Spherical black holes have a minimum temperature in AdS set by the cosmological constant: $T_{min} \sim 1/L$, where $\Lambda = -3/L^2$. (L is called the radius of curvature of AdS.) Small black holes act just like the asymptotically flat Schwarzschild solution. Their temperature is $T \sim 1/r_0$, where r_0 is the horizon radius, and they can evaporate. However, large black holes are different and have a temperature which grows linearly with r_0. This means that large black holes have positive specific heat: Their temperature increases when we add mass. So the equilibrium configurations of black holes and radiation are stable. For $T < 1/L$, the equilibrium configuration is just a gas of particles in AdS. There is a (first order) phase transition between this gas of particles at low T and a large black holes at high T, which is called the Hawking-Page transition [1].

We are more interested in the planar black hole. It has a metric, which is static and translationally invariant. Setting $L = 1$ for convenience, it is given by:

$$ds^2 = r^2[-f(r)dt^2 + dx_i dx^i] + \frac{dr^2}{r^2 f(r)}, \tag{1}$$

where $i = 1,\ 2$ and

$$f(r) = 1 - \frac{r_+^3}{r^3}. \tag{2}$$

This is a solution to Einstein's equation with negative cosmological constant. The horizon is at $r = r_+$ and $r_+ = 0$ is AdS itself. Dropping numerical factors, the Hawking temperature is $T \sim r_+$ (for all r_+). The total energy is $E \sim r_+^3 V \sim T^3 V$ and the entropy is $S \sim A \sim r_+^2 V \sim T^2 V$. So, not only is the specific heat positive, the energy and entropy behave exactly like a thermal gas in 2+1 dimensions! This is the first indication that general relativity with AdS boundary conditions can be related to non-gravitational systems.

Charged (planar) black holes in AdS are similar to Reissner-Nordström. The metric is of the same form as in Eq. (1) with:

$$f(r) = 1 - (1+b)\frac{r_+^3}{r^3} + b\frac{r_+^4}{r^4}. \tag{3}$$

The vector potential is:

$$A_t = \mu - \frac{\rho}{r}, \tag{4}$$

where ρ is the charge density, and μ is chosen so that A_t vanishes at the horizon. (This is required in order for $A_\nu A^\nu$ to remain finite there.) The parameters b, r_+ in the metric are related to μ and ρ, so the solutions are labelled by only two parameters, which can be thought of as the mass and charge. There is an extremal limit corresponding to the maximum charge for a given mass, which has $T = 0$ and non-zero entropy. Having $S \neq 0$ at $T = 0$ is uncommon for a thermodynamic system and seems to imply a highly degenerate ground state. We will see that the Reissner-Nordström AdS solution is often unstable at low temperature when coupled to other matter (e.g., a charged scalar field), and the new black hole has $S = 0$ at $T = 0$.

3. Gauge/gravity duality

We begin with a few comments about gauge theory and string theory. Gauge theories are generalizations of electromagnetism in which the familiar U(1) gauge symmetry is replaced by a non-commuting group such as SU(N). Our standard model of particle physics is based on a gauge theory. In particular, QCD (the theory of the strong interactions) has SU(3) gauge symmetry. The interactions are weak at high energy but become strong at low energy causing quark confinement.

String theory is based on the idea that elementary particles are not pointlike, but excitations of a one dimensional string. Strings interact with a simple splitting and joining interaction. It has been shown that string theory includes gravity and in fact reduces to general relativity (with certain matter) in a classical limit.

't Hooft argued in the 1970's that a 1/N expansion of an SU(N) gauge theory would resemble a theory of strings [2]. This was based on the structure of the perturbation theory. This turns out to be correct, but it took more than twenty years for this idea to be made precise. The precise formulation takes the form of a duality [3]:

Gauge/gravity duality: With anti-de Sitter boundary conditions, string theory (which includes gravity) is completely equivalent to a (non-gravitational) gauge theory living on the boundary at infinity.

Of course, these theories look very different at weak coupling, but the point is that when the string theory is weakly coupled, the gauge theory is strongly coupled, and vice versa. This duality is a manifestation of the idea that quantum gravity (at least with AdS boundary conditions) is holographic: It can be completely described by degrees of freedom living in a lower dimensional space. This idea was first discussed by 't Hooft [4] and Susskind [5]. Even though this duality has not been rigorously proved, there is by now very strong evidence that it is true.

Setting $r_+ = 0$ in Eq. (1) we can write anti-de Sitter space in planar coordinates as:

$$ds^2 = r^2(-dt^2 + dx_i dx^i) + \frac{dr^2}{r^2}. \tag{5}$$

Conformally re-scaling this metric by $1/r^2$, one can smoothly extend the geometry to the boundary at $r = \infty$. The metric on the boundary is just Minkowski spacetime, and the gauge theory lives on this spacetime. (The physical spacetime away from the boundary is often referred to as the "bulk".) The metric in Eq. (5) is invariant under a scaling symmetry:

$$r \to ar, \quad (t, x^i) \to (t/a, x^i/a). \tag{6}$$

It follows that small r corresponds to large distances or low energy in the gauge theory.

One of the main applications of gauge/gravity duality has been to gain new insight into strongly coupled gauge theories. In particular, it provides a simple geometric picture of confinement, which can be seen as follows: The potential between two quarks separated by Δx^i is obtained from the length of a string in the bulk with end points at the two quarks on the boundary[1]. Since the string is under tension, it wants to minimize its length, and the geometry is warped, as Δx^i increases, the minimal curve extends farther into smaller radii. In fact, in AdS itself, the length of this curve is independent of Δx^i due to the scaling symmetry given in Eq. (6). Thus, AdS describes the vacuum of a scale invariant theory, and there is no confinement.

The gravity dual of a confining vacuum differs from AdS, and in that the spacetime ends at non-zero radius. (Recall that small radius describes low energy.) This can happen, e.g., if a compact extra dimension smoothly caps off. In this case, when Δx^i becomes large, the string drops down to the minimum radius keeping x^i nearly constant. It then moves a distance Δx^i

[1] Since the boundary is at infinity, the actual length is infinite, but this just corresponds to an infinite self-energy of the point-like quarks. There is a well defined way to subtract this divergence, and the resulting finite quantity defines the potential.

and returns to the boundary again along a curve of nearly constant x^i. So, the length of the string grows linearly with Δx^i when it becomes large. This gives a linearly growing potential between quarks, i.e., confinement! In this way, one obtains a simple geometric description of a complicated quantum field theory phenomenon.

It is easy to construct a solution describing a confining vacuum. We start with the five dimensional analogue of the planar AdS black hole. The metric is again given by Eq. (1) with now $i = 1, 2, 3$ and

$$f(r) = 1 - \frac{r_+^4}{r^4}. \tag{7}$$

We now analytically continue $t = i\chi$. As usual, when constructing a Euclidean black hole metric, the space only exists for $r \geq r_+$ and χ must be periodic with period $\beta = 1/T$ in order to avoid a conical singularity at $r = r_+$. We now analytically continue $x_3 = it$ to obtain a new Lorentzian solution. The result is:

$$ds^2 = r^2[f(r)d\chi^2 + dx_i dx^i - dt^2] + \frac{dr^2}{r^2 f(r)}, \tag{8}$$

with $i = 1, 2$. This metric is globally static and non-singular. It is called the AdS soliton [6, 7]. The dual theory is a four dimensional gauge theory compactified on a circle (parameterized by χ). At low energies, it reduces to a $2+1$ gauge theory which has a confining vacuum. The AdS soliton is the dual gravitational description of this state.

One can also turn the duality around, and use the gauge theory to gain insight into quantum gravity. In particular, it provides answers to fundamental questions about quantum properties of black holes. For example, it has been shown that the gauge theory has enough micro-states to reproduce the Bekenstein-Hawking entropy of black holes in AdS. This explains the origin of the enormous entropy arising in black hole thermodynamics. Furthermore, since the entire process of forming a small black hole in AdS and letting it evaporate can be represented by standard Hamiltonian evolution in the gauge theory, it is clear that there is no loss of information. Black hole evaporation does not violate quantum mechanics. After arguing against this for thirty years, Hawking finally conceded this point in 2005 [8]. However, the actual mechanism by which the information gets out of the black hole is still under active investigation.

So far our discussion has included quantum gravity and string theory. However, in a certain (large N) limit, all stringy and quantum effects are suppressed and the bulk theory is just general relativity (with AdS boundary conditions and possibly higher dimensions). If gauge/gravity duality is correct, general relativity must be able to reproduce other areas of physics.

One check comes from considering the renormalization group (RG). The RG flow in a quantum field theory corresponds to obtaining an effective low energy action by integrating out high energy modes. This corresponds to radial dependence on the gravity side. One can now compare two separate calculations. On the gauge theory side, one adds a mass term to the action and follows the RG flow to low energies to obtain a new field theory. On the gravity side, adding a mass term to the dual field theory correspond to modifying the boundary conditions for certain matter fields coupled to gravity. So, one looks for static solutions to Einstein's equation with these modified boundary conditions. Remarkably, one finds detailed numerical agreement between the small r behaviour of the gravity solution and the end point of the RG flow [9, 10].

Recently, people have started to extend the rules of gauge/gravity duality to other non-gravitational systems besides gauge theories. The correspondence is not as well established in this case, but there is growing evidence that it still works. Starting in 2007, it was shown that there were applications of gauge/gravity duality to condensed matter physics. The Hall effect and Nernst effect were reproduced using general relativity [11, 12]. (The Nernst effect is the existence of a transverse voltage arising from an applied temperature gradient in the presence of a magnetic field.) The basic idea is that a non-gravitational system in thermal equilibrium at temperature T is described by a black hole in AdS with Hawking temperature T. Using linear response theory, the electrical and thermal transport properties in the non-gravitational system are obtained from linear perturbations of the black hole. Given this early success, it was natural to ask: Is there a gravitational description of superconductivity?

4. Superconductivity

In conventional superconductors (e.g., Al, Nb, Pb), pairs of electrons with opposite spin can bind to form a charged Boson, called a Cooper pair. Below a critical temperature T_c, there is a second order phase transition and these Bosons condense. The DC conductivity becomes infinite, resulting in superconductivity. This is well described by BCS theory [13], in which the binding is due to interactions with lattice vibrations or phonons. This interaction is weak and a typical Cooper pair is much larger than the lattice spacing.

A new class of high T_c superconductors were discovered in 1986 [14]. They are cuprates, and the superconductivity is along the CuO_2 planes. The highest T_c known today (at atmospheric pressure) is $T_c = 134^o K$ for a mercury, barium, copper oxide compound. If we apply pressure, T_c climbs to about $160^o K$. Another class of superconductors were discovered in 2008 based on iron, and not copper [15]. The highest T_c so far is $56^o K$. These materials are also layered and the superconductivity is again associated with the two dimensional planes. They are called *iron pnictides*, since they involve other elements like arsenic in the nitrogen group of the periodic table. There is evidence that electron pairs still form in these high T_c materials, but the pairing mechanism is not well understood. Unlike BCS theory, the coupling is not weak.

One might hope that gauge/gravity duality can be used to gain some insight into these high temperature superconductors. The first step is to find a gravitational dual of a superconductor. This was found in [16] (for more complete reviews, see [17]). The minimal ingredients are the following: In a superconductor, we need the notion of temperature and condensate. On the gravity side, the role of temperature is played by a black hole with the Hawking temperature of the black hole equal to the field theory temperature. The role of a charged condensate will be played by a charged scalar field. So we need to find a black hole that has scalar hair at low temperatures, but no hair at high temperatures. At first sight, this does not sound like an easy task. There are no hair theorems, which say that certain matter fields must be trivial outside a black hole. The field usually wants to fall into the black hole or radiate out to infinity. But these theorems usually apply to linear fields coupled to gravity. It is known that we can have hair for certain non-linear theories.

A surprisingly simple solution to this problem was found by Gubser [18]. He has argued that a charged scalar field around a *charged* black hole in AdS would have the desired property. Consider:

$$S = \int d^4x\sqrt{-g}\left(R + 6 - \frac{1}{4}F_{\mu\nu}F^{\mu\nu} - |\nabla\Psi - iqA\Psi|^2 - m^2|\Psi^2|\right). \tag{9}$$

This is just general relativity with a negative cosmological constant $\Lambda = -3$ (again setting the AdS radius to one), coupled to a Maxwell field and charged scalar with mass m and charge q. It is easy to see why black holes in this theory might be unstable to forming scalar hair: For an electrically charged black hole, the effective mass of Ψ is:

$$m_{eff}^2 = m^2 + q^2 g^{tt} A_t^2. \tag{10}$$

The last term is negative, so there is a chance that m_{eff}^2 becomes sufficiently negative near the horizon to destabilize the scalar field.

We have seen that black holes in this theory indeed develop scalar hair at low temperature. These four dimensional hairy black holes are dual to $2+1$ dimensional field theories, so it is the right context to try to understand the superconductivity along the two dimensional planes. One might wonder why such a simple type of hair was not noticed earlier. A possible explanation is that this does not work for asymptotically flat black holes. The AdS boundary conditions are crucial. One way to understand the difference is by the following quantum argument: Let Q_i be the initial charge on the black hole. If qQ_i is large enough, even maximally charged black holes with zero Hawking temperature create pairs of charged particles. This is simply due to the fact that the electric field near the horizon is strong enough to pull pairs of oppositely charged particles out of the vacuum via the Schwinger mechanism of ordinary field theory. The particle with opposite charge to the black hole falls into the horizon, reducing Q_i, while the particle with the same sign charge as the black hole is repelled away. In asymptotically flat spacetime, these particles escape to infinity, so the final result is a standard Reissner-Nordström black hole with final charge $Q_f < Q_i$. In AdS, the charged particles cannot escape, since the negative cosmological constant acts like a confining box, and they settle outside the horizon. This gas of charged particles is the quantum description of the hair. This quantum process has an entirely classical analogue in terms of super-radiance of the charged scalar field.

To find the hairy black holes, we start with the ansatz for the metric:

$$ds^2 = -g(r)e^{-\chi(r)}dt^2 + \frac{dr^2}{g(r)} + r^2(dx^2 + dy^2), \tag{11}$$

and the following ansatz for the matter fields:

$$A = \phi(r)\; dt, \quad \Psi = \psi(r). \tag{12}$$

The field equations require that the phase of Ψ is constant, so we can choose ψ to be real. Substituting into the field equations yields four coupled non-linear ordinary differential equations, which can be solved numerically (for some analytic results, see [19, 20]). At the horizon, $r = r_+$, g and ϕ vanish, and χ and ψ are constant. Asymptotically, the metric approaches as in Eq.(5), and the Maxwell potential approaches:

$$\phi = \mu - \frac{\rho}{r} + \cdots . \tag{13}$$

The asymptotic behaviour of the scalar field depends on the mass. For $m^2 = -2$, we have:

$$\psi = \frac{\psi^{(2)}}{r^2} + \cdots . \tag{14}$$

Gauge/gravity duality relates this asymptotic behaviour to properties of the dual field theory: μ is the chemical potential and ρ is the charge density. There is a dimension two operator dual to ψ, and its expectation value is $O_2 = \psi^{(2)}$.

We want to know how the condensate O_2 behaves as a function of temperature. But, there is a scaling symmetry analogous to Eq. (6), which applies to the hairy black holes. This will trivially re-scale the temperature and condensate. It is convenient to use the chemical potential to fix a scale and consider $\sqrt{O_2}/\mu$ as a function of T/μ. When one does this, one finds that the condensate is non-zero only when T/μ is small enough. Setting T_c equal to the critical temperature when the condensate first turns on, we get Figure 1. This curve is qualitatively similar to that obtained in BCS theory, and observed in many materials, where the condensate rises quickly and goes to a constant at zero temperature.

The critical temperature is proportional to μ, which is not surprising, since it is the only other scale in the problem. Near T_c, there is a square root behaviour $O_2 \sim (1 - T/T_c)^{1/2}$. This is the standard mean field behaviour predicted by Landau-Ginzburg theory. As $T \to 0$, one finds that the area of the horizon of the hairy black hole vanishes, consistent with a unique ground state [21].

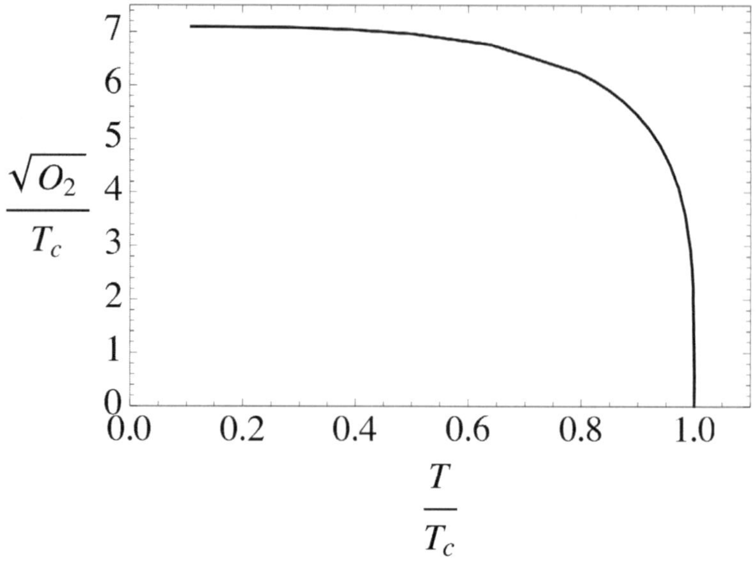

Figure 1. The condensate as a function of temperature. The critical temperature is proportional to the chemical potential.

One can compute the free energy (Euclidean action) of these hairy configurations and compare with the solution without hair: $\psi = 0$, $A_t = \mu - \rho/r$. The free energy is always lower for the hairy configurations and scales like $(T_c - T)^2$ near T_c, showing that this is a second order phase transition. Actually, we can compare the free energy at fixed charge or fixed chemical potential. These correspond to two different ensembles, but in both cases, the free energy is lower for the hairy configuration. The net result is that Reissner-Nordström AdS is unstable at low temperatures in the theory given by Eq. (9) and develops scalar hair.

We have next wanted to compute the conductivity as a function of frequency in the dual theory. According to the gauge/gravity duality dictionary, this is obtained by perturbing the Maxwell field around the black hole. We have imposed ingoing wave boundary conditions at the horizon, since that corresponds to causal propagation in the dual theory. Assuming harmonic time dependence $e^{-i\omega t}$, we get an ordinary differential equation which can be solved numerically.

Asymptotically,

$$A_x = A_x^{(0)} + \frac{A_x^{(1)}}{r} + \cdots \,. \tag{15}$$

The gauge/gravity dictionary says that the limit of the electric field in the bulk is the electric field on the boundary: $E_x = -\dot{A}_x^{(0)}$, and the expectation value of the induced current is the first sub-leading term: $J_x = A_x^{(1)}$. From Ohm's law, we get:

$$\sigma(\omega) = \frac{J_x}{E_x} = -\frac{J_x}{\dot{A}_x^{(0)}} = -\frac{iJ_x}{\omega A_x^{(0)}} = -\frac{iA_x^{(1)}}{\omega A_x^{(0)}} \,. \tag{16}$$

So the conductivity is directly related to the ratio of the leading two terms at large r of a perturbation of the Maxwell field. This is typical of how gauge/gravity duality works. The leading term often corresponds to adding a source (in this case, the electric field), and the first

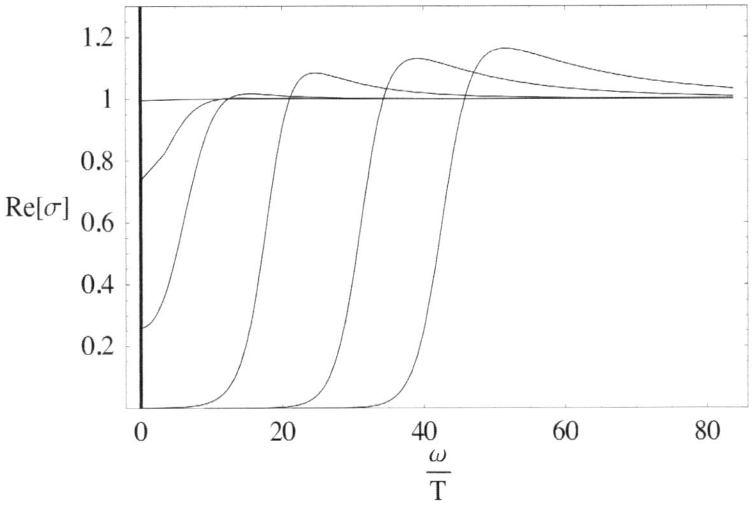

Figure 2. The formation of a gap in the real part of the conductivity as the temperature is lowered below the critical temperature. The curves describe successively lower temperatures. There is also a delta function at $\omega = 0$.

sub-leading term corresponds to the response (the current).[2]

Figure 2 shows the real part of the optical conductivity for a variety of temperatures. As one lowers the temperature below T_c, a pronounced gap appears at low frequency.[3] There is also a delta function at $\omega = 0$ for all $T < T_c$ corresponding to infinite DC conductivity. This cannot be seen by numerically solving for the real part of the conductivity, but can be inferred from the behaviour of the imaginary part. The Kramers-Kronig relations relate the real and imaginary parts of any causal quantity, such as the conductivity, when expressed in frequency space. They imply that the real part has a delta function if and only if the imaginary part has a pole. One, indeed, finds a pole in $\mathrm{Im}[\sigma(\omega)]$ at $\omega = 0$ for all $T < T_c$.

At very low temperatures, one can compare the width of the gap in the optical conductivity ω_g to the critical temperature. Even though the model has two free parameters (the mass m and charge q of the scalar field), one finds that this ratio is approximately constant and given by [22]:

$$\frac{\omega_g}{T_c} \approx 8. \tag{17}$$

This holds provided q is not too small, in which case the gap becomes much less pronounced and ω_g is ill defined. (It is also modified by higher curvature modifications to general relativity [20, 23].) In BCS theory, this ratio is about 3.5, but in some of the high T_c cuprates, this ratio has been measured and is, indeed, about eight [24].

5. Josephson junctions

A Josephson junction consists of two superconductors separated by a weak link. There are different types of Josephson junctions depending on the nature of this link. If the weak link is

[2] The perturbation of the Maxwell field induces a perturbation of the metric whose dual interpretation is simply that the current carries non-zero momentum.

[3] It was originally thought that the conductivity was behaving like $\sigma \sim e^{-\Delta/T}$ as expected for theories with a mass gap Δ. However, it is now known that the conductivity remains non-zero, although exponentially small, at $T = 0$ and low frequency.

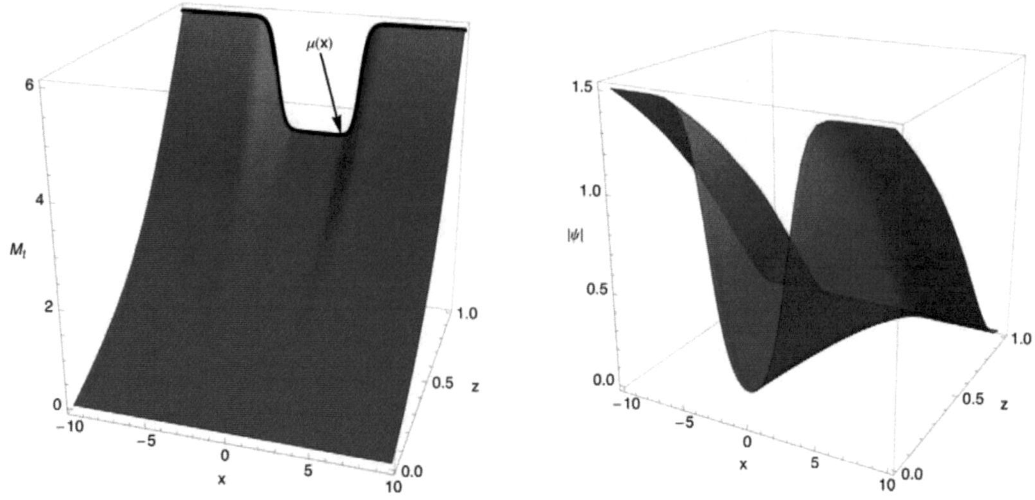

Figure 3. *Left*: $M_t = A_t$ as a function of x and z, showing the boundary profile $\mu(x)$ imposed at $z = 1$. *Right*: The absolute value of the condensate as a function of x and z.

an insulator, it is called an SIS junction. If it is a normal conductor, it is called an SNS junction. It can even be a narrow superconducting bridge. Josephson showed that in all of these cases, even without a voltage difference across the junction, there is a current flow:

$$J = J_{\max} \sin \gamma, \tag{18}$$

where γ is the phase difference between the condensates in the two superconductors[4]. Josephson junctions are used in numerous applications of superconductivity.

Since our holographic superconductors act like ordinary conductors above the critical temperature, we can construct an SNS junction if we can make the critical temperature of the superconductor vary as a function of position. Recall that T_c is proportional to the chemical potential μ. So, we can model this using exactly the same gravitational theory as before, and simply let μ be position dependent [25]. More precisely, we choose $\mu(x)$ to be almost constant, $\mu(x) \approx \mu_\infty$, except for a narrow gap of width ℓ, where it drops down to $\mu_0 < \mu_\infty$. Then there is a range of temperatures, which are above the critical temperature in the gap, but below the critical temperature everywhere else. This is precisely an SNS Josephson junction. Since the bulk solution now depends on x as well as r, we need a more general ansatz for our fields. We set:

$$\Psi = |\psi| e^{i\varphi} , \qquad A = A_t \, dt + A_r \, dr + A_x \, dx . \tag{19}$$

Previously, we have fixed a gauge with $A_r = 0$. Here, it is more convenient to include A_r and work with gauge invariant variables $M = A - d\varphi$.

We have worked in a "probe" approximation, in which we can ignore the backreaction of the matter fields on the spacetime geometry. This still captures the essential physics of the superconductor. The Maxwell and scalar equations reduce to non-linear two dimensional PDEs, which we have solved numerically. At large $|x|$, the solution must approach the homogeneous

[4] This is the DC Josephson effect. If one puts a voltage difference across the junction, one gets an AC current.

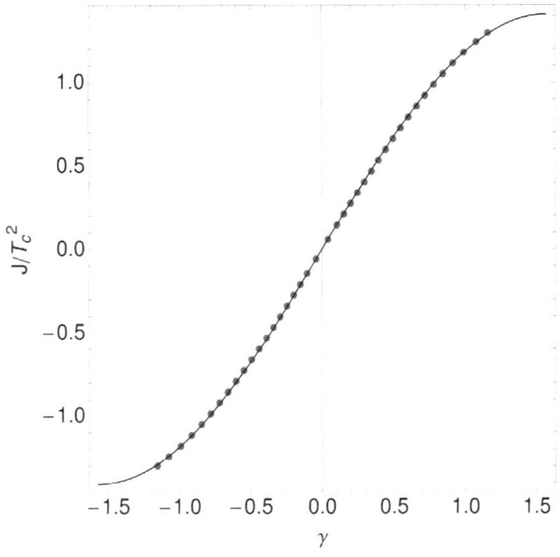

Figure 4. The current as a function of the phase difference. The solid line is the best fit sine curve.

superconductor (now with non-zero current). Near $r = \infty$, the general solution is given by:

$$M_t = \mu(x) - \frac{\rho(x)}{r} + O\left(\frac{1}{r^2}\right), \tag{20}$$

$$M_r = O\left(\frac{1}{r^3}\right), \quad and \tag{21}$$

$$M_x = \nu(x) + \frac{J}{r} + O\left(\frac{1}{r^2}\right). \tag{22}$$

We specify $\mu(x)$ as above and must choose either $\nu(x)$ or J as a boundary condition for M_x. J must be independent of x by virtue of the field equations. Since J represents the current in the dual theory, this can be interpreted as current conservation. The gauge invariant phase difference is:

$$\gamma = \Delta\varphi - \int_{r=\infty} A_x \, dx = -\int \nu(x) \, dx. \tag{23}$$

Since the phase difference γ between the two superconductors is a non-local quantity, we check using Eq. (18), by first picking J and then computing γ. To display the results, it is convenient to compactify the r direction by defining $z = 1 - r_+/r$. We also suppress the y direction, since nothing depends on y. In Figure 3, on the left, we have shown $M_t(x, z)$. Since the phase φ is independent of t, $M_t = A_t$. At infinity ($z = 1$), it shows the form of the chemical potential $\mu(x)$ that we have chosen with the gap. Regularity requires that M_t vanishes at the horizon ($z = 0$). On the right, we have shown the magnitude of the scalar field $|\psi(x, z)|$. Note that it is highly suppressed inside the gap. In effect, ψ turns on outside the gap and leaks in. (The phase, which is not displayed, is essentially constant outside the gap and only changes inside.)

Given the solutions, one can now study the properties of the junction [25]. Fixing a series of different values for the current J and computing the corresponding phase difference γ yields the plot in Figure 4. The solid line is the best fit to $J = J_{\max} \sin\gamma$. So, indeed, general relativity reproduces Eq. (18). The dependence of the maximum current $J_{\max}$ on the width of the gap

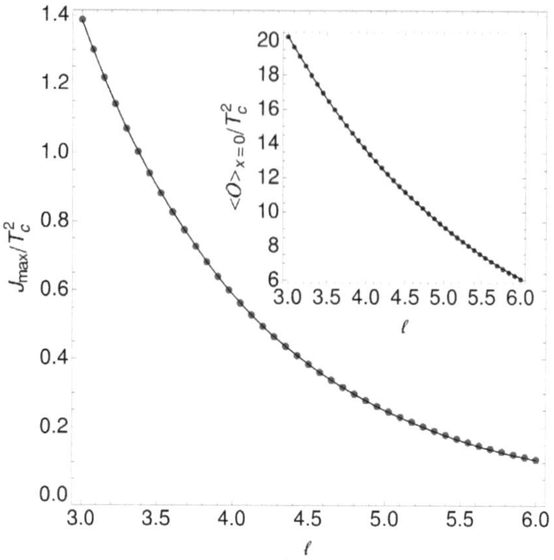

Figure 5. The behaviour of the maximum current as a function of the width of the gap. The solid line is the best fit exponential. The inset shows that the condensate at the middle of the gap falls off in a similar way.

ℓ is shown in Figure 5. Once again, this curve is in good agreement with condensed matter expectations [26], which for SNS junctions, predicts an exponential decay with growing ℓ in the maximum current, given by:

$$J_{\rm max}/T_c^2 = A_0\, e^{-\frac{\ell}{\xi}}. \tag{24}$$

The constant ξ represents a correlation length. This universal law is only expected to hold as long as $\xi \ll \ell$. Furthermore, a similar behaviour is expected of the condensate within the barrier at zero current, given by:

$$\langle \mathcal{O} \rangle_{x=0,J=0}/T_c^2 = A_1\, e^{-\frac{\ell}{2\xi}}. \tag{25}$$

Note that the value of the coherence length ξ should be the same in Eqs. (24) and (25). The gravitational calculations, indeed, show the same value $\xi \approx 1.2$, arises in both fits. This is illustrated by the inset in Figure 5.

6. Discussion

To summarize, we have seen that black holes in anti-de Sitter space behave exactly like a thermal system in one lower dimension. Gauge/gravity duality predicts that this is not a coincidence, but part of a much deeper connection between general relativity and non-gravitational physics. We have seen that one can, indeed, recover the main properties of superconductors, including Josephson junctions, from general relativity.

This is encouraging, but the simple gravitational model discussed here certainly does not capture the whole story. One problem is that it has the wrong symmetry. The symmetry of a superconductor refers to the energy gap near the Fermi surface. The fact that our bulk solution is rotationally invariant, means that it describes an s-wave superconductor. High T_c cuprate superconductors are known to be d-wave. There has been some recent work trying to construct a d-wave holographic superconductor [27], but this remains one of the main open problems in the field. Interestingly enough, there is evidence that the new iron based superconductors may be s-wave.

There have been many further developments over the past few years. For example, the response to magnetic fields has been studied. One of the characteristic properties of superconductors is that they expel magnetic fields. However, a superconductor can expel a magnetic field only up to a point. A sufficiently strong field will destroy the superconductivity. Superconductors are divided into two classes depending on how they make the transition from a superconducting to a normal state as the magnetic field is increased. In type I superconductors, there is a first order phase transition at a critical field B_c, above which magnetic field lines penetrate uniformly and the material is no longer a superconductor. In type II superconductors, there is a more gradual second order phase transition. The magnetic field starts to penetrate the superconductor in the form of vortices with quantized flux at a lower critical field B_{c1}. The vortices become more dense as the magnetic field is increased, and at an upper critical field strength $B_{c2} > B_{c1}$, the material stops being a superconductor. It was shown that holographic superconductors are type II, just like the high T_c cuprates. A vortex solution has been constructed [28, 29].

The author has focussed on superconductivity, but it has been shown recently that general relativity can also describe other aspects of condensed matter physics such as Fermi surfaces, Fermi and non-Fermi liquids, and quantum critical points. We know that several condensed matter physicists who are now learning general relativity in order to use this new tool.

Carrying this further, one might speculate that someday one might use gravity to predict new exotic states of matter and then try to look for them in the lab. One can also use the duality in reverse. Since classical gravity corresponds to a certain (large N) limit of the dual field theory, which is hard to realize in condensed matter systems, a connection with quantum gravity might be easier to arrange. It is certainly a striking thought that we might someday be able to learn about quantum gravity by doing condensed matter experiments!

Acknowledgements

The author would like to thank the organizers of the ICGC-2011, for a very stimulating meeting. This is an updated version of a talk he gave at GR-19, which appeared in *Class. Quant. Grav.* **28** 114008 (2011). The work was supported in part by NSF grant PHY-0855415.

References

[1] Hawking S W and Page D N 1983 *Commun. Math. Phys.* **87** 577
[2] 't Hooft G 1974 *Nucl. Phys. B* **72** 461
[3] Maldacena J M 1998 *Adv. Theor. Math. Phys.* **2** 231, 1999 *Int. J. Theor. Phys.* **38** 1113 [arXiv:hep-th/9711200], Gubser S S, Klebanov I R and Polyakov A M 1998 *Phys. Lett. B* **428** 105 [arXiv:hep-th/9802109], Witten E 1998 *Adv. Theor. Math. Phys.* **2** 253 [arXiv:hep-th/9802150]
[4] 't Hooft G [arXiv:gr-qc/9310026]
[5] Susskind L 1995 *J. Math. Phys.* **36** 6377 [arXiv:hep-th/9409089]
[6] Horowitz G T and Myers R C 1998 *Phys. Rev. D* **59** 026005 [arXiv:hep-th/9808079]
[7] Witten E 1998 *Adv. Theor. Math. Phys.* **2** 505 [arXiv:hep-th/9803131]
[8] Hawking S W 2005 *Phys. Rev. D* **72** 084013 [arXiv:hep-th/0507171]
[9] Freedman D Z, Gubser S S, Pilch K and Warner N P 1999 *Adv. Theor. Math. Phys.* **3** 363 [arXiv:hep-th/9904017]
[10] de Boer J, Verlinde E P and Verlinde H L 2000 *JHEP* **0008** 003 [arXiv:hep-th/9912012]
[11] Hartnoll S A and Kovtun P 2007 *Phys. Rev. D* **76** 066001 [arXiv:0704.1160[hep-th]]
[12] Hartnoll S A, Kovtun P K, Muller M and Sachdev S 2007 *Phys. Rev. B* **76** 144502 [arXiv:0706.3215[cond-mat.str-el]]
[13] Bardeen J, Cooper L N and Schrieffer J R 1957 *Phys. Rev.* **108** 1175
[14] Bednorz J G and Muller K A 1986 *Z. Phys. B* **64** 189
[15] Kamihara Y, Watanabe T, Hirano M and Hosono H 2008 *J. Am. Chem. Soc.* **130** 3296
[16] Hartnoll S A, Herzog C P and Horowitz G T 2008 *Phys. Rev. Lett.* **101** 031601 [arXiv:0803.3295[hep-th]], 2008 *JHEP* **0812**, 015 [arXiv:0810.1563[hep-th]]
[17] Horowitz G T 2011 *From Gravity to Thermal Gauge Theories: The AdS/CFT Correspondence*, (ed

E Papantonopoulos) Springer [arXiv:1002.1722[hep-th]], Hartnoll S A 2009 *Class. Quant. Grav.* **26** 224002 [arXiv:0903.3246[hep-th]], Herzog C P 2009 *J. Phys. A* **42** 343001 [arXiv:0904.1975[hep-th]], McGreevy J 2010 *Adv. High Ener. Phys.* **2010** 723105 [arXiv:0909.0518[hep-th]]

[18] Gubser S S 2008 *Phys. Rev. D* **78** 065034 [arXiv:0801.2977[hep-th]]

[19] Siopsis G and Therrien J 2010 *JHEP* **1005** 013 [arXiv:1003.4275[hep-th]], Herzog C P 2010 *Phys. Rev. D* **81** 126009 [arXiv:1003.3278[hep-th]]

[20] Gregory R, Kanno S and Soda J 2009 *JHEP* **0910** 010 [arXiv:0907.3203[hep-th]], Barclay L, Gregory R, Kanno S and Sutcliffe P 2010 *JHEP* **1012** 029 [arXiv:1009.1991[hep-th]]

[21] Horowitz G T and Roberts M M 2009 *JHEP* **0911** 015 [arXiv:0908.3677[hep-th]]

[22] Horowitz G T and Roberts M M 2008 *Phys. Rev. D* **78** 126008 [arXiv:0810.1077[hep-th]]

[23] Siani M 2010 *JHEP* **1012** 035 [arXiv:1010.0700[hep-th]]

[24] Gomes K K, Pasupathy A N, Pushp A, Ono S, Ando Y and Yazdani A 2007 *Nature* **447** 569

[25] Horowitz G T, Santos J E and Way B 2011 *Phys. Rev. Lett.* **106** 221601 [arXiv:1101.3326[hep-th]]

[26] Tinkham M 1996 *Introduction to Superconductivity* (McGraw Hill)

[27] Benini F, Herzog C P, Rahman R and Yarom A 2010 *JHEP* **1011** 137 [arXiv:1007.1981[hep-th]]

[28] Montull M, Pomarol A and Silva P J 2009 *Phys. Rev. Lett.* **103** 091601 [arXiv:0906.2396[hep-th]], Domenech O, Montull M, Pomarol A, Salvio A and Silva P J 2010 *JHEP* **1008** 033 [arXiv:1005.1776[hep-th]]

[29] Albash T and Johnson C V [arXiv:0906.0519[hep-th]], 2009 *Phys. Rev. D* **80** 126009 [arXiv:0906.1795[hep-th]]

Proceedings of the 7th International Conference on Gravitation and Cosmology (ICGC2011) IOP Publishing
Journal of Physics: Conference Series **484** (2014) 012003 doi:10.1088/1742-6596/484/1/012003

Connections between gravitational dynamics and thermodynamics

R-G Cai

State Key Laboratory of Theoretical Physics, Institute of Theoretical Physics, Chinese Academy of Sciences, P.O. Box 2735, Beijing 100190, China

E-mail: cairg@itp.ac.cn

Abstract. We have briefly summarized three aspects of connections between gravitational dynamics and thermodynamics by focusing on three kinds of spacetime horizon: Causal Rindler horizon of spacetime, black hole event horizon, and apparent horizon in a Friedmann-Robertson-Walker (FRW) universe. For the causal Rindler horizon, we have derived the Einstein's field equations from the Clausius relation. For black hole horizon in a static spherically symmetric spacetime, we have shown that at the black hole horizon, the Einstein's field equations can be cast to a form of the first law of thermodynamics. Applying to the Clausius relation to apparent horizon of a FRW universe, one is able to derive the Friedmann equations, not only in Einstein gravity, but also in Lovelock gravity. We have shown that there exists Hawking radiation associated with the apparent horizon in the FRW universe.

1. Introduction

According to Einstein, general relativity describes dynamics of spacetime by use of spacetime metric, while thermodynamics is another subject which describes macroscopic properties of thermal system in terms of energy, pressure, entropy, and temperature, etc. of the system. The first evidence for the deep connections between gravitational dynamics and thermodynamics comes from black hole thermodynamics. Black hole as a fantastic object is predicted by general relativity. Black hole thermodynamics tells us that black hole has a temperature proportional to its surface gravity, and an entropy proportional to its horizon area, and the temperature, entropy, and black hole mass satisfies the first law of thermodynamics. Not only the first law, other three laws of thermodynamics are also obeyed for black hole. In this sense, black hole is nothing but an ordinary thermodynamic system. However, black hole is a special thermal system because its entropy is proportional to its horizon area, while entropy of ordinary thermal system is proportional to its volume. Another feature of black hole thermodynamics is that heat capacity of some black holes may be negative, for example, Schwarzschild black hole.

The geometric feature of black hole temperature and entropy leads one to conjecture that gravity might be an emergent phenomenon, and is a coarse graining description of some underlying microscopic degrees of freedom. In fact, this idea was first proposed by Sakharov in 1967 [1], before black hole thermodynamics was set up. According to Sakharov, spacetime background emerges as a mean field approximation of underlying microscopic degrees of freedom, similar to hydrodynamics or continuum elasticity theory from molecular physics. In recent reviews [2, 3], Padmanabhan has summarized some aspects to understand gravity from thermodynamical viewpoint.

In this paper, we have studied some connections between gravitational dynamics and thermodynamics, in particular, focus on the relation between Einstein's field equations and the first law of thermodynamics. We have investigated the relation by using the three kinds of spacetime horizons: Rindler horizon, black hole horizon, and apparent horizon in a Friedmann-Robertson-Walker (FRW) universe.

The organization of this paper is as follows: In the next Section, we have briefly repeated the derivation process of Einstein's field equations from the first law of thermodynamics made by Jacobson by applying the latter to Rindler horizon of spacetime [4]. In Section 3, we have focused on black hole horizon and shown that the Einstein's field equations can be cast to a form of the first law of thermodynamics, not only in general relation, but also in Horava-Lifshitz gravity [5, 6]. In Section 4, we have focused on apparent horizon of FRW universe, and investigated the relation between Friedmann equations and the first law [7, 8, 9], and shown that there exists Hawking radiation associated with apparent horizon [10].

2. Rindler horizon: From the first law to Einstein's field equations

In the Rindler chart, the Minkowski space can be written as:

$$ds^2 = -x^2dt^2 + dx^2 + dy^2 + dz^2, \tag{1}$$

where $0 < x < \infty$, $-\infty < t,\ y,\ z < \infty$. There is a coordinate singularity at $x = 0$, where the acceleration of the Rindler observers diverges. The locus $x = 0$ in the Rindler chart corresponds to the locus $X^2 - T^2 = 0$ with $X > 0$ in the Cartesian chart of Minkwoski space, the latter consists of two null half-planes. The locus $x = 0$ is just the Rindler horizon, and the Rindler observer cannot see any information outside the Rindler wedge.

According to Davies [11] and Unruh [12], an observer, who is with acceleration a in Minkowski space, will detect a temperature given by $T = \frac{a}{2\pi}$, where the so-called geometric units have been taken: $c = \hbar = k_B = 1$. In the Rindler coordinates given in Eq. (1), for an observer with a fixed $x = x_0$, the acceleration $a = 1/x_0$.

Now, consider a certain event P in any spacetime, by equivalent principle, one can introduce a local inertial frame around P with Riemann normal coordinates. One further transforms the local inertial frame to a local Rindler frame by accelerating along an axis with acceleration κ. Then, there is the Unruh temperature associated with the local Rindler horizon as:

$$T = \kappa/2\pi. \tag{2}$$

Suppose the matter in the spacetime is described by the stress-energy tensor T_{ab}. Then the heat flux across the Rindler horizon $\mathcal{H}$ is given by:

$$\delta Q = \int_{\mathcal{H}} T_{ab}\chi^a d\Sigma^b, \tag{3}$$

where χ^a is an approximation boost Killing vector on $\mathcal{H}$. Note that the relations: $\chi^a = -\kappa\lambda k^a$ and $d\Sigma^a = k^a d\lambda dA$, where k^a is the tangent vector to the horizon generators for the fine parameter λ, which vanishes at P, and dA is the area element on a cross section of the horizon. Then, we have:

$$\delta Q = -\kappa \int_{\mathcal{H}} \lambda T_{ab}k^a k^b d\lambda dA. \tag{4}$$

Now, assume that the entropy is proportional to the horizon area, so that the entropy variation associated with a piece of horizon is given by:

$$dS = \eta\delta\mathcal{A}, \tag{5}$$

where η is a constant and $\delta\mathcal{A} = \int_{\mathcal{H}} \theta d\lambda dA$. Using the Raychaudhuri equation:

$$\frac{d\theta}{d\lambda} = -\frac{1}{2}\theta^2 - \sigma^2 - R_{ab}k^a k^b, \tag{6}$$

and assuming the vanishing of expansion θ and shear σ at P, to the leading order of λ, one has $\theta = -\lambda R_{ab}k^a k^b$, which leads to:

$$\delta\mathcal{A} = -\int_{\mathcal{H}} \lambda R_{ab}k^a k^b d\lambda dA. \tag{7}$$

By the Clausius relation $\delta Q = TdS$, one can have:

$$R_{ab} + f g_{ab} = \frac{2\pi}{\eta} T_{ab}, \tag{8}$$

where f is an arbitrary function. To determine the function f, we have employed the conservation law of the stress-energy tensor: $T^{ab}{}_{;b} = 0$. This leads to $f = -R/2 + \Lambda$, where Λ is a constant, which is nothing but the cosmological constant, as one will see shortly. Put f back to the above equation, one arrives at:

$$R_{ab} - \frac{1}{2}Rg_{ab} + \Lambda g_{ab} = \frac{2\pi}{\eta} T_{ab}. \tag{9}$$

One can see immediately from Eq. (9) that it is nothing but the Einstein's field equations, once $G = 1/4\eta$ is identified.

Thus, we have simply repeated the process to derive the Einstein's field equations from the Clausius relation made by Jacobson [4]. The key idea is to demand that this relation holds for all the local Rindler causal horizons through each spacetime point, with δQ and T interpreted as the energy flux and Unruh temperature seen by an accelerated observer just inside the horizon. In this sense, the Einstein's field equations are nothing but equations of state of spacetime. If this viewpoint is correct, it has significant implication for quantum gravity theory: It may be no more appropriate to canonically quantize the Einstein's field equations than it would be to quantize the wave equation for sound in air.

Further remarks: (i) For $f(R)$ gravity [13] and scalar-tensor gravity [9], it turns out that a non-equilibrium thermodynamic set up has to be employed in order to produce corresponding gravitational field equations. Namely, one needs an entropy production term added to the Clausius relation. (ii) It is further shown that assuming the non-vanishing of the shear for Einstein gravity, the non-equilibrium set up is needed, $dS = \delta Q/T + d_i S$, where the entropy production term is proportional to the squared shear of the horizon [14]. And it leads to a universal ratio of the shear viscosity η to entropy density of the horizon: $\eta/s = 1/4\pi$. (iii) For any diffeomorphism invariant theory, however, it has been shown recently that given Wald's entropy formula, by the Clausius relation, it is possible to derive the gravitational field equations [15, 16]. However, an issue exists that to have the Wald's entropy formula, one has to know first the gravity theory. In this sense, the derivation is not natural and satisfied [17].

3. Black hole horizon: Equivalence between Einstein's field equations and the first law of thermodynamics

We now move to black hole horizon and discuss the relation between the Einstein's field equations and the first law of thermodynamics [5, 6].

Let us first consider the Einstein's field equations:

$$G_{ab} \equiv R_{ab} - \frac{1}{2}Rg_{ab} = 8\pi G T_{ab}, \tag{10}$$

where G_{ab} is the Einstein tensor. Consider a generic static, spherically symmetric spacetime:

$$ds^2 = -f(r)dt^2 + f^{-1}(r)dr^2 + b^2(r)(d\theta^2 + \sin^2\theta d\phi^2), \tag{11}$$

where $f(r)$ and $b(r)$ are two continuous functions of r. Suppose the metric in Eq. (11) describes a black hole with a non-degenerated horizon at r_+. Then, the horizon has a Hawking temperature given by:

$$T = \frac{1}{4\pi}f'(r_+), \tag{12}$$

where a prime denotes the derivative with respect to r. The Einstein's field equations for the metric in Eq. (11) turn to be:

$$G^t_t = \frac{1}{b^2}\Big(-1 + fb'^2 + b(f'b' + 2fb'')\Big), \quad and \quad G^r_r = \frac{1}{b^2}(-1 + bf'b' + fb'^2). \tag{13}$$

At the black hole, where $f(r_+) = 0$, G^t_t and G^r_r become identical as:

$$G^t_t = G^r_r = \frac{1}{b^2}(-1 + bfb')|_{r_+}. \tag{14}$$

The $r - r$ component of the Einstein's field equations at the horizon can be written as:

$$-1 + bf'b' = 8\pi b^2 P, \tag{15}$$

where $P = T^r_r$ is the radial pressure of matter at the horizon. Now, we multiply a displacement dr_+ of the horizon on both sides of the above equation, the resulting equation can be cast as:

$$Td\left(\frac{4\pi b^2}{4G}\right) - d\left(\frac{r_+}{2G}\right) = PdV, \tag{16}$$

where V is the black hole volume and $dV = 4\pi b^2 dr_+$. This equation can be clearly rewritten as:

$$TdS - dE = PdV, \tag{17}$$

with identifications: $S = 4\pi b^2/4G = A/4G$ and $E = r_+/2G$. Note that Eq. (17) is nothing but the first law of thermodynamics. Here, T is the Hawking temperature of the black hole horizon and S is the Bekenstein-Hawking entropy of the black hole. Note that E is the Misner-Sharp energy at the horizon, and is not the ADM mass of the black hole.

Thus, we have shown that at a black hole horizon, the Einstein's field equations can be cast to the form of the first law of thermodynamics, which implies that there exists a deep connection between gravity and thermodynamics. It is further shown that even for Horava-Lifshitz gravity, which is not a diffeomorphism invariant theory, its field equation at black hole horizon also can be cast to the form of the first law of thermodynamics [6]. In addition, it is found that the story goes on for Lovelock gravity [18], BTZ black hole spacetime [19], stationary black holes and evolving spherically symmetric horizons [20]. However, a non-equilibrium thermodynamics setting is needed for $f(R)$ gravity [21].

4. Apparent horizon: Friedmann equation, the first law of thermodynamics and Hawking radiation

Now, we move to apparent horizon in a FRW universe.

4.1. From first law to Friedmann equation

Let us start with an $(n+1)$-dimensional FRW metric:

$$ds^2 = -dt^2 + a^2(t)\left(\frac{dr^2}{1-kr^2} + r^2 d\Omega_{n-1}^2\right), \tag{18}$$

where a is the scale factor and $d\Omega_{n-1}^2$ denotes an $(n-1)$-dimensional sphere with unit radius. Without loss of generality, one can take $k = 1$, 0 or -1, corresponding to a closed, flat or open universe respectively. Einstein's field equations in the FRW metric in Eq. (18) take the form:

$$H^2 + \frac{k}{a^2} = \frac{16\pi G}{n(n-1)}\rho, \quad and \tag{19}$$

$$\dot{H} - \frac{k}{a^2} = -\frac{8\pi G}{n-1}(\rho + p), \tag{20}$$

where ρ and p are energy density and pressure of the perfect fluid in the spacetime, $H = \dot{a}/a$ is the Hubble parameter, and the overdot stands for the derivative with respect to cosmic time t. Introducing the physical radius $\tilde{r} = ar$ and the metric in Eq. (18) can be rewritten as:

$$ds^2 = h_{ab}dx^a dx^b + \tilde{r}^2 d\Omega_{n-1}^2, \tag{21}$$

where $x^0 = t$, $x^1 = r$, and $h_{ab} = \mathrm{diag}(-1, a^2/(1-kr^2))$. By definition of apparent horizon, $h^{ab}\partial_a\tilde{r}\partial_b\tilde{r} = 0$, we have the apparent horizon radius as:

$$\tilde{r}_A = \frac{1}{\sqrt{H^2 + k/a^2}}. \tag{22}$$

Now, we have applied the first law of thermodynamics (the Clausius relation) $dE = TdS$, to the apparent horizon of the FRW universe. To do that, we have assumed that there exist temperature and entropy associated with the apparent horizon as:

$$T = \frac{1}{2\pi\tilde{r}_A}, \quad and \quad S = \frac{A}{4G}, \tag{23}$$

where $A = \tilde{r}_A^{n-1}\Omega_{n-1}$ is the apparent horizon area. Suppose the matter source in the FRW universe is a perfect fluid, its stress-energy tensor is given by: $T_{ab} = (\rho + p)U_a U_b + p g_{ab}$, where U_a is the 4-velocity. Following [22], defining two physical quantities, the energy supply vector Ψ_a, and work density W as:

$$\Psi_a = T_a^b\partial_b\tilde{r} + W\partial_a\tilde{r}, \quad and \quad W = -\frac{1}{2}T^{ab}h_{ab}, \tag{24}$$

one, then, can calculate the amount of energy across the apparent horizon within the time interval dt as:

$$dE = A(\rho + p)H\tilde{r}_a dt. \tag{25}$$

Using the energy dE together with the temperature T, and entropy S in Eq. (23), the Clausius relation leads to:

$$\dot{H} - \frac{k}{a^2} = -\frac{8\pi G}{n-1}(\rho + p). \tag{26}$$

Clearly, this is nothing but the second Friedmann Eq. (20). Further, using the continuity equation $\dot{\rho} + n(\rho + p) = 0$, and integrating Eq. (26) yields:

$$H^2 + \frac{k}{a^2} = \frac{16\pi G}{n(n-1)}\rho. \tag{27}$$

Here, an integration constant has been put into the energy density ρ. In fact, the integration constant is just the cosmological constant. This way the cosmological constant appears as an integration constant, whose exact value should be given by initial condition. Thus, the so-called cosmological constant problem does not appear here.

Thus, we have derived the Friedmann equations governing the dynamics of the FRW universe by applying the first law to the apparent horizon of the spacetime with assumption in Eq. (23). In fact, replacing the entropy formula in Eq. (23) by corresponding ones in Gauss-Bonnet gravity and more general Lovelock gravity, in the same way we are also able to derive corresponding Friedmann equations in those gravity theories [7].

The above approach can be used in more general cases [23], in fact. For example, given a relation between entropy and apparent horizon area, we are able to obtain corresponding modified Friedmann equation. (For details, see [24]).

4.2. Equivalence between the Friedmann equation and the first law

Now, we have proved that the Friedmann equation at the apparent horizon can be cast to the form of the first law of thermodynamics [8].

Let us start with the FRW metric in Eq. (18), by definition, $\kappa = \frac{1}{2\sqrt{-h}}\partial_a(\sqrt{-h}h^{ab}\partial_b\tilde{r})$, one has the surface gravity at apparent horizon, given by:

$$\kappa = -\frac{1}{\tilde{r}_A}(1 - \frac{\dot{\tilde{r}}_A}{2H\tilde{r}_A}). \tag{28}$$

In terms of the apparent horizon radius, the first Friedmann Eq. (19) can be expressed as:

$$\frac{1}{\tilde{r}_A^2} = \frac{16\pi G}{n(n-1)}\rho. \tag{29}$$

Taking derivative on both sides of the above equation and using the continuity equation, one has:

$$\frac{1}{\tilde{r}_A^3}d\tilde{r}_A = \frac{8\pi G}{n-1}(\rho + p)H dt. \tag{30}$$

With identification $T = \kappa/2\pi$, and $S = A/4G$, this equation can be rewritten as:

$$dE = TdS + WdV, \tag{31}$$

where $E = \rho V$, V is the volume of the region inside the apparent horizon, and $W = (\rho - p)$. This is nothing but the form of the so-called unified first law at the apparent horizon of the FRW universe. It is also shown that the form in Eq. (31) also holds for the Gauss-Bonnet gravity and Lovelock gravity. Further, it is found to hold in brane world scenarios [25]. With this, we can obtain the entropy expression for apparent horizon in brane world scenarios. (For a review, see [26]).

It is interesting to note that with the first law given in Eq. (31), starting from some modified Friedmann equation, one is able to get entropy expression of apparent horizon in some quantum corrected gravity. For example, one can obtain the entropy expression associated with the apparent horizon in loop quantum cosmology [24].

4.3. Hawking radiation of apparent horizon

In the process to derive the Friedmann equation, we have assumed that there is a Hawking temperature given in Eq. (23) associated with apparent horizon in a FRW universe. In this subsection, we have proved this by using the quantum tunnelling approach, which is first employed by Parikh and Wilczek [27] to discuss Hawking radiation for black hole horizon.

In terms of the physical radial coordinate $\tilde{r} = ar$, the 4-dimensional FRW metric can be expressed as:

$$ds^2 = -\frac{1-\tilde{r}^2/\tilde{r}_A^2}{1-k\tilde{r}^2/a^2}dt^2 - \frac{2H\tilde{r}}{1-k\tilde{r}^2/a^2}dtd\tilde{r} + \frac{1}{1-k\tilde{r}^2/a^2}d\tilde{r}^2 + \tilde{r}^2 d\Omega_2^2. \tag{32}$$

In the spherically symmetric metric in Eq. (32), one can define a Kodama vector $K^a = -\epsilon^{ab}\nabla_b\tilde{r} = \sqrt{1-k\tilde{r}^2/a^2}(\partial/\partial t)^a$. The norm of the Kodama vector is $K^2 = -(1-\tilde{r}^2/\tilde{r}_A^2)$. One sees that the Kodama vector is time-like, null and spacelike, inside the apparent horizon, at the apparent horizon, and outside the apparent horizon respectively. As a result, the Kodama vector can play the role as a Killing vector does in a de Sitter space in static coordinates.

Within WKB approximation, a particle with mass m in the metric in Eq. (32) satisfies the Hamilton-Jacob equation:

$$g^{\mu\nu}\partial_\mu \mathbf{S}\partial_\nu \mathbf{S} + m^2 = 0, \tag{33}$$

where $\mathbf{S}$ is its action. In the s-wave approximation, one can define the energy, and the radial wave-number of the particle as:

$$\omega = -K^a\partial_a\mathbf{S} = -\sqrt{1-k\tilde{r}^2/a^2}\partial_t\mathbf{S}, \quad and \quad k_{\tilde{r}} = (\partial/\partial\tilde{r})^a\partial_a\mathbf{S} = \partial_{\tilde{r}}\mathbf{S}. \tag{34}$$

And the action can be expressed as:

$$\mathbf{S} = -\int \frac{\omega}{\sqrt{1-k\tilde{r}^2/a^2}}dt + \int k_{\tilde{r}}d\tilde{r}. \tag{35}$$

Then, the Hamilton-Jacob Eq. (33) turns to be:

$$-\frac{\omega^2}{1-k\tilde{r}^2/a^2} + \frac{2H\tilde{r}\omega}{\sqrt{1-k\tilde{r}^2/a^2}}k_{\tilde{r}} + (1-\frac{\tilde{r}^2}{\tilde{r}_A^2})k_{\tilde{r}}^2 + m^2 = 0, \tag{36}$$

which has the solution as:

$$k_{\tilde{r}} = \frac{-H\tilde{r} \pm \sqrt{H^2\tilde{r}^2 + (1-\tilde{r}^2/\tilde{r}_A^2)[1-m^2(1-k\tilde{r}^2/a^2)/\omega^2]}}{(1-\tilde{r}^2/\tilde{r}_A^2)\sqrt{1-k\tilde{r}^2/a^2}}\omega, \tag{37}$$

where the plus/minus sign corresponds to an outgoing/incoming mode. Now, we have considered an incoming mode, since the observer is inside the apparent horizon, like the case of particle tunnelling for the cosmological event horizon in de Sitter space [28]. It is obvious that the action $\mathbf{S}$ has a pole at the apparent horizon. Through a contour integral, we have obtained an imaginary part of the action as:

$$\begin{aligned} \text{Im}\,\mathbf{S} &= -\text{Im}\int \frac{H\tilde{r} + \sqrt{H^2\tilde{r}^2 + (1-\tilde{r}^2/\tilde{r}_A^2)[1-m^2(1-k\tilde{r}^2/a^2)/\omega^2]}}{(1-\tilde{r}^2/\tilde{r}_A^2)\sqrt{1-k\tilde{r}^2/a^2}}\omega d\tilde{r} \\ &= \pi\tilde{r}_A\omega. \end{aligned} \tag{38}$$

In the WKB approximation, the emission rate Γ is the square of the tunnelling amplitude (here, the particle tunnels from outside to inside the apparent horizon), and is given by:

$$\Gamma \propto \exp(-2\ \text{Im}\,\mathbf{S}). \tag{39}$$

Combining Eq. (39) with Eq. (38), one can see clearly that the emission rate can be cast in a form of thermal spectrum $\Gamma \sim \exp(-\omega/T)$, with temperature as:

$$T = \frac{1}{2\pi\tilde{r}_A}. \tag{40}$$

Thus, we have finished the proof that an observer inside the apparent horizon will see a thermal spectrum with temperature given by Eq. (40) when particles tunnel from outside the apparent horizon to inside the apparent horizon. This can be explained as Hawking radiation of apparent horizon in the same spirit in the tunnelling approach proposed by Parikh and Wilczek that the Hawking radiation of black hole is expressed as a tunnelling phenomenon. Furthermore, at this level of approximation, the mass of particle does not enter the emission rate. This is just the remarkable feature of thermal spectrum.

5. Conclusion

A surge of evidence shows that there exist deep connections between gravity and thermodynamics. In this paper, we have just mentioned pieces of them. The intrinsic relation between gravity and thermodynamics is required to further study. If gravity is, indeed, not a fundamental interaction in Nature, instead, it is an induced coarse graining description of some microscopic degrees of freedom of spacetime, it is natural to see the deep connection between gravity and thermodynamics, and even further to see the relation between gravity and hydrodynamics. The latter is a very active issue under study recently in the framework of AdS/CFT correspondence.

Acknowledgements

The author thanks the organizers of ICGC-2011 for kind invitation and for warm hospitality during the conference. This work was supported in part by the National Natural Science Foundation of China (No. 10821504, No. 10975168, and No. 11035008), and by the Ministry of Science and Technology of China under Grant No. 2010CB833004.

References

[1] Sakharov A D 1968 *Sov. Phys. Dokl.* **12** 1040
[2] Padmanabhan T 2010 *Report Prog. Phys.* **73** 046901
[3] Padmanabhan T 2011 *J. Phys. Conf. Ser.* **306** 012001
[4] Jacobson T 1995 *Phys. Rev. Lett.* **75** 1260
[5] Padmanabhan T 2002 *Class. Quant. Grav.* **19** 5387
[6] Cai R G and Ohta N 2010 *Phys. Rev.D* **81** 084061
[7] Cai R G and Kim S P 2005 *JHEP* **0502** 050
[8] Akbar M and Cai R G 2007 *Phys. Rev. D* **75** 084003
[9] Cai R G and Cao L M 2007 *Phys. Rev. D* **75** 064008
[10] Cai R G, Cao L M and Hu Y P 2009 *Class. Quant. Grav.* **26** 155018
[11] Davies P C W 1975 *J. Phys. A* **8** 609
[12] Unruh W G 1976 *Phys. Rev. D* **14** 870
[13] Eling C, Guedens R and Jacobson T 2006 *Phys. Rev. Lett.* **96** 121301
[14] Eling C 2008 *JHEP* **0811** 04
[15] Brustein R and Hadad M 2009 *Phys. Rev. Lett.* **103** 101301, 2010 *Erratum-ibid* **105** 239902
[16] Parikh M K and Sarkar S 2009 [arXiv:0903.1176[hep-th]]
[17] Padmanabhan T 2009 [arXiv:0903.1254[hep-th]]
[18] Paranjape A, Sarkar S and Padmanabhan T 2006 *Phys. Rev. D* **74** 104015
[19] Akbar M and Siddiqui A A 2007 *Phys. Lett. B* **656** 217
[20] Kothawala D, Sarkar S and Padmanabhan T 2007 *Phys. Lett. B* **652** 338
[21] Akbar M and Cai R G 2007 *Phys. Lett. B* **648** 243
[22] Hayward S A, Mukohyama S and Ashworth M C 1999 *Phys. Lett. A* **256** 347
[23] Akbar M and Cai R G 2006 *Phys. Lett. B* **635** 7

[24] Cai R G, Cao L M and Hu Y P 2008 *JHEP* **0808** 090
[25] Cai R G and Cao L M 2007 *Nucl. Phys. B* **785** 135, Sheykhi A, Wang B and Cai R G 2007 *Nucl. Phys. B* **779** 1, *Phys. Rev. D* **76** 023515
[26] Cai R G 2008 *Prog. Theor. Phys. Supp.* **172** 100
[27] Parikh M K and Wilczek F 2000 *Phys. Rev. Lett.* **85** 5042
[28] Parikh M K 2002 *Phys. Lett. B* **546** 189, Medved A 2002 *Phys. Rev. D* **66** 124009

Expressing entropy globally in terms of (4D) field-correlations

R D Sorkin

Perimeter Institute, 31 Caroline Street North, Waterloo ON, N2L 2Y5 Canada, and Department of Physics, Syracuse University, Syracuse, NY 13244 - 1130, USA

E-mail: rsorkin@perimeterinstitute.ca

Abstract. We have expressed the entropy of a scalar field directly in terms of its spacetime correlation function, assuming that the higher correlators are of "Gaussian" form. The resulting formula associates an entropy to any spacetime region, and when this region is globally hyperbolic with Cauchy surface then the entropy can be interpreted as the entropy of the reduced density-matrix belonging to the Cauchy surface. One acquires, in particular, a new expression for the entropy of entanglement across an event-horizon. Thanks to its spacetime character, this expression makes sense in a causal set as well as in a continuum spacetime.

1. Introduction

As usually conceived of, entropy is local in time, being defined relative to a hypersurface Σ. To compute such an entropy (at least in its "Gibbsian" guise), one must be able to identify a density-matrix $\rho(\Sigma)$, which one can plug into the formula $S = \mathrm{Tr}\,\rho\log\rho^{-1}$. But according to our current understanding, quantum fields are, in general, too singular to admit meaningful restriction to lower dimensional subsets of spacetime. If it is, therefore, doubtful whether a concept like "state on a hypersurface" can be well defined, it is virtually certain that this concept will break down in the discrete context of a causal set. For reasons such as these, it would be desirable to define entropy in a more global manner by associating it with a spacetime region rather than a sub-manifold of co-dimension one. A more global notion of entropy would also seem fitting in connection with black holes, whose very definition is global in character.

Of course, black hole entropy will not be understood fully except against the background of a theory of quantum gravity and quantum spacetime. Nevertheless, it seems fair to say that quantum correlations between the hole's interior and its exterior must contribute to its entropy. Perhaps, when suitably understood, some such entropy of entanglement will even turn out to tell the whole story. Be that as it may, a portion of the entanglement entropy can be identified without invoking full quantum gravity, namely that portion belonging to whatever "matter" fields are present in the neighbourhood of the horizon.

When we compute such an entropy in the continuum, we obtain an infinite answer, as is well known. Introducing a cutoff ℓ on the other hand, we obtain $S_{entanglement} \sim \frac{A}{\ell^2}$, which compares favourably with the exact formula $S_{BH} = \frac{2\pi A}{\ell_p^2}$, where $\ell_p = \sqrt{8\pi G}$ is the rationalized Planck length [1]. We have recognized here the familiar area law, but with a coefficient that is not easily determined because it is not "universal".

On one hand, this sensitivity of S to the details of the cutoff is to be welcomed, because it lets us learn something about the magnitude of ℓ and the nature of Planck-scale physics. (Although they are sometimes viewed as uninteresting; non-universal quantities are actually the most interesting if we are after the microscopic physics!)

But on the other hand, the very fact that the answer is not universal also creates a difficulty. When we try to compute $S_{entanglement}$ with finite ℓ, the answer will depend on the details of how ℓ is introduced (sometimes called, rather misleadingly, "scheme dependence"). In itself, this ambiguity might not be a problem, but for the fact that it tends to call into question the covariance of S. For example, in computing entanglement entropy for a Schwarzschild black hole, we might think to neglect all contributions from within a distance ℓ of the horizon, but what does this mean? Do we neglect modes within a shell $\Delta r = \ell$, where r is the Schwarzschild radial coordinate? Or should we be using instead of r, proper distance along the spacelike surface Σ, whose entropy we want (proper distance to a lightlike surface being not in itself well-defined)? But then the answer will depend on how we extend Σ away from the horizon H. If, for example, we define Σ as a surface of constant Schwarzschild-time t then it will go through the bifurcation 2-sphere. But realistic black holes do not have such bifurcation surfaces; and even if they did, our attention when we came to consider the crucial question of entropy *increase* would have to focus on portions of the horizon which were farther to the future.

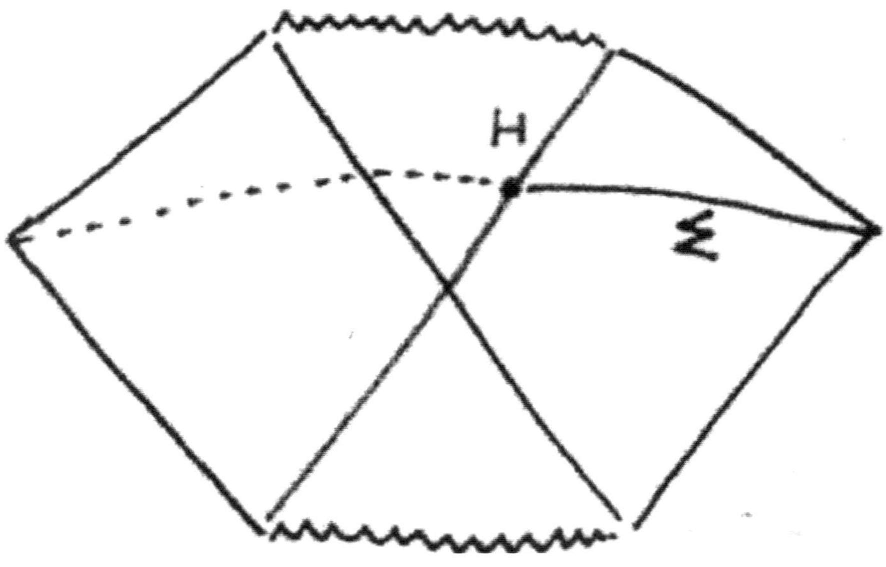

Figure 1. The black hole horizon H and a hypersurface Σ, whose entropy is desired. The dotted line extends Σ to a Cauchy surface of the full spacetime.

2. Entanglement entropy in a causal set

The difficulties we have just reviewed make it hard to define an entropy of entanglement that is at the same time covariant, but if one replaces spacetime by a causal set, the difficulties disappear because no cutoff is needed. That, of course, is one reason for thinking of the causal set as more fundamental than the continuum [2, 3, 4]. But, even if one is seeking no more than a "phenomenological cutoff" to render the entropy finite, a causet sprinkled into the spacetime in question offers (uniquely as far as we know) a covariant way to obtain one. For this reason alone, it is worth asking whether a natural definition of entanglement entropy can be found within causal set theory.

Let us consider then, the problem of defining a horizon-entropy in a fixed, background causal set. First of all, one must attach a meaning to the words "event horizon". Although, this might seem to pose a problem because a causet contains no subset that could play the role of a null surface, the distinction between the interior and exterior of a black hole still makes sense because it depends solely on the causal order. Since quantum entanglement concerns only the interior and exterior regions, and not the horizon per se, there is, in fact, no problem.

The next step, if we were to mimic the continuum story perfectly, would be to define the concept of a spacelike (or achronal) surface Σ, after which we would try to associate with each suitable Σ some sort of "momentary state" or density matrix ρ_Σ. In fact, there does exist a natural choice for Σ, since we can define it as an anti-chain, or better a maximal anti-chain (or "slice"). That which is not readily available in a causet, however, is anything like ρ_Σ: a causet-based field theory admits no obvious notion of "state on a hypersurface".

More generally, none of the surface-quantities which one routinely defines in the continuum seems to carry over naturally to an anti-chain, except possibly a measure of (spatial) volume. In fact, it seems to be a rule of thumb that "purely spatial" concepts do not live happily in a causet[1] including, for example, the concepts of induced metric g_{jk} and extrinsic curvature $K_{jk} \sim \dot{g}_{jk}$. For this reason, a causet is not congenial to canonical quantum gravity, albeit one *can* define many surface-quantities in an approximate sense, including both metrical ones and topological ones like homology groups [5].

The question then is whether we can free $S_{entanglement}$ from reference to a density matrix localized to a hypersurface. In seeking the answer, we have limited ourself to what is perhaps the simplest case, that of a free scalar field $\hat{\phi}(x)$. For this case, a corresponding theory exists in the causet, and it has been given both algebraic and path-integral (or "quantum measure") formulations. In as much as the path-integral formulation extends to the interacting case, it is arguably preferable to the operator formulation. Nevertheless, we have worked in the latter context, first of all because more tools are available there, and more importantly, we seem to lack any definition of entropy couched in the language of histories[2].

Could it be that this lack betokens an inherent dependence of entropy on some notion of "state localized in time"? Were that true, the search for a histories-based definition of entropy would necessarily be futile, but such pessimism was called into question by the global definition of entropy, we have arrived at below in the algebraic context. If indeed, a more (Lorentzian and non-equilibrium) histories-based concept of entropy is out there somewhere, its discovery would, to our mind, mark important progress in the path-integral formulation of quantum theories.

3. Entropy from the spacetime correlator

Let us recall some possible definitions of entropy in the context of quantum field theory. Conceived algebraically (as it commonly is), a quantum field is a collection of operators[3] $\phi(x)$ acting irreducibly in a Hilbert space $\mathfrak{H}$, together with a "global state", represented by another operator ρ (called "the density-matrix"), such that $\mathrm{Tr}\,\rho A$ yields the "expectation value" $\langle A \rangle$ of

[1] More generally, there is evidence for the rule that it is the continuum concepts, which are "C^0-stable" that have simple causet counterparts.

[2] We are ignoring here the trick that computes the thermodynamic partition function via a Wick-rotated path integral. That procedure is indeed histories-based, but the histories in question develop in imaginary time, and more importantly, the entropy defined thereby is limited to states of thermal equilibrium.

[3] In the continuum, the symbol $\phi(x)$ must, of course, be interpreted formally, since, the field ϕ is at best an operator-valued distribution. In the following lines, we have ignored this technicality and whatever difficulties flow from it. They are peculiar to continuous spacetime and do not seem to be germane to the causal set context of primary interest in this paper, within which, the symbol ϕ^j belonging to element j of the causet will literally denote an operator, albeit an unbounded one.

A. One can then define the *entropy* as $S = S(\rho) = \mathrm{Tr}\,\rho\log\rho^{-1}$. (In practice the ρ that enters this definition will seldom be the "exact microscopic density matrix". Rather it will be some coarse-grained version thereof, whose entropy is non-zero even when the underlying "microscopic state" is pure.)

Now, let R be an arbitrary spacetime region. The operators $\phi(x)$ for $x \in R$ generate a sub-algebra $\mathfrak{A}_R$ of $\mathfrak{A}$, and the expectation-value for $\mathfrak{A}$ restricts to a similar functional on $\mathfrak{A}_R$. In general, the sub-algebra $\mathfrak{A}_R$ will no longer act irreducibly in $\mathfrak{H}$. Assuming, however, that we can represent $\mathfrak{A}_R$ in some other Hilbert space $\mathfrak{H}_R$, in which it does act irreducibly, and assuming further that we can find in $\mathfrak{H}_R$ a density matrix ρ_R, such that $\langle A\rangle = \mathrm{Tr}\,\rho_R A$ for operators $A \in \mathfrak{A}_R$, we can go on to define $S(R) = -\,\mathrm{Tr}\,\rho_R\log\rho_R$, *the entropy of ρ relative to the region R.*

Remark:
Let $\mathfrak{A}$ be the algebra generated by the $\phi(x)$. As such, it is a concrete algebra of linear operators in $\mathfrak{H}$, but in what might be called the "strictly algebraic" formulations of quantum field theory, $\mathfrak{A}$ is regarded as an abstract $\star$-algebra, while a density-matrix ρ is regarded simply as an expectation-value-functional on $\mathfrak{A}$. Since the set of all such functionals is convex, any given ρ will be (either exactly or to a close approximation) a convex combination $\sum_\alpha p_\alpha\rho_\alpha$ of extremal (or "pure") states ρ_α. In this abstract setting, $S(\rho)$ could perhaps be defined as the infimum[4] over all such sums of the quantity $\sum_\alpha p_\alpha\log p_\alpha^{-1}$. In this way, one could imagine defining the entropy $S(R)$ directly in terms of the expectation-value functional, without recourse to either $\mathfrak{H}_R$ or ρ_R.

The definition we have just given of the entropy of an arbitrary spacetime region R is the basis of the causal set construction we are seeking. First, though, we need to relate the entropy of a region to the entropy of a hypersurface. To define the latter itself, we can imagine repeating the same steps that led to $S(R)$, only with R shrunken down to a hypersurface Σ, and with $\mathfrak{A}_\Sigma$ then being the algebra generated (formally) by the "initial data" operators $\phi(x)$ and $\dot\phi(x)$ for $x \in \Sigma$.

The point now is that when Σ is a Cauchy surface for the region R, we have $\mathfrak{A}_\Sigma = \mathfrak{A}_R$, and therefore, $S(\Sigma)$ coincides with the more globally defined entropy $S(R)$. By taking for R, the so-called *domain of dependence* of Σ, we can, thus, express the entropy of any desired hypersurface as the entropy of a spacetime region. And this in turn will let us express it directly in terms of the correlation function $\langle\phi(x)\,\phi(y)\rangle$.

In application to a black hole spacetime, Σ would be the exterior portion of the hypersurface, for which we desired the entanglement entropy, and the region R would then be given by:

$$R = \mathrm{future}(\Sigma) \cap \mathrm{past}(H) = D^+(\Sigma). \tag{1}$$

Remark:
Time-reversing the above, we could also have taken $R = D^-(\Sigma)$, but the analogue of Eq. (1) would not hold with that choice, making it apparently less convenient when transposed to a causet.

In summary, *entanglement entropy is a special case of the entropy $S(R)$ of a spacetime region.* Our task now is to find a simple formula for $S(R)$, assuming, as always, that ϕ is free and "Gaussian".

4. Sketch of a derivation of a formula

What follows is in parts still a "work in progress"; not every loose end has been tied up: We have arrived at an algebra $\mathfrak{A} = \mathfrak{A}(R)$ and an expectation-functional or "state" $\langle\cdot\rangle$ thereon. In order to derive from these structures an entropy $S(R)$, we will seek to represent $\mathfrak{A}(R)$ irreducibly in a

[4] For $\dim\mathfrak{H} < \infty$ this is proved in [6].

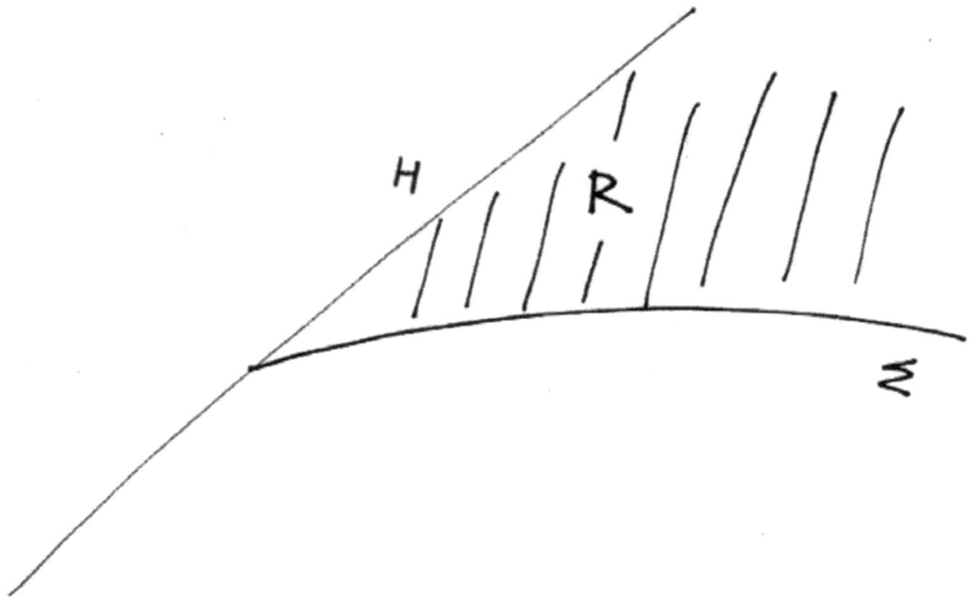

Figure 2. The entropy of the surface Σ can be identified with the entropy of the region R

Hilbert space $\mathfrak{H} = \mathfrak{H}(R)$, and to find therein a density-matrix $\rho : \mathfrak{H} \to \mathfrak{H}$, such that $\langle A \rangle = \mathrm{Tr}\, \rho A$ for all $A \in \mathfrak{A}$. In what follows, we have omitted the reference to R, as if R were the full spacetime or causet as the case may be. There is no loss of generality in doing so, since we do not need to refer again to spacetime points (respectively causet elements) outside of R. Also, we have written "ϕ^j" for an individual field value. This notation is more convenient than "$\phi(x)$", and it is also more apt for a causet C, where the index j runs literally from 1 to N, $N = |C|$, being the cardinality of C.

Remark:
The condition that $\mathfrak{A}$ acts irreducibly in $\mathfrak{H}$ is crucial here. If we drop it, we could, by "purification", always find a representation in which $\mathrm{Tr}\, \rho \log \rho$ would vanish, because ρ would be the projector onto a single vector in $\mathfrak{H}$. This would be the case in the so-called GNS representation, and for that matter, in the representation of the ϕ^j in the original, global Hilbert space, assuming the original global state to have been pure. In view of these observations, it is somewhat disconcerting that, strictly speaking, the existence of an irreducible representation is not always guaranteed. Consider, for example, the "region" of a causet consisting of a single element e. The corresponding sub-algebra of $\mathfrak{A}$ will be generated by a single field-value $\phi(e) =: q$. Now by assumption, $\langle q \rangle = 0$, but in general $\langle qq \rangle$ will not vanish. If it does not, then q cannot act irreducibly in any representation (because then Schur's lemma $\Rightarrow q = c \in \mathbb{R} \Rightarrow q = \langle q \rangle = 0 \Rightarrow \langle qq \rangle = \langle 0 \rangle = 0$). In such a case, the entropy could not be defined by the means we have adopted. However, the "strictly algebraic" definition we have remarked earlier would still seem to apply and to yield $S = \infty$. In the following, we have simply assumed that an irreducible representation does exist. The formula in Eq. (9), which we will derive, will not actually depend on this assumption, if interpreted as in Eq. (8). It would, however, yield $S = 0$ in our one-element example, rather than $S = \infty$.

Because, we are dealing with a free field, the commutator of ϕ^j with ϕ^k is a c-number and we can write:

$$[\phi^j, \phi^k] = i\Delta^{jk}, \tag{2}$$

for some real, skew matrix Δ^{jk}. We have defined similarly the *Wightman function*[5] as the Hermitian matrix:

$$W^{jk} \equiv \langle \phi^j \phi^k \rangle.$$

Notice that Δ is twice the imaginary part of W.

Assume now that our theory is "Gaussian" (or "Wickian") in the sense that the Bosonic form of *Wick's rule* holds with $\langle \phi \rangle = 0$:

$$\langle \phi\phi \cdots \phi \rangle = \sum \langle \phi\phi \rangle \cdots \langle \phi\phi \rangle.$$

(The sum is taken over all ways of pairing the ϕ's, and within each pairing, the original odering must be preserved.) Thanks to this assumption the "two-point function" W^{jk} determines the theory fully.

For example, its imaginary part determines the "equations of motion" for ϕ. Thus (thinking for a moment of the theory on the entire spacetime or causet), if the sequence α_j is in the kernel of Δ, in the sense that $\sum \Delta^{kj}\alpha_j = 0$, then $A = \sum \alpha_j \phi^j$ vanishes as well. (Proof: When $\alpha \in \ker \Delta$, A commutes with every ϕ^j, as follows from the calculation $[\phi^j, A] = [\phi^j, \alpha_k \phi^k] = i\Delta^{jk}\alpha_k = 0$. But, since the ϕ^j act irreducibly in $\mathfrak{H}$, this implies that $A = c\mathbf{1}$, and then $c = \langle A \rangle = \alpha_j \langle \phi^j \rangle = 0$, whence $A = 0$.) This last fact yields a set of linear relations among the operators ϕ. In the continuum, they are just the equations of motion for $\phi(x)$ (in our case, the Klein-Gordon equation), but in the causet, they yield only relatively a few conditions, because the *exact* kernel of Δ is typically rather small.

Now, our algebra $\mathfrak{A}$ is generated by the individual field-variables ϕ^j, and we can, therefore, characterize it fully by specifying the relations among these generators. But, since we have taken our field to be non-interacting, this is quite simple to do. In addition to the linear relations just described, there are only the bilinear relations in Eq. (2) (the "canonical commutation relations"). Moreover, we have just seen that any zero eigenvector of the commutator-form Δ^{ij} results in a linear dependence among the ϕ^j. Hence, by passing to a linearly independent set of generators, we can assume that Δ^{ij} is invertible.

It follows that there exists a basis for the ϕ^j of the form q^α, p_α ($\alpha = 1, 2, \cdots, n$), such that:

$$[q^\alpha, p_\beta] = i\delta^\alpha_\beta \ .$$

By construction, this basis block-diagonalizes Δ, meaning it block-diagonalizes the skew part of W (Recall that $W^{jk} - W^{kj} = i\Delta^{jk} = [\phi^j, \phi^k]$). We claim that one can also choose it to diagonalize the symmetric part:

$$R^{jk} \equiv Re\ W^{jk} = \frac{1}{2}\langle \{\phi^j, \phi^k\} \rangle \ ,$$

hence, to block-diagonalize W^{jk} itself [6]. With this, our problem splits up into a product of individual problems, each concerning a single pair $(q,\, p)$ - a single degree of freedom - that one might imagine as position and momentum for a free particle or harmonic oscillator (or as the field-components ϕ^j for a 2-element causet, a 2-chain). It, thus, suffices to evaluate the entropy in that special case.

[5] Some connections between the Wightman function and the entropy of a Gaussian density matrix are also discussed in [7].

[6] In fact, diagonalization is possible in general, without assuming that the ϕ^j are linearly independent. Given any positive-semidefinite matrix W, with associated real matrices R and Δ, such that $W = R + i\Delta/2$, one can introduce a (real) basis in which R is diagonal and Δ is block-diagonal with 2×2 blocks. Moreover, one can arrange that the non-zero matrix elements $R^{jj} = 1$. The proof is slightly too long to reproduce here, but see [8] for a similar result that applies when Δ is invertible.

4.1. The entropy for a single degree of freedom

Our task is now as follows: Given a conjugate pair of variables q and p such that $[q,p]=i$, and given the correlators:

$$\langle qq\rangle \quad \langle pp\rangle \quad Re\,\langle qp\rangle$$

for a Gaussian density-matrix ρ, to find $S(\rho)=\mathrm{Tr}\,\rho\log\rho^{-1}$. ρ is Gaussian, which means that in a q-basis it takes the form:

$$\rho(q,q')\equiv\langle q|\rho|q'\rangle=(cst)\exp\left(-\frac{A}{2}(q^2+q'^2)+\frac{iB}{2}(q^2-q'^2)-\frac{C}{2}(q-q')^2\right). \tag{3}$$

Then, $S(\rho)$ must be some function of the parameters A, B, and C, and our task is to determine this function.

To that end, notice first that S must be dimensionless and invariant under unitary transformations. From this, it can be shown that S can depend on the correlators only in the combination $\langle qq\rangle\langle pp\rangle-(Re\,\langle qp\rangle)^2=\frac{detR}{det\Delta}$, where in this simple 2×2 situation, Δ and R reduce to:

$$\Delta=2\,Im\begin{pmatrix}\langle qq\rangle & \langle qp\rangle\\ \langle pq\rangle & \langle pp\rangle\end{pmatrix}=\begin{pmatrix}0 & 1\\ -1 & 0\end{pmatrix},$$

and

$$R=Re\begin{pmatrix}\langle qq\rangle & \langle qp\rangle\\ \langle pq\rangle & \langle pp\rangle\end{pmatrix}=\begin{pmatrix}\langle qq\rangle & Re\,\langle qp\rangle\\ Re\,\langle qp\rangle & \langle pp\rangle\end{pmatrix}.$$

It turns out that (as a short calculation will confirm):

$$\langle qq\rangle\langle pp\rangle-(Re\,\langle qp\rangle)^2=\frac{C}{2A}+\frac{1}{4},$$

whence the entropy depends only on the ratio $\frac{C}{A}$, while B drops out entirely. But, with B set to 0, we can take over from [9] the entropy calculation there, with the result[7]:

$$-S=\frac{\mu\log\mu+(1-\mu)\log(1-\mu)}{1-\mu},$$

where

$$\mu=\frac{\sqrt{1+2C/A}-1}{\sqrt{1+2C/A}+1}.$$

Putting the pieces together, yields now:

$$S=\left(\sigma+\frac{1}{2}\right)\log\left(\sigma+\frac{1}{2}\right)-\left(\sigma-\frac{1}{2}\right)\log\left(\sigma-\frac{1}{2}\right), \tag{4}$$

where we have defined the eigenvalues of $\Delta^{-1}R$ (which are purely imaginary) to be $\pm\, i\sigma$: spectrum $(\Delta^{-1}R)=\pm\, i\sigma$.

This nice symmetrical expression is already quite simple, but we can simplify it still further by swapping the eigenvalues of $\Delta^{-1}R$ for those of $\Delta^{-1}W=\Delta^{-1}R+\frac{i}{2}$ (Recall that $W=R+\frac{i\Delta}{2}$). Writing these latter eigenvalues as $\pm\, i\omega_\pm=i(\frac{1}{2}\pm\sigma)$, brings S finally to the convenient form:

$$S=\omega_+\log\omega_+-\omega_-\log\omega_-. \tag{5}$$

Notice in this connection that positivity of the correlation matrix (also known as the uncertainty principle) implies that $\sigma\geq\frac{1}{2}$, whence ω_+ and $\omega_-=\omega_+-1$ are both ≥ 0.

[7] Might there be a more direct route from Eq. (3) to $S(\rho)$, perhaps via a diagonalization of ρ or via a use of the "replica trick"?

5. The entropy in full

The formula in Eq. (5) belongs to a single degree of freedom, corresponding to one block of our block-diagonalized matrix W. Summing of Eq. (5) over all the blocks then yields the total entropy in the form of a sum over the spectrum of the full matrix $\Delta^{-1}W$, which for short, we will call L, or more conveniently iL in oder that its eigenvalues be real, and therefore, we can write:

$$\Delta^{-1}W = iL \ . \tag{6}$$

Denoting the eigenvalues of L by λ, we have then:

$$S = \sum \lambda \log |\lambda| \ . \tag{7}$$

We have seen that the eigenvalues λ are all real, even though L is in general neither real nor symmetric. We have also seen that each negative λ is paired with a positive eigenvalue $1 - \lambda$.

As derived, this last formula relies on the fact that the q^α and p^α are all linearly independent. When we revert to the original matrices W^{jk} and Δ^{jk}, we will encounter (after their diagonalization) a block submatrix consisting entirely of zeroes. In this subspace, Δ is obviously not invertible, but neither is there any contribution to S. To adapt our formula in Eq. (7) to this circumstance, we can rewrite the eigenvalue equation for $\Delta^{-1}W$, so that it ignores this block of zeroes. To wit, we define the eigenvalues λ as the solutions of the equation $W^{jk}v_k = i\lambda\Delta^{jk}v_k$, where it is understood that $\Delta^{jk}v_k$ must not vanish (just as v must not vanish in the more traditional eigenvalue equation $Lv = \lambda v$). In other words, we have defined the eigenvalues of L as the solutions of the equation:

$$Wv = i\lambda\Delta v \qquad\qquad (\Delta v \neq 0) \ . \tag{8}$$

With this understanding (and with the usual convention that $0 \log 0 = 0$), we have finally:

$$S = \mathrm{Tr}\ L \log |L| \equiv \sum_\lambda \lambda \log |\lambda| \ . \tag{9}$$

A remarkably simple formula!

Remark:
Derived under the assumption that the ϕ^j can be represented irreducibly in some Hilbert space, Eqs. (8) and (9) give a good account of the entropy when the kernels of Δ and R (or equivalently W) coincide. From the condition $W \geq 0$ that W is positive semidefinite, it follows that $ker\Delta \supseteq kerR$, but the converse inclusion is not guaranteed in general. When it fails, no irreducible representation exists (as seen in an earlier remark), and our entropy is to that extent ill-defined. If we adopt the more general "strictly algebraic" definition described earlier, the resulting entropy will diverge. By way of illustration, consider our earlier example of a single "field operator" $\phi^1 = q$ with $\langle q \rangle = 0$ and $\langle qq \rangle = s^2$ (and with $\mathfrak{A}$ being the algebra of polynomials in q). One can think of q in this case as a classical random variable with Gaussian distribution function $\frac{1}{\sqrt{2\pi s^2}} \exp(\frac{-x^2}{2s^2})$, the pure states corresponding to specific sharp values $q = x$ for $x \in \mathbb{R}$. Since these states are uncountably infinite in number, our "strictly algebraic" entropy will be infinite, absent some sort of short-distance cutoff or regulator.

Remark:
Given the commutator function Δ^{jk}, the construction in [10] and [11] produces a distinguished "vacuum" by taking the two-point function W to be the positive projection of $i\Delta$, i.e., by setting $W = \sum'_n w_n |n\rangle\langle n|$, where the sum is over the postive eigenvalues w_n of the matrix

$i\Delta = \sum_n w_n |n\rangle\langle n|$. Evidently, the positive eigenvalues produce solutions of Eq. (8) with $\lambda = 1$, while the negative eigenvalues produce solutions with $\lambda = 0$. The net entropy, therefore, vanishes, as it must, since the vacuum in question is the "ground state" of a Fock-type representation, hence, a pure state.

Acknowledgements
Research at Perimeter Institute for Theoretical Physics is supported in part by the Government of Canada, through NSERC, and by the Province of Ontario, through MRI, and NSF grant PHY-0404646.

References

[1] Sorkin R D 1983 *On the Entropy of the Vacuum Outside a Horizon* eds B Bertotti, F de Felice and A Pascolini *Tenth International Conference on General Relativity and Gravitation* **II** 734 [http://www.pitp.ca/personal/rsorkin/some.papers/31.padova.entropy.pdf]
[2] Bombelli L, Lee J, Meyer D and Sorkin R D 1987 *Phys. Rev. Lett.* **59** 521
[3] Surya S *Recent Research in Quantum Gravity* ed A Dasgupta (Nova Science Publishers, New York) [arXiv:1103.6272v1 [gr-qc]]
[4] Dowker F 2005 *100 Years of Relativity - Spacetime Structure: Einstein and Beyond* ed Abhay Ashtekar (World Scientific) [gr-qc/0508109]
[5] Surya S 2008 *Theor. Compu. Sci.* **405** 188
[6] Uhlmann A 1970 *Reports Math. Phys.* **1 (2)** 147
[7] Elze H-T 1996 *Phys. Lett. B* **369** 295 [hep-th/9406085]
[8] Dhar A, Saito K and Peter Hanggi [arXiv:1106.3207]
[9] Bombelli L, Koul R K, Lee J and Sorkin R D 1986 *Phys. Rev. D* **34** 373
[10] Johnston S 2009 *Phys. Rev. Lett* **103** 180401 [arXiv:0909.0944 [hep-th]]
[11] Sorkin R D 2011 *J. Phys. Conf. Ser.* **306** 01201 [arXiv:1107.0698] [http://www.pitp.ca/personal/rsorkin/some.papers/142.causet.dcf.pdf]

Effects of inhomogeneities on the expansion of the Universe: A challenge to dark energy?

M-N Célérier

Laboratoire Univers et Théories (LUTH), Observatoire de Paris, CNRS, Université Paris Diderot, 5 place Jules Janssen, 92190 Meudon, France

E-mail: marie-noelle.celerier@obspm.fr

Abstract. The current standard model of cosmology, the ΛCDM model, is based on the homogeneous FLRW solutions of the Einstein's equations to which some perturbations are added to account for the CMB features and structure formation at large scales. This model fits rather well to the observations, provided 95% of the energy density budget of the Universe should be of an unknown physical nature, i.e., dark matter and dark energy. Now, the aim of a cosmological model is not merely to reproduce the observations, but also to give a physical understanding of the Universe we live in. Moreover, even if the assumption of homogeneity seems to be more or less valid at large scales, it appears to be in contradiction with observations at intermediate scales (between the scale of non-linear structure formation and that where structures virialize). This is the reason why, during the last decade, a community of researchers has been formed, whose aim has been to look for the best way to take into account the influence of the inhomogeneities seen in the Universe, and to construct accurate cosmological models, which could possibly get rid of the dark components. This task, which is still in its infancy, is currently progressing towards promising results. Two types of methods can be found in the literature: Spatial averaging of scalar quantities, and use of exact inhomogeneous solutions of general relativity. We have given, here, a brief report of the second one.

1. Introduction

The current standard model of cosmology is based on an hypothesis, the cosmological "Principle", stating that the Universe is spatially homogeneous beyond some large scale. However, even if the value of this scale is usually claimed to be of the order of 100 Mpc, this assumption has never been proved and is still under debate. A mathematical simplification describing the Universe by a FLRW model *at all scales* is, therefore, commonly used. Up to 1998, this model was Einstein-de Sitter (EdS).

Then, in 1998, it was found that the Supernova Ia (SN Ia) luminosity is dimmer than predicted by the EdS model [1]. As a consequence, this model was ruled out. However, contrary to what was widely advocated at the time, and still is, one cannot put forward accelerated expansion at this stage. Actually, all current evidence for dark energy is indirect, because the only effect of dark energy is on the expansion history of the Universe.

What usually done in cosmology is: First assume a cosmological model, then fit it to data measuring observables sensitive to the expansion history through the redshift, thus, obtain the values of the cosmological parameters *of the assumed model*.

In an FLRW cosmology, the expansion rate $H = \dot{a}/a$ is given by:

$$H^2(z) = H_0^2 \left\{ \Omega_\Lambda \exp\left[\int_0^z \frac{\mathrm{d}z_1}{1+z_1} 3\left[1 + w_\Lambda(z_1)\right] \right] + \Omega_k(1+z)^2 + \Omega_M(1+z)^3 + \Omega_R(1+z)^4 \right\}. \tag{1}$$

It depends on four constants: the fraction of dark energy Ω_Λ, the fraction of curvature Ω_k, the fraction of matter, including dark matter Ω_M, the fraction of radiation Ω_R, and one function w_Λ, which is the equation of state parameter of the dark energy. Since $H^2(0) = H_0^2$, $\Sigma_i \Omega_i = 1$, which removes one constant. $\Omega_k \sim 0$ from the WMAP experiment measuring the CMB power spectrum, and Ω_R is also small from the CMB (5×10^{-5}). The two other constants Ω_Λ and Ω_M, and the function w_Λ are rather well constrained by their effect on the expansion history, which is manifest through various cosmological observables. The result is the "concordance" model with $\Omega_\Lambda \sim 0.7$, $\Omega_M \sim 0.3$ (with ~ 0.25 from dark matter) and $w = -1$.

Is it a good cosmological model? It seems to work, since it reproduces nearly all the cosmological data. But, 95% is "dark", i.e., not explained. What about a model whose components are so fundamentally not understood, and could the accelerated expansion be a mere mirage?

It has been shown shortly after the discovery of the SN Ia dimming that inhomogeneities in our local Universe can *in principle* mimic an accelerated expansion. A general demonstration for models spherically symmetric around the observer has been provided in [2]. Various inhomogeneous models with no cosmological constant reproducing the supernova data have also been proposed in [3, 4, 2, 5].

To understand intuitively how such an effect can arise, it is worth turning to the very pedagogical model proposed in [5]. This model is composed of two regions. An inner homogeneous region with a given density is matched at some redshift to another homogeneous region with a higher density. Both the regions *decelerate*, but since the inner void expands faster than the outside region because it contains less matter, the observer inside the void, observing a source in the outer region, experiences *an apparent acceleration* of the Universe expansion.

Three methods have been proposed to take inhomogeneities into account: (i) Linear perturbation theory, which is valid when *both* the curvature and density contrasts remain small (this is not the case in the non-linear regime of structure formation and where the SN Ia are observed), (ii) Averaging of scalar quantities, promising and currently improving, and (iii) Exact inhomogeneous solutions, which are valid at all scales and are exact perturbations of the Friedmann background, which they can reproduce as a limit with any precision. This last one is the method we have considered in the following:

2. One patch Lemaître-Tolman (LT) models with $\Lambda = 0$

2.1. The LT solution

The Lemaître[6]-Tolman[7] (LT) solution is a spherically symmetric and non-static solution of Einstein's equations, with a dust gravitational source. Its metric, in comoving coordinates and synchronous time gauge, in units where $c = 1$, is:

$$\mathrm{d}s^2 = \mathrm{d}t^2 - \frac{R'^2}{1+2E(r)}\mathrm{d}r^2 - R^2(t,r)(\mathrm{d}\vartheta^2 + \sin^2\vartheta\, \mathrm{d}\varphi^2), \tag{2}$$

where the prime denotes derivation with respect to the radial coordinate r. A first integral of the field equations gives a dynamical equation for R as:

$$\dot{R}^2 = 2E + \frac{2M}{R} + \frac{\Lambda}{3}R^2, \tag{3}$$

where the dot denotes derivation with respect to the time coordinate t. The mass density in energy units, coming from another field equation is:

$$8\pi G\rho = \frac{2M'}{R^2 R'}. \tag{4}$$

Integrating Eq. (3) gives:

$$\int_0^R \frac{\mathrm{d}\widetilde{R}}{\sqrt{2E + 2M/\widetilde{R} + \frac{1}{3}\Lambda\widetilde{R}^2}} = t - t_B(r). \tag{5}$$

This solution exhibits three integration functions: The energy per unit mass of the particles within the comoving shell of radius r, $E(r)$, the gravitational mass of the particles in that shell $M(r)$, and the "bang time" $t_B(r)$ meaning that the singularity (bang or crunch) is not simultaneous for different shells.

To reproduce the cosmological data with no dark energy, the cosmological constant Λ is set to zero. Then Eq. (5) can be solved explicitly and the solutions are:

- when $E < 0$ (elliptic evolution):

$$R(t,r) = \frac{M}{(-2E)}(1 - \cos\eta), \quad and \tag{6}$$

$$\eta - \sin\eta = \frac{(-2E)^{3/2}}{M}(t - t_B(r)), \tag{7}$$

  where $\eta(t,r)$ is a parameter.
- when $E = 0$ (parabolic evolution):

$$R(t,r) = \left[\frac{9}{2}M(t - t_B(r))^2\right]^{1/3}. \tag{8}$$

- when $E > 0$ (hyperbolic evolution):

$$R(t,r) = \frac{M}{2E}(\cosh\eta - 1), \quad and \tag{9}$$

$$\sinh\eta - \eta = \frac{(2E)^{3/2}}{M}(t - t_B(r)). \tag{10}$$

Note, however, that a given model can exhibit an $E(r)$ function whose sign changes with r while going smoothly through zero. Even if, for simplification sake, the parametric solutions are commonly used with a fixed sign for $E(r)$, one must be aware of this possible property of the LT models.

Since all the above formulas are covariant under coordinate transformations of the form $\tilde{r} = g(r)$, one of the three integration functions can be fixed at our convenience by the choice of g. Therefore, each LT solution is fully determined by two arbitrary functions of r and a coordinate choice.

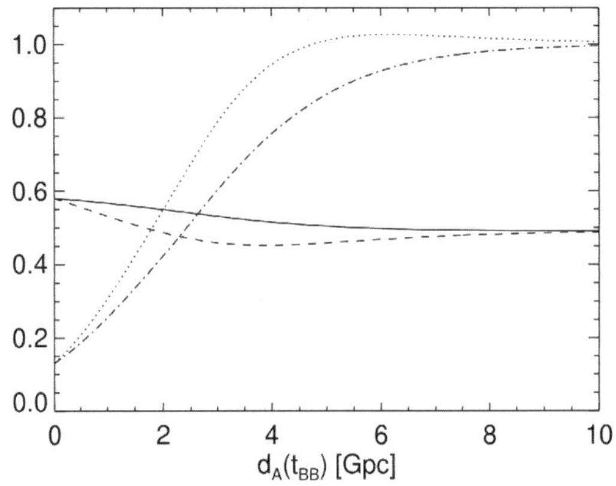

Figure 1. The void for the GBH central observer. *Dotted curve*: The matter density in units of the critical density. *Dash-dot curve*: Ω_M. *Solid curve*: H_T, the transverse expansion rate. *Dashed curve*: H_L, the longitudinal expansion rate.

2.2. Examples of one patch LT models with $\Lambda = 0$

This class of solutions has been first used to construct void models in which a local LT under-density centred on the observer and a priori parametrized is matched at some redshift to an EdS model. We have discussed below two examples of such models.

In [8], the so called GBH models are specified by their matter content $\Omega_M(r)$ and their expansion rate $H(r)$, and are governed by five free constant parameters. As we have said above, they are smoothly matched to an EdS universe at large redshift. These models are fitted to a series of observations (CMB first peak, Large Scale Structures (LSS), Baryon Acoustic Oscillations (BAO), SN Ia data, Hubble Space Telescope measure of H_0, age of the globular clusters, gas fraction in clusters) to constrain their parameters. They suggest the possibility that the Universe spherically smoothened around us might be a large (around 2.5 Gpc) void within an EdS universe, with no dark energy. This void is depicted in Figure 1, with the matter content Ω_M and the two expansion rates of the LT model, the transverse expansion rate $H_T \equiv \dot{R}/R$ and the longitudinal expansion rate $H_L \equiv \dot{R}'/R'$. However, even if they are able to reproduce most of the observations, they exhibit a value of H_0 too low as regards to the latest measurements, which give $H_0 \sim 74$ (km/s)/Mpc [9]. This can be seen in Figure 2, where $H_0 \sim 64$ (km/s)/Mpc.

To try to get rid of this drawback, a set of five LT void models have been proposed in [10]. They have been tested against the full CMB spectrum, BAO, SN, and H_0. The inclusion of a non-zero overall curvature drastically improves the goodness of fit of these models, bringing them very close to that of the ΛCDM model. And, by varying the density profile, the value of H_0 has been increased. However, these models have still difficulty exhibiting a value of H_0 sufficiently high as regards that of [9].

In another work [11], an LT model has been fitted to two sets of observables: The angular diameter distance together with the redshift-space mass density, and the angular diameter distance together with the expansion rate, assumed to have the same form on our past light cone as in the ΛCDM model. The two left arbitrary LT functions have been determined and have given the mass distribution in spacetime. Without any priori assumption on its form, the current density profile does not exhibit a void but a hump, and the observer is located in a shallow and wide funnel on top of this hump, as shown in Figure 3. Moreover, in this model,

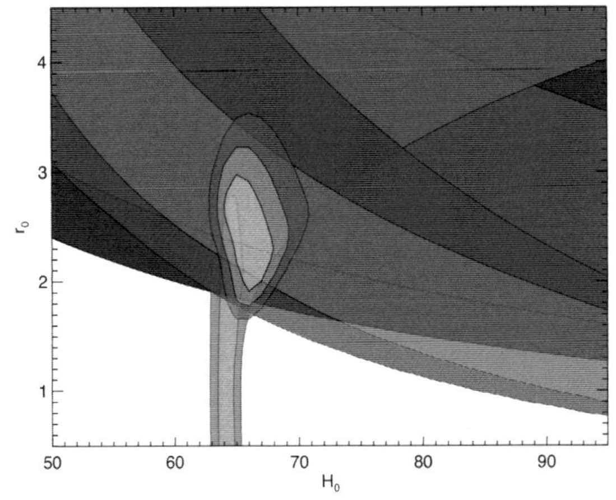

Figure 2. One of the likelihood for the most simple GBH model (r_0 characterizes the size of the void): in yellow with 1-, 2-, and 3-σ contours = the likelihood for the combined data set; in blue, purple and green respectively with 1- and 2-σ contours = the likelihood for the individual SN Ia, BAO and CMB data sets. The value of H_0 is too low, around 64 (km/s)/Mpc.

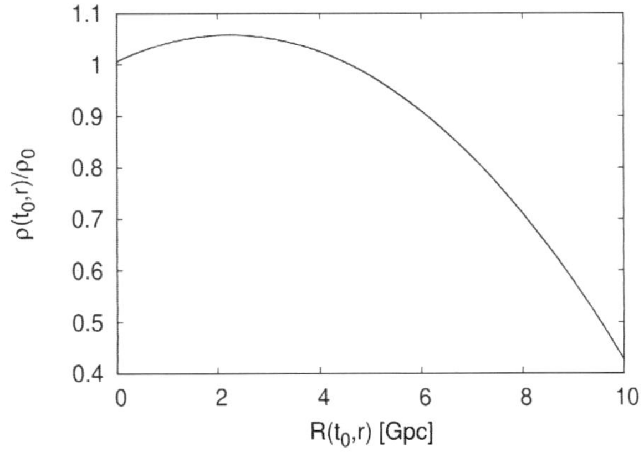

Figure 3. The current density profile of [11] does not exhibit a void but a hump. The observer is located in a shallow and wide funnel on top of this hump.

the value of H_0 is that measured, by construction.

However, single patch LT models with a central observer have been criticized on the ground of being in contradiction with the Copernican Principle. Now, when using such models in cosmology, one should not claim that the observer is at the very centre of any actual spherically symmetric universe. Rather, these models must be considered as mathematical simplifications or preliminary steps. Here, the inhomogeneities are smoothened over angles around us and only their radial component is considered. Note that this is an improvement as regards to FLRW models, where the inhomogeneities are smoothened over the whole space. Since, the exact central location of the observer is mandatory by construction, these models cannot take anisotropies into account. Effects merely designed to test them, such as CMB dipole and low multipoles, kSZ effect, CMB spectral distortion, cosmic parallax, etc., can in no way be used to test these models [12].

3. Swiss-cheese and meat-ball models

3.1. Example of an LT Swiss-cheese model

For a more physical representation of our observed Universe, Swiss-cheese models with no dark energy have also been proposed. In [13], the model is a lattice of LT bubbles with radius ≥ 350 Mpc in an EdS background. Initially, the void at the centre of each hole is dominated by negative curvature and a compensating over-density matches smoothly the density and curvature EdS values at the border of the hole. Since the voids expand faster than the cheese, the over-dense regions contract, and become thin shells at the borders of the bubbles while under-dense regions turn into emptier voids, eventually occupying most of the volume. An extra very important result is that *the evolution* of the voids, bends the photon paths, and affects more the photon physics than the geometry of the inhomogeneities.

3.2. Quasi-spherical Szekeres Swiss-cheese models

Another exact solution of Einstein's equation has been used to build Swiss-cheese models, the quasi-spherical Szekeres (QSS) solution. The Szekeres solutions [14] are dust solutions of Einstein's equations with no symmetry at all, i.e., no Killing vector. Their metric in comoving coordinates and synchronous time gauge is given as:

$$\mathrm{d}s^2 = \mathrm{d}t^2 - \mathrm{e}^{2\alpha}\mathrm{d}r^2 - \mathrm{e}^{2\beta}(\mathrm{d}x^2 + \mathrm{d}y^2), \tag{11}$$

where α and β are functions of (t, r, x, y) to be determined from the Einstein's equations. The $\beta' = 0$ sub-family has not been considered so far in cosmology, since it does not contain the LT solution as a spherically symmetric limit.

The $\beta' \neq 0$ sub-family can be parametrized so that a set of 5 independent functions of r determines each solution. The sign of one of them fixes the geometry of the $t = constant$ 3-surfaces and the type of evolution (elliptic, parabolic, hyperbolic). The sign of another function determines the geometry of the $t = constant r = constant$ 2-surfaces, which can be quasi-spherical, quasi-plane or quasi-hyperbolic. Only the quasi-spherical solution has been found useful in cosmology, since it is the only one which exhibits the three Friedmann limits (hyperbolic, flat and spherical). This solution possesses an origin, not a centre.

Its metric in comoving coordinates and synchronous time gauge can be written, with the parametrization proposed in [15] as:

$$\mathrm{d}s^2 = \mathrm{d}t^2 - \frac{(\Phi' - \Phi E'/E)^2}{1-k}\mathrm{d}r^2 - \frac{\Phi^2}{E^2}(\mathrm{d}x^2 + \mathrm{d}y^2), \tag{12}$$

where Φ is a function of t and r, k is a function of r, and E is a function of r, x and y, given as:

$$E = \frac{S}{2}\left[\left(\frac{x-P}{S}\right)^2 + \left(\frac{y-Q}{S}\right)^2 + 1\right], \tag{13}$$

with $S(r)$, $P(r)$, $Q(r)$ are functions of r.

Applying the Einstein's equations to the metric in Eq. (12) and assuming that the energy-momentum tensor is that of dust, these equations reduce to the following two:

$$\dot{\Phi}^2 = \frac{2M}{\Phi} - k + \frac{1}{3}\Lambda\Phi^2, \quad \textit{and} \tag{14}$$

$$\kappa\rho = \frac{2M' - 6ME'/E}{\Phi^2(\Phi' - \Phi E'/E)}, \tag{15}$$

where $\kappa = 8\pi G$, Λ is the cosmological constant, and M is an arbitrary function of r.

As in the LT solution, the bang time function $t_B(r)$, follows from Eq. (14) as:

$$\int_0^\Phi \frac{\mathrm{d}\widetilde{\Phi}}{\sqrt{-k + 2M/\widetilde{\Phi} + \frac{1}{3}\Lambda\widetilde{\Phi}^2}} = t - t_B(r). \tag{16}$$

Note that Eq. (16) exhibits the same form as Eq. (5) of the LT solution. Therefore, for $\Lambda = 0$, the parametric solutions are the same, with the Szekeres function $k(r)$ corresponding to the LT function $-2E(r)$, i.e.,

- when $k > 0$ (elliptic evolution):

$$\Phi(t,r) = \frac{M}{k}(1 - \cos\eta), \quad \textit{and} \tag{17}$$

$$\eta - \sin\eta = \frac{k^{3/2}}{M}(t - t_B(r)). \tag{18}$$

- when $k = 0$ (parabolic evolution):

$$\Phi(t,r) = \left[\frac{9}{2}M(t - t_B(r))^2\right]^{1/3}. \tag{19}$$

- when $k < 0$ (hyperbolic evolution):

$$\Phi(t,r) = \frac{M}{k}(1 - \cosh\eta), \quad \textit{and} \tag{20}$$

$$\sinh\eta - \eta = \frac{(-k)^{3/2}}{M}(t - t_B(r)). \tag{21}$$

Remember that, as for the LT case, the sign of $k(r)$ can change in a given model.

Since all the formulas given so far are covariant under coordinate transformations of the form $\tilde{r} = g(r)$, this means that one of the functions $k(r)$, $S(r)$, $P(r)$, $Q(r)$, $M(r)$ or $t_B(r)$ can be fixed at our convenience by the choice of g. Hence, each QSS solution is fully determined by only five functions of r and a coordinate choice.

To begin with and for simplicity sake, axially symmetric classes of QSS models with $\Lambda = 0$ have been used to construct Swiss-cheese models, which are able to reproduce the supernova data [16]. Here, the holes are no more spheres, as in LT Swiss-cheeses, but axially symmetric patches and the cheese is either an EdS or an open FLRW background. The luminosity distance is computed on axially directed null geodesics. Compared to those obtained with corresponding LT Swiss-cheese models, the results of these are quantitatively different but qualitatively comparable. To reproduce the observational data, structures of at least 500 Mpc are necessary.

3.3. Meat ball models

The meat ball models are the opposite of the Swiss-cheese models. The holes are replaced by balls and the cheese by voids, which occupy more volume than the over-densities [17]. Most of the photons travel through the voids, missing these localized over-densities. The most interesting finding is that the lensing probability distribution function is skewed with a mode at de-magnified values. It is, thus, possible that these lensing effects, possibly increased by a selection effect, and also because they are stronger in an homogeneous background without cosmological constant than in a ΛCDM model, might explain part of the "dark energy". But these lensing effects alone cannot explain the full dark energy. However, added to other effects, they can boost LT models or make smaller the volume of voids needed in Swiss-cheese models.

4. Conclusion and prospects

The first very simple inhomogeneous models used to deal with the dark energy problem and proposed very soon after the discovery of the dimming of the luminosity of the supernovae have been with time replaced by more and more sophisticated ones. These are able to reproduce to a better accuracy, not only the supernova observations but nearly all the available cosmological data.

To try to improve the reproduction of our observed local Universe, from FLRW models to axially symmetric Szekeres Swiss-cheese models, the number of symmetries has decreased, aiming at the removal of any of them. It has been claimed that the geometric, dynamical, lensing effects of inhomogeneities act in the same direction [17]. Their addition can, therefore, modify markedly our view of the ΛCDM universe.

Even if at the end of the day, the effects of inhomogeneities should prove insufficient to explain out dark energy, the current results obtained with exact models show that their impact on the value of the cosmological parameters will have to be considered as non-negligible. Now, to be able to take properly into account all the data available to constrain the models, and in particular LSS, BAO and CMB, perturbation theories applicable to the LT and Szekeres models will have to be developed.

The final aim is to construct the metric of the cosmos directly from observations. Preliminary codes have been developed for LT models [18, 19, 20], but two main problems remain: The choice of the scale, and to get rid of spherical symmetry. This implies a jump to full numerical relativity, which might be the future aim of cosmological investigations.

References

[1] Riess A G *et al* 1998 *Astron. J.* **116** 1009
[2] Célérier M-N 2000 *A. & A.* **353** 63
[3] Dabrowski M P and Hendry M A 1998 *Ap. J.* **498** 67
[4] Pascual-Sánchez J F 1999 *Mod. Phys. Lett.* **A14** 1539
[5] Tomita K 2000 *Ap. J.* **529** 382011, 2001 *MNRAS* **326** 287, 2001 *Prog. Theor. Phys.* **106** 929
[6] Lemaître G 1933 *Ann. Soc. Sci. Brux. A* **53** 51 (English translation, with historical comments: 1997 *Gen. Rel. Grav.* **29** 637)
[7] Tolman R C 1934 *Proc. Nat. Acad. Sci. USA* **20** 169 (Reprinted, with historical comments: 1997 *Gen. Rel. Grav.* **29** 931)
[8] García-Bellido J and Haugbølle T 2008 *JCAP* **04** 003
[9] Riess A G *et al* 2011 *Ap. J.* **730** 119 [Erratum-ibid **732** 129]
[10] Biswas T, Notari A and Valkenburg W A 2010 *JCAP* **11** 030
[11] Célérier M-N, Bolejko K and Krasiński A 2010 *A. & A.* **518** A21
[12] Célérier M-N [arXiv:1108.1373]
[13] Marra V, Kolb E W, Matarrese S and Riotto A 2007 *Phys. Rev. D* **76** 123004
[14] Szekeres P 1975 *Commun. Math. Phys.* **41** 55
[15] Hellaby C 1996 *Class. Quant. Grav.* **13** 2537
[16] Bolejko K and Célérier M-N 2010 *Phys. Rev. D* **82** 103510
[17] Kainulainen K and Marra V 2009 *Phys. Rev. D* **80** 127301

[18] Lu T H-C and Hellaby C 2007 *Class. Quant. Grav.* **24** 4107
[19] McClure M L and Hellaby C 2008 *Phys. Rev. D* **78** 044005
[20] Bolejko K, Hellaby C and Alfedeel A H A 2011 *JCAP* **09** 011

The direction of gravity

E V Linder

Berkeley Lab and University of California, Berkeley, CA 94720 USA, and
Institute for the Early Universe WCU, Ewha Womans University, Seoul, Korea

E-mail: evlinder@lbl.gov

Abstract. Gravity directs the paths of light rays and the growth of structure. Moreover, gravity on cosmological scales does not simply point down: It accelerates the universal expansion by pulling outward, either due to a highly negative pressure dark energy or an extension of general relativity. We have examined methods to test the properties of gravity through cosmological measurements. We have then considered specific possibilities for a sound gravitational theory based on the Galilean shift symmetry. The evolution of the laws of gravity from the early universe to the present acceleration to the future fate - the paths of gravity - carries rich information on this fundamental force of physics, and on the mystery of dark energy.

1. Introduction

The direction of gravity, in both senses of the word, is a key characteristic of the universe. The revolution of the late 1990s was the discovery that gravity did not exclusively pull down in the conventional, attractive way but that cosmic expansion showed an acceleration equivalent to gravity pulling outward. Secondly, gravitation plays the major role in directing the contents of the universe, bending the paths of light and governing the growth of large scale structure.

A new frontier has opened in exploring the force of gravity, understanding not only its spatial direction but its direction of energy and mass in the sense of a conductor directing an orchestra. Some outstanding questions now being addressed include: Why is gravity pulling outward, due to an unknown dark energy component or a new law? Is gravity (e.g., Newton's constant) constant, or strengthening or weakening with time? Does gravity govern the growth of cosmic structure in exactly as the manner it governs cosmic expansion, as in the laws of general relativity, or are there further degrees of freedom? Does gravity behave the same on all scales?

At its most basic, we should test our understanding of cosmic gravity because we can, as a reality check. Observations of the deflection of light by gravity (lensing), and the growth of matter clustering by gravity are reaching the stage where they can provide incisive information. Moreover, we should justify the long extrapolation of general relativity from the small scales, where it is tested to the cosmic scales where it is applied (more than a factor of billion in length), and from high curvature to low curvature regions. Tests of cosmic gravity are further motivated by the fact that the first two precision observational tests have basically shown gaping holes in our understanding: General relativity plus attractive matter fails to predict acceleration in the cosmic expansion, and fails to agree with the growth and clustering of large scale structure.

In Section 2, we have explored methods for testing gravity in phenomenological and model independent approaches. Adopting a specific model showing desirable characteristics, Galilean gravity, and in Section 3, we have illustrated the theoretical constraints, an alternate theory of gravity must satisfy and handles by which future data will be able to test stringently the nature of gravity.

2. The differences of gravity

To test general relativity, one can either simply look for consistency with observations or one can measure parameters or quantities that would differ from the general relativity prediction if deviations exist. (For an early work addressing some implications of deviations, e.g., from the inverse square law of attraction, see [1].) We have concentrated, here, on cosmic scales, those above galaxy cluster lengths but well within the horizon 5 − 500 Mpc.

General relativity says that effects on the cosmic expansion are mirrored exactly in the growth of structure, with no extra gravitational degrees of freedom to cause an offset. Therefore, comparing the expansion history to the growth history is one of the major tests of the physics. This implies that it is crucial to fit for the expansion and growth simultaneously, to be alerted to any deviations. Should the gravity theory be assumed (e.g., as general relativity) incorrectly? This will bias the fits of both the expansion and cosmological model. The converse is true as well (e.g., assuming ΛCDM). Thus, even if one is not interested in gravity, one still must fit for it or the quantities one is interested in will be biased.

To separate cleanly the gravity effects on growth from the expansion effects on growth, the gravitational growth index parameter γ was developed in [2]. Recall that linear growth of a fractional density perturbation $\delta\rho/\rho$ as a function of expansion factor a can be well approximated by:

$$\frac{\delta\rho}{\rho}(a) = \frac{\delta\rho}{\rho}(a_i) \times e^{\int_{a_i}^{a} (da'/a')\,\Omega_m^{\gamma}(a')}, \quad and \tag{1}$$

$$\Omega_m(a) = \frac{\Omega_m a^{-3}}{\Omega_m a^{-3} + (1-\Omega_m)e^{3\int_a^1 (da'/a')\,[1+w(a')]}}, \tag{2}$$

where Ω_m is the present matter density, in units of the critical density and the dark energy equation of state $w(a) = w_0 + w_a(1-a)$ to an excellent approximation. This accurately deconvolves the two influences on the growth, the expansion history in the form of $\Omega_m(a)$ from the gravity law in the form of γ, providing a fit to better than 0.1% in many of the standard cases. Suppose one considered growth measurements arising from weak gravitational lensing surveys. The weak lensing signal involves a convolution of the gravitational growth of massive structures at various redshifts along the line of sight and the geometric distance factors serving as a focal length. If one assumes general relativity (GR), omitting to fit for gravity, then the dark energy parameters one derives from the (GR) growth and distances could be strongly biased. The fit for the time variation of the dark energy equation of state w_a will be incorrect by $\Delta w_a \approx 8\Delta\gamma$, where $\Delta\gamma$ is the amount by which the true gravity deviates from GR. That is, failing to fit for gravity also induces failure to fit expansion accurately. The converse is true as well. Fitting for gravity is, therefore, necessary.

If we want to go beyond the coarse grained approach of γ, how should we do so in a practical manner? Adding parameters without careful thought will merely blow up the uncertainties in both gravity and expansion. As with the gravitational growth index γ, one must carefully study how to separate cleanly the different physical influences. One could work within a specific model that determines the gravity and expansion behaviours but such one by one comparisons are time consuming and often give little general insight. Instead, we have considered various approaches to a more model independent analysis.

General relativity, as summarized by John Wheeler, is that "matter tells spacetime how to curve, and spacetime tells matter how to move". These statements can be thought of mathematically as Poisson's equation $\nabla^2\phi = 4\pi G_N a^2\delta\rho$, and Newton's first law $-\vec{\nabla}\psi = \ddot{x}$.

We have suggestively written the spacetime potentials with two different symbols: ϕ and ψ. In GR, they are one and the same, but in other theories of gravity they can differ. To tie them closely to the observations, we consider two modified Poisson equations:

$$\nabla^2(\phi+\psi) = 8\pi G_N a^2 \delta\rho \times G_{\rm light}, \quad \textit{and} \tag{3}$$

$$\nabla^2\psi = 4\pi G_N a^2 \delta\rho \times G_{\rm matter}. \tag{4}$$

The function $G_{\rm light}$ tests how light responds to gravity, and is central to gravitational lensing and integrated Sachs-Wolfe measurements. The function $G_{\rm matter}$ tests how matter responds to gravity, and is central to growth of massive structures and peculiar velocities; the growth index γ is closely related. (Also, see [3] for a more detailed discussion of these modified Poisson equations.)

In general, $G_{\rm light}$ and $G_{\rm matter}$ will be functions of time and space, e.g., redshift $z = a^{-1}-1$ and length scale r or wavemode k. One approach to parametrizing them with a practical number of degrees of freedom is to be guided by classes of gravity theories. Looking at both higher dimension gravity and scalar-tensor theories, one sees a deviation starting from early universe GR behaviour and involving either a scale independent or characteristic k^2 dependence (arising from the Laplacian in the equations of motion). Thus, a useful form is [4]:

$$G_{\rm matter} = 1 + \frac{ca^s(k/H_0)^n}{1+3|c|a^s(k/H_0)^n}, \tag{5}$$

where $n = 0$, 2 respectively, and a similar form could hold for $G_{\rm light}$. Note that leaving out the denominator, so G simply deviates from the GR value of 1 as a^s is dangerous as it unfairly weights high vs low redshift observations and can easily lead to bias. The Padé form of Eq. (5) is accurate to $\sim$1% for DGP and $f(R)$ gravity.

For current data, involving weak gravitational lensing, galaxy peculiar velocities, CMB power spectra, supernova distances, and baryon acoustic oscillations [4] simultaneously fits for gravity (the amplitude c and time dependence s) and expansion (matter density Ω_m and dark energy equation of state w_0, w_a). They have found that the gravity constraints are nearly unaffected by the expansion fit, and the expansion constraints are nearly unaffected by the gravity fit. To a significant extent, however, this reflects the status of current data: almost no constraint exists on s. As data improves and becomes more incisive in testing gravity, we do expect some degradation in expansion constraints when also fitting vs fixing, gravity - perhaps a factor of ~ 2 in confidence contour area.

Both the above theories of gravity mentioned are simple models with restricted parameters. In [5], the authors have given one gravity parameter, the 5D crossover scale that determines both the amplitude and time dependence of the deviations from GR. The $f(R)$ family of scalar-tensor models has more freedom, but can successfully be treated in terms of the scalar on mass [6], and this is well fit by an amplitude (of order $\sim 100\ H_0$) and a power law time dependence (i.e., s) - basically two parameters.

Indeed, if one considers the phase-space, or "paths of gravity" of the deviations from GR, the authors of [5] have shown a definite, near parabolic track in the $G'_{\rm matter}$– $G_{\rm matter}$ plane, and the family of $f(R)$ tracks can be calibrated through a re-scaling of s into a single, also near parabolic track [7]. (See Figure 3 of [7] for an illustration and further details.) These two classes diverge from each other, and GR, having today $\Delta G_{\rm matter} \approx \pm 0.3$. This suggests, in these cases at least, that we should strive for observations capable of testing this beyond Einstein's parameter at the 10% level or better.

Not every viable theory of gravity may be so simple, however. It is useful to have another, more model independent approach in our toolkit. A reasonable choice is to test for the values of the G functions in bins of redshift and wavenumber, i.e., asking whether gravity behaves the same (and like GR) at high and low redshift, and for high and low wavenumbers (small and large scales). While basis functions or principal components are other possible choices, even next generation data will not have the leverage to fit more than 2 modes each with reasonable signal to noise (not just low noise). Recall that N modes in each of redshift and wavenumber gives rise to $2N^2$ parameters (for $G_{\rm matter}$ and $G_{\rm light}$), and hence, $N^2(2N^2+1) \sim 2N^4$ correlation functions, so attempting to fit more than 2 modes is impractical. We, thus, have considered "$2 \times 2 \times 2$" gravity: high/low redshift, large/small scales, $G_{\rm matter}$ and $G_{\rm light}$ [8].

Such an approach is fully model independent, and we can investigate the leverage of current and future observations to test cosmic gravity. In Figure 1, we have shown the results of Markov Chain Monte Carlo fits to data and simulations in the two bins each of wavenumber and redshift. Galaxy redshift surveys such as BigBOSS [9] will be powerful "gravity machines", capable of mapping the density and velocity fields over a wide range of redshifts. This is particularly useful at constraining $G_{\rm matter}$, the function sensitive to growth. Gravitational lensing information and the integrated Sachs-Wolfe effect of the cosmic microwave background act to constrain $G_{\rm light}$. We have seen that current data are weak in the ability to test $G_{\rm matter}$, with uncertainties of order one, while $G_{\rm light}$ can be determined to roughly 10% from the Canada-France-Hawaii Telescope Legacy Survey lensing data [10] and the WMAP CMB temperature power spectrum [11]. However, note that this precision does not indicate accuracy; the deviation from GR seen in the bottom right panel, for example, arises from acknowledged observational systematics [8]. BigBOSS will bring $G_{\rm matter}$ also to the 10% precision level. Recall that this is just what the "paths of gravity" phase-space approach above argued was necessary to test GR, so we have anticipated that next generation observations will reach interesting leverage in understanding cosmic gravity.

For the galaxy surveys, one must marginalize over astrophysical effects such as galaxy bias. Here, we have taken care of that by allowing the bias to float freely in each 0.1 bin of redshift. Due to the complementarity with the other techniques included in Stage III experiments, this marginalization causes little degradation ($\lesssim$ 10%) in the final constraints. To reduce the uncertainties on $G_{\rm matter}$ further, one would need to increase the survey volume or use other growth probes such as CMB lensing. To show the effect of a next generation galaxy weak lensing survey on $G_{\rm light}$, we have included a rather optimistic survey in the inner red contours, with the result that $G_{\rm light}$ could be determined in some bins at the 1% level. In any case, the future of testing gravity looks extremely promising.

3. The paths of gravity

Exploring beyond the simplest models of gravity is useful to understand what evolutionary signatures of the gravitational modifications might exhibit. This allows us to investigate how robust are the well defined phase-space tracks predicted by those particular models, and to look for characteristics that might be smoothed over by the model independent binned approach of the previous section.

Moreover, $f(R)$ models are arbitrary and unnatural in the same sense that a quintessence potential $V(\phi)$ is: We have no clear physics guidance as to the appropriate functional form and high energy physics corrections will in any case alter it. Two ways to protect against these problems are to avoid completely any potential, using only kinetic terms, and to look for a theory where the degrees of freedom arise geometrically, e.g., from higher dimensions, protecting against corrections. Galilean gravity incorporates both these solutions.

Galilean gravity is based on a shift symmetry in the effective field π, and involves non-linear kinetic terms that are allowed (and only these are allowed) in order to guarantee second order field equations [13, 14, 15, 16]. These also provide a Vainshtein screening that restores the theory to general relativity on small scales. Thus, this is a well defined, robust theory that we can explore as a possible extension to GR.

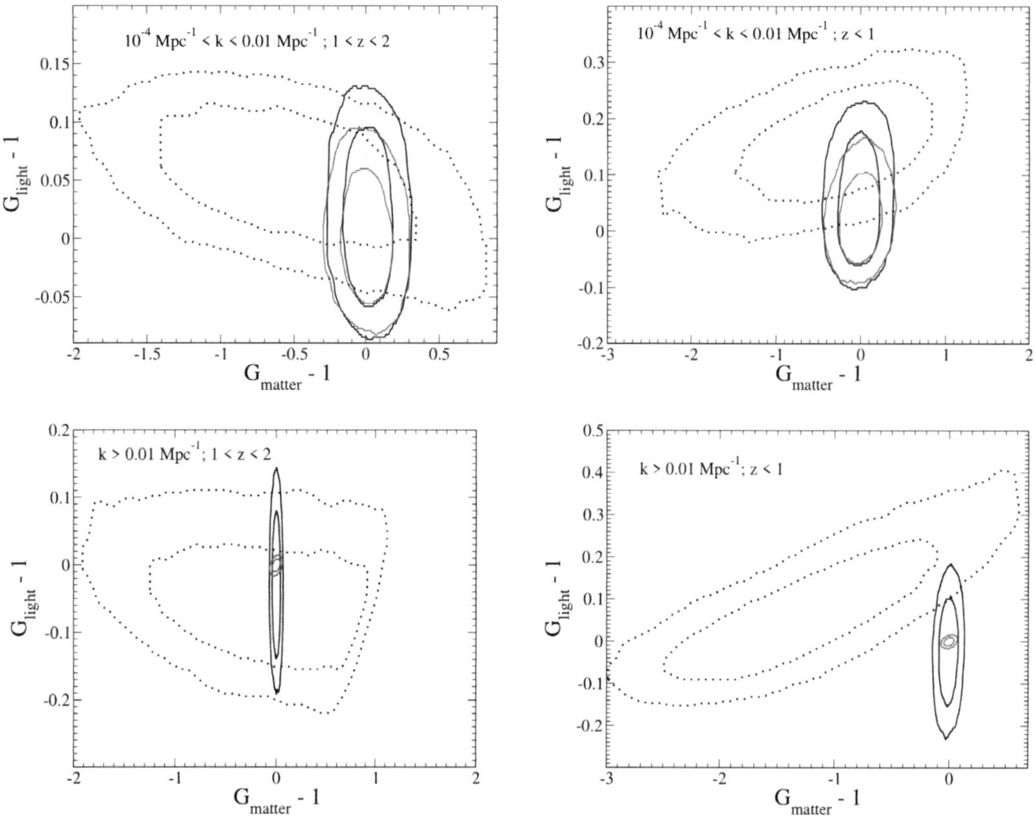

Figure 1. Constraints are shown on the 8 parameters of post-GR model independent gravity, from current and simulated future data. Dotted contours show the 68% and 95% confidence regions from current WMAP CMB [11], CFHTLS lensing [10], and Union2.1 supernova distance data [12], while solid black contours show constraints from simulated data of the next generation BigBOSS galaxy redshift survey (marginalized over galaxy bias) plus Stage III experiments. Small red contours include as well a highly optimistic galaxy weak lensing survey covering 4000 deg^2 with a galaxy number density of 55 arcmin^{-2}.

We can write the action for the Galilean theory as:

$$\begin{aligned} S &= \int d^4x\,\sqrt{-g}\Bigg[\left(1-2c_0\frac{\pi}{M_{\rm pl}}\right)\frac{M_{\rm pl}^2R}{2}-\frac{c_2}{2}(\partial\pi)^2-\frac{c_3}{M^3}(\partial\pi)^2\Box\pi-\frac{c_4\mathcal{L}_4}{2}-\frac{c_5\mathcal{L}_5}{2} \\ &\quad -\frac{M_{\rm pl}}{M^3}c_G G^{\mu\nu}\partial_\mu\pi\partial_\nu\pi-\mathcal{L}_{\rm m}\Bigg]. \end{aligned} \tag{6}$$

General relativity arises simply from the "1" inside the circular brackets. The terms involving c_2 through c_5 comprise the standard Galilean extension [13, 14] (see [16] promoting the coefficients to functions), and one can consider additional couplings to matter given by the c_0 term for a linear coupling and c_G for a derivative coupling. Again, these terms guarantee second order field equations.

From the action, one can derive the background equations of motion and find under what circumstances one gets late time acceleration. Note that acceleration occurs despite the absence of any cosmological constant or indeed any scalar field potential. Galilean cosmology has attractor solutions in the radiation and matter eras, avoiding fine tuned sensitivities to initial conditions. The effective equation of state w at early times can be close to that of matter, allowing for a non-negligible contribution of early dark energy density. The field generically becomes phantom $w < -1$, at some epoch near today, and has a future attractor to a de Sitter state. These are all interesting properties for a theory to possess, and are discussed in some detail in [17].

Perturbing the equations of motion leads to the modified Poisson equations for $G_{\rm matter}$ and $G_{\rm light}$, and the growth of structure. We can then study the paths of gravity, the evolution of these $G(a)$ functions. In the early universe, Galilean gravity acts as a thawing field, locked to general relativity at high redshift and then gradually deviating. The deviation increases linearly as the effective dark energy density does: $G \sim 1 + b\,\Omega_\pi$. However, unlike the authors in [5] and $f(R)$ cases, the evolution is not a simple, quasi-parabolic trajectory from GR to a frozen new value of the gravitational strength. The interaction between the multiple Galilean terms leads to non-monotonic behaviour, strengthening gravity at $z \sim 10$ then restoring to GR before deviating again toward a de Sitter attractor. Interestingly, in this asymptotic state, the gravitational slip, the difference between the metric potentials ϕ and ψ with which we started our exploration of extensions to GR, vanishes (and hence, $G_{\rm matter} = G_{\rm light}$). But, although $\phi = \psi$, this is not general relativity, i.e., the strength of gravity differs from Newton's constant $G \neq 1$.

Galilean cosmology, thus, shows some fascinating properties and signatures by which we can hope to distinguish it observationally from GR. Figure 2 shows the paths of gravity, the evolution of $G_{\rm matter}(a)$ and $G_{\rm light}(a)$ for the uncoupled Galilean for different values of the initial dark energy density. The larger the initial density, the stronger the deviation from GR at $z \sim 10$, and closer to the present bump occurs; a low initial density is still capable of giving the same late time behaviour but with an earlier and milder first deviation from GR. One might conjecture whether the early enhancement in the gravitational strength could play a role in the development of high redshift structure and early massive galaxy clusters.

Detailed comparison of the predictions of Galilean cosmology for the expansion and growth history with observational data is in progress [18]. However, physical soundness of the theory in terms of lacking pathologies or instabilities already puts important constraints on the Galilean parameters [17]. The most important conditions are the no-ghost condition, preventing the energy from being unbounded from below, and stability of perturbations, which can be written in terms of the sound speed as $c_s^2 \geq 0$. These requirements not only restrict the viable parameter space, but in the linear and derivative coupling cases force those contributions to be sub-dominant to the standard Galilean terms at high redshift.

Galilean cosmology has a much richer phenomenology than the simple extensions to GR previously considered and we are still exploring its implications. It has a sounder theoretical foundation than many other theories (including most quintessence models) and the combination of theoretical and observational constraints may soon give rise to definite predictions for how to find deviations from general relativity.

4. Conclusions

Gravitation is a ubiquitous and fundamental force, but one that has not yet been rigorously tested, especially over the immense extrapolation to cosmic scales. The "direction of gravity" is an open, active area of research, and may hold key insights into cosmic acceleration and the growth of large scale structure.

Measuring the expansion history alone, i.e., the dark energy equation of state $w(a)$, is not sufficient to reveal the physics; rather, the combination of the expansion history and growth

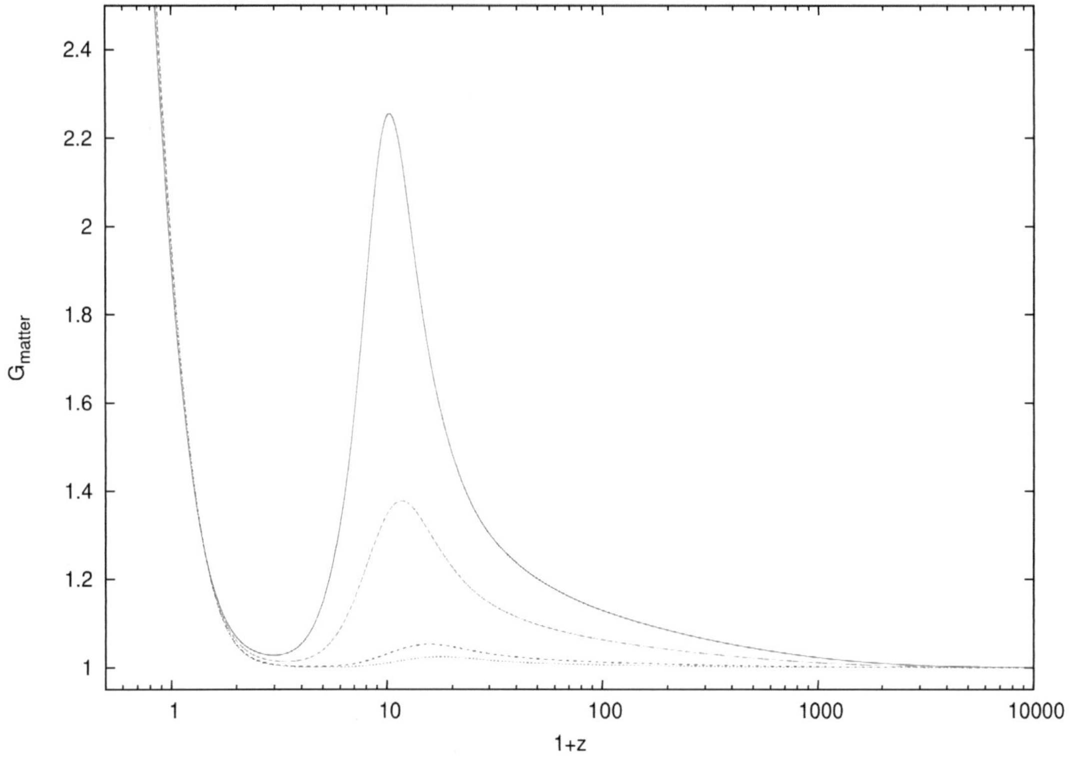

Figure 2. The gravitational strength evolution varies with the initial dark energy density. Here, the curves have $\rho_{de}(z=10^6)=10^{-4}$, 5×10^{-5}, 10^{-5}, and 5×10^{-6} from top to bottom. While the bump in strength grows and shifts to later times for higher density, the early time and late time behaviours are on attractors. An interesting complementarity exists with the effective dark energy equation of state such that the less deviation from GR in $G_{\rm matter}$, the stronger phantom deviation from $w=-1$.

history, or testing gravity as well as the equation of state, is essential. This introduces the functions $G_{\rm matter}$ and $G_{\rm light}$ from the modified Poisson equations, detailing gravity's direction of matter growth and of light deflection. Failure to fit simultaneously for expansion and gravity runs the risk of substantially biasing both results.

Fortunately, simultaneous fitting is straightforward (with proper definition of parameters to separate the physics) and does not substantially degrade the constraints. Next generation observations will carry a wealth of information on the cosmic density and velocity fields, and enable model independent gravity constraints on 8 post-GR parameters at the 10% or better level. For the simplest extensions to general relativity, this is within the precision necessary to distinguish the correct theory.

Galilean cosmology offers a more robust foundation with respect to physical naturalness. It also provides a rich phenomenology with attractor solutions in the early and late universe and multiple signatures in both the expansion and gravity behaviour. The parameter space can be constrained through both theoretical considerations and comparison to observations. Although, the exploration of the paradoxical topic of the direction of gravity is still in early days, the results so far are exciting - robust, testable alternatives to general relativity. The next generation of cosmology surveys and theoretical studies should finally test gravity diligently on cosmic scales and address the mystery of cosmic acceleration.

Acknowledgements

The author gratefully acknowledges his collaborators Scott Daniel and Stephen Appleby for Figures 1 and 2 respectively, and thank the ICGC-2011 organizers for the invitation to speak, and the wonderful hospitality. This work has been supported in part by the Director, Office of Science, Office of High Energy Physics, of the U.S. Department of Energy under Contract No. DE-AC02-05CH11231, and by World Class University grant R32-2009-000-10130-0 through the National Research Foundation, Ministry of Education, Science and Technology of Korea.

References

[1] Peebles P J E [arXiv:astro-ph/0208037]
[2] Linder E V 2005 *Phys. Rev. D* **72** 043529
[3] Bertschinger E 2011 *Phil. Trans. Roy. Soc. A* **369** 4947
[4] Zhao G-B, Li H, Linder E V, Koyama K, Bacon D J and Zhang X [arXiv:1109.1846]
[5] Dvali G R, Gabadadze G and Porrati M 2000 *Phys. Lett. B* **485** 208
[6] Appleby S A and Weller J 2010 *JCAP* **12** 006
[7] Linder E V 2011 *Phil. Trans. Roy. Soc. A* **369** 4985
[8] Daniel S F and Linder E V 2010 *Phys. Rev. D* **82** 103523
[9] Schlegel D *et al* [arXiv:1106.1706]
[10] Fu L *et al* 2008 *A. & A.* **479** 9
[11] Jarosik N *et al* 2011 *Ap. J. Supp. Ser.* **192** 14
[12] Suzuki *et al* 2011 *Ap. J. In press* [arXiv:1105.3470]
[13] Nicolis A, Rattazzi R and Trincherini E 2009 *Phys. Rev. D* **79** 064036
[14] Deffayet C, Esposito-Farese G and Vikman A 2009 *Phys. Rev. D* **79** 084003
[15] Deffayet C, Deser S and Esposito-Farese G 2009 *Phys. Rev. D* **80** 064015
[16] Deffayet C, Pujolas O, Sawicki I and Vikman A 2010 *JCAP* **10** 026
[17] Appleby S and Linder E V 2011 *JCAP Submitted* [arXiv:1112.1981]
[18] Appleby S and Linder E V *In preparation*

Journal of Physics: Conference Series **484** (2014) 012007 doi:10.1088/1742-6596/484/1/012007

Improved source localization with LIGO-India

S Fairhurst

School of Physics and Astronomy, Cardiff University, UK

E-mail: `stephen.fairhurst@astro.cf.ac.uk`

Abstract. A global network of advanced gravitational wave interferometric detectors is under construction. These detectors will offer an order of magnitude improvement in sensitivity over the initial detectors, and will usher in the era of gravitational wave astronomy. Gravitational wave sources can be localized on the sky, primarily through time of arrival measurements at the detectors, allowing for followup observations by telescopes. In this paper, we have evaluated the improvement in sky localization obtained by relocating one of the Advanced LIGO detectors to India.

There is currently, a proposal to situate one of the Advanced LIGO [1] detectors in India [2]. This would entail installing one detector at each at Livingston, LA; Hanford, WA (rather than two at Hanford), and the third detector in India. The global network will be augmented by the Advanced Virgo detector in Cascina, Italy [3] and the KAGRA detector in Kamioka, Japan [4]. One of the main motivations for the move to India is the improvement in localization of sources afforded by a network with an additional site remote from the other sites. Localization of gravitational wave sources to (relatively) small regions of the sky will enable wide field electromagnetic telescopes, such as LOFAR [5], Palomar Transient Factory [6], Pan Starrs [7], and SkyMapper [8] to perform followup observations of gravitational wave events. The joint electromagnetic and gravitational wave observation of sources will provide new insight into the central engines powering electromagnetic transients [9, 10], and allow for independent measurements of distance and redshift sources [11].

In this paper, we have investigated in detail the improvement in localization afforded by the installation of a LIGO India detector. At this stage, a site for a detector in India has not been fixed, so we consider two locations. The first, as introduced by Schutz [12], has the detector placed at the location of the Giant Metrewave Radio Telescope (GMRT) [13] at 74 deg 02' 59" E, 19 deg 05' 47" N. The second is a site close to Bangalore, at 76 deg 26' E, 14 deg 14' N, in an area that is seismically quiet [14]. In both cases, the arms are chosen (arbitrarily) to be oriented North and West, although the orientation of the detector will not greatly affect the localization [15, 16]. Examining two locations gives some insight into the effect of moving the detector. Of course, a more detailed study (such as the one described in [15]) would be required before fixing the location and orientation of a detector in India.

The primary method by which a gravitational wave source will be localized on the sky is by triangulation using measured time delays between the sites [17]. Thus, the localization accuracy will be dominated by the timing accuracy in the detectors as well as the separation between sites. In [17], the timing accuracy in a given detector was shown to be:

$$\sigma_t = \frac{1}{2\pi\rho\sigma_f}. \tag{1}$$

Timing accuracy is inversely proportional to both the SNR ρ, and effective bandwidth σ_f of the source, defined as:

$$\begin{aligned} \rho^2 &= 4\int_0^\infty \frac{|h(f)|^2}{S(f)}df\,, \quad \textit{and} \\ \sigma_f^2 &= \left(\frac{4}{\rho^2}\int_0^\infty f^2\frac{|h(f)|^2}{S(f)}df\right) - \left(\frac{4}{\rho^2}\int_0^\infty f\frac{|h(f)|^2}{S(f)}df\right)^2\,, \end{aligned} \tag{2}$$

where $h(f)$ is the Fourier transform of the signal, and $S(f)$ denotes the one sided noise power spectral density of the detector. The approximations used to obtain Eq. (1) break down at low SNR, where second order effects become important [18]. Also, we have neglected the effect of correlation between timing and other parameters. These correlations will affect the observed end time in all sites in a similar way and will, thus, not have a significant effect on the recorded time delays between sites.

We consider localization of a source using only observed time delays between the sites. From a set of observed time delays, the sky location of the source can be inferred. The errors on the arrival times determine the uncertainties in the location. For two sites, a single time delay suffices to localize the source to a ring in the sky. In principle, it may be possible to break the degeneracy on the ring based on additional information, such as spin or other features in the signal [19]. Even then, the localization area will be large, and we will not consider localization regions for a two detector network here. For three sites, the observed time delays lead to two possible locations of the source, which are mirror images in the plane of the detectors. In many cases, the observed amplitudes of the signal in the three detectors will enable the correct choice of a single location, although this is not always possible [20, 16]. Here, we assume that the degeneracy can be broken and for each source consider only the 90% confidence localization region around the true sky location.

There are numerous proposed configurations for the advanced detectors [1, 21], and it is likely that several configurations will be used as the sensitivity of the detectors develop. For this work, we are primarily interested in the sensitivity of the detectors and the frequency bandwidth of the signal. For simplicity, we have restricted attention to Binary Neutron Star (BNS) systems and characterise the sensitivity by the *BNS horizon*: The distance at which an optimally oriented and located BNS system would be observed with a signal to noise ratio of 8. The *BNS range*: The volume and orientation averaged distance at which a BNS gives SNR 8, is a factor of 2.26 smaller than the *horizon*. In [22], we have listed the sensitivity and frequency bandwidth of a number of different configurations of the advanced detectors with various sensitivities and bandwidths. In this paper, we have used a fiducial detector with a BNS range of 160 Mpc (360 Mpc horizon), and a frequency bandwidth of 100 Hz. It is important to note that the results obtained are just illustrative of what might be achievable by an advanced detector network. The actual localization of sources will depend upon the sensitivities of the detectors and may also be influenced by non-stationary features in the noise. Furthermore, it is very likely that the source localization of the detector networks will evolve over time as the detectors approach their final, design sensitivities.

In [22], we have introduced several characterizations of detector networks with regard to localization. Here, we have presented the results for networks containing a detector in India. Where appropriate, we also have shown the comparable network with two detectors in Hanford, instead of one in Hanford and one in India, to give a sense of the improvement. First, we have examined the localization for a source at a fixed distance. We simulate the signal for a BNS merger, oriented face on, at a distance of 160 Mpc (equal to the BNS range of a single detector), and compute the 90% localization area. We require the expected network SNR of the source to be greater than 12 and require a single detector SNR above 5 in at least two detectors. This

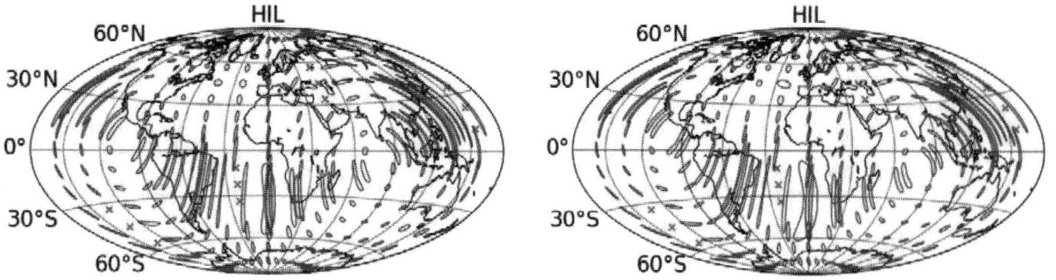

Figure 1. Localization with a three site LIGO network. The plots show the 90% localization ellipses for a face on BNS merger at 160 Mpc. A $\times$ is plotted at points where the network would not confidently detect the system. The left plot shows the localization ellipses with a detector in India at the location of GMRT. The right plot shows localization for a site close to Bangalore. The qualitative features of the plots are similar but the localization ellipses do vary substantially in some parts of the sky. This is most evident close to the plane defined by the three detectors. Moving the detector changes this plane and consequently the regions of the sky for which sources cannot be well localized.

excludes regions of the sky, where a source at 160 Mpc would not be confidently detected. These thresholds are motivated by results of recent searches of LIGO and Virgo data [23].

In Figure 1, we have shown the localization of face on BNS sources at 160 Mpc that is possible using three LIGO detectors located at Hanford, Livingston, and India. We have shown the results for the two different LIGO India locations described above. The detector orientation will not have any effect on these results, as we are considering only face on systems, which give a circularly polarized gravitational wave signal, although the orientation will have an effect on localization of generic binary systems. In both cases, there are large areas of the sky for which the localization is good, but there is a clear band where localization is rather poor and the ellipses become elongated. This corresponds to sources lying in (or close to) the plane formed by the three detectors. For sources close to the plane, the localization out of the plane is poor as a large change in position corresponds to only a small change in time delays between sites. We have not shown the corresponding plots for a Hanford–Hanford–Livingston network, as the two site network would only serve to localize sources to a ring on the sky.

Next, we have considered the global network comprising three LIGO detectors, the Virgo detector, and the KAGRA detector. In Figure 2, we have shown the localization afforded by a network comprising four or five detectors (three LIGO plus one or both of Virgo and KAGRA). For each network, we have shown localizations with either two US LIGO sites (with two Hanford detectors) or three LIGO sites including a LIGO-India detector located near Bangalore. The three site networks (HHKL and HHLV) provide comparable localizations to the three site LIGO network HIL, in that there are regions of the sky where sources are well localized, other regions where the localization is poor and even a few areas where the source would not be confidently detected. The results for HHKL show poorer localization than the other three site networks, as these sites are rather close to being co-linear (as discussed in [22]), so there is only really one good direction of localization. The four site networks show good localization over the majority of the sky with slightly worse performance for the HIKL network, as the four sites are nearly co-planar and sources lying close to this plane are not well localized. The five site network shows good localization over the entire sky.

Next, we would like to investigate the distribution of 90% localization areas for a realistic catalogue of detectable sources. We choose the sources to be uniformly distributed in volume

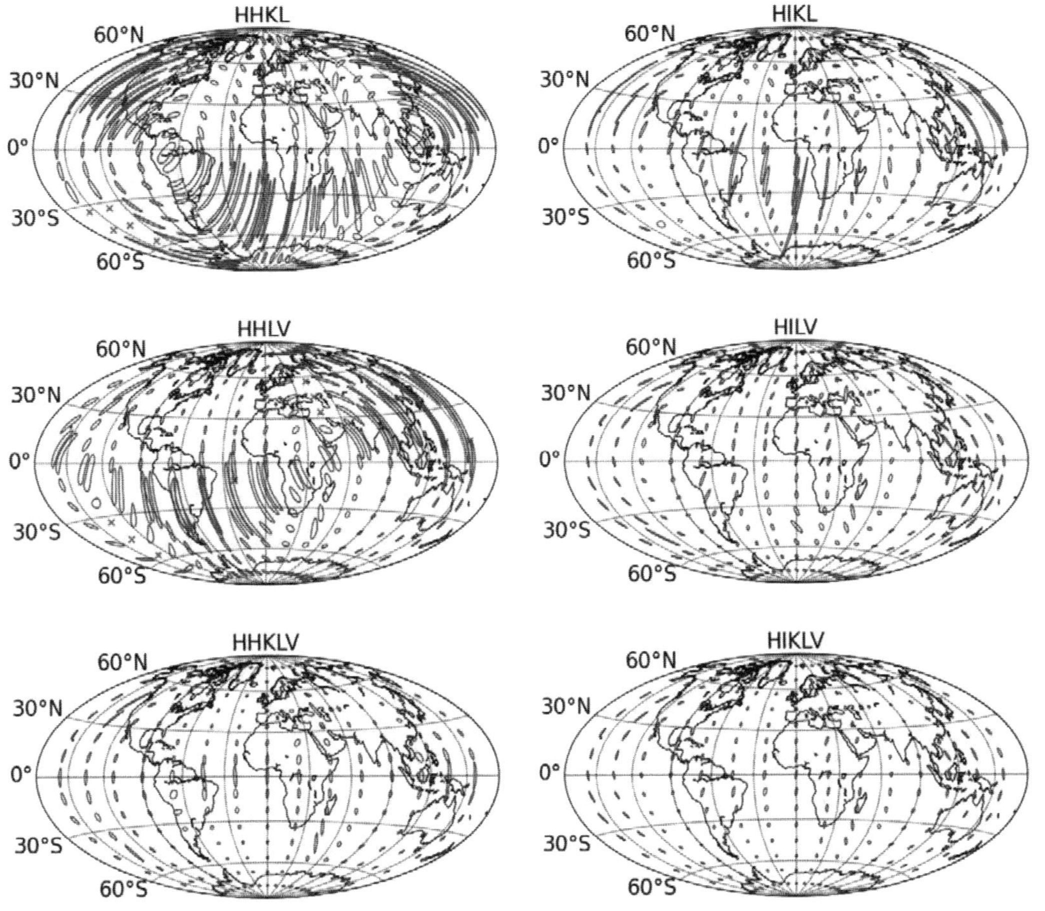

Figure 2. The localization accuracy for face on BNS at 160 Mpc in various networks of advanced detecors. The ellipses contain the 90% localization regions for sources from various points in the sky. A × is plotted at points where the network would not confidently detect the system. The plots show the localization for six different networks: (i) Hanford–Hanford–KAGRA–Livingston (HHKL), (ii) Hanford–India–KAGRA–Livingston (HIKL), (iii) Hanford–Hanford–Livingston–Virgo (HHLV), (iv) Hanford–India–Livingston–Virgo (HILV), (v) Hanford–Hanford–KAGRA–Livingston–Virgo (HHKLV), and (vi) Hanford–India–KAGRA–Livingston–Virgo (HIKLV).

and source orientation and impose the same thresholding criteria introduced above to restrict to detectable signals. The rate of BNS mergers in the nearby universe is uncertain to about two orders of magnitude [24]. For illustrative purposes, we have chosen a rate that would give 40 sources per year with an SNR greater than 8 in a single detector. This corresponds to the "realistic" astrophysical rate of [24]. Since, we have used the same rate for all the networks, the results are directly comparable, but the precise numbers should not be interpreted too strictly as the true rate could reasonably be an order of magnitude larger or smaller than the rate used here.

The number of sources that will be detected by the various networks increases significantly with the addition of a detector, but varies little between the networks. With our assumed rate, the three detector networks would observe 55 - 60 sources annually, four detector networks 82-87,

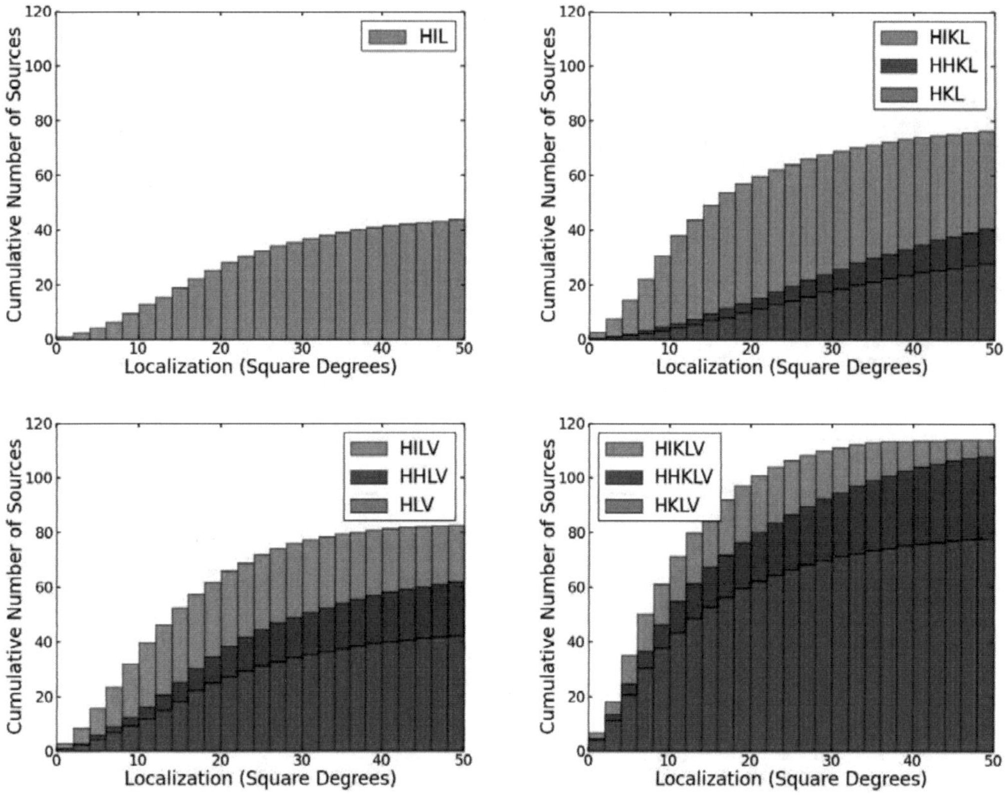

Figure 3. The number of sources localized within a given area in the sky for various networks of detectors. LIGO only networks (upper left): HIL shown, HHL and HL omitted, as no sources are well localized. LIGO-KAGRA networks (upper right): HIKL, HHKL, HKL shown (from top to bottom). LIGO-Virgo networks (lower left): HILV, HHLV and HLV shown. LIGO-KAGRA-Virgo networks (lower right): HIKLV, HHLKV and HKLV shown. In all cases, the presence of a detector in India (rather than two at Hanford) improves the source localization dramatically, although, it does not significantly affect the total number of observable signals (not shown on the plots). A single Hanford detector (and no detector in India) leads to a 25 - 40% reduction in well localized sources, and indeed in the overall number of observed sources. All results assume a rate of 40 events per year at SNR 8 or higher in a single detector.

and five detector networks 110 - 115. However, the fraction of sources that are well localized will vary significantly with detector locations. Figure 3 shows how well sources are localized with the various three, four and five detector networks. In all cases, there is a clear improvement in localization between a network with a LIGO detector in India over the corresponding network with two detectors at Hanford. This is particularly evident for a LIGO only network, where the (two) Hanford–Livingston network is incapable of localizing sources to better than a ring on the sky, while almost half of the sources observed by Hanford–Livingston–India are localized to within $20\,\mathrm{deg}^2$. For the four detector networks, an detector in India provides two or three times as many sources localized within $20\,\mathrm{deg}^2$ compared to having two LIGO detectors at Hanford. The five site network opens the possibility of the majority of sources to be localized within $10\,\mathrm{deg}^2$. These localizations are of interest as the current and future generations of wide field, transient telescopes have field of views from 3 to 20 $\deg^2$ [5, 6, 7, 8]. Thus, it will be feasible to perform follow-up observations of the majority of gravitational wave sources.

In the long term, we expect to be in a position where five advanced gravitational wave detectors are operating at their design sensitivity. Even at this stage, there will be a considerable advantage to having each detector at a unique site. It seems un-reasonable to assume that the detectors will operate with 100% duty cycle, both due to the need to perform maintenance on the detectors, and the inevitable disturbances from local environmental factors. Indeed, it is typical to assume that each detector will have around an 80% duty cycle [25], with the downtime in the two Hanford detectors being highly correlated. The five site network provides considerable redundancy and, assuming 80% duty cycle in each detector, would give nearly 75% duty cycle for a four or more detector network and 95% duty cycle for a three or more detector network. This compares to a 40% four site duty cycle and 80% three or more site duty cycle for the global network with two LIGO Hanford detectors.

Finally, we must consider the fact that the re-location of a detector to India will delay its installation by several years [2]. Consequently, there will be a time when the network is operating with only two Advanced LIGO detectors. In Figure 3, we have also shown localization of sources using networks with one LIGO detector at Hanford and one at Livingston, but no LIGO-India detector. These show a reduction in the number of well localized sources by between 25 and 40%, and indeed a reduction in the overall number of observed sources by a similar amount. For example, with the realistic astrophysical rate and the above duty cycle assumptions, each year the HLV network provides 13 sources localized to better than 20 $\deg^2$, and 40 total sources compared to 18 well localized and 60 total sources for the HHLV network. Thus, the re-location of a detector to India would cause a reduction in the number of sources detected (and localized) in the short term, but lead to significant long term benefits.

In summary, we have discussed in detail how the re-location of one of the LIGO detectors to India provides a significant improvement to the ability of the global gravitational wave network to localize sources, as well as providing a greater redundancy to the network. This improvement will come at the cost of a delay in the completion of the third Advanced LIGO detector, but the long term benefits make this worthwhile. The ability of four and five site gravitational wave detector networks to accurately localize sources will be critical to enabling gravitational wave observations to play their full part in the upcoming era of multi-messenger astronomy.

Acknowledgements

This work has benefitted from discussions with numerious colleagues including Duncan Brown, Ray Frey, Bala Iyer, Dave Reitze, Bangalore Sathyaprakash, Patrick Sutton and Stan Whitcomb. This research was made possible thanks to support from the Royal Society.

References

[1] Harry G M and the LIGO Scientific Collaboration 2010 *Class. Quant. Grav.* **27** 084006
[2] Iyer B, Souradeep T, Unnikrishnan C, Dhurandhar S, Raja S, Kumar A and Sengupta A S 2011 *LIGO-India Tech. Report* [https://dec.ligo.org/cgi-bin/DocDB/ShowDocument?docid=75988]
[3] Advanced Virgo [http://wwwcascina.virgo.infn.it/advirgo/]
[4] KAGRA [http://gwcenter.icrr.u-tokyo.ac.jp/en/]
[5] LOFAR [http://www.lofar.org/]
[6] Rau A *et al* 2009 *PASP* **121** 1334 [ISSN 00046280]
[7] Pan-STARRS [http://pan-starrs.ifa.hawaii.edu/public/]
[8] Sky Mapper [http://rsaa.anu.edu.au/skymapper/]
[9] Nakar E 2007 *Phys. Report* **442** 166
[10] Ott C D 2009 *Class. Quant. Grav.* **26** 204015
[11] Schutz B F 1986 *Nature* **323** 310
[12] Schutz B F 2011 *Class. Quant. Grav.* **28** 125023
[13] Giant Metrewave Radio Telescope [http://gmrt.ncra.tifr.res.in/]

Journal of Physics: Conference Series **484** (2014) 012007 doi:10.1088/1742-6596/484/1/012007

[14] Iyer B *Private Communication*
[15] Klimenko S, Saulson P, Sathyaprakash B, Fritschel P, Raab F, Weiss R and Finn L S 2010 *LIGO Document: LIGO-T1000251-v1* [https://dcc.ligo.org/cgi-bin/DocDB/ShowDocument?docid=11604]
[16] Veitch J *et al* 2012 *Phys. Rev. D* **85** 104045
[17] Fairhurst S 2009 *New J. Phys.* **11** 123006
[18] Vitale S and Zanolin M 2010 *Phys. Rev. D* **82** 124065
[19] Van der Sluys M *et al* 2008 *Ap. J.* **688** L61
[20] Abadie J *et al* 2012 *A. & A.* **541** A155
[21] Advanced LIGO Anticipated Sensitivity Curves *LIGO Document: LIGO-T0900288-v3* [https://dcc.ligo.org/cgi-bin/DocDB/ShowDocument?docid=2974]
[22] Fairhurst S 2011 *Class. Quant. Grav.* **28** 105021
[23] Abadie J *et al* 2012 *Phys. Rev. D* **85** 82002
[24] Abadie J *et al* 2010 *Class. Quant. Grav.* **27** 173001
[25] Sathyaprakash B *et al Scientific Benefits of Moving One of LIGO Hanford Detectors to India* [http://www.ligo.caltech.edu/NSF/related/10.2011/ligo-india-110923.pdf]

Astrophysics, cosmology, and fundamental physics with compact binary coalescence and the Einstein Telescope

C Van Den Broeck

Nikhef – National Institute for Subatomic Physics, Science Park 105, 1098 XG Amsterdam, The Netherlands

E-mail: vdbroeck@nikhef.nl

Abstract. The second-generation interferometric gravitational wave detectors, currently under construction are expected to make their first detections within this decade. This will firmly establish gravitational wave physics as an empirical science, and will open up a new era in astrophysics, cosmology, and fundamental physics. Already with the first detections, we will be able to, among other things, establish the nature of short-hard gamma ray bursts, definitively confirm the existence of black holes, measure the Hubble constant in a completely independent way, and for the first time gain access to the genuinely strong-field dynamics of gravity. Hence, it is time to consider the longer-term future of this new field. The Einstein Telescope (ET) is a concrete conceptual proposal for a third-generation gravitational wave observatory, which will be ~ 10 times more sensitive in strain than the second-generation detectors. This will give access to sources at cosmological distances, with a correspondingly higher detection rate. We have given an overview of the science case for ET, with a focus on what can be learned from signals emitted by coalescing compact binaries. Third-generation observatories will allow us to map the coalescence rate out to redshifts $z \sim 3$, determine the mass functions of neutron stars and black holes, and perform precision measurements of the neutron star equation of state. ET will enable us to study the large-scale structure and evolution of the Universe without recourse to a cosmic distance ladder. Finally, we have discussed how it will allow for high-precision measurements of strong-field, dynamical gravity.

1. Introduction

A network of second-generation, kilometre scale interferometric gravitational wave detectors is currently under construction: Advanced LIGO in the US [1], Advanced Virgo in Europe [2], and KAGRA in Japan [3]; a further large interferometer might be built in India [4]. In Germany, GEO-HF is operational [5]. The first direct detection of gravitational waves (GWs) is expected before the end of the decade. Potential sources include pulsars, and indeed, the (now decommissioned) first-generation detectors already put interesting bounds on the non-axisymmetry of the Crab pulsar [6]. Next, there is the possibility of observing a stochastic gravitational wave background; here too, the initial observatories are already able to put relevant constraints [7]. Among the most promising sources are coalescing compact binaries composed of neutron stars and/or black holes [8]. Observing these will reveal the nature of short-hard gamma ray bursts (GRBs) [9], establish once and for all the existence of black holes or uncover the presence of alternative compact objects such as Boson stars [10], and enable us to independently study the expansion of the Universe by using coalescing binaries as 'standard sirens' [11]. They may also give us empirical access to the neutron star equation of state (EOS) [12, 13, 14, 15, 16, 17, 18]. Last but not the least, they will allow us to probe the genuinely strong-field dynamics of spacetime [19, 20, 21, 22, 23, 24, 25]. (For a recent review, see [26].) Thus, an exciting field of observational gravitational physics is about to open up, with implications for astrophysics, cosmology, and fundamental physics.

It is appropriate to start considering the longer-term future of this new field. In 2008-2011, a conceptual design study was conducted for a third-generation gravitational wave observatory called Einstein Telescope (ET) [27]. ET is envisaged to be a large ($\sim$ 10 km), underground, partially cryogenic facility housing several interferometers arranged in a triangular configuration. It will be about a factor of ten more sensitive in strain than the second-generation detectors. While the second-generation observatories may see up to hundreds of sources per year [8], ET might have as many as a million detections. A first study of the considerable data analysis challenges associated with such a high detection rate was presented in [28]. In this paper, we have investigated in general terms what kind of novel science can be expected from ET. In doing so, we have not attempted to be exhaustive and focus on compact binary coalescence. (For a more complete overview, see the recent paper [29].)

This paper is structured as follows: In Section 2, we have discussed the detailed re-construction of the compact binary coalescence rate as a function of redshift, which (a network of) third-generation detectors will allow for the accuracy with which the mass function of neutron stars and black holes can be determined, and the possibility of measuring the EOS of neutron stars. In Section 3, we have turned to cosmology, in particular the potential of coalescing binaries as the so-called standard sirens. Finally, in Section 4, we have described how high-precision tests of the strong field dynamics of general relativity could be performed with ET. A summary is given in Section 5.

Throughout this paper, we have used units such that $G = c = 1$, unless stated otherwise. We have denoted binary neutron stars by BNS, neutron star-black hole binaries by NSBH, and binary black holes by BBH.

2. Astrophysics

2.1. Re-construction of the coalescence rate

ET will be able to see coalescing binaries out to cosmological distances. The coalescence rate as a function of redshift is expected to rise with increasing redshift z, reaching a peak around $z \sim 1.5$, after which it must decrease again and eventually reach zero at some large redshift value where no binaries had yet formed. The coalescence rate per redshift bin is given by [30]:

$$\frac{dR}{dz}(z) = \dot{\rho}_c(z)\frac{dV}{dz}, \tag{1}$$

where dV/dz is the derivative of comoving volume with respect to redshift, and

$$\dot{\rho}_c(z) \propto \int_{t_d^{\rm min}}^{\infty} \frac{\dot{\rho}_*(z_f(z,t_d))}{1+z_f(z,t_d)} P(t_d)\, dt_d, \tag{2}$$

with $\dot{\rho}_*$, the star formation rate, z the redshift at which merger takes place, z_f the redshift at which the binary is formed, and $P(t_d)$ a probability distribution of the delay time t_d between formation and merger. The factor $(1+z_f)^{-1}$ within the integral takes into account the difference between intrinsic and observed rates. The advanced detectors, which will see mergers out to $z \simeq 0.3$, will be able to measure the coalescence rate density at $z = 0$, $\dot{\rho}_c(0)$; even knowing this one number would already be valuable in understanding the fate of extremely massive stars. At larger redshifts, there is significant disagreement between various predictions for the *shape* of

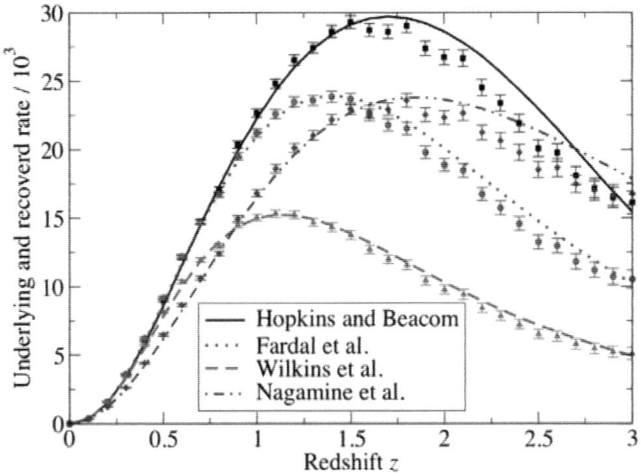

Figure 1. Four different models for the coalescence rate as a function of redshift [32, 33, 34, 35]. Assuming each of these in turn to be the underlying rate, one can simulate to what extent ET will be able to map it. The results are given by the 'data points' with $1-\sigma$ error bars.

the function $dR/dz(z)$, as illustrated in Figure 1, where coalescence rates are normalized such that $\dot{\rho}_c(0)$ is the same for all of them.

In order to arrive at a crude idea of how well a *network* of ETs, *e.g.*, with locations at the two LIGO sites and the Virgo site, would be able to map the coalescence rate, one can proceed as follows: With such a network, one would be able to partially dis-entangle the luminosity distance $D_{\rm L}$ from the angles appearing in the GW amplitude (see Eq. (3) below). Fisher matrix estimates indicate that distance could be estimated with $1-\sigma$ relative error $\sigma_0 \sim 1/\rho$, with ρ being the network SNR, if detector noise were the only source of uncertainty. However, distance measurements will also be affected by weak lensing. Modelling the associated relative distance error as $\sigma_w = 0.05\,z$ [31], one can take the combined uncertainty to be $\Delta D_{\rm L}/D_{\rm L} = (\sigma_0^2 + \sigma_w^2)^{1/2}$. Next, consider a large number of simulated 'catalogues' of $\mathcal{O}(10^5)$ sources each, distributed in redshift according to one of four different underlying true rate distributions $dR/dz(z)$ taken from recent literature [32, 33, 34, 35]. Within each catalogue, take the 'measured' distance for each source to be Gaussian distributed, centred around the true distance and with spread $\Delta D_{\rm L}/D_{\rm L}$. A measured distance $\hat{D}_{\rm L}$ can be converted into a measured redshift $\hat{z}$ using a cosmological model, *e.g.*, a ΛCDM cosmology with parameter values from the recent WMAP results [36], and by counting the number of sources in each redshift bin, one arrives at a 'measured' coalescence rate as a function of redshift, for a given catalogue. Starting from $z \sim 0.7$, this rate needs to be corrected for loss of GW detection efficiency due to inconvenient positioning and/or orientation of the source. Finally, in each redshift bin, one can compute the average and variance of the measured $d\hat{R}/dz(z)$ over all the simulated catalogues. The results are shown as the 'data points' and $1-\sigma$ error bars in Figure 1. For redshifts up to $z \sim 1.5$, the uncertainty on the rate is $\sim 1\%$. In particular, one would be able to tell the difference between the four theoretical distributions considered here.

2.2. Re-construction of the neutron star and black hole mass functions

Current measurements of neutron star masses pertain to a small sample of neutron stars within our own galaxy (see *e.g.*, [37]). Accordingly, the true distribution of neutron star masses, and in particular, the mass region where the transition to black holes occurs, remains unknown. The possible time evolution of these mass functions over cosmological timescales is also something we currently have no access to. In Figure 2, we have shown Fisher matrix estimates of how accurately the mass of a canonical ($1.4\,M_\odot$) neutron star in a BNS or NSBH system can be determined from the inspiral signal only, as a function of the companion's mass, and for $z = 1, 2, 3$. (The results suggest that in the near-equal mass case, it will be difficult to estimate the mass; however, this is due to a singularity of the Fisher matrix itself.) As we have seen, star formation models suggest that the rate at which coalescences occur will peak at $z \simeq 1 - 3$, a range where ET can measure masses with high accuracy. Thus, ET will be able to make a detailed map of the neutron star and black hole mass functions during the most interesting part of star formation history.

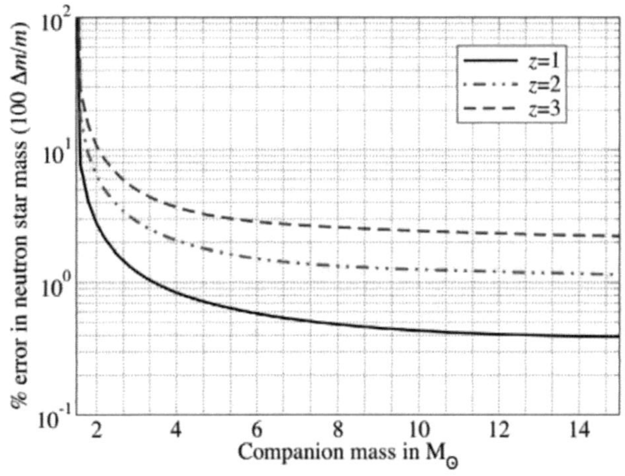

Figure 2. The percentage uncertainty in measuring the mass of a canonical ($1.4\,M_\odot$) neutron star in a BNS or NSBH system, as a function of the companion's mass, for sources at $z = 1, 2, 3$. Comparing with Figure 1, we have seen that ET would be able to measure component masses with $0.5 - 3\%$ uncertainty within the redshift interval where most of the inspirals take place.

2.3. Determining the equation of state of neutron stars

The later part of BNS or NSBH inspiral, as well as the merger, will bear imprints of the neutron star EOS. To the linear order, the tidal field $\mathcal{E}_{ij}$ of the companion object will cause a spherically symmetric neutron star to acquire a quadrupole moment $Q_{ij} = -\lambda\mathcal{E}_{ij}$. The tidal deformability parameter λ will be a function of the star's mass m. The neutron star's deformation affects the orbital motion and enters the inspiral waveform's phase starting from 5PN order (in the usual notation), where it contributes $\propto \lambda/M^5\,(v/c)^5$, with M, the total mass of the binary and v, the orbital velocity. Since, $\lambda \propto R^5$, with R, the neutron star radius, and $R/M = \mathcal{O}(10)$ for a neutron star, the tidal contribution to the phase will be non-negligible despite occurring at such high order in v/c. The function $\lambda(m)$ is set by the neutron star EOS.

In [12], a variety of EOS were considered, from very stiff to very soft, for different neutron star compositions (npeμ only, π/hyperon/quark matter, and strange quark matter). At $m = 1.4\,M_\odot$, the resulting values for $\lambda(m)$ vary over an order of magnitude. Fisher matrix results indicate that in the gravitational wave frequency range 10 Hz - 450 Hz, with the advanced detector network, and a single high-SNR source, we will only have access to the stiffest EOS leading to the largest values of the deformability λ. (For a more recent assessment which also includes the plunge, see [16].) On the other hand, ET will have access to the entire range considered in [12].

In [13], merger and post-merger BNS waveforms were numerically computed for a range of EOS with varying stiffness. For very soft EOS, the merger is followed by the prompt formation of a black hole. By constrast, for a very stiff EOS, an unstable bar is formed, which spins down over a timescale of milliseconds before also collapsing to a black hole. With second-generation detectors, in the post-merger, it is again the stiffest EOS that might cause a detectable signal because of the unstable bar. Thus, already with the upcoming advanced detector network, there is good hope of gaining information about the EOS of neutron stars in this way. However, also in the post-merger phase, ET will enable far more accurate measurements due to better sensitivity in the kHz region, where the bar mode signal, if present, will have the most power.

In [14, 15], NSBH signals were considered, in which case larger induced neutron star quadrupole moments are to be expected. It was found that, for mass ratios of 2 or 3, advanced detectors would be able to extract λ/M^5 with $10-40\%$ accuracy for a single source at a distance of 100 Mpc, and in particular, they would be able to distinguish between a neutron star and a black hole of the same mass. ET would be able to do better by an order of magnitude.

Finally, as shown in [17], combining information from *multiple* BNS inspirals observed with an advanced detector network will allow us to at least tell the difference between a soft, moderate, and stiff EOS, and constrain the EOS to 10% accuracy with as few as $\mathcal{O}(10)$ sources near the SNR threshold.

In summary, the prospects are good for empirically studying the EOS of neutron stars already with advanced detectors, but detailed measurements may have to wait until the ET era.

3. Cosmology

3.1. Compact binary coalescences as standard sirens

As noted by Schutz as early as 1986 [11], compact binary coalescences can be used as 'standard sirens', a designation based on 'standard candles', but referring to GW rather than electromagnetic sources. The observed strain amplitude of inspiral signals goes like:

$$\mathcal{A}(t) \propto \frac{1}{D_{\rm L}} \mathcal{M}_{\rm obs}^{5/3} \mathcal{F}(\theta, \phi, \iota, \psi) F^{2/3}(t). \tag{3}$$

Here, $\mathcal{M}_{\rm obs}$ is the observed chirp mass (which is related to the *physical* chirp mass $\mathcal{M}_{\rm phys}$ by $\mathcal{M}_{\rm obs} = (1+z)\,\mathcal{M}_{\rm phys}$), and $\mathcal{F}$ is a known function of sky position and orientation. The chirp mass can be measured very accurately from the phase. Thus, if some information is available on sky position and orientation, one can infer the luminosity distance $D_{\rm L}$ directly from the gravitational wave signal, with no need for calibration through a cosmic distance ladder, unlike the currently used standard candles. This opens up the possibility of completely independent cosmological measurements.

Assuming that at sufficiently large scales, the Universe can be modelled as a Friedman-Lemaître-Robertson-Walker (FLRW) spacetime, its evolution is governed by the Hubble constant H_0, the density of matter $\Omega_{\rm M}$ relative to the critical density, the density of dark energy $\Omega_{\rm DE}$, the contribution from spatial curvature Ω_k, and the dark energy EOS parameter w. The latter is defined as $w = p_{\rm DE}/\rho_{\rm DE}$, with $\rho_{\rm DE}$ and $p_{\rm DE}$ are the density and pressure of dark energy respectively; note that we have allowed for the possibility that w is time dependent. Given a number of sources for which the luminosity distance $D_{\rm L}$ can be measured, as well as the redshift z, the cosmological parameters $\vec{\Omega} = (H_0, \Omega_{\rm M}, \Omega_{\rm DE}, \Omega_k, w)$ can be inferred by fitting the distance-redshift relationship $D_{\rm L}(\vec{\Omega}; z)$.

This means that information about the redshifts of sources will also be required. In recent years, a variety of methods, as given below, have been developed for utilizing binary inspiral signals as standard sirens. The first three will already allow for an independent measurement of H_0 with an advanced detector network.

(i) Some fraction of inspirals will have electromagnetic counterparts, such as GRBs, which may allow for the identification of the host galaxy, and hence the redshift [38].

(ii) Assuming knowledge of the distribution of *physical* neutron star masses $m_{\rm phys}$, the *observed* masses $m_{\rm obs} = (1+z)\, m_{\rm phys}$ in BNS sources lead to a redshift estimate without the need for electromagnetic counterparts [39].

(iii) Approximate knowledge of sky position by using multiple detector sites together with the distance uncertainty leads to a 3-dimensional volume within which the inspiral will have occurred. Using a galaxy catalogue, each event can be associated with a list of possible redshifts, which for moderate distances can be translated into a list of possible values of H_0. A second inspiral event will lead to a different list of H_0 values, which will only be partially consistent with the first. As more events are observed, the true value of H_0 will quickly emerge. This method allows the use of *all* inspiral events, including BBH [40].

(iv) If the EOS of neutron stars has already been determined (*e.g.,* using some subset of the observed sources), one can extract the redshift directly from the GW waveform through the effect of tidal deformations on the orbital motion, which involves the *intrinsic* masses [41].

In [31] and [42], detailed studies were made of how accurately the cosmological parameters $\vec{\Omega}$ could be measured with the use of electromagnetic counterparts, *i.e.*, using the method (i) above. With $\mathcal{O}(1000)$ such events over the course of 5 – 10 years, $\Omega_{\rm M}$ and $\Omega_{\rm DE}$ could be measured with an accuracy comparable to what is possible with CMB measurements, but in a completely independent way. One can also consider the possible time dependence of the EOS parameter w, taking the CMB measurements of $\Omega_{\rm M}$, $\Omega_{\rm DE}$, and Ω_k as priors. It is convenient to approximate $w \simeq w_0 + (1-a)\, w_a$, with w_0 the present-day value, and a, the scale factor. In Figure 3, we have compared accuracies in measuring w_0 and w_a, on the one hand using standard sirens seen by ET, and on the other hand, considering the SNAP Type Ia supernova survey, which may be available on the same timescale as ET [43]. The measurement quality is comparable in the two cases, but we stress once again that standard sirens allow for an *independent* measurement; in particular, there will be no need for a cosmic distance ladder. More recent work by Taylor and Gair [44] based on method (ii), which allows the use of *all* BNS inspiral events seen in gravitational waves, has demonstrated that a network of ETs would, in fact, be able to beat the predicted accuracy of the future supernova surveys.

4. Probing the strong-field dynamics of spacetime

Until the 1970s, all tests of general relativity (GR) took the form of observing test particles (photons, planets modelled as test masses,...) moving in a weak, stationary gravitational field [45]. This changed with the discovery of tight binary pulsars [46, 47, 48], whose orbital parameters were found to be changing in accordance with the emission of gravitational radiation as predicted by GR. Even so, the most interesting aspect of classical GR, namely the strong-field dynamics of the gravitational field itself, remains out of reach. In the foreseeable future, only the direct detection of gravitational radiation, in particular from coalescing compact binaries, will enable significant progress in this direction.

The authors of [49, 50, 20] have first pointed out the possibility of a very broad test of the dynamics of GR, based on a consistency requirement on the post-Newtonian coefficients appearing in the phase of an inspiral waveform. In the stationary phase approximation, the phase as a function of frequency, takes the form [51, 52]:

$$\Psi(f) = 2\pi f t_c - \phi_c + \sum_{j=0}^{7} \left[\psi_j + \psi_j^{(l)} \ln f\right] f^{(j-5)/3}, \tag{4}$$

where t_c and ϕ_c are respectively the time and phase at coalescence. Now, according to general relativity, in the non-spinning case, all of the coefficients ψ_j and $\psi_j^{(l)}$ only depend on the two

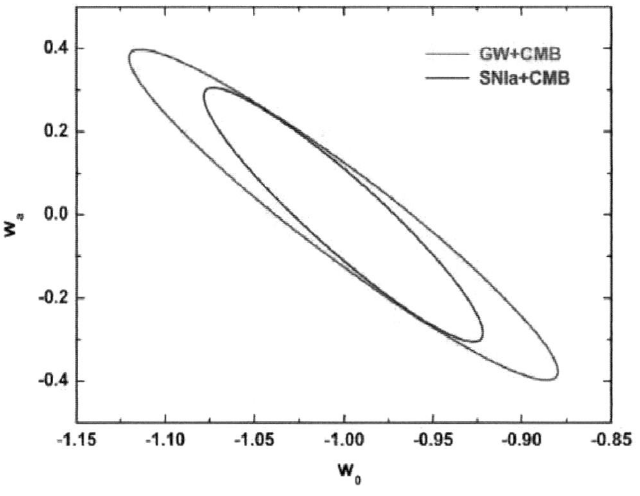

Figure 3. Measurement uncertainties for the possible time dependence in the dark energy EOS parameter w, modelled as $w(a) \simeq w_0 + (1-a)\, w_a$, with a as the scale factor. The slightly larger, red error ellipse is for standard sirens as seen with ET, the blue one for the possible future SNAP Type Ia supernova survey. In both cases, prior information from the CMB is assumed for $\Omega_{\rm M}$, $\Omega_{\rm DE}$, and Ω_k. (Adapted from [42].)

component masses (m_1, m_2), and GR makes a specific prediction for this dependence. Li *et al* [23, 24] subsequently developed a test of GR in the context of Bayesian inference, which computes an odds ratio based on the data d, and whatever background information I one may hold:

$$\mathcal{O}_{\rm GR}^{\rm modGR} = \frac{P(\mathcal{H}_{\rm modGR}|d, I)}{P(\mathcal{H}_{\rm GR}|d, I)}. \tag{5}$$

Here, $\mathcal{H}_{\rm GR}$ is the hypothesis that GR is correct, while $\mathcal{H}_{\rm modGR}$ is the hypothesis that *one or more* of the ψ_j, $\psi_j^{(l)}$ do not have the functional dependence on component masses as welll as spins that GR predicts, without specifying which ones. Computing the posterior probability $P(\mathcal{H}_{\rm modGR}|d, I)$ amounts to testing a large number of independent sub-hypotheses, in each of which a particular subset of the phase coefficients are allowed to vary freely, while the others have the dependences on masses and spins as in GR. This has been dubbed the TIGER method. The framework is well suited to a situation, where almost all sources have low SNR, as will be the case with the advanced detectors. It also trivially allows for combining information from multiple sources: For each source, the same 'yes/no' question is asked, whose correct answer is the same for all sources, so that evidence for or against GR has a tendency to build up.

As an example, one can consider a possible violation of GR in the coefficient ψ_3, which is of particular interest, because 1.5 PN is the lowest post-Newtonian order at which so-called 'tail effects' become apparent, which roughly speaking result from the interaction of gravitational waves with the rest of spacetime [53, 54][1]. As shown in [23], with $\mathcal{O}(10)$ BNS detections, the Advanced LIGO and Virgo network will be able to pick up a 10% deviation in this coefficient with essentially zero false alarm probability, and *constraining* it will be possible at the $\sim$ 1%

[1] This is an example of an essentially field-theoretic effect, rather than just pertaining to the behaviour of test masses in a given gravitational field. Such effects cannot be probed with the currently observed binary neutron stars, due to their low orbital velocity ($v/c \sim 10^{-3}$) and orbital compactness ($GM/c^2R \sim 10^{-6}$) compared with a binary on the verge of merger (which has $v/c \sim 0.4$ and $GM/c^2R \sim 0.2$ respectively). More generally, terms in Eq. (4) with $j > 0$ are only accessible with direct gravitational wave detection.

level. In the case of ET, the Fisher matrix results of [20] suggest that already with a *single* source, we will be able to measure ψ_3 with an accuracy better than 0.1% using the inspiral signal only. Presumably, these numbers will be even better for BBH when dynamical spins as well as merger and ringdown are taken into account.

It should be emphasized that a much wider class of GR violations can be picked up by using the TIGER method, not just simple shifts in post-Newtonian coefficients. (See [23, 24] for some striking examples.) A likely condition for a *generic* violation to be detectable in this way is that, for fixed coalescence phase, it induces a shift in the waveform of half a cycle around the frequency where the detector is the most sensitive; both for the advanced detectors and for ET this is ~ 150 Hz. Finally, we have noted that although TIGER will be very efficient in *finding* a deviation from GR, it will not necessarily allow for a determination of its precise nature. For the latter, one may want to use one of the general parameterized post-Einsteinian (ppE) waveform families of [19, 21, 25].

As shown in [55], observing the ringdown signal with ET will allow for a powerful test of the No Hair Theorem. As the black hole resulting from merger evolves towards a quiescent state, it is well modelled as a perturbed Kerr black hole. The Einstein's equations dictate that the frequencies and damping times of the quasi-normal modes only depend on the mass M and spin J of the underlying Kerr geometry. For a $500\,M_\odot$ black hole at 6 Gpc, a departure as small as 8% in the frequency of the dominant mode could be detected. Kamaretsos *et al* went a step further by relating ringdown mode amplitudes to the progenitor masses and spins [56, 57]. It would be of great interest to cast these ideas into the language of TIGER, in which case smaller deviations can presumably be found even for a single source, and information from multiple sources can be combined.

5. Summary

The advanced interferometric gravitational wave detectors that are currently being built have the potential to settle important scientific questions: What is the local coalescence rate, and what does this imply for our understanding of the evolution of massive stars? What is the nature of short-hard gamma ray bursts? Do black holes really exist, or is there some alternative kind of very compact object? What does the EOS of neutron stars approximately look like? Are estimates of the present-day expansion of the Universe correct? Is general relativity really the correct theory of gravity in the strong-field, dynamical regime?

Einstein Telescope will take us to much larger detection rates and out to cosmological distances, where new questions can be addressed. What is the mass distribution of neutron stars and black holes, and how does it evolve in time? What is the true EOS of neutron stars (and does this, too, have time dependence)? In cosmology, we will arrive at a comprehensive, completely independent determination of all the basic parameters governing the Universe. Is the EOS of dark energy time dependent? Last but not the least, ET will probe the strong-field dynamics of spacetime in an unprecedented way.

Acknowledgements

CVDB is supported by the research programme of the Foundation for Fundamental Research on Matter (FOM), which is partially supported by the Netherlands Organisation for Scientific Research (NWO). It is a pleasure to thank the organizers of ICGC-2011 for their kind invitation. The author would also like to thank the contributors to the Einstein Telescope Conceptual Design Study, and in particular B S Sathyaprakash, for many fruitful discussions.

References

[1] Harry G M 2010 (for the LIGO Scientific Collaboration) *Class. Quant. Grav.* **27** 084006

[2] Virgo Collaboration *Advanced Virgo Baseline Design* note VIR027A09 May 16, 2009 `https://tds.ego-gw.it/itf/tds/file.php?callFile=VIR-0027A-09.pdf`
[3] Kuroda K and the LCGT Collaboration 2010 *Class. Quant. Grav.* **27** 084004
[4] Sathyaprakash B S (for the LIGO Scientific Collaboration) 2011 *Scientific Benefits of LIGO-India*, LSC internal report G1100991
[5] Grote H (for the LIGO Scientific Collaboration) 2008 *Class. Quant. Grav.* **25** 114043, Grote H (and the LIGO Scientific Collaboration) 2010 *Class. Quant. Grav.* **27** 084003
[6] Abbott B *et al* 2008 *Ap. J.* **683** L45
[7] Abbott B *et al* 2009 *Nature* **460** 990
[8] Abadie J *et al* (for the LIGO Scientific Collaboration and Virgo Collaboration) 2010 *Class. Quant. Grav.* **27** 173001
[9] Nakar E 2007 *Phys. Report* **442** 166
[10] Palenzuela C, Lehner L and Liebling S L 2008 *Phys. Rev. D* **77** 044036
[11] Schutz B 1986 *Nature* **323** 310
[12] Hinderer T, Lackey B D, Lang R N and Read J S 2010 *Phys. Rev. D* **81** 123016
[13] Read J S, Markakis C, Shibata M, Uryu K, Creighton J E D and Friedman J L 2009 *Phys. Rev. D* **79** 124033
[14] Lackey B D, Kyutoku K, Shibata M, Brady P R and Friedman J L 2012 *Phys. Rev. D* **85** 044061
[15] Lackey B D, Kyutoku K, Shibata M, Brady P R and Friedman J L [arXiv::1303.6298 [gr-qc]]
[16] Damour T, Nagar A and Villain L 2012 *Phys. Rev. D* **85** 123007
[17] Del Pozzo W, Li T G F, Agathos M, Van Den Broeck C and Vitale S 2013 *In preparation*
[18] Yagi K and Yunes N 2013 [arXiv:1303.1528 [gr-qc]]
[19] Yunes N and Pretorius F 2009 *Phys. Rev. D* **80** 122003
[20] Mishra C K, Arun K G, Iyer B R and Sathyaprakash B S 2010 *Phys. Rev. D* **82** 064010
[21] Cornish N, Sampson L, Yunes N and Pretorius F 2011 *Phys. Rev. D* **84** 062003
[22] Del Pozzo W, Veitch J and Vecchio A 2011 *Phys. Rev. D* **83** 082002
[23] Li T G F, Del Pozzo W, Vitale S, Van Den Broeck C, Agathos M, Veitch J, Grover K, Sidery T, Sturani R and Vecchio A 2012 *Phys. Rev. D* **85** 082003
[24] Li T G F, Del Pozzo W, Vitale S, Van Den Broeck C, Agathos M, Veitch J, Grover K, Sidery T, Sturani R and Vecchio A 2012 *J. Phys. Conf. Ser.* **363** 012028
[25] Chatziioannou K, Yunes N and Cornish N 2012 *Phys. Rev. D* **86** 022004
[26] Van Den Broeck C 2013 *Probing dynamical spacetimes with gravitational waves*, in Springer Handbook of Spacetime, eds A Ashtekar and V Petkov, Springer Verlag (In press) [arXiv:1301.7291 [gr-qc]]
[27] Abernathy M *et al Einstein Gravitational Wave Telescope: Conceptual Design Study* (Available from *European Gravitational Observatory*, document number ET-0106A-10)
[28] Regimbau T, Dent T, Del Pozzo W, Giampanis S, Li T G F, Robinson C, Van Den Broeck C, Meacher D, Rodriguez C, Sathyaprakash B S and Wjcik K 2012 *Phys. Rev. D* **86** 122001
[29] Sathyaprakash, B S *et al* 2012 *Class .Quant. Grav.* **29** 124013
[30] Regimbau T and Hughes S A 2009 *Phys. Rev. D* **79** 062002
[31] Sathyaprakash B S, Schutz B and Van Den Broeck C 2010 *Class. Quant. Grav.* **27** 215006
[32] Hopkins A and Beacom J 2006 *Ap. J.* **651** 142
[33] Fardal M, Katz N, Weinberg D and Davé R 2007 *MNRAS* **379** 985
[34] Wilkins S, Trentham N and Hopkins A 2007 *MNRAS* **379** 985
[35] Nagamine K, Ostriker J, Fukugita M and Cen R 2006 *Ap. J.* **653** 881
[36] Komatsu K *et al.* 2011 *Ap. J. Supp.* **192** 18
[37] Ozel F, Psaltis D, Narayan R and Santos Villarreal A 2012 *Ap. J.* **757** 55
[38] Nissanke S, Holz D E, Hughes S A, Dalal N and Sievers J L 2010 *Ap. J.* **725** 496
[39] Taylor S R, Gair J R and Mandel I 2012 *Phys. Rev. D* **85** 023535
[40] Del Pozzo W 2012 *Phys. Rev. D* **86** 043011
[41] Messenger C and Read J S 2012 *Phys. Rev. Lett.* **108** 091101
[42] Zhao W, Van Den Broeck C, Baskaran D and Li T G F 2011 *Phys. Rev. D* **83** 023005
[43] SNAP Collaboration *Supernova Acceleration Probe: A Satellite Experiment to Study the Nature of the Dark Energy* [arXiv:astro-ph/0405232]
[44] Taylor S R and Gair J R 2012 *Phys. Rev. D* **86** 023502
[45] Misner C W, Thorne K S and Wheeler J A 1973 *Gravitation* (W H Freeman and Company, New York)
[46] Hulse R A and Taylor J H 1975 *Ap. J.* **195** L51
[47] Burgay M *et al* 2003 *Nature* **426** 531
[48] Kramer M *et al* 2006 *Science* **314** 97
[49] Arun K G, Iyer B R, Qusailah M S S and Sathyaprakash B S 2006 *Class. Quant. Grav.* **23** L37

[50] Arun K G, Iyer B R, Qusailah M S S and Sathyaprakash B S 2006 *Phys. Rev. D* **74** 024006
[51] Blanchet L, Faye G, Iyer B R and Joguet B 2002 *Phys. Rev. D* **65** 061501(R), 2005 *Erratum ibid* **71** 129902(E)
[52] Blanchet L, Damour T, Esposito-Fareèse G and Iyer B R 2004 *Phys. Rev. Lett.* **93** 091101
[53] Blanchet L and Sathyaprakash B S 1994 *Class. Quant. Grav.* **11** 2807
[54] Blanchet L and Sathyaprakash B S 1995 *Phys. Rev. Lett.* **74** 1067
[55] Gossan S, Veitch J and Sathyaprakash B S 2012 *Phys. Rev. D* **85** 124056
[56] Kamaretsos I 2012 *J. Phys. Conf. Ser.* **363** 012047
[57] Kamaretsos I, Hannam M and Sathyaprakash B S 2012 *Phys. Rev. Lett.* **109** 141102

Exact solutions for isometric embeddings of pseudo-Riemannian manifolds

G Amery and J Moodley

Astrophysics and Cosmology Research Unit, School of Mathematics, Statistics and Computer Science, University of KwaZulu-Natal, Private Bag X54001, Durban 4000, South Africa

E-mail: garethamery@gmail.com, sanjana.jm@gmail.com

Abstract. Embeddings into higher dimensions are of direct importance in the study of higher dimensional theories of our Universe, in high energy physics and in classical general relativity. Theorems have been established that guarantee the existence of local and global codimension-1 embeddings between pseudo-Riemannian manifolds, particularly for Einstein embedding spaces. A technique has been provided to determine solutions to such embeddings. However, general solutions have not yet been found and most known explicit solutions are for embedded spaces with relatively simple Ricci curvature. Motivated by this, we have considered isometric embeddings of 4-dimensional pseudo-Riemannian spacetimes into 5-dimensional Einstein manifolds. We have applied the technique to treat specific 4-dimensional cases of interest in astrophysics and cosmology (including the global monopole exterior and Vaidya-de Sitter-class solutions), and provided novel physical insights into, for example, Einstein-Gauss-Bonnet gravity. Since difficulties arise in solving the 5-dimensional equations for given 4-dimensional spaces, we have also investigated embedded spaces, which admit bulks with a particular metric form. These analyses help to provide insight to the general embedding problem.

1. Introduction

Inspired by string theory, D-brane theory and Horava-Witten theory, there has been much interest in extra dimensional effective theories, evidenced by phenomenological models such as the braneworld scenarios [1], which attempt to tackle problems such as the mass-hierarchy problem, the origins of the primordial spectrum, the inflationary field and dark energy. Einstein-Gauss-Bonnet (EGB) gravity [2, 3] takes a complementary approach, building a natural five-dimensional (5D) theory by considering additional terms in the action that only contribute for more than four dimensions. Induced matter theory [4] attempts to describe energy-momentum on the brane in terms of purely geometric effects in higher dimensions. This higher dimensional physics is complemented by the recent proof of several existence theorems for both local [5, 6, 7] and global embeddings [8].

The Dahia-Romero theorem [5] proves that a 4D analytic pseudo-Riemannian manifold M with metric $ds^2 = g_{ik}dx^i dx^k$ has a local analytic isometric embedding into a 5D Einstein manifold N with metric (in Gaussian normal coordinates, without loss of generality):

$$d\tilde{s}^2 = \tilde{g}_{\alpha\beta}(x^j, y)dx^\alpha dx^\beta = \bar{g}_{ik}(x^j, y)dx^i dx^k + \epsilon dy^2, \quad \epsilon^2 = 1, \quad \bar{g}_{ik}(x^j, 0) = g_{ik}, \tag{1}$$

along the hypersurface $y = 0$. Here, y denotes the fifth coordinate, $\tilde{R}_{\alpha\beta} = \frac{-2\Lambda}{3}\tilde{g}_{\alpha\beta}$ and Λ is the 5D cosmological constant. The metric in Eq. (1) is a solution to the Codazzi, Gauss and propagation equations (equivalent to the field equations) are given respectively by:

$$\tilde{R}_{in} = \bar{g}^{jk}(\bar{\nabla}_j\bar{\Omega}_{ik} - \bar{\nabla}_i\bar{\Omega}_{jk}) = 0\,, \tag{2}$$

$$\tilde{G}^n{}_n = -\frac{1}{2}\bar{g}^{ik}\bar{g}^{jm}[\bar{R}_{ijkm} + \epsilon(\bar{\Omega}_{ik}\bar{\Omega}_{jm} - \bar{\Omega}_{jk}\bar{\Omega}_{im})] = \Lambda\,, \quad and \tag{3}$$

$$\tilde{R}_{ik} = \bar{R}_{ik} + \frac{\epsilon}{2}\frac{\partial^2\bar{g}_{ik}}{\partial y^2} + \frac{\epsilon\bar{g}^{jm}}{4}\left(\frac{\partial\bar{g}_{ik}}{\partial y}\frac{\partial\bar{g}_{jm}}{\partial y} - 2\frac{\partial\bar{g}_{im}}{\partial y}\frac{\partial\bar{g}_{jk}}{\partial y}\right) = \frac{-2\Lambda}{3}\,\bar{g}_{ik}\,, \tag{4}$$

with the condition $\bar{g}_{ik}(x^j,0) = g_{ik}$. The term $\bar{\Omega}_{ik} = -\frac{1}{2}\frac{\partial\bar{g}_{ik}}{\partial y}$ is the extrinsic curvature of the surface $y = 0$ with $\bar{\Omega}_{ik}(x^j,0) = \Omega_{ik}$. By appealing to the Cauchy-Kowalewskaja theorem, it can be shown that a unique analytic solution $\bar{g}_{ik}$ exists, and that Eqs. (2) and (3) need only be solved at $y = 0$ [5]. The propagation Eq. (4) is used to specify the rest of the bulk. The embedded and embedding spaces are well-behaved with regards to stability and causality [9].

Despite these existing results, it has been difficult to obtain general solutions to the embedding equations as well as explicit solutions for chosen embedded spaces, particularly those with non-trivial energy-momentum and/or Ricci curvature. This motivates the investigation of embeddings for 4D non-vacuum spacetimes relevant in astrophysics and cosmology. We have focused on 5D Einstein spaces, because of their role in high-energy physics and their geometric simplicity.

Above and in the following, we have consistently adopted the following notational conventions: Roman lower case indices label the coordinates of the embedded (lower dimensional) space, Roman upper case indices label its spatial coordinates, and Greek indices label the coordinates of the embedding (higher dimensional bulk) space. A tilde denotes quantities pertaining to the bulk, an overbar denotes quantities obtained from the n-dimensional component of the bulk metric, and $d\Psi^2 = d\theta^2 + \sin^2\theta d\phi^2$.

2. Results

We have applied the above methodology to investigate embeddings of 4D spherically symmetric spaces into 5D Einstein bulks [10, 11, 12]. Here, we have presented a summary of our findings. First, we have examined an embedding of the 4D exterior global monopole [10], which is a pointlike topological defect formed during symmetry breaking phenomena in the early universe [13], and occurs as a limit in EGB Kaluza-Klein black holes [3]. The gravitational field exterior to a 4D global monopole is a non-vacuum space with simple metric:

$$ds^2 = -dt^2 + K^{-1}dr^2 + r^2d\Psi^2,\ K = 1 - 8\pi G\eta^2,\ \eta \sim 10^{16}\text{GeV}, \quad and \quad r > r_c,$$

but non-trivial Ricci curvature $R = \frac{2(K-1)}{r^2}$, and energy-momentum $T_{00} = \frac{-\eta^2}{r^2}$, $T_{11} = \frac{\eta^2}{Kr^2}$. The embedding Eqs. (2), (3) and (4) with $\Lambda = 0$ can be solved to obtain a Ricci flat bulk:

$$d\tilde{s}^2 = -dt^2 + K^{-1}dr^2 + (r - \alpha y)^2 d\Psi^2 + \epsilon dy^2,\ \alpha^2 = \epsilon(1 - K), \tag{5}$$

that embeds the exterior global monopole at $y = 0$. Considering 5D Ricci flat bulks of the form:

$$\bar{g}_{ik} = \text{diag}\ [-e^{-B(y,r)},\ e^{B(y,r)},\ (r - \sigma y)^2,\ (r - \sigma y)^2\sin^2\theta], \quad \sigma \in \mathbb{R}, \tag{6}$$

we have found only the 4D solutions $ds^2 = -(1 - \epsilon\sigma^2)dt^2 + (1 - \epsilon\sigma^2)^{-1}dr^2 + r^2d\Psi^2$. Thus, the global monopole metric is the canonical example for embeddings of type given in Eq. (6). Applying the transformation $R = r - \alpha y$, $Y = \alpha(\epsilon K)^{-1/2}r + (\epsilon K)^{1/2}y$ to Eq. (5) yields 5D Minkowski space with no deficits, where the surface $Y = \alpha(\epsilon K)^{-1/2}R$ corresponds to the monopole metric. This transformation involves a rotation of (r, y) to (R, Y) by different small angles for the r and y axes, which are related to the energy scale η. By the global existence theorems [8], we have deduced that the monopole exterior can be embedded into a 5D globally flat space. With the imposition of a deficit angle, various 5D cosmic string solutions may also be constructed. Since 5D Minkowski space is stable, this potentially alleviates the stability problem for the 4D monopole exterior metric, provided we regard gravity as being five-dimensional.

Hypersurfaces of constant y, which correspond to different global monopoles, and so the embedding space appears to have a 'linelike' structure, although it is not a product metric. This facilitates a comparison with the $r \to 0$ limit of the EGB black hole [3], in which we have observed how the Gauss-Bonnet term increases the number of permissible (product) vacuum topologies. Furthermore, our (locally flat) metric, with appropriate compactifications, can be viewed as a local embedding into the 5D EGB Kaluza-Klein black hole exterior. This can be made global [10, 8], thereby regaining the model of [3]. The energy-momentum tensor $\bar{T}_{00} = -\eta^2/(r-\alpha y)^2$, $\bar{T}_{11} = \eta^2/K(r-\alpha y)^2$ for $\bar{g}_{ik}$ reduces to the matter tensor of the global monopole when $y=0$. This is relevant to induced matter theory [4] as our bulk is empty, but its hypersurfaces of constant y contain matter.

We also have investigated [11, 10] what 4D spherically symmetric spaces in the usual form $ds^2 = -e^{2\nu(t,r)}dt^2 + e^{2\lambda(t,r)}dr^2 + r^2 d\Psi^2$ can embed into 5D Einstein bulks of particular metric forms:

$$d\tilde{s}^2 = A(y)g_{00}dt^2 + B(y)g_{11}dr^2 + C(y)g_{22}d\theta^2 + D(y)g_{33}d\phi^2 + \epsilon dy^2, \quad \textit{and} \tag{7}$$

$$d\tilde{s}^2 = A(y,r)ds^2 + \epsilon dy^2. \tag{8}$$

For Eq. (7), the only possible embedded solutions are the general Schwarzschild-de Sitter space, and the Einstein universe given respectively by (with f, h arbitrary functions):

$$ds^2 = -f(t)\left(1 + \frac{k}{r} + \frac{\alpha_1}{3}r^2\right)dt^2 + \left(1 + \frac{k}{r} + \frac{\alpha_1}{3}r^2\right)^{-1} dr^2 + r^2 d\Psi^2, \; \alpha_1, k \in \mathbb{R}, \tag{9}$$

$$ds^2 = -h(t)dt^2 + \left(1 + \frac{\alpha_2}{2}r^2\right)^{-1} dr^2 + r^2 d\Psi^2, \; \alpha_2 \in \mathbb{R}, \alpha_2 < 0. \tag{10}$$

A 5D solution in Eq. (7) for embedding metric in Eq. (9) is already known for both $\Lambda = 0$ and $\Lambda \neq 0$ [7], and has $A = B = C = D$. We partially solve the 5D embedding space given in Eq. (7) for the Einstein universe given in Eq. (10) as:

$$A = 1, \quad B = C = D = \left(1 + \sqrt{\frac{-\epsilon\alpha_2}{2}}\, y\right)^2, \quad \Lambda = 0, \tag{11}$$

$$A = \left[\cosh\left(\sqrt{\frac{-2\epsilon\Lambda}{3}}\, y\right) - a\sqrt{\frac{-3\epsilon}{2\Lambda}} \sinh\left(\sqrt{\frac{-2\epsilon\Lambda}{3}}\, y\right)\right]^2, \; B = C = D = 1, \; \Lambda = \frac{-3\alpha_2}{2}, \tag{12}$$

where Eq. (11) is Ricci flat, and Eq. (12) is a particular non-Ricci-flat solution with $a \in \mathbb{R}$ (obtained via Lie symmetry analysis with collaborators [12]). The solution in Eq. (11) can be transformed to Minkowski form so that the embedding space is flat, regaining the result given in Eq. [14]. The bulk metric in Eq. (10) cannot have $A = B = C = D$. For ansatz in Eq. (8), we have obtained $\frac{\partial A}{\partial r} = 0$, so it reduces to the case of embedding solution in Eq. (9).

Finally, we have investigated [10] the 4D spherically symmetric spaces in retarded time coordinates $ds^2 = \sigma(v,r)dv^2 - 2\mu(v,r)dvdr + r^2 d\Psi^2$, which can embed into Einstein bulks given in Eq. (1) with:

$$\bar{g}_{ik} = A(y,v)g_{ik}, \tag{13}$$

$$\bar{g}_{ik} = A(y)g_{00}\delta_i^0\delta_k^0 + A(y)g_{22}\delta_i^2\delta_k^2 + A(y)g_{33}\delta_i^3\delta_k^3 + B(y)g_{01}\delta_i^0\delta_k^1 + B(y)g_{10}\delta_i^1\delta_k^0. \tag{14}$$

We have determined that $\frac{\partial A}{\partial v} = 0$ for ansatz in Eq. (13), $A = B$ for Eq. (14), and the only 4D solution is the general Vaidya-de Sitter space with constant mass. This is a vacuum space, and so its Einstein bulk is already known [7]. Work on the non-constant mass case is ongoing.

3. Conclusion and future work

We have successfully embedded the exterior global monopole metric, general Schwarzschild-de Sitter space, Einstein universe and general constant mass Vaidya-de Sitter space as the canonical embeddings of types given in Eqs. (6), (7), (8), (13) and (14). These may be promoted to global results in various ways. Full solutions for the exterior global monopole metric and Einstein universe into non-Ricci-flat Einstein spaces have yet to be determined. It would be interesting to consider embedding a 4D global monopole with core mass and/or a cosmological constant term, and to embed global cosmic strings. Further work would include embedding the Reissner-Nordström black hole as another non-vacuum test case, and investigating what spaces can embed into 5D Einstein metrics of different forms. Other points of interest are the comparison of embeddings in general relativity with EGB gravity cases, and the relationship between the conformal geometries of embedding and embedded spaces.

Acknowledgements

The authors wish to thank S D Maharaj, P G L Leach, N K Dadhich and the faculty at the Inter-University Centre for Astronomy and Astrophysics (IUCAA, India) for useful discussions and insightful comments. GA would like to thank IUCAA for their hospitality during a research visit, which was the genesis of this project. JM was supported by a graduate bursary from the National Institute for Theoretical Physics in South Africa.

References

[1] Randall L and Sundrum R 1999 *Phys. Rev. Lett.* **83** 3370, Randall L and Sundrum R 1999 *Phys. Rev. Lett.* **83** 4690, Arkani-Hamed N, Dimopoulos S and Dvali G 1998 *Phys. Lett. B* **429** 263, Arkani-Hamed N, Dimopoulos S and Dvali G 1999 *Phys. Rev. D* **59** 086004, Dvali G, Gabadadze G and Porrati M 2000 *Phys. Lett. B* **485** 208
[2] Dadhich N K 2005 Probing universality of gravity *Proc. 11th Reg. Conf. Math. Phys.* (Singapore: World Scientific), Dadhich N K 2007 On the Gauss-Bonnet gravity *Proc. 12th Reg. Conf. Math. Phys.* (Singapore: World Scientific)
[3] Maeda H and Dadhich N K 2006 *Phys. Rev. D* **74** 021501
[4] Wesson P S 1999 *Spacetime Matter* (Singapore: World Scientific)
[5] Dahia F and Romero C 2002 *J. Math. Phys.* **43** 5804
[6] Dahia F and Romero C 2002 *J. Math. Phys.* **43** 3097
[7] Lidsey J E, Romero C, Tavakol R and Rippl S 1997 *Class. Quant. Grav.* **14** 865, Anderson E and Lidsey J E 2001 *Class. Quant. Grav.* **18** 4831
[8] Katzourakis N I 2005 [math-ph/0407067v4], Moodley J 2008 *M.Sc. Thesis, University of KwaZulu-Natal*, Moodley J and Amery G 2012 *To be submitted to J. Math. Phys.*
[9] Dahia F and Romero C 2005 *Class. Quant. Grav.* **22** 5005
[10] Moodley J 2012 *Ph.D. Thesis, University of KwaZulu-Natal*, Moodley J and Amery G 2012 *To be submitted to Commun. Math. Phys.*, Moodley J and Amery G 2012 *In preparation*
[11] Moodley J and Amery G 2011 *Pramana: J. Phys.* **77** 533
[12] Okelola M 2011 *M.Sc. Thesis, University of KwaZulu-Natal*, Okelola M, Govinder K S, Moodley J and Amery G 2012 (In preparation)
[13] Barriola M and Vilenkin A 1989 *Phys. Rev. Lett.* **63** 341
[14] Wesson P S 1994 *Ap. J.* **436** 547

Symmetry reduction on non-expanding horizons

R Basu[1], A Chatterjee[2] and A Ghosh[3]

[1] S.N. Bose National Centre for Basic Sciences, Sector-III, Block - JD, Salt Lake, Kolkata 700098, India
[2] The Institute of Mathematical Sceinces, CIT Campus, Taramani, Chennai 600113, India
[3] Saha Institute of Nuclear Physics, Sector-1, Block-AF, Bidhannagar, Kolkata 700064, India

E-mail: [1]rudranil@bose.res.in, [2]achatterjee@imsc.res.in, [3]amit.ghosh@saha.ac.in

Abstract. Local symmetries of a non-expanding horizon have been investigated in the 1st order formulation of gravity. When applied to a spherically symmetric horizons, only a U(1) subgroup of the Lorentz group survives as residual local symmetry that one can make use of in constructing an effective theory on the horizon.

In this paper, we have explored the local symmetries of a non-expanding horizon (NEH), a 3-dimensional null hypersurface, which imitates properties of a black hole horizon. Precise definition of NEH can be found in [1, 2], sufficient to characterize an NEH to be a lightlike hypersurface Δ imbedded in spacetime, such that the unique (up to scaling by a function) lightlike, real vector field l tangential to Δ is expansion, shear and twist-free. Since l is also normal to Δ, it is geodesic as well. These properties of Δ are independent of the scaling of l [1, 3]. Let us further assume that Δ is topologically equivalent to $\mathbb{S} \times \mathbb{R}$, where $\mathbb{S}$ is a 2-sphere. To understand local symmetries, it is imperative that the gravitational dynamics be described by first order formulation of gravity, so that we are also able to decipher those Lorentz transformations, which survive as symmetries on this null surface. Incidentally, since the definition of NEH is very general with minimal number of local conditions, our symmetry analysis, as done in the following paragraphs will survive even for Killing horizons and event horizons.

Gravity is invariant, apart from diffeomorphisms, under the local Lorentz group $SL(2,\mathbb{C})$, which is explicit in the first order formulation only. Here, our specific interest is to investigate whether the definition of NEH (more precisely the boundary conditions on it) leads to breaking of the bulk local Lorentz symmetries on Δ. This suspicion is motivated by the very well known examples of breaking of other local symmetries such as diffeomorphisms through boundary conditions in general relativity.

We have systematically analyzed to find out if there is indeed breaking of local Lorentz symmetry on NEH. Based on the residual gauge group that we have found through a kinematical analysis, we have proposed an effective theory on the horizon. This is worked out using the pre-symplectic structure on the covariant phase-space of the first order theory. It is supposed that subsequent quantization of that theory, with loop quantum gravity in the bulk would yield the quantum states of a black hole. There is a recent upsurge of interest in such effective theories, where an $SU(2)$ Chern-Simons theory has been proposed [4, 5, 6] in correlation with previous works [7, 8, 9, 10] as the effective quantum theory on the horizon in contrast to a $U(1)$ theory [11, 12, 13]. However, all of these analyses are based on isolated horizon definitions, which need some more geometric structures over our prime interest, the NEH conditions. In this medley, we have put the relevance of our paper through the symmetry reduction mechanism on NEH as a conclusive answer from $SL(2,\mathbb{C}) \rightarrow ISO(2) \ltimes \mathbb{R}$. A farther, rather dramatic reduction to $U(1)$ follows due to some special properties of the lie algebra $\mathfrak{iso}(2)$.

First, let us see how a NEH Δ reduces the local Lorentz symmetry. In order to facilitate our study, we have classified the Lorentz transformation matrices into the conjugacy classes of $SL(2,\mathbb{C})$ as:

$$\Lambda_{IJ} = -\xi l_I n_J - \xi^{-1} n_I l_J + 2m_{(I}\bar{m}_{J)}, \tag{1}$$
$$\Lambda_{IJ} = -2l_{(I}n_{J)} + (e^{i\theta} m_I \bar{m}_J + c.c.), \tag{2}$$
$$\Lambda_{IJ} = -l_I n_J - (n_I - cm_I - \bar{c}m_I + |c|^2 l_I)l_J + (m_I - \bar{c}l_I)\bar{m}_J + (\bar{m}_I - cl_I)m_J, \quad and \tag{3}$$
$$\Lambda_{IJ} = -l_I n_J - (l_I - bm_I - \bar{b}m_I + |b|^2 n_I)n_J + (m_I - \bar{b}n_I)\bar{m}_J + (\bar{m}_I - bn_I)m_J. \tag{4}$$

Let us consider the Palatini connection $\mathbb{A}_{IJ}$, and in the interior of the spacetime, let us expand $\mathbb{A}_{IJ}$ in the internal Lorentz basis as:

$$\mathbb{A}_{IJ} = -2\mathbb{W}l_{[I}n_{J]} + 2\mathbb{V}m_{[I}\bar{m}_{J]} + 2(\bar{\mathbb{N}}n_{[I}m_{J]} + c.c.) + 2(\bar{\mathbb{U}}l_{[I}m_{J]} + c.c.), \tag{5}$$

where $\mathbb{W}$, $\mathbb{V}$, $\mathbb{N}$, $\mathbb{U}$ are connection 1-forms; as defined, $\mathbb{W}$ is real, $\mathbb{V}$ is imaginary and $\mathbb{N},\mathbb{U}$ are complex (in all, there are six of them associated with the six generators). For the rest of our analysis, we have fixed an internal Lorentz frame for which $l_I, n_I, m_I, \bar{m}_I$ are constants. However, our results will be unaffected by such a choice.

The pull-back of the Palatini connection to the NEH Δ is of the form:

$$A_{IJ} \triangleq -2Wl_{[I}n_{J]} + 2Vm_{[I}\bar{m}_{J]} + 2(\bar{U}l_{[I}m_{J]} + c.c.), \tag{6}$$

where W, V, U are respectively the pull-backs of $\mathbb{W}, \mathbb{V}, \mathbb{U}$. Clearly, the 1-form N, which is the pull-back of $\mathbb{N}$, vanishes on Δ by the NEH boundary conditions. Proof: The simplest way to show this is to relate the connection 1-forms to the Newman-Penrose coefficients (the constant $l_I, n_I, m_I, \bar{m}_I$ basis simplifies these relations):

$$\mathbb{W} = -(\gamma+\bar{\gamma})l - (\epsilon+\bar{\epsilon})n + (\alpha+\bar{\beta})m + (\bar{\alpha}+\beta)\bar{m}\ , \tag{7}$$
$$\mathbb{V} = -(\gamma-\bar{\gamma})l - (\epsilon-\bar{\epsilon})n + (\alpha-\bar{\beta})m + (\beta-\bar{\alpha})\bar{m}\ , \tag{8}$$
$$\mathbb{U} = -\bar{\nu}l - \bar{\pi}n + \bar{\mu}m + \bar{\lambda}\bar{m}\ , \quad and \tag{9}$$
$$\mathbb{N} = \tau l + \kappa_{\text{NP}} n - \rho m - \sigma\bar{m}\ . \tag{10}$$

l_a pulled back to Δ vanishes. κ, ρ and σ also vanishe on Δ, as a consequence of the non-expanding nature of Δ; hence, does N. So, only three independent connection 1-forms W, V, U survive on Δ. This indicates that the pulled back connection given in Eq. (6) does not take the value in the full $\mathfrak{sl}(2,\mathbb{C})$, but rather in a sub-algebra of it. The following analysis puts it into firm ground. Under the local Lorentz transformations given in Eqs. (1), (2), (3) and (4), the Palatini connection in Eq. (5) transforms as:

$$\mathbb{A}_{IJ} \mapsto \Lambda_I{}^K \mathbb{A}_{KL} \Lambda_J{}^L + \Lambda_{IK} d\Lambda_J{}^K. \tag{11}$$

A lengthy but straightforward calculation shows that under the Lorentz transformations given in Eqs. (1), (2) and (3), the connection 1-forms transform as:

$$\mathbb{W} \mapsto \mathbb{W} - d\ln\xi, \quad \mathbb{V} \mapsto \mathbb{V}, \quad \mathbb{U} \mapsto \xi\mathbb{U}, \quad \mathbb{N} \mapsto \xi^{-1}\mathbb{N}. \tag{12}$$

$$\mathbb{W} \mapsto \mathbb{W}, \quad \mathbb{V} \mapsto \mathbb{V} - id\theta, \quad \mathbb{U} \mapsto e^{-i\theta}\mathbb{U}, \quad \mathbb{N} \mapsto e^{-i\theta}\mathbb{N}. \tag{13}$$

$$\begin{aligned} &\mathbb{W} \mapsto \mathbb{W} - c\mathbb{N} - \bar{c}\bar{\mathbb{N}}, \quad \mathbb{V} \mapsto \mathbb{V} - c\mathbb{N} + \bar{c}\bar{\mathbb{N}}, \\ &\qquad \mathbb{U} \mapsto \mathbb{U} - d\bar{c} + \bar{c}(\mathbb{W} - \mathbb{V}) - \bar{c}^2\bar{\mathbb{N}}, \quad \mathbb{N} \mapsto \mathbb{N}. \end{aligned} \tag{14}$$

Since $\mathbb{N}$ transforms homogeneously, its pull-back $N \triangleq 0$ in one frame implies that it vanishes in all Lorentz frames related by Eqs. (1), (2) and (3). However, under Eq. (4), the connection 1-forms transform as:

$$\begin{aligned} &\mathbb{W} \mapsto \mathbb{W} + b\mathbb{U} + \bar{b}\bar{\mathbb{U}}, \quad \mathbb{V} \mapsto \mathbb{V} - b\mathbb{U} + \bar{b}\bar{\mathbb{U}}\ , \\ &\mathbb{U} \mapsto \mathbb{U}, \quad \mathbb{N} \mapsto \mathbb{N} + d\bar{b} - \bar{b}(\mathbb{W} + \mathbb{V}) - \bar{b}^2\bar{\mathbb{U}}. \end{aligned} \tag{15}$$

Clearly, in this case, $N \triangleq 0$ if and only if b satisfies the equation $db \triangleq b(W - V + b\bar{U}) =: bY$, where Y is a 1-form. This equation has a non-trivial solution if and only if Y is a closed 1-form. However, we have shown that the equation admits only the trivial solution, $b = 0$. (For a detailed proof, see [14].) This proves conclusively that out of the four transformations given in Eqs. (1), (2), (3) and (4), the fourth one is not allowed on Δ, due to non-trivial boundary conditions on it. Apart from the fourth transformation, the rest three are generated by a Borel sub-algebra [15] of $\mathfrak{sl}(2,\mathbb{C})$ and the corresponding group is $ISO(2) \ltimes \mathbb{R}$ connection [14]. It is interesting to note that the Cartan Killing metric of this Lie algebra has two zero modes and two non-zero modes, which makes it non semi-simple.

For later convenience, we have fixed a basis for the Lie algebra $\mathfrak{iso}(2) \ltimes \mathbb{R}$ as:

$$B_{IJ} = (\partial\Lambda_{IJ}/\partial\xi)_{\xi=1} = -2l_{[I}n_{J]}, \tag{16}$$

$$R_{IJ} = (\partial\Lambda_{IJ}/\partial\theta)_{\theta=0} = 2im_{[I}\bar{m}_{J]}, \tag{17}$$

$$P_{IJ} = (\partial\Lambda_{IJ}/\partial \mathrm{Re}\, c)_{c=0} = 2m_{[I}l_{J]} + 2\bar{m}_{[I}l_{J]}, \quad \textit{and} \tag{18}$$

$$Q_{IJ} = (\partial\Lambda_{IJ}/\partial \mathrm{Im}\, c)_{c=0} = 2im_{[I}l_{J]} - 2i\bar{m}_{[I}l_{J]}. \tag{19}$$

Hence, we can expand A in this basis as:

$$A_{IJ} = 2A_B B_{IJ} + 2A_R R_{IJ} + 2A_P P_{IJ} + 2A_Q Q_{IJ}, \tag{20}$$

where $2A_B = W$, $2A_R = -iV$, $2A_P = -\mathrm{Re}\, U$, and $2A_Q = \mathrm{Im}\, U$. The connection 1-forms A_B, A_R, A_P, A_Q will turn out to be more useful in the context of an effective theory on the horizon.

Let us now turn our attention to the symplectic structures. The Hölst action [16] gives rise to the symplectic current 3-form (in units of $4\pi G\gamma_B = 1$ and $\mathbb{E}^I$ is the spacetime tetrad 1-form):

$$\mathbb{J}(\delta_1, \delta_2) = -\frac{1}{4}\mathrm{Tr}\left(\delta_1(\mathbb{E} \wedge \mathbb{E}) \wedge \delta_2\mathbb{H} - (1 \leftrightarrow 2)\right). \tag{21}$$

Here, the trace involves the $\mathfrak{sl}(2,\mathbb{C})$ Cartan-Killing metric. The expansion of the tetrad in the null tetrad basis is $\mathbb{E}^I = -nl^I - ln^I + m\bar{m}^I + \bar{m}m^I$. So the two-form $\mathbb{E}^I \wedge \mathbb{E}^J$ pulled back onto Δ, and expanded in the $\mathfrak{iso}(2) \ltimes \mathbb{R}$ basis is given by:

$$E^I \wedge E^J \triangleq {}^2\epsilon R^{IJ} + \mathrm{Re}(n \wedge m)P^{IJ} - \mathrm{Im}(n \wedge m)Q^{IJ}, \tag{22}$$

where ${}^2\epsilon = im \wedge \bar{m}$. Now, the symplectic current given in Eq. (21) is a closed spacetime 3-form $d\mathbb{J} = 0$. Now, consider a region $\mathcal{M}$ of the spacetime, with the partially Cauchy surfaces $M_\pm$ and the horizon Δ enclosing it. Moreover, let $M_\pm$ intersect Δ at circles $S_\pm$. Integrating $d\mathbb{J}$ over $\mathcal{M}$, we have found that the sum-total contribution of the symplectic current from the boundaries of $\mathcal{M}$ must vanish as:

$$\int_{M_+ \cup M_- \cup \Delta \cup i^0} \mathbb{J}(\delta_1, \delta_2) = 0. \tag{23}$$

We have assumed that the boundary conditions at infinity are such that the contribution of i^0 to the integral in Eq. (23) vanishes. We must also ensure that the symplectic structure is independent of the choice of our foliation by the partial Cauchy slices. Using Eqs. (20) and (22), and the fact that the trace in Eq. (21) is taken over a degenerate Killing metric, the pull-back of the symplectic current in Eq. (21) is:

$$J(\delta_1, \delta_2) \triangleq \frac{1}{2}\delta_1{}^2\epsilon \wedge \delta_2(iV + \gamma_B W) - (1 \leftrightarrow 2). \tag{24}$$

It is easy to see why only the combination $iV + \gamma_B W$ survives the pull-back: The pull-back of the connection A, hence also of H, have all the $\mathfrak{iso}(2) \ltimes R$ components. However, the pull-back $E^I \wedge E^J$ is only $\mathfrak{iso}(2)$-valued, as is obvious from Eq. (22). Furthermore, only the RR and BB components survive the tracing, because of the degeneracy of the metric. Since $E^I \wedge E^J$ has no B-component, only the RR components survive in the symplectic current, which gives rise to the combination $iV + \gamma_B W$ in Eq. (24), where ${}^2\epsilon$ is the area 2-form of some spherical cross-section of Δ. In the derivation of the symplectic current, it is sufficient to assume that the spherical cross-section foliates Δ and is not necessarily a geometric 2-sphere. However, for the rest of our analysis, we have restricted ourselves to the unique foliation of Δ, in which each leaf is a geometric 2-sphere; this is possible if and only if the horizon Δ is spherically symmetric. For such a horizon with a fixed area $\mathcal{A} = \int {}^2\epsilon$, the 1-form W is closed and dV is proportional to ${}^2\epsilon$ [17, 18], and

$$dW \triangleq 0, \quad and \quad dV \triangleq \frac{4\pi i}{\mathcal{A}}\,{}^2\epsilon, \tag{25}$$

where d is the exterior derivative intrinsic to Δ. Using Eq. (25), we have found that the symplectic current 3-form is exact on Δ, given by:

$$\begin{aligned} J(\delta_1, \delta_2) &\triangleq dj(\delta_1, \delta_2), \text{ where} \\ j(\delta_1, \delta_2) &= -\frac{\mathcal{A}}{8\pi}\delta_1(iV + \gamma_B W) \wedge \delta_2(iV + \gamma_B W). \end{aligned} \tag{26}$$

It is to be noted that in the $\mathfrak{iso}(2) \ltimes \mathbb{R}$ basis, the 1-form $iV + \gamma_B W = -2(A_R - \gamma_B A_B) =: -2A_{\mathrm{CS}}$.

We have now chosen a particular orientation of the relevant spacetime boundaries M_+, M_- and Δ, such that the current conservation given in Eq. (23) reduces to:

$$\left(\int_{M_+} - \int_{M_-}\right)\mathbb{J}(\delta_1, \delta_2) = \frac{\mathcal{A}}{2\pi}\left(\int_{S_-} - \int_{S_+}\right)(\delta_1 A_{\mathrm{CS}} \wedge \delta_2 A_{\mathrm{CS}}).$$

This gives a foliation independent symplectic structure, whose boundary part is given by (putting back $4\pi G\gamma_B = 1$):

$$\Omega(\delta_1, \delta_2) = -\frac{\mathcal{A}}{8\pi^2 G\gamma_B}\int_{\mathbb{S}} \delta_1 A_{\mathrm{CS}} \wedge \delta_2 A_{\mathrm{CS}}, \tag{27}$$

where $\mathbb{S}$ is the unique spherical cross-section of Δ and $A_{\rm CS} = A_R - \gamma_B A_B$.

The form in Eq. (27) suggests that on a spherically symmetric NEH, one can take the effective boundary theory as a $U(1)$ Chern-Simons theory. Two distinct cases of $U(1)$ arise: (i) If either the pull-back of A_B vanishes on $\mathbb{S}$ [19] or one restricts the gauge freedom to a constant class (ξ = constant, as has been the original choice [11]) then one gets a compact $U(1)$, and (ii) In general, if no restrictions are imposed, then one gets a non-compact $U(1)$.

Acknowledgements

We thank Abhay Ashtekar for correspondence, Parthasarathi Majumdar and Ramesh Kaul at the initial stage of collaboration and Parthasarathi Mitra for discussion. RB thanks Council for Scientific and Industrial Research (CSIR), Government of India, for support through the SPM Fellowship SPM-07/575(0061)/2009-EMR-I.

References

[1] Ashtekar A, Fairhurst S and Krishnan B 2000 *Phys. Rev. D* **62** 104025 [arXiv:gr-qc/0005083]
[2] Ashtekar A and Krishnan B 2004 *Living Rev. Rel.* **7** 10 [arXiv:gr-qc/0407042]
[3] Chatterjee A and Ghosh A 2006 *Class. Quant. Grav.* **23** 7521 [arXiv:gr-qc/0603023]
[4] Engle J, Perez A and Noui K [arXiv:0905.3168[gr-qc]], Engle J, Noui K, Perez A and Pranzetti D 2010 *Phys. Rev. D* **82** 044050 [arXiv:1006.0634[gr-qc]], Engle J, Noui K, Perez A and Pranzetti D 2011 *JHEP* **1105** 016 [arXiv:1103.2723[gr-qc]]
[5] Basu R, Kaul R K and Majumdar P [arXiv:0907.0846[gr-qc]]
[6] Kaul R K and Majumdar P 2011 *Phys. Rev. D* **83** 024038 [arXiv:1004.5487[gr-qc]]
[7] Krasnov K V 1997 *Phys. Rev. D* **55** 3505 [arXiv:gr-qc/9603025]
[8] Rovelli C 1996 *Phys. Rev. Lett.* **77** 3288 [arXiv:gr-qc/9603063]
[9] Kaul R K and Majumdar P 2000 *Phys. Rev. Lett.* **84** 5255 [arXiv:gr-qc/0002040]
[10] Agullo I, Fernando Barbero G J, Borja E F, Diaz-Polo J and Villasenor E J S 2009 *Phys. Rev. D* **80** 084006 [arXiv:0906.4529[gr-qc]]
[11] Ashtekar A, Corichi A and Krasnov K 2000 *Adv. Theor. Math. Phys.* **3** 419 [arXiv:gr-qc/9905089]
[12] Ashtekar A, Baez J, Corichi A and Krasnov K 1998 *Phys. Rev. Lett.* **80** 904 [arXiv:gr-qc/9710007]
[13] Ashtekar A, Baez J C and Krasnov K 2000 *Adv. Theor. Math. Phys.* **4** 1 [arXiv:gr-qc/0005126]
[14] Basu R, Chatterjee A and Ghosh A [arXiv:1004.3200[gr-qc]]
[15] Private communication with Samir K Paul
[16] Hölst S 1996 *Phys. Rev. D* **53** 5966 [arXiv:gr-qc/9511026]
[17] Ashtekar A, Beetle C and Fairhurst S 2000 *Class. Quant. Grav.* **17** 253 [arXiv:gr-qc/9907068]
[18] Chatterjee A and Ghosh A 2009 *Phys. Rev. D* **80** 064036 [arXiv:0812.2121[gr-qc]]
[19] Lewandowski J 2000 *Class. Quant. Grav.* **17** L53 [arXiv:gr-qc/9907058]

Localizing the energy and momentum of linear gravity

L M Butcher

Kavli Institute for Cosmology, Madingley Road, Cambridge, CB3 0HA, UK

E-mail: l.butcher@mrao.cam.ac.uk, lmb62@mrao.cam.ac.uk

Abstract. Although there is little doubt that gravitational waves exist and carry energy as they propagate, it has been notoriously difficult to explain where in spacetime this energy resides. We have summarized a new approach to the localization of gravitational energy-momentum, valid within the linear approximation to general relativity . Built around a local description of the *exchange* of energy-momentum between matter and linear gravity, the framework defines a *unique* symmetric gravitational energy-momentum tensor, free of second derivatives, and motivates a natural gauge-fixing programme, which renders the description unambiguous. Once the gauge has been fixed according to this programme, the gravitational energy-momentum tensor obeys the dominant energy condition: gravitational energy-density is never negative, and gravitational energy-flux is never spacelike.

1. Introduction

Half a century ago, a simple argument established that gravitational waves carry energy and can exchange this energy with matter. Often attributed to Feynman [1] (certainly popularized by Bondi [2]), the argument asked us to imagine a gravitational detector comprising a rigid rod along which two "sticky beads" are threaded. A passing gravitational wave then acts to alter the proper distance between the beads, and this motion, opposed by friction, heats the detector, and thus, mediates a transfer of energy from gravity to matter. Despite the simplicity of this idea, even after fifty years, it has not been possible to explain *where* in spacetime this gravitational energy resides, and it is generally accepted that attempts to do so are "looking for the right answer to the wrong question" [3].

The elusiveness of the "right answer", and the wrongness of the question, are often identified as arising from gravity's gauge freedom, the consequence of which is a one-to-many mapping between physical spacetime and whatever localization of gravitational energy-momentum might be proposed. However, there is no reason *a priori* that gauge dependence should preclude the construction of a physically unambiguous tensor, provided we are prepared to remove the gauge freedom in some well-defined way. Unfortunately, no previous approach has supplied instructions of this nature, and more importantly, neither the construction of these energy-momentum objects, nor their key properties, appear to favour one gauge over another; thus, it appears impossible to justify any of these seemingly arbitrary choices as *natural*.

In spite of this difficulty, one aspect of this enduring problem stands opposed to conventional wisdom, and motivates further consideration: when gravity and matter interact, the *exchange* of energy can be localised! To see this, we need look no further than the sticky bead detector: here, the energy exchange is certainly localized in so far as it takes place only within the confines of the detector. Furthermore, we can imagine a very small detector, much smaller than a wavelength of the incident gravitational radiation, and observe that at each instant, a well-defined power is developed in the detector as heat; thus, at least in this case, the rate of energy exchange is associated with a particular point in spacetime. One might hope, therefore, that consistency with this phenomenon would be enough to localize the energy and momentum of the gravitational field outside the detector, or even when no detector is present.

Based on these ideas, a framework has been developed to localize the energy and momentum of gravity, within the linear approximation to general relativity [4]; what follows is a summary of this work.[1]

2. Derivation

We define the gravitational field h_{ab} on a *flat* background spacetime $(\check{\mathcal{M}}, \check{g}_{ab})$ by a diffeomorphism $\phi : \mathcal{M} \to \check{\mathcal{M}}$ that maps the physical spacetime $(\mathcal{M}, g_{ab})$ onto the background[2]:

$$\phi^* g_{ab} = \check{g}_{ab} + h_{ab}. \tag{1}$$

The physical spacetime is assumed to be "nearly flat", and ϕ is chosen such that h_{ab} is small everywhere, so that the linear approximation to the Einstein's field equations is valid:

$$\widehat{G}_{ab}{}^{cd} h_{cd} = \kappa \check{T}_{ab} + O(h^2), \tag{2}$$

where $\check{T}_{ab} \equiv \phi^* T_{ab} \sim O(h)$ is the matter energy-momentum tensor T_{ab} mapped onto the background, and

$$\widehat{G}_{ab}{}^{cd} h_{cd} \equiv \check{\nabla}_c \check{\nabla}_{(a} h_{b)}{}^c - \tfrac{1}{2}\check{\nabla}^2 h_{ab} - \tfrac{1}{2}\check{\nabla}_a \check{\nabla}_b h + \tfrac{1}{2}\check{g}_{ab}\left(\check{\nabla}^2 h - \check{\nabla}_c \check{\nabla}_d h^{cd}\right) \tag{3}$$

is the linearized Einstein tensor $G^{(1)}_{ab}$.

The gravitational energy-momentum tensor τ_{ab} is defined by seeking a symmetric tensor, quadratic in $\check{\nabla}_c h_{ab}$, which solves:

$$\check{\nabla}_a j_\mu{}^a + \phi^*(\nabla_a J_\mu{}^a) = 0, \tag{4}$$

neglecting terms $O(h^3)$. In the above equation, $J_\mu{}^a \equiv T^a{}_b e_\mu{}^b$ are the (1 energy, 3 momentum) current-densities of matter, associated with the (1 timelike, 3 spacelike) vector fields $e_\mu{}^a \equiv (\phi^{-1})^* \check{e}_\mu{}^a$, the images of the Lorentzian coordinate basis $\check{e}_\mu{}^a \equiv (\partial/\partial x^\mu)^a$ that generate the translational symmetries of the background; the $j_\mu{}^a \equiv \tau^a{}_b \check{e}_\mu{}^b = \tau^a{}_\mu$ constitute the energy-momentum current-densities of the gravitational field. Consequently, Eq. (4) indicates that the extent to which material energy-momentum fails to be conserved at a point in the physical spacetime is exactly equal and opposite to the extent to which gravitational energy-momentum fails to be conserved at the corresponding point in the background. Interactions between matter and gravity can then be understood in terms of a *local exchange* of energy and momentum between the two.

It is not possible to construct a τ_{ab} to solve Eq. (4) for all gravitational fields, so a condition must be placed on h_{ab} in order to proceed. Of all possible symmetric tensors τ_{ab}, quadratic in

[1] We have worked in units where $c = 1$, write $\kappa \equiv 8\pi G$, and use the sign conventions of Wald [5]: $\eta_{\mu\nu} \equiv \mathrm{diag}(-1,1,1,1)$, $[\nabla_c, \nabla_d]v^a \equiv 2\nabla_{[c}\nabla_{d]}v^a \equiv R^a{}_{bcd}v^b$, and $R_{ab} \equiv R^c{}_{acb}$; Roman letters (except i, j, k, l) are used as abstract tensor indices and Greek letters as numerical indices running from 0 to 3. The indices i, j, k, l are reserved for spatial components, and run from 1 to 3.

[2] As usual, fields defined on $\mathcal{M}$ have their indices raised and lowered with g_{ab}, and those on $\check{\mathcal{M}}$ with $\check{g}_{ab}$. Lorentzian coordinates $\{x^\mu\}$ are commonly deployed in $\check{\mathcal{M}}$, for which $\check{g}_{\mu\nu} = \eta_{\mu\nu}$.

$\check{\nabla}_c h_{ab}$, and all (non-trivial, linear and Lorentz invariant) field conditions, only one combination solves the Eq. (4) as:

$$\kappa\bar{\tau}_{ab} = \tfrac{1}{4}\check{\nabla}_a h_{cd}\check{\nabla}_b\bar{h}^{cd}, \quad \textit{and} \tag{5}$$

$$\check{\nabla}^a\bar{h}_{ab} = 0, \tag{6}$$

where the overbars signify trace-reversal. Because Eq. (6) is simply the equation of *harmonic gauge*, which can always be satisfied through a choice of ϕ, the field condition does not restrict the physical applicability of our approach in any respect. In fact, the only effect of the field condition is to vastly reduce the gauge freedom in our description of gravitational energy-momentum given in Eq. (5). Essentially, Eq. (6) indicates that ϕ is to be chosen such that it maps Lorentzian coordinates $\{x^\mu\}$ of the background onto harmonic coordinates $y^\mu(p) \equiv x^\mu(\phi(p))$ of the physical spacetime. This ensures that the energy-momentum currents ${J_\mu}^a$ are defined by the generators of a harmonic coordinate system; these represent the approximate translational symmetries of the physical spacetime (present due to its small curvature), and give a sensible replacement for Killing vectors in the absence of an exact symmetry.

3. Infinitesimal probe

A small amount of gauge freedom remains after the harmonic condition in Eq. (6) has been enforced; this ambiguity is extinguished by considering the local exchange of energy-momentum between gravity and an *infinitesimal detector*, as discussed in the introduction. The detector takes the form of a matter "point-source", with an energy-momentum tensor:

$$\begin{aligned}\check{T}_{00} &= M\delta(\vec{x}) + \tfrac{1}{2}I_{ij}\partial_i\partial_j\delta(\vec{x}),\\ \check{T}_{0i} &= \tfrac{1}{2}(\dot{I}_{ij} - L_{ij})\partial_j\delta(\vec{x}), \quad \textit{and}\\ \check{T}_{ij} &= \tfrac{1}{2}\ddot{I}_{ij}\delta(\vec{x}),\end{aligned} \tag{7}$$

derived by shrinking a compact source down to a point[3]. The exchange is rendered completely gauge invariant by the *monopole-free micro-average*: The incoming wave is split into a sum of Heaviside step-functions, and the energy-momentum delivered by each is integrated over a vanishingly small 4-volume centred on the probe[4]. The result:

$$\langle\partial^\mu\tau_{\mu\nu}\rangle^{\not{M}}_{\int} = -\tfrac{1}{4}\delta(\vec{x})\ddot{I}_{ij}\partial_\nu h^{\mathrm{TT}}_{ij} \tag{8}$$

is equal to the bare (i.e., not micro-averaged) energy-momentum delivered by the incident field in transverse-traceless gauge. This motivates the programme of fixing the final piece of gauge freedom by insisting that the incident h_{ab} be transverse-traceless; consequently, τ_{ab} represents the gauge-invariant gravitational energy-momentum that is accessible to an infinitesimal probe at rest in the transverse-traceless frame.

4. Local positivity

The use of transverse-traceless gauge is further motivated by the *positivity property* of τ_{ab}: Whenever the gravitational field is transverse-traceless, τ_{ab} describes positive energy-density

[3] M, I_{ij} and L_{ij} are the mass, moment of inertia, and angular momentum of the detector respectively. Overdots indicate differentiation with respect to $t \equiv x^0$, and the three spatial coordinates are abbreviated $\vec{x} = (x^1, x^2, x^3)$.
[4] Details are to be found in Sections IV B, and IV C of [4].

and causal energy-flux, for all observers. To state this more rigorously: If, at some point $p \in \check{\mathcal{M}}$, the gravitational field h_{ab} obeys the transverse-traceless conditions:

$$\check{\nabla}^a h_{ab} = 0, \quad h = 0, \quad and \quad u^a h_{ab} = 0, \tag{9}$$

for some timelike vector u^a, then τ_{ab} satisfies the following inequalities:

$$v^a \tau_{ab} v^b \geq 0, \quad and \tag{10a}$$

$$v^a \tau_{ac} \tau^c{}_b v^b \leq 0, \tag{10b}$$

at p, for any timelike vector v^a. This is a remarkable and exceptionally valuable property that has not been seen in any previous localization of gravitational energy-momentum. It extends the *dominant energy condition* to include the energy-momentum of the gravitational field, and ensures that τ_{ab} provides a description of gravitational energy-momentum that makes intuitive sense on local scales.

5. Closing remarks

By considering the local exchange of energy-momentum between matter and linear gravity in Eq. (4), a unique symmetric gravitational energy-momentum tensor τ_{ab}, free of second derivatives, was derived in Eq. (5); the harmonic gauge condition given in Eq. (6) arose as a natural consequence of this derivation, and extinguished the vast majority of the description's gauge freedom. The last trace of gauge freedom was then fixed by insisting that the gravitational field be transverse-traceless. This programme, not only guarantees an agreement with the gauge-invariant energy-momentum transferred onto a infinitesimal probe given in Eq. (8), but also ensures that τ_{ab} encodes positive energy-density in Eq. (10a) and causal energy flux given in Eq. (10b).

References

[1] Preskill J and Thorne K S 2002 *Foreword to Feynman Lectures on Gravitation, Feynman et al* (Westview Press) p xxv
[2] Bondi H 1957 *Nature* **179** 1072
[3] Misner C, Thorne K S and Wheeler J A 1973 *Gravitation* (W H Freeman) p 466
[4] Butcher L M, Hobson M P and Lasenby A N 2010 *Phys. Rev. D* **82** 104040
[5] Wald R M 1984 *General Relativity* (University of Chicago Press)

Proceedings of the 7th International Conference on Gravitation and Cosmology (ICGC2011) IOP Publishing
Journal of Physics: Conference Series **484** (2014) 012012 doi:10.1088/1742-6596/484/1/012012

Spherical gravitational wave detectors: MiniGRAIL and Mario Schenberg

C F Da Silva Costa and O D Aguiar
USP and INPE, São Paulo, Brazil

E-mail: filipe.dasilva@usp.br, odylio.aguiar@inpe.br

Abstract. Spherical gravitational wave detectors allow the analysis of multiple independent channels and, therefore, are able to determine gravitational wave directions and polarizations. There are two spherical detectors being developed now: MiniGRAIL (Netherlands) and Mario Schenberg (Brazil). Both share the same principle of detection and main features. They have done commissioning runs and shown progress in their development. We have presented here the status of Mario Schenberg. Its transducers have been redesigned for sensitivity improvements. While an offline analysis was already developed for MiniGRAIL, we have investigated a low latency data analysis technique for Mario Schenberg. Both analysis are based on directional detection.

1. Introduction

Two spherical resonant mass detectors are being developed: MiniGRAIL (Netherlands) [1] and Mario Schenberg (Brazil) [2, 3] (see Section 2). Their last commissioning runs were respectively, in autumn 2010 and in 2008. Similar to bars, surface oscillations are measured in order to look for gravitational waves signatures. Unlike bars that only monitor one mode, the spherical shape allows an isotropic and a multichannel analysis of signals. Actually, gravitational waves (GW) couple to the five quadrupolar modes of the sphere (see Section 3). The spherical shape also provides a better effective mass than a bar [4]. Despite their narrow band, spherical gravitational detectors are a low cost solution that could play a significant role in gravitational astrophysics, since one detector can determine alone the direction of a GW source.

An offline analysis has been already developed for MiniGRAIL (see Section 4), and a low latency analysis using similar principle as the offline analysis is under development (see Section 5). These analysis techniques could be applied to both detectors with minor changes.

2. Experiment description

Both spheres of MiniGRAIL and Mario Schenberg are made of CuAl (6%) alloy, which provides a high mechanical $Q \sim 10^7$. Their respective masses and diameters are 1400 kg/68 cm, and 1150 kg/65 cm, which give resonant frequencies $f_0 \sim 3$ kHz. External vibrations (seismic noise) are attenuated by a dumping system composed of hanging masses: by ~ 350 dB@3 kHz (MiniGRAIL), and ~ 400 dB@3.2 kHz (Mario Schenberg). Both antennas aim to operate at temperature of ~ 50 mK in order to reduce thermal noise.

The mechanical vibrations of the sphere are converted into electromagnetic (EM) signals by six transducers. MiniGRAIL uses capacitive transducers amplified by 2-stage SQUID systems

Figure 1. Scheme of the 5th transducer generation. The lower ring (left) is introduced in the sphere holes. The middle part works as a spring, which will be tuned to the sphere resonant frequency at low temperature. The upper part contains the electromagnetic resonant cavity, which is closed by a 50 μm membrane thick. This transducer is $\sim$ 30 mm diameter [7].

(dc-SQUIDs from Quantum Design) [5]. Instead of passive transducers, Mario Schenberg is the first spherical detector to use parametric transducers with resonant cavities of klystron type [6] (see Figure 1). These transducers could allow quantum squeezing. The resonant cavity is pumped with a microwave signal of 10 GHz, which is modulated by the oscillations of the transducer ($df/dx \sim 0.5$ GHz/μm). These modulations are thus, analyzed. The contact between transducers and the sphere is done by thermal contraction of the sphere. Bodies and membranes of the transducers are both made of niobium, which minimizes the differential thermal contraction between parts after cooling down. Since their first design [2], a large effort has been done to increase the transducer sensitivity. Instead of microstrip antennas (commonly used), a new technique has been developed to couple directly the EM probe to the cavity and moreover from outside of it. In this way, it avoids thermic and seismic noise transmission. A commissioning run to test the new technologies is planned by the end of 2012.

2.1. Data acquisition system

We have written our own software to control and manage the acquisition (ADC, GPS, VXI-PC cards). We have tested the synchronization between the GPS Pulse Per Second (PPS) and the data acquisition. For a given sampling frequency f_s, we get a timing error of $1/f_s$ [8]. Scintillators are also being installed and calibrated to veto high energetic astroparticle showers, which could interact with the sphere.

3. Mode channels

GW couple with the five quadrupolar modes of the sphere [9, 10]. These quadrupolar modes have frequencies around 2980 Hz (MiniGRAIL) within a bandwidth of $\sim$ 230 Hz, and 3200 Hz (Mario Schenberg) within a bandwidth of $\sim$ 80 Hz. In order to monitor these normal modes, six transducers are disposed following the TIGA configuration [11, 12]. Transducers signals are converted into the quadrupolar components h_m (m mode channels) by:

$$h_m(\omega) = T_{mk}(\omega)V_k(\omega), \qquad m = 1, ..., 5 \text{ and } k = 1, ..., 6, \tag{1}$$

where T_{mk} is the system transfer function (calibrated), and V_k are the tensions provided by the k transducers. After solving analytically T_{mk} using the TIGA configuration, we have determined the GW tensor h_{ij}, which are directly related to the quadrupolar mode h_m by:

$$h_{ij} = h_m Y_{ij}^m, \tag{2}$$

where Y_{ij}^m is a quadrupolar tensor basis (i.e., traceless and symmetric tensors) used to express the spherical harmonic Y_2^m [9, 13]. This basis is explicitly defined in [14].

Thus, these five quadrupolar modes provide the complete information about the gravitational waves: direction and polarization.

4. Offline analysis

A complete data analysis pipeline has been developed for MiniGRAIL [15, 16]. It uses the Coherent Waveburst for the Sphere (CWS), an adaptation of the Klimenko's Waveburst [17], to generate lists of triggers. This method presents the advantage of not depending on the signal waveform and direction as the Matched Filter (MF) technique does.

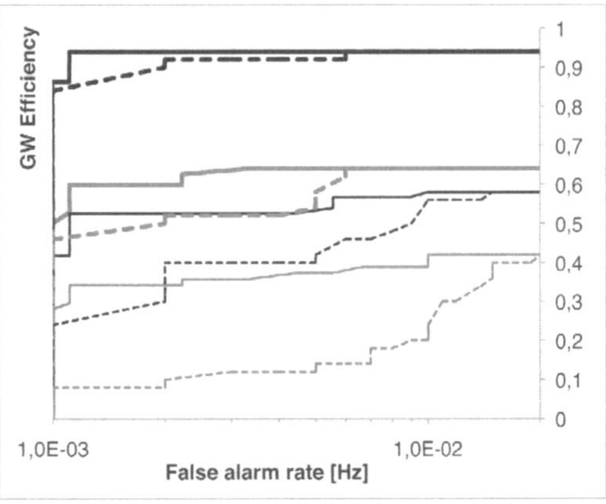

Figure 2. ROC: Detection efficiency vs. false alarm rate for signal injections (Sine-Gaussian) of increasing amplitude: $h_{rss} = \{3, 5, 7, 10\} \times 10^{-21}\text{Hz}^{-1/2}$ (respectively, lines of increasing thickness). Broken and solid lines refer to the angular distance respectively between the CWS and the determinant method, and between CWS and MF methods.

The triggers are vetoed by two independent direction reconstructions: A likelihood method and an algebraic method based on the transverse traceless properties of the GW matrix. Using a Receiver Operation Characteristic (ROC), we select a threshold of 0.2 radians between the two directions. All triggers with a higher difference are discarded. After the CWS, we have used a linear MF that takes into account noise correlation among the five channels and the noise evolution. The MF maximizes the SNR and thus, compared to the CWS, it increases the time precision by one order of magnitude, giving an error of 10 ms. The signal amplitude is estimated with an accuracy of 10%. The MF allows a third direction reconstruction ((θ or ϕ) = 15 - 60 mrad) reducing again by a factor ~ 3 the false alarm rate. In Figure 2, we have shown the ROC of GW efficiency to false alarm rate. It is obtained by varying the maximum allowed distance between the direction reconstruction methods. We can exclude 90% of non-GW signals.

5. Low latency detection

GW triggers of high SNR could be confronted with other event messengers like neutrinos, gamma rays, and other signals, if a fast identification of these triggers is provided. We have been developing and testing a low latency analysis pipeline. In order to reduce the computational time, the transfer function is computed before data acquisition and it is stored on disk. It relies on an analytic solution and the parameters of the present status of the detector.

Triggers are defined by a SNR threshold. The SNR is calculated considering previous noise levels, but for simplification, the SNR calculation does not take into account the correlations between channels. As the detector is narrow band, we have used a pass band filter, which is continuously adjusted to the best strain sensitivity zone.

Generated triggers are vetoed primarily by checking for glitches or cosmic rays. To confirm that our trigger could be a possible event, we have checked if its energy is deposited following a spherical quadrupolar pattern. An indirect way to test this deposition is, thus, to reconstruct the GW arrival direction for each bin in the frequency window. We have constructed two histograms, one for each direction angle. A signal having a quadrupolar distribution generates a Gaussian distribution around the direction with a low variance. On the other hand, noise produces a false direction with a large variance. Evidently, this also depends on the SNR level. In Figure 3, we have shown variances for different signal injections amplitudes. Like the offline analysis, a follow up MF is being tested.

In this method, the computational time is maintained below the acquisition time. For example, on a 2.4 GHz processor, it needs $\sim$ 2 s to process for 4 s of data and $\sim$ 9 s for 32 s of data ($f_s = 16384$ Hz). The most time consuming parts are FFT transformations.

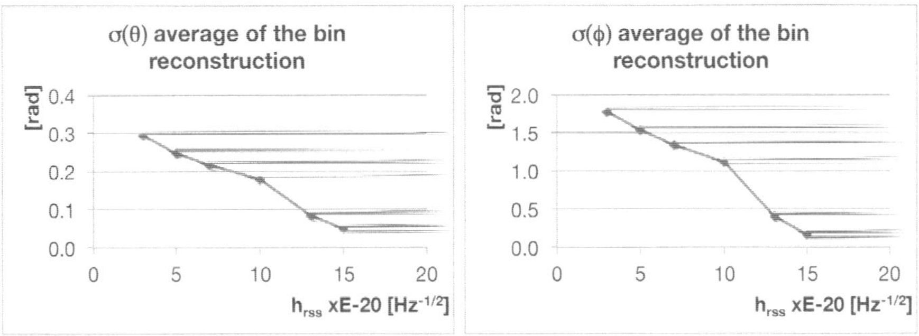

Figure 3. We inject Sine-Gaussian signals covering the entire sensitive frequency band with direction: $\theta = 0.62$ and $\phi = -2.60$ (Euler angles), and six different amplitudes h_{rss}. For each amplitude, we do an average over 25 injections. Here, we have shown the direction reconstruction variances for each signal amplitude. Signal injections are done in our simulator to include signal deformations due to mechanical and electrical attenuations.

6. Conclusion

This low latency method will be able to provide fast direction re-construction with a good confidence for high SNRs. In order to increase the GW detection efficiency of the offline analysis, we have been also investigating the use of Bayesian method for un-modelled bursts as described in [18]. Resonant spheres are a promising technique, and obtaining a better strain sensitivity is purely a technological problem. Another transducer technology for monitoring more than the fundamental modes will turn spherical detectors into large band detectors.

Acknowledgements

This work has been supported by FAPESP (under grants No. 2010/09101-6 and project grants 2006/56041-3). To C A Costa and S Foffa for their help.

References

[1] de Waard A, Bassan M, Benzaim Y, Fafone V *et al* 2006 *Class. Quant. Grav.* **23** S79
[2] Aguiar O D, Andrade L A, Barroso J J *et al* 2006 *Class. Quant. Grav.* **23** S239
[3] Costa C A *et al* 2008 *Class. Quant. Grav.* **25** 184002
[4] Lenzi C H, Magalhães N S *et al* 2008 *Phys. Rev. D* **78** 062005
[5] Gottardi L 2007 *Phys. Rev. D* **75** 022002
[6] Pimentel G L, Aguiar O D, Barroso J J and Tobar M E 2008 *J. Phys. Conf. Ser.* **122** ed Science I
[7] Aguiar O D *et al* 2012 *J. Phys. Conf. Ser.* **363** 012003
[8] Da Silva Costa C F *et al* 2012 *Gravitational wave antenna Mario Schenberg: Data acquisition system manual sid.inpe.br/mtc-m19/2012/04.17.15.03 INPE, Brazil*
[9] Zhou C Z and Michelson P F 1995 *Phys. Rev. D* **51**(6) 2517
[10] Costa C A, Aguiar O D and Magalhães N S 2003 [gr-qc/0312035]
[11] Johnson W W and Merkowitz S M 1993 *Phys. Rev. Lett.* **70** 2367
[12] Merkowitz S M and Johnson W W 1997 *Phys. Rev. D* **56** 7513
[13] Maggiore M 2008 *Gravitational Waves* **1** *Theory and Experiments* (Oxford University Press)

[14] Da Silva Costa C F 2010 *Data Analysis Pipeline for the Spherical Gravitational Wave Antenna MiniGRAIL* thesis nb. 4210 *Faculte des sciences de l'Universite de Geneve*
[15] Foffa S and Sturani R 2009 *Class. Quant. Grav.* **26** 105013
[16] Da Silva Costa C F, Foffa S and Sturani R 2009 *Class. Quant. Grav.* **26** 204019
[17] Klimenko S and Mitselmakher G 2004 *Class. Quant. Grav.* **21** S1819
[18] Searle A C, Sutton P J and Tinto M 2009 *Class. Quant. Grav.* **26** 155017

Spherical gravitational collapse in 5D Einstein-Gauss-Bonnet gravity

S G Ghosh[1], **S Jhingan**[1] **and D W Deshkar**[2]

[1]Centre for Theoretical Physics, Jamia Millia Islamia, New Delhi 110025, India
[2] S.S.E.S. Amravati's Science College, Congress Nagar, Nagpur 440012, India

E-mail: sgghosh@gmail.com,sanjay.jhingan@gmail.com,dwdeshkar@gmail.com

Abstract. We have considered a spherical gravitational collapse of inhomogeneous dust (and null dust) in Einstein gravity with the Gauss-Bonnet (GB) combination of quadratic curvature terms. It turns out that the presence of the coupling constant of the GB terms completely changes the causal structure of the singularities from the analogous general relativistic case. The gravitational collapse of inhomogeneous dust in the five-dimensional GB extended Einstein equations leads to formation of a massive, but weak, timelike singularity, which is forbidden in general relativity. The apparent horizons of two different collapsing solutions have shown interesting mathematical similarity.

1. Introduction

The Lemaitre-Tolman-Bondi (LTB) [1] and Vaidya [2] solutions have been extensively used, not only to study the formation of naked singularities and black holes in spherical collapse, but in cosmology as well. It is well known that these two solutions admit both naked and covered singularities depending upon the choice of initial data, and there is a smooth transition from one phase to the other [3]. These results have led to strong evidence against the cosmic censorship conjecture (CCC) [4], which asserts that *there can be no singularity visible from future null infinity.* In recent years, a renewed interest has grown in higher order gravity, which involves higher derivative curvature terms, and amongst the most extensively studied theory is the so-called Einstein-Gauss-Bonnet (EGB) gravity. It appears naturally in the low-energy effective action of heteroclite string theory [5]. Boulware and Deser [6] have found exact black hole solutions in D ($\geq$ 5)-dimensional gravitational theories with a Gauss-Bonnet (GB) term modifying the usual Einstein-Hilbert action. Recently, Maeda [7] has considered the spherically symmetric gravitational collapse of a inhomogeneous dust with the D ($\geq$ 5)-dimensional action including the GB term, and investigated its effects on the final fate of gravitational collapse without finding the explicit form of the solution.

In this paper, we have considered the 5-dimensional (5D) action with the GB terms for gravity and given an *exact model* of the gravitational collapse of a inhomogeneous dust including the second order perturbative effects of quantum gravity. We also have considered the effects of GB terms on the structure of apparent horizons in dust solutions as well as in null dust solutions. The nature of singularities of such a spacetime in terms of its being hidden within a black hole, or whether it would be visible to outside observers, and the consequence of GB term on 5D-LTB collapse have been analyzed.

2. Basic equations in EGB

The gravitational part of the 5D action that we have considered is [8, 9]:

$$S = \int d^5x\sqrt{-g}\left[\frac{1}{2\kappa_5^2}(R - 2\Lambda + \alpha L_{GB})\right] + S_{\text{matter}}, \tag{1}$$

where R and Λ are the 5D Ricci scalar and the cosmological constant respectively. $\kappa_5 \equiv \sqrt{8\pi G_5}$, where G_5 is the 5D gravitational constant. The GB Lagrangian is the combination of the Ricci scalar, Ricci tensor R_{ab}, and Riemann tensor $R^a{}_{b\rho\sigma}$ as $L_{GB} = R^2 - 4R_{ab}R^{ab} + R_{ab\rho\sigma}R^{ab\rho\sigma}$. In the 4-dimensional spacetime, the GB terms do not contribute to the field equations. The α is the coupling constant of the GB terms. From the action given in Eq. (1), we have derived the following field equations:

$$G_{ab} - \alpha H_{ab} = T_{ab}, \quad \text{with} \quad G_{ab} = R_{ab} - \frac{1}{2}g_{ab}R, \quad and \tag{2}$$

$$H_{ab} = 2\Big[RR_{ab} - 2R_{a\alpha}R^{\alpha}_{b} - 2R^{\alpha\beta}R_{a\alpha b\beta} + R_{a}{}^{\alpha\beta\gamma}R_{b\alpha\beta\gamma}\Big] - \frac{1}{2}g_{ab}L_{GB},$$

where G_{ab} is the Einstein tensor, and H_{ab} is the Lanczos tensor.

3. Null dust collapse

The energy-momentum tensor of a null fluid is $T_{ab} = \psi(v,r)l_a l_b$, where $\psi(v,r)$ is the non-zero energy density and l_a is a null vector, such that $l_a = \delta^0_a$, with $l_a l^a = 0$. Expressed in terms of Eddington advanced time coordinate (ingoing coordinate) v, the metric of general spherically symmetric spacetime [8, 9] is given by:

$$ds^2 = -A(v,r)^2 f(v,r)\, dv^2 + 2A(v,r)\, dv\, dr + r^2 d\Omega_3^2, \tag{3}$$

where $d\Omega_3^2 = d\theta^2 + \sin^2\theta d\phi^2 + \sin^2\theta \sin^2\phi\, d\psi^2$. Here, $A(v,r)$ is an arbitrary function. It is the field equation $G^0_1 = 0$ leads to $A(v,r) = g(v)$. This could be absorbed by writing $d\tilde{v} = g(v)dv$. Then, $f(v,r)$ is obtained by solving field equation to give the general solution as:

$$f(v,r) = 1 + \frac{r^2}{4\alpha}\left[1 \pm \sqrt{1 + \frac{8\alpha m(v)}{r^4}}\right]; \; \psi(v,r) = \frac{3}{2r^3}\dot{m}(v), \tag{4}$$

where $m(v)$ is an arbitrary function of v. There are two families of solutions, which correspond to the $\pm$ in Eq. (4), and they are respectively for the plus (minus) branch solutions. The energy density of the null fluid is given by $\psi(v,r)$ for both branches, where the dot denotes the derivative with respect to v. In the general relativistic limit $\alpha \to 0$, the minus-branch solution reduces to the 5D Vaidya solution in Einstein gravity. In the static case $\dot{m} = 0$, this solution reduces to the solution which was independently discovered by Boulware and Deser [6] and Wheeler [10]. It has been shown that a timelike naked singularity is formed from gravitational collapse of null dust in EGB, which does not appear in the general relativistic case [9].

The apparent horizon (AH) is the outermost marginally trapped surface for the outgoing photons. The AH is defined as the surface such that $\Theta \simeq 0$, which implies that $f = 0$. It is clear that AH [8] is the solution of:

$$\left[1 + \frac{r^2}{4\alpha}\left[1 - \sqrt{1 + 8\alpha\left(\frac{m(v)}{r^4}\right)}\right]\right] = 0, \Rightarrow r_{AH} = \sqrt{m(v) - 2\alpha}. \tag{5}$$

In the relativistic limit $\alpha \to 0$, $r_{AH} \to \sqrt{m(v)}$. One sees that $g_{vv} = 0$ implies that $r = \sqrt{m(v) - 2\alpha}$ is timelike surface [8].

4. Inhomogeneous dust collapse

The solution we seek is collapse of a spherical dust in 5D-EGB. The energy-momentum tensor for dust is $T_{ab} = \epsilon(t,r)\delta^t_a\delta^t_b$, where $u_a = \delta^a_t$ is the 5D velocity. The metric for the 5D case, in comoving coordinates is [11, 12, 13]:

$$ds^2 = -dt^2 + A(t,r)^2 dr^2 + R(t,r)^2 d\Omega_3^2 \,. \tag{6}$$

The coordinate r is the comoving radial coordinate, t is the proper time of freely falling shells, R is a function of t and r with $R \geq 0$, and A is also a function of t and r. The master equation of the system is given by:

$$\dot{R}^2\left[1 - 4\alpha\frac{W^2-1}{R^2}\right] = (W^2 - 1) + \frac{F}{R^2} - 2\alpha\frac{\dot{R}^4}{R^2}. \tag{7}$$

Here, $F = F(r)$ is an arbitrary function of r and is referred to as mass function. Indeed, the energy density $\epsilon = 3F'/2R^3R'$, must be non-negative. It is easy to see that as $\alpha \to 0$ the master solution given in Eq. (7) of the system reduces to the corresponding 5D-LTB solution in [13]. In the present discussion, we are concerned with gravitational collapse, which requires $\dot{R}(t,r) < 0$. Eq. (7), for $W(r) = 1$, can be integrated to [11]:

$$t_\varsigma(r) - t = \frac{\sqrt{\alpha}}{2\sqrt{2}}\tan^{-1}\left[\frac{3R^2 - \sqrt{R^4 + 8\alpha F}}{2\sqrt{2}R[\sqrt{R^4+8\alpha F} - R^2]^{1/2}}\right] + \sqrt{\frac{\alpha R^2}{\sqrt{R^4+8\alpha F} - R^2}}\,, \tag{8}$$

where $t_\varsigma(r)$ is an arbitrary function of integration. The mass function F can be related with initial data (density) at the scaling surface $t = 0$ ($R = r$), reduces to form $F(r) = \frac{2}{3}\int_0^r \epsilon(0,r)r^3 dr$, which completely specifies the mass function in terms of the initial density profile. The function F must be positive, because $F < 0$ implies the existence of negative mass. This can be seen from the mass function $m(t,r)$ [13], which in the 5D-LTB-EGB case is given by:

$$m(t,r) \;=\; R^2\left(1 - g^{ab}R_{,a}R_{,b}\right) = R^2\left(1 - \frac{R'^2}{A^2} + \dot{R}^2\right). \tag{9}$$

Using Eq. (7) into Eq. (9), we get $m(t,r) = F(r) - 2\alpha\dot{R}^4$. It may be noted that one can also calculate mass using the formula proposed by Maeda [7] for the generalized mass function in the EGB. The mass function $F(r) = m(t,r) + 2\alpha\dot{R}^4$, is equivalent, up to a constant factor, to the generalized mass function in EGB [7].

The main advantage of working with the AH is that it is local in time and can be located at a given spacelike hypersurface. Considering Eq. (7), the AH condition [11] becomes:

$$R(t_{AH}(r), r) = \sqrt{F(r) - 2\alpha}. \tag{10}$$

It is clear that the presence of the coupling constant of the GB terms α produces a change in the location of these horizons. In the relativistic limit $\alpha \to 0$, $R_{AH} \to \sqrt{F(r)}$ [12]. For non-zero α, the structure of the AH is non-trivial. Interestingly, the theory demands α to be a positive number, which forbids AH from reaching the centre, thereby making the singularity massive and eternally visible, which is forbidden in the corresponding general relativistic scenario. In general relativity, non-central singularity is always covered [14] (see also [15]). However, in the presence of the GB term, we have found that even the non-central singularity is naked, in spite of being massive ($F(r > 0) > 0$). Further, Eq. (10) has a mathematical similarity for the AH given by Eq. (5) in null fluid collapse [8].

Eq. (10) implicitly defines a curve $t_{ah}(r)$ and represents the AH, i.e., the time at which the shell gets trapped. To further analyze the horizon curve, we have used the singularity condition in Eq. (8) to get:

$$t_c(r) - t = \frac{\pi\sqrt{\alpha}}{4\sqrt{2}} + \sqrt{\frac{\alpha R^2}{\sqrt{R^4+8\alpha F} - R^2}} + \frac{\sqrt{\alpha}}{2\sqrt{2}}\tan^{-1}\left[\frac{3R^2 - \sqrt{R^4+8\alpha F}}{2\sqrt{2}R[\sqrt{R^4+8\alpha F} - R^2]^{1/2}}\right] \quad (11)$$

Then, the AH condition in Eq. (10) reduces Eq. (11) to the form:

$$t_c(r) - t_{AH}(r) = \frac{\pi\sqrt{\alpha}}{4\sqrt{2}} + \frac{\sqrt{\alpha}}{2\sqrt{2}}\tan^{-1}\left[\frac{F-4\alpha}{2\sqrt{2\alpha(F-2\alpha)}}\right] + \frac{1}{2}\sqrt{F-2\alpha}. \quad (12)$$

Clearly, for a positive α, the central shell does not get trapped, and the untrapped region around the centre increases with increasing α, for both homogeneous and inhomogeneous models. Thus, the singularities are always naked as they are formed prior to the formation of apparent horizon.

5. Discussion

In this paper, we have found exact spherically symmetric 5D-LTB solutions, for $W(r) = 1$ case, to GB extended Einstein equations, namely 5D-LTB-EGB, which describes gravitational collapse of spherically symmetric inhomogeneous dust in a 5D spacetime. While, there may be an AH about this singularity, for $\alpha > 0$, the singularity always remains visible to any observer as the AH lies beyond singularity, which is actually not in the spacetime. It is interesting to note that the coupling constant of the GB terms produces a change in the location of the AH by the factor 2α, and is exactly the same as in the case of 5D null fluid collapse in EGB. The final fate of gravitational collapse is quite different in the sense that a massive naked singularity is formed, which is not allowed in 5D-LTB. However, the strength of singularity is weaker as compared to the corresponding 5D-LTB [11], and this may not be a serious threat to CCC. It is seen that the time for which the occurrence of the central shell focusing singularity for the collapse is increased as compared to the analogous 5D-LTB case. The reason may be, there is relatively less mass-energy collapsing in the 5D-LTB-EGB spacetime as compared to the 5D-LTB case. In particular, our results in the limit $\alpha \to 0$ reduce *vis-à-vis* to 5D relativistic case.

Acknowledgements

SGG thanks the University Grants Commission (UGC), Government of India, for the major research project grant F. No. 39-459/2010 (SR).

References

[1] Lemaitre G 1933 *Ann. Soc. Sci. Brux., Ser. I: Sci. Math. A.& A.* **53** 51, Tolman R C 1934 *Proc. Nat. Acad. Sci. USA* **20** 169, Bondi H 1948 *MNRAS* **107** 410
[2] Vaidya P C 1951 *Proc. Ind. Acad. Sci. A* **33** 264
[3] Joshi P S 1993 *Global Aspects in Gravitation and Cosmology* (Clendron, Oxford), Joshi P S 2007 *Gravitational Collapse and Spacetime Singularities* (Cambridge University Press)
[4] Penrose R 1969 *Riv. Nuovo Cimento* **1** 252, Penrose R 1979 *General Relativity - An Einstein Centenary Volume*, ed S W Hawking and W Israel (Cambridge University Press) 581
[5] Gross D J and Sloan J H 1987 *Nucl. Phys. B* **291** 41, Bento M C and Bertolami O 1996 *Phys. Lett. B* **368** 198
[6] Boulware D G and Deser S 1985 *Phys. Rev. Lett.* **55** 2656
[7] Maeda H 2006 *Phys. Rev. D* **73** 104004
[8] Ghosh S G and Deshkar D W 2008 *Phys. Rev. D* **77** 047504
[9] Maeda H 2006 *Class. Quant. Grav.* **23** 2155
[10] Wheeler J T 1986 *Nucl. Phys. B* **268** 737

[11] Jhingan S and Ghosh S G 2010 *Phys. Rev. D* **81** 024010
[12] Ghosh S G and Beesham A 2001 *Phys. Rev. D* **64** 124005
[13] Ghosh S G and Banerjee A 2003 *Int. J. Mod. Phys. D* **12** 639
[14] Christodoulou D 1984 *Commun. Math. Phys.* **93** 171
[15] Cooperstock F, Jhingan S, Joshi P S and Singh T P 1997 *Class. Quant. Grav.* **14** 2195

Journal of Physics: Conference Series **484** (2014) 012014 doi:10.1088/1742-6596/484/1/012014

Generalized junction conditions for collapsing models

G Govender, S D Maharaj and M Govender

Astrophysics and Cosmology Research Unit, School of Mathematics, Statistics and Computer Science, University of KwaZulu-Natal, Private Bag X54001, Durban 4000, South Africa

E-mail: maharaj@ukzn.ac.za

Abstract. We have constructed the general junction conditions on the surface of a dissipating relativistic star. The stellar exterior is a spacetime described by the generalised Vaidya metric and a two-fluid energy-momentum tensor, and therefore, defines the local atmosphere, which must be a super-position of standard null radiation and a general null fluid. We have highlighted briefly that our result will effect the physics of the dissipation at the stellar boundary.

1. Introduction

A compact star is formed when a massive star ($M_* \geq 8\ M_\odot$) breaks away from a state of hydrostatic equilibrium and collapses under its own gravity. The stellar object that is formed at the end of the collapse process is usually very dense and small in size (typical radius: $R \sim 7-20\ km$) as compared to the initial configuration. Consequently, the gravitational field in the interior and exterior of the star is significantly strong, and this is reflected in the curvature of the spacetime geometries, so general relativity is required. During the contraction of the massive star and the evolution of the compact object, gravitational binding energy is converted into heat energy, which is used for ionization and dissociation in the dense interior. The excess heat energy is dissipated as radiation across the stellar surface. This process is described adequately for a relativistic star by constructing the junction conditions at the surface. These equations relate the interior and exterior matter variables as well as the respective geometries.

We have extended and generalized the Santos [1] junction condition for relativistic dissipation, by matching the spacetime geometry for an interior stellar fluid with isotropic pressure to that of the local exterior, which is defined by the generalised Vaidya radiating metric. The physical consequence, here, is that the local atmosphere must be a two-fluid system consisting of standard null radiation (photons), and an additional null fluid that has a characteristic non-zero pressure and energy density. The effect of our result on the surface radiation parameters is also considered.

2. The interior and exterior stellar geometry

The dynamics of the gravitational field in the stellar interior is given by the shear-free metric:

$$ds^2 = -A^2(r,t)dt^2 + B^2(r,t)[dr^2 + r^2(d\theta^2 + \sin^2\theta d\phi^2)], \tag{1}$$

where $A(r,t)$ and $B(r,t)$ are metric potentials and the material inside the star is defined as a relativistic fluid with heat conduction by:

$$T^{-}_{ab} = (\mu + p)u_a u_b + p g_{ab} + q_a u_b + q_b u_a, \tag{2}$$

where μ and p are the fluid energy density and isotropic pressure respectively, g_{ab} is the metric tensor field and $\boldsymbol{u}$ and $\boldsymbol{q}$ are the fluid four-velocity and the radial heat flux. The relevant Einstein's field equations $G^{-}_{ab} = T^{-}_{ab}$ for the interior are:

$$p = \frac{1}{A^2}\left(-2\frac{\ddot{B}}{B} - \frac{\dot{B}^2}{B^2} + 2\frac{\dot{A}}{A}\frac{\dot{B}}{B}\right) + \frac{1}{B^2}\left(\frac{B'^2}{B^2} + 2\frac{A'}{A}\frac{B'}{B} + \frac{2}{r}\frac{A'}{A} + \frac{2}{r}\frac{B'}{B}\right), \quad and \tag{3a}$$

$$q = -\frac{2}{AB^2}\left(-\frac{\dot{B}'}{B} + \frac{B'\dot{B}}{B^2} + \frac{A'}{A}\frac{\dot{B}}{B}\right), \tag{3b}$$

where dots and primes denote differentiation with respect to coordinate time t and radial distance r respectively.

The dynamics of the gravitational field in the local region outside the star is described by the generalised Vaidya outgoing radiation metric:

$$ds^2 = -\left(1 - 2\frac{m(v,\mathsf{r})}{\mathsf{r}}\right)dv^2 - 2dvd\mathsf{r} + \mathsf{r}^2(d\theta^2 + \sin^2\theta d\phi^2), \tag{4}$$

where $m(v, \mathsf{r})$ is the mass function across the surface, and is related to the gravitational energy within a given radius r. The characteristic feature about the metric given in Eq. (4) is that the mass function contains a spatial dependence in the radial direction during dissipation, and this is significantly different from the standard scenario, in which the mass at the boundary only has a time dependance. Husain [2], and Wang and Wu [3] have shown that an appropriate energy momentum tensor for a two-fluid atmosphere is:

$$T^{+}_{ab} = \varepsilon l_a l_b + (\rho + p)(l_a n_b + l_b n_a) + p g_{ab}, \tag{5}$$

which represents a super-position of null radiation and an arbitrary null fluid. ε is the energy density of the photon radiation and ρ and p are the energy density and pressure of the additional null fluid, respectively. In general, T^{+}_{ab} represents a Type II fluid as defined by Hawking and Ellis [4]. The null vector $\boldsymbol{l}$ is a double null eigenvector of the energy-momentum tensor in Eq. (5) and the vector $\boldsymbol{n}$ is normal to the spatial hypersurface.

For the spacetime metric in Eq. (4) and the energy momentum tensor in Eq. (5), the Einstein's field equations for the local two-fluid stellar atmosphere can be written as:

$$\varepsilon = -2\frac{m_v}{\mathsf{r}^2}\tilde{v}^2, \tag{6a}$$

$$\rho = 2\frac{m_{\mathsf{r}}}{\mathsf{r}^2}, \quad and \tag{6b}$$

$$p = -\frac{m_{\mathsf{rr}}}{\mathsf{r}}. \tag{6c}$$

It is interesting to note that in the standard framework, the Santos [1] junction condition tells us that on the boundary of the star, the pressure of the stellar fluid is proportional to the magnitude of the heat flux given as:

$$(p_r)^{s}_{\Sigma} = (qB)_{\Sigma}, \tag{7}$$

where the superscript s denotes a quantity as defined in the standard Santos framework. Furthermore, the local atmosphere outside the star contains only pure radiation in the form of null photons that have a characteristic non-zero energy density given by the Eq. (6a). In our new framework, the additional null fluid component defined by the two extra field Eqs. (6b) and (6c) should significantly transform Eq. (7), and consequently change the way in which the heat energy is dissipated as radiation across the surface.

3. The matching process

The interior and exterior spacetimes given in Eqs. (1) and (4) are matched across the spatial hypersurface (stellar boundary) Σ. The intrinsic metric to the hypersurface Σ is defined by:

$$ds^2_\Sigma = -d\tau^2 + \mathcal{Y}^2\left(d\theta^2 + \sin^2\theta d\phi^2\right), \tag{8}$$

where τ is an intermediate timelike variable defined on the surface, $\mathcal{Y} = \mathcal{Y}(\tau)$ is a metric function and $\xi^\alpha = (\tau, \theta, \phi)$ are coordinates on the surface.

The first fundamental form $(ds^2_+)_\Sigma = (ds^2_-)_\Sigma = ds^2_\Sigma$ gives:

$$A(\mathsf{r}_\Sigma, t)dt = \left(1 - \frac{2m}{\mathsf{r}_\Sigma} + 2\frac{d\mathsf{r}_\Sigma}{dv}\right)^{1/2} dv, \quad and \tag{9a}$$

$$\mathsf{r}_\Sigma(v) = rB(\mathsf{r}_\Sigma, t), \tag{9b}$$

and the second fundamental form $(K^+_{\alpha\beta})_\Sigma = (K^-_{\alpha\beta})_\Sigma$ yields the following equations:

$$m(v,\mathsf{r}) = \left[\frac{rB}{2}\left(1 + r^2\frac{\dot{B}^2}{A^2} - \frac{1}{B^2}(B + rB')^2\right)\right]_\Sigma, \quad and \tag{10}$$

$$\left(-\frac{1}{B}\frac{A'}{A}\right) = \left(1 + r\frac{B'}{B} + r\frac{\dot{B}}{A}\right)^{-1} \times \left[\frac{m_\mathsf{r}}{\mathsf{r}} + \frac{B'}{B^2} + \frac{r}{2}\frac{B'^2}{B^3} - \frac{r}{2}\frac{\dot{B}^2}{BA^2} - \frac{1}{A}\left(r\frac{\dot{B}'}{B} + r\frac{\ddot{B}}{A} - r\frac{B'\dot{B}}{B^2} - rB\frac{\dot{A}}{A^2}\right)\right]_\Sigma, \tag{11}$$

where Eq. (10) is the mass profile at the boundary. With some algebraic simplification and using the interior field Eqs. (3a), (3b) and the exterior field Eq. (6b), Eq. (11) reduces to:

$$(p)_\Sigma = (qB - \rho)_\Sigma, \tag{12}$$

which is a generalization of the Santos junction condition given in Eq. (7).

4. The effect of the exterior null fluid energy density on the dissipation parameters

Since our generalised junction condition in Eq. (12) holds on the surface of the dissipating star, it is expected that the additional null fluid in the local stellar atmosphere should have a significant effect on the radiation parameters associated with the dissipation. This may be important to our understanding of stellar atmospheres for relativistic stars. Shapiro and Teukolsky [5], and Kippenhan and Weigert [6] have discussed the atmospheres of compact objects extensively. Furthermore, investigating the effect of the null fluid energy density, which in our case could also be crucial for modelling the final collapse phase leading up to the point of black hole formation as demonstrated in a recent investigation by Sarwe and Tikekar [7]. We have reconstructed the general profiles of some of the relevant radiation quantities, taking into consideration our new result in Eq. (12) as:

$$(p_r)^g_\Sigma = \left[-\frac{2}{B}\left(-\frac{\dot{B}'}{B}+\frac{B'\dot{B}}{B^2}\right)-\rho(r,t)\right]_\Sigma, \tag{13}$$

$$L^g_\Sigma = \frac{1}{2}r^2B^2\left[p\left(r\frac{\dot{B}}{A}+\frac{(rB)'}{B}\right)+\rho\left(1+r\frac{B'}{B}\right)\right]\left(r\frac{\dot{B}}{A}+\frac{(rB)'}{B}\right)^{-1}, \tag{14}$$

$$L^g_\infty = \frac{1}{2}r^2B^2\left[p\left(r\frac{\dot{B}}{A}+\frac{(rB)'}{B}\right)+\rho\left(1+r\frac{B'}{B}\right)\right]\left(r\frac{\dot{B}}{A}+\frac{(rB)'}{B}\right), \tag{15}$$

$$(\overline{T})^g_\Sigma = \left[\left(\frac{1}{r^2B^2}\right)\left(\frac{L^g_\Sigma}{4\pi\delta}\right)\right]^{\frac{1}{4}}, \quad and \tag{16}$$

$$z^s_\Sigma = \left[\left(r\dot{B}+\frac{(rB)'}{B}\right)^{-1}\right]-1. \tag{17}$$

In the Eqs. (13) through (17), the superscript g denotes the quantity as defined or measured in the generalized framework with the metric in Eq. (4), $(p_r)^g_\Sigma$ represents the pressure of the stellar fluid on the surface, and is obtained from the boundary condition given in Eq. (12). L^g_Σ and L^g_∞ are the luminosity of the emitted radiation at the surface of the star and the luminosity as observed at infinity, respectively. $(\overline{T})^g_\Sigma$ is the effective surface temperature of the radiation and z^s_Σ is the gravitational redshift. All of the above dissipation parameters are constrained and generalized by the energy density ρ, of the exterior null fluid.

5. Discussion

We have generalized the Santos junction condition for dissipation in a compact relativistic star. The inclusion of a radial spatial dependence in the mass function across the stellar boundary leads to the local atmosphere consisting of a two-fluid system comprised of null radiation (photons) as in the standard scenario and an additional null fluid having a characteristic non-vanishing energy density and pressure. Our new boundary condition on the surface of the star shows that the stellar fluid pressure is now constrained by the energy density of the exterior null fluid and that the physics of the relativistic dissipation process may consequently change. This is clear, since in the standard Santos framework, the exterior of the star contains only null radiation but in the generalized framework the atmosphere contains an additional null fluid, which can be interpreted as another matter field that arises due to the dissipation. Our result is an improvement on the work of Santos, since the outside of the relativistic star is more general and more realistic. The matter field mentioned above has been identified as a string fluid but may also be a field of particles like neutrinos. Consequently, the radiation parameters at the surface could be increased or decreased depending on the form of the metric potentials, which have to be obtained by solving the junction condition in Eq. (12). This work is currently in progress.

The profiles for the radiation parameters in our generalized scenario, have been generated, and we have demonstrated that each of these quantities associated with the dissipation is, indeed, affected by the null fluid density. This was not the case in the standard framework. We conclude that our results may be significant for understanding the atmospheres of relativistic stars, and are relevant to previous investigations by Glass and Krisch [8, 9], and Krisch and Glass [10].

Acknowledgements

MG and GG thank the National Research Foundation and the University of KwaZulu-Natal for financial support. SDM acknowledges that this work is based upon research supported by the South African Research Chair Initiative of the Department of Science and Technology and the National Research Foundation. The authors would also like to thank Sanjay Wagh and Subharthi Ray for constructive criticisms and useful discussions.

References

[1] Santos N O 1985 *MNRAS* **216** 403
[2] Husain V 1996 *Phys. Rev. D* **53** 1759
[3] Wang A and Wu Y 1999 *Gen. Rel. Grav.* **31** 107
[4] Hawking S W and Ellis G F R 1973 *The Large Scale Structure of Spacetime* (Cambridge University Press)
[5] Shapiro S L and Teukolsky S A 1983 *Black Holes, White Dwarfs and Neutron Stars* (New York: John Wiley and Sons)
[6] Kippenhan R and Weigert A 1994 *Stellar Structure and Evolution* (Berlin: Springer Verlag)
[7] Sarwe S and Tikekar R 2010 *Int. J. Mod. Phys. D* **19** 1889
[8] Glass E N and Krisch J P 1998 *Phys. Rev. D* **57** 5945
[9] Glass E N and Krisch J P 1999 *Class. Quant. Grav.* **16** 1175
[10] Krisch J P and Glass E N 2005 *J. Math. Phys.* **46** 062501

Gravity in five-dimensional warped product spacetimes with time-dependent warp factor

S Guha[1] and P Bhattacharya[2]

[1] Department of Physics, St. Xavier's College (Autonomous), Kolkata 700016, India
[2] Gopal Nagar High School, Singur 712409, West Bengal, India

E-mail: `srayguha@yahoo.com`

Abstract. We have considered gravity in a 5-dimensional warped product spacetime with a time-dependent warp factor and time-dependent extra dimension. The braneworld has been described by a spatially flat FRW-type metric. The five-dimensional field equations have been constructed and solved, and the status of the energy conditions have been examined. In the high energy regime, the bulk is assumed to be sourced by a scalar field. The effective cosmological constant of the 4-dimensional universe is a variable quantity monitored by the time-dependent warp factor, and leads to a geometric interpretation of dynamical dark energy.

1. Introduction

From the higher-dimensional theories of gravity [1], it appears that extra-dimensions play crucial role in a dynamical spacetime. The simplest example is available in the framework of a five-dimensional (5D) spacetime with a single warped extra dimension, in which the 4D cosmological constant arises from the curvature related to the expansion of a (3+1) spacetime.

In the RS model with a single extra dimension [2], matter fields are localized on a 4D hypersurface in a constant curvature 5D bulk with a constant extra-dimensional scale factor. The exponential warp factor is a function of the extra coordinate, and reflects the confining role of the negative bulk cosmological constant to localize gravity on the 3-brane through the curvature of the bulk [3]. However, the process of localization of gravity may include some time-dependence during a particular phase of evolution [4] and may be related to a time-dependent warp factor, which in turn may be related to a time-varying bulk cosmological constant. Hence, we have considered RS-type braneworlds in a 5D warped product spacetime having a non-compact fifth dimension and an exponential warp factor, which depends both on time as well as on the extra-dimensional coordinate. The braneworld is chosen to be spatially flat FRW-type, as it can be embedded in any constant curvature bulk [5]. A stabilized bulk [6] with constant curvature, and characterized by a negative cosmological constant removes the appearance of non-conventional cosmologies [7], and the ordinary FRW equations are recovered at low energies [8]. In the high energy regime, gravity along with scalar fields and particles are able to access the extra dimensions. We, therefore, have considered a bulk sourced by a scalar field minimally coupled to gravity in the high energy regime [9]. The interaction between the bulk and the brane introduces a new term for the change in the extrinsic curvature of the brane into the EFE on the brane [5]. This leads to a geometric interpretation of dynamical dark energy: The effective cosmological constant of the 4D universe is a variable quantity monitored by the time-dependent warp factor. The gravitational force law is sensitive to the background cosmological expansion at early times and is related to the time-dependent extrinsic curvature of the brane [10]. The zero-mode graviton fluctuation is not guaranteed to be localized on the brane and the effective 4D Newton's constant may not be finite. However, the effect of this time-dependence dies down at later times.

The five-dimensional action is of the form [11]:

$$S = \int d^5x\sqrt{\bar{g}}\left[\frac{1}{2\bar{\kappa}^2}(\bar{R} - 2\bar{\Lambda}) + \frac{1}{2}g^{AB}\nabla_A\psi\nabla_B\psi - V(\psi)\right] + \int d^4x\sqrt{-g}L_m, \tag{1}$$

and supplemented with the brane curvature term:

$$-M_P^2\int d^4x\sqrt{-g}R\ . \tag{2}$$

The 5D metric is given by:

$$dS^2 = e^{2f(t,y)}\left(dt^2 - R^2(t)(dr^2 + r^2d\theta^2 + r^2sin^2\theta d\phi^2)\right) - A^2(t,y)dy^2. \tag{3}$$

As the graviton crosses the brane, it produces a 'bending effect', which varies with time, and is expressed in terms of the extrinsic curvature of the brane, representing the tangential components of the local variation of the normal unit vector at that point [12]. This bending gives rise to an observable result in the form of a smooth scalar function represented by the warping function, which is, therefore, a function of both t and y. The 4D effective action has terms depending on the time-dependent warp factor, and thus, the 4D Newton's constant is time-dependent [13].

2. Five-dimensional field equations

The non-vanishing components of the five-dimensional Einstein tensor are:

$$\bar{G}_t^t = \frac{3}{e^{2f}}\left(\frac{\dot{R}^2}{R^2} + \frac{2\dot{R}\dot{f}}{R} + \dot{f}^2 + \frac{\dot{f}\dot{A}}{A} + \frac{\dot{R}}{R}\frac{\dot{A}}{A}\right) - \frac{3}{A^2}\left(2f'^2 + f'' - \frac{f'A'}{A}\right), \tag{4}$$

$$\bar{G}_y^t = -\frac{3}{e^{2f}}\left((\dot{f})' - \frac{\dot{A}}{A}f'\right), \qquad \bar{G}_t^y = \frac{3}{A^2}\left((\dot{f})' - \frac{\dot{A}}{A}f'\right), \tag{5}$$

$$\bar{G}_y^y = \frac{3}{e^{2f}}\left(\frac{\ddot{R}}{R} + \frac{\dot{R}^2}{R^2} + \frac{3\dot{R}\dot{f}}{R} + \dot{f}^2 + \ddot{f}\right) - \frac{6f'^2}{A^2}, \quad and \tag{6}$$

$$\bar{G}_j^i = \frac{1}{e^{2f}}\left(\frac{2\ddot{R}}{R} + \frac{\dot{R}^2}{R^2} + \frac{4\dot{R}\dot{f}}{R} + \dot{f}^2 + 2\ddot{f} + \frac{\dot{f}\dot{A}}{A} + \frac{2\dot{R}}{R}\frac{\dot{A}}{A} + \frac{\ddot{A}}{A}\right) - \frac{3}{A^2}\left(2f'^2 + f'' - \frac{f'A'}{A}\right). \tag{7}$$

The bulk stress-energy tensors are $\bar{T}_t^t = \rho_B$, $\bar{T}_j^i = -P_B$, $\bar{T}_y^y = -P_y$, *and* $\bar{T}_y^t = -Q_B$. The conservation of the energy-momentum tensor $T_{b;a}^a = 0$ leads to the two equations:

$$\dot{\rho}_B + 3(\rho_B + P_B)(\dot{f} + \frac{\dot{R}}{R}) + (\rho_B + P_y)\frac{\dot{A}}{A} = 0, \quad and \tag{8}$$

$$P_y' + (\rho_B - 3P_B + 4P_y)f' = 0. \tag{9}$$

3. Results

For an isotropic bulk, $P_B = P_y$, and we have $(\rho_B + P_B) \geq |2Q_B|$, where Q_B may be positive or negative, but $(\rho_B + P_B)$ must be non-negative so as not to violate the WEC. The condition $(\rho_B + P_B) > 0$, corresponds to the *high energy condition* for which $|Q_B| \neq 0$. At low energies, $|Q_B| = 0$ and hence, $(\rho_B + P_B) = 0$. Assuming positive flux for Q_B, we can write $\dot{\rho_B} = -2Q_B\left(3\dot{f} + \frac{3\dot{R}}{R} + \frac{\dot{A}}{A}\right)$, which represents the equation of continuity for this isotropic bulk.

3.1. *Low energy regime*

To prevent matter or energy flowing out of the brane along the fifth dimension, we require that $\bar{T}_y^t = 0$, which implies that $\bar{G}_y^t = 0$ and hence, we obtain:

$$\dot{f}' = \frac{\dot{A}}{A}f'. \tag{10}$$

Assuming that A, f and their first order derivatives are continuous, Eq. (10) can be easily integrated to give the result:

$$A(t,y) = \chi(y)f'(t,y). \tag{11}$$

Thus, the extra-dimensional scale factor at a given y and t, depends on the way the warping function varies along the extra dimension at that instant at the specific location. The localization of gravity at different locations in the bulk will be different at different times. Since there are six unknowns (f, R, χ, ρ_B, P_B and P_y), but only three independent field equations as they include the conservation equations, we need to impose additional constraints to solve the field equations.

3.1.1. Conformally flat bulk: The non-zero components of the Weyl tensor for the case in Eq. (11) are equated to zero and solved for $f(t,y)$. This leads us to the condition [14]:

$$f(t,y) = F_1(t)F_2(y), \tag{12}$$

where F_1 and F_2 are arbitrary functions such that:

$$\frac{dF_2(y)}{dy} = 0, \tag{13}$$

which, therefore, represents the unwarped case. Thus, we cannot have a conformally flat bulk with a generalized warp factor and extra-dimensional scale factor.

3.1.2. A stabilized bulk: For this condition, we must have $\dot{A} = 0$, so that $A = A(y)$. Re-scaling $A(y) \to 1$, we find that for $T_y^t = 0$, we need $\dot{f}' = 0$. Thus, either $f = f(y)$ or $f = f(t)$.

Case (i): $f = f(y)$**:** This corresponds to the usual RS scenario. For $f'' = 0$, we have $f = const \times y$, which represents a RS 3-brane embedded into 5D AdS. For a constant curvature bulk, the bulk Weyl tensor vanishes and the bulk is characterized by a negative cosmological constant with $\rho_B = -P_B$. The observed universe will have $q = -1$.

Case (ii): $f = f(t)$**:** This is the trivial case. We can absorb $f(t)$ inside the 4D metric.

3.1.3. Specific choice $f(t,y) = \frac{1}{2}(ln\tau(t) + ln\Gamma(y))$: In this case, the validity of the strong energy condition in the bulk is governed by the nature of warping, as well as by the effect of the extra dimension. For an isotropic bulk with $\tau(t) = t^{-1/2}$ and $R(t) = t$, the weak energy condition is found to be valid.

3.2. High energy regime

Here, the energy scale is in the GeV range. The bulk energy-momentum tensor for a non-self-interacting massless scalar field source [15] is given by:

$$\bar{T}_{IJ}^{scalar} = -\partial_I\phi\partial_J\phi + \frac{1}{2}g_{IJ}\partial_K\phi\partial^K\phi. \tag{14}$$

The components of the bulk energy-momentum tensor for the given metric are:

$$\bar{T}_t^t = \frac{1}{2}\dot{\phi}^2 e^{-2f} + \frac{\phi'^2}{2A^2} = -\bar{T}_y^y, \quad \bar{T}_i^i = \frac{\phi'^2}{2A^2} - \frac{1}{2}\dot{\phi}^2 e^{-2f}, \quad and \quad \bar{T}_y^t = \dot{\phi}\phi' e^{-2f}. \tag{15}$$

Assuming that $\phi(t,y) = \phi_1(t) + \phi_2(y)$, we find that $\dot{\phi}_1 = k\frac{\dot{A}}{A}$ and $\phi_2' = \frac{3}{k}f'$, where k is a scalar. Consequently, the solution for the bulk scalar field turns out to be:

$$\phi(t,y) = klnA + \frac{3}{k}f. \tag{16}$$

The effective Newton's constant and the cosmological constant on the brane will depend on the scalar field in the bulk.

4. Four-dimensional cosmology

The extrinsic curvature of the brane is governed by the time-dependent warp factor and the field equations for the effective matter has an extra term arising out of the junction conditions. It is geometrical in origin, which modifies the dynamics of the gravitational field compared to Einstein gravity. Consequently, the effective cosmological constant of the 4D universe is a variable quantity monitored by time-dependent warp factor. For an isotropic bulk with $f(t,y) = \frac{1}{2}(ln\tau(t) + ln\Gamma(y))$, the universe is initially decelerated, but makes a transition to an accelerated phase at later times.

5. Conclusions

The time-dependent warp factor monitors the localization of gravity and the evolution of the 4-dimensional universe up to the radiation-dominated phase, leading to a geometric interpretation of a dynamical dark energy. At high energies, the effective Newton's constant and the effective cosmological constant on the brane depend on the scalar fields in the bulk.

Acknowledgements

SG thanks IUCAA, Pune, India, for the Visiting Associateship.

References

[1] Arkani-Hamed N, Dimopoulos S and Dvali G 1998 *Phys. Lett. B* **429** 263, Antoniadis I, Arkani-Hamed N, Dimopoulos S and Dvali G 1998 *Phys. Lett. B* **436** 257, Khoury J, Ovrut B A, Steinhardt P J and Turok N 2001 *Phys. Rev. D* **64** 123522, Akama K 1982 *Lect. Notes Phys.* **176** 267 [arXiv:hep-th/0001113]
[2] Randall L and Sundrum R 1999 *Phys. Rev. Lett.* **83** 3370, *ibid* 4690
[3] Maartens R and Koyama K [arXiv:hep-th/1004.3962]
[4] Guha S and Chakraborty S 2012 *Int. J. Theor. Phys.* **51** 55
[5] Maia M D, Monte E M, Maia J M F and Alcaniz J S 2005 *Class. Quant. Grav.* **22** 1623, Maia M D, Monte E M and Maia J M F 2004 *Phys. Lett. B* **585** 11
[6] Goldberger W D and Wise M B 1999 *Phys. Rev. D* **60** 107505
[7] Binetruy P, Deffayet C and Langlois D 2000 *Nucl. Phys. B* **565** 269
[8] Kanti P, Kogan I, Olive K A and Pospelov M 1990 *Phys. Lett. B* **468** 31
[9] Maeda K and Wands D 2000 *Phys. Rev. D* **62** 124009
[10] Deffayet C 2001 *Phys. Lett. B* **502** 199, Dvali G, Gabadadze G and Porrati M 2000 *Phys. Lett. B* **484** 112, *ibid* **485** 208

[11] Lue A, Scoccimarro R and Starkmann G D 2004 *Phys. Rev. D* **69** 124015
[12] Maia M D and Silva G S *Phys. Rev. D* **50** 7233
[13] Neupane I P *Phys. Lett. B* **683** 88, *Phys. Rev. D* **83** 086004
[14] We have used GRTensor package: K. Lake and P. J. Musgrave, (Queen's University, Kingston, 2003).
[15] Ghosh S and Kar S 2009 *Phys. Rev. D* **80** 064024, Koyama K and Takahashi K 2003 *Phys. Rev. D* **68** 103512

Journal of Physics: Conference Series **484** (2014) 012016 doi:10.1088/1742-6596/484/1/012016

High-velocity collision of particles around a rapidly rotating black hole

T Harada

Department of Physics, Rikkyo University, Toshima, Tokyo 171-8501, Japan

E-mail: harada@rikkyo.ac.jp

Abstract. We have derived a general formula for the centre-of-mass (CM) energy for the near-horizon collision of two general geodesic particles around a Kerr black hole. We have found that if the angular momentum of the particle satisfies the critical condition, the CM energy can be arbitrarily high. We have then applied the formula to the collision of a particle orbiting an innermost stable circular orbit (ISCO) and another generic particle near the horizon, and found that the CM energy is arbitrarily high if we take the maximal limit of the black hole spin. In view of the astrophysical significance of the ISCO, this implies that particles can collide around a rapidly rotating black hole with a very high CM energy without any artificial fine-tuning. We have next applied the formula to the collision of general inclined geodesic particles and shown that in the direct collision scenario, the collision with an arbitrarily high CM energy can occur near the horizon of maximally rotating black holes, not only at the equator but also on a belt centred at the equator between two latitudes. This is also true in the scenario through the collision of a last stable orbit particle. This strongly suggests that if signals due to high-energy collision are to be observed, such signals will be generated primarily on this belt.

1. Introduction

Many black hole candidates have been observed in X-ray binaries and galactic centres. In the former case, electromagnetic radiation emitted from accretion disks around black holes of mass $\sim 10\ M_\odot$ is observed. In the latter case, the central objects are believed to be supermassive black holes of mass $\sim 10^6 - 10^9 M_\odot$. However, no direct signature of black hole horizons has been observed yet. In the near future, the observation with the resolution of the black hole horizon scale might catch the direct image of the black hole horizon, which is called a *black hole shadow*, in terms of the millimetre electromagnetic waves. Moreover, when gravitational radiation from black holes is observed, it will directly reveal the spacetime geometry in the vicinity of black hole horizons.

Astrophysical black holes are generally rotating and uniquely described by a Kerr metric, in which the line element is given by:

$$\begin{aligned} ds^2 &= -\left(1-\frac{2Mr}{\rho^2}\right)dt^2 - \frac{4Mar\sin^2\theta}{\rho^2}d\phi dt + \frac{\rho^2}{\Delta}dr^2 + \rho^2 d\theta^2 \\ &\quad + \left(r^2+a^2+\frac{2Mra^2\sin^2\theta}{\rho^2}\right)\sin^2\theta d\phi^2, \qquad (1) \end{aligned}$$

where $\rho^2 = r^2 + a^2\cos^2\theta$ and $\Delta = r^2 - 2Mr + a^2$. If $0 \le |a| \le M$, Δ vanishes at $r = r_\pm = M \pm \sqrt{M^2 - a^2}$. The horizon radius is given by $r_H = r_+$. The angular velocity of the horizon is given by:

$$\Omega_H = \frac{a}{r_H^2 + a^2} = \frac{a}{2M(M + \sqrt{M^2 - a^2})}. \quad (2)$$

We define the non-dimensional Kerr parameter $a_* = a/M$. We can assume $a \geq 0$ without loss of generality.

Bañados *et al* [1] have proposed a scenario, where rapidly rotating black holes may act as particle accelerators. More precisely, they have considered the following scenario: Let us drop two particles of the same rest mass at rest at infinity on the equatorial plane of a rotating black hole and assume that the two particles collide near the horizon. If the maximum spin limit $a_* \to 1$ is taken and the angular momentum of either particle is fine-tuned, then the centre-of-mass (CM) energy, which is the invariant measure of the collision energy, can be arbitrarily high. This article is based on our recent papers [2, 3].

2. The collision of general geodesic particles

The CM energy of particles 1 and 2 at the same spacetime point is defined by:

$$E_{\rm cm}^2 = -(p_1 + p_2)^a (p_1 + p_2)_a = m_1^2 + m_2^2 - 2g_{ab}p_1^a p_2^b, \quad (3)$$

where p_i^a and m_i are the four-momentum and the rest mass of particle i $(= 1, 2)$ respectively. This is a scalar quantity, therefore coordinate-independent and in principle observable. This is the energy of the two colliding particles observed by a local observer, which is at rest in the CM frame of the two particles.

The general geodesic equations are integrable with four conserved quantities [4]. The conserved quantities are the rest mass m, the energy E, the angular momentum L, and the Carter constant $\mathcal{Q}$. The geodesic orbits are given by the following set of first-order ordinary differential equations:

$$\begin{aligned} \rho^2 \dot{t} &= -a(aE\sin^2\theta - L) + (r^2 + a^2)P/\Delta, & (4)\\ \rho^2 \dot{\phi} &= -\left(aE - L/\sin^2\theta\right) + aP/\Delta, & (5)\\ \rho^2 \dot{r} &= \sigma_r \sqrt{R}, \quad \textit{and} & (6)\\ \rho^2 \dot{\theta} &= \sigma_\theta \sqrt{\Theta}, & (7) \end{aligned}$$

where dot $(.) = d/d\lambda$, $\sigma_r = \pm 1$, $\sigma_\theta = \pm 1$,

$$\begin{aligned} P &= P(r) = (r^2 + a^2)E - aL, & (8)\\ R &= R(r) = P^2 - \Delta[m^2 r^2 + (L - aE)^2 + \mathcal{Q}], \quad \textit{and} & (9)\\ \Theta &= \Theta(\theta) = \mathcal{Q} - \cos^2\theta\left[a^2(m^2 - E^2) + L^2/\sin^2\theta\right]. & (10) \end{aligned}$$

The condition $\dot{t} > 0$ must be satisfied for physical particle orbits, which is called the 'forward-in-time' condition. In the near-horizon limit, this reduces to:

$$L \leq \Omega_H^{-1} E \equiv L_c\ . \quad (11)$$

We have considered the collision of two particles outside the event horizon. Then, we have shown that the CM energy is bounded except in the limit to the horizon. For the near-horizon collision, the CM energy is given by:

$$\begin{aligned} E_{\rm cm}^2 &= m_1^2 + m_2^2 + \frac{1}{r_H^2 + a^2\cos^2\theta}\left[(m_1^2 r_H^2 + \mathcal{K}_1)\frac{E_2 - \Omega_H L_2}{E_1 - \Omega_H L_1} + (m_2^2 r_H^2 + \mathcal{K}_2)\frac{E_1 - \Omega_H L_1}{E_2 - \Omega_H L_2}\right.\\ &\left. - \frac{2(L_1 - a\sin^2\theta E_1)(L_2 - a\sin^2\theta E_2)}{\sin^2\theta} - 2\sigma_{1\theta}\sqrt{\Theta_1}\sigma_{2\theta}\sqrt{\Theta_2}\right], \quad (12) \end{aligned}$$

where $\mathcal{K}_i \equiv \mathcal{Q}_i + (L_i - aE_i)^2$. Therefore, if the angular momentum of either of the particles is fine-tuned so that $L_i \to \Omega_H^{-1} E_i$, the CM energy $E_{\rm cm}$ can be arbitrarily high. We have called particles with $L = \Omega_H^{-1} E$ as *critical particles.*

We have defined the effective potential as follows:

$$\frac{1}{2}\dot{r}^2 + \frac{r^4}{\rho^4} V(r) = 0, \tag{13}$$

where

$$V(r) = -\frac{R(r)}{2r^4}. \tag{14}$$

It is useful to classify critical particles in terms of the behaviour of the effective potential at the horizon. The classification is summarized in Table 1.

Table 1. The classification of critical particles in terms of $V(r_H)$, $V'(r_H)$ and $V''(r_H)$.

Class	$V(r)$ at $r = r_H$	BH spin	Scenario
I	$V = V' = 0$, $V'' < 0$	$\lvert a\rvert = M$	Direct collision
II	$V = V' = V'' = 0$	$\lvert a\rvert = M$	LSO (ISCO) collision
III	$V = V' = 0$, $V'' > 0$	$\lvert a\rvert = M$	Multiple scattering
IV	$V = 0$, $V' > 0$	$0 < \lvert a\rvert < M$	Multiple scattering

3. The collision of an ISCO particle

The Kerr black hole has an innermost stable circular orbit (ISCO) and its generalization to non-equatorial orbits, which is called a last stable orbit (LSO). The ISCO is important in astrophysical contexts. The inner edge of the standard accretion disk is usually given by the ISCO. In the extreme-mass-ratio-inspirals (EMRIs), the transition of the adiabatic inspiral phase to the plunge phase occurs at the ISCO. We have seen that the fine-tuning is naturally realised by an ISCO particle as: $a_* \to 1$, $r_{\rm ISCO} \to r_H$ and $L \to \Omega_H^{-1} E$.

We have considered the collision of an ISCO particle with another generic particle, and two distinct kinds of collision of an ISCO particle. The one is the near-horizon collision, where the ISCO particle plunges to the vicinity of the horizon and collide with a counterpart particle there. The other is the on-ISCO collision, where the ISCO particle collide with a counterpart at the ISCO radius. The schematic figures for these two kinds of collision are shown in Figure 1. For the near-horizon collision of an ISCO particle with a generic counterpart, we obtain:

$$\frac{E_{\rm cm}}{2m_0} \approx \frac{1}{2^{1/2}3^{1/4}} \frac{\sqrt{2e_2 - l_2}}{\sqrt[4]{1 - a_*^2}}, \tag{15}$$

for $a_* \approx 1$. For the collision of an ISCO particle with a generic counterpart at the ISCO radius, we obtain:

$$\frac{E_{\rm cm}}{2m_0} \approx \frac{1}{2^{1/6}3^{1/4}} \frac{\sqrt{2e_2 - l_2}}{\sqrt[6]{1 - a_*^2}}, \tag{16}$$

for $a_* \approx 1$.

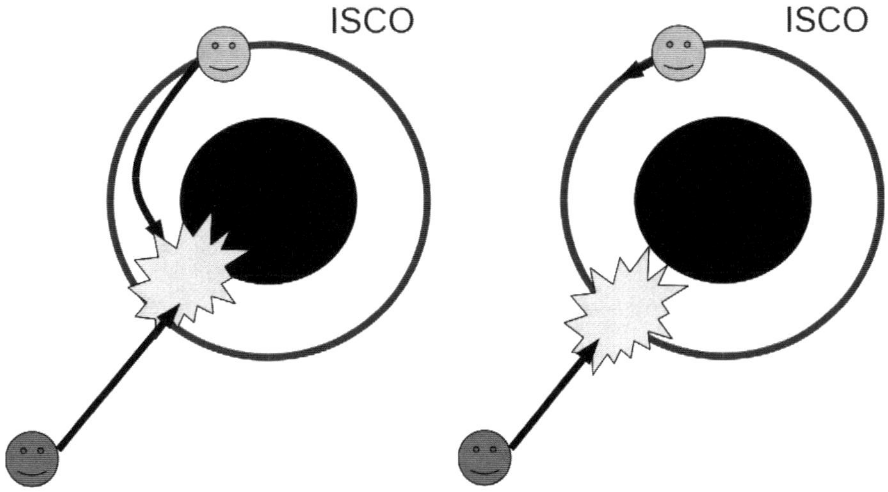

Figure 1. *Left panel*: The near-horizon collision of an ISCO particle plunging from the ISCO. *Right panel*: The collision of an ISCO particle at the radius of the ISCO

4. High-velocity collision apart from the equatorial plane

It is intriguing whether the high-velocity collision is restricted to that on the equatorial plane or not. Here, we have concentrated ourselves on the direct collision scenario and the ISCO collision scenario, for which the high-velocity collision with the critical particles is of classes I and II respectively, because the physical relevance of the other cases is not clear. This implies that we can assume $V = V' = 0$ and $V'' \leq 0$ at $r = r_H$. Consequently, the black hole must be maximally rotating and θ is restricted. The allowed region is the belt centred at the equator between latitudes $\pm 42.94^\circ$.

5. Summary

The CM energy of two colliding geodesic particles near the horizon can be arbitrarily high in the maximal rotation limit with the fine-tuning of the angular momentum. The required fine-tuning of the angular momentum is naturally realised by a particle orbiting the ISCO or LSO. The high-velocity collision can occur, not only at the equator but also at the latitude between $\pm 42.94^\circ$ near the horizon in the maximal rotation limit of the black hole.

As future prospects, we would like to raise the investigation of the effects of gravitational radiation reaction and self-force, and the theoretical predictions to astrophysical observation.

References

[1] Bañados M, Silk J and West S M 2009 *Phys. Rev. Lett.* **103** 111102
[2] Harada T and Kimura M 2011 *Phys. Rev. D* **83** 084041
[3] Harada T and Kimura M 2011 *Phys. Rev. D* **83** 024002
[4] Carter B 1968 *Phys. Rev.* **174**, 1559

Optical analogues of spherically symmetric black hole spacetimes

S S Hegde[1] **and C V Vishveshwara**[2]

[1]Research Education Advancement Programme, Bangalore Association for Science Education, Bangalore, India

[2]Formerly with the Indian Institute of Astrophysics, Bangalore 560034, India

E-mail: `physicshegde@gmail.com`

Abstract. We have given an analytical formalism for developing optical analogues of spherically symmetric black hole spacetimes, and demonstrated the exact similarity between the electromagnetic wave equations in an inhomogeneous medium in flat spacetime and in a general relativistic curved spacetime. The permittivity and permeability of the inhomogeneous optical medium act as the metric components of an effective optical spacetime. Manifest properties of black holes, like curved trajectories of light rays and quasi-normal modes follow directly from our formalism. It is also applied to the specific case of Schwarzschild spacetime. The formalism would facilitate table-top experiments to investigate black hole phenomena.

1. Introduction

Advances in Transformation Optics and Analogue Gravity have led to systems that closely resemble those engendered by the general theory of relativity such as the black holes. Models that have come under the category of Analogue Gravity [1] make use of sound waves in flowing medium [2] like super fluids in BEC [3, 4], to mimic certain aspects of a black hole. Such systems developed until now are not exact analogues of any real spacetimes with corresponding metric tensors. On the other hand, Transformation Optics has utilized the formalism of general relativity [5] to build gradient index static media in order to gain excellent control over the light paths in those media [6]. Although, much of the efforts in Transformation Optics have been directed towards developing technological applications, the results can be used effectively to investigate and construct models of optical analogues of gravity. In this context, we have given a general analytical formalism for developing optical black hole analogues using static optical media. Our purpose is to simulate, in as much detail as possible, a gravitational black hole and its manifestly characteristic phenomena, such as the existence of a 'no-return' surface, namely, the event horizon, the bending of light rays, and the quasi-normal modes. To achieve this, we must have: (i) A region outside the horizon that replicates the black hole spacetime, and (ii) A perfect light absorber, which can act as the event horizon.

In the following sections, we have described how a particular profile of permittivity and/or permeability of the optical analogue can lead to an 'optical spacetime' that has the black hole metric, and therefore, reproduces the gravitational phenomena outside the event horizon. Within the framework of our formalism, we further have discussed the broadband omni-directional light absorber developed by Narimanov and Khildishev [7] that serves exactly as a perfect absorber, or a black hole.

2. Electromagnetic wave equation in an inhomogeneous medium

The macroscopic electromagnetic field strength tensor in a continuous medium is defined as [2]:

$$I^{\mu\nu} = Z^{\mu\nu\sigma\tau} F_{\sigma\tau}, \tag{1}$$

where $F_{0\mu} = -F_{\mu 0} = -E_\mu$, $F_{ij} = \epsilon_{ijk} B^k$, $I^{0\nu} = I^{\nu 0} = D^\nu$, and $I^{ij} = \epsilon^{ijk} H_k$. Here, $F_{\sigma\tau}$ is the field strength tensor in free space and $Z^{\mu\nu\sigma\tau}$ is a fourth rank tensor whose components are the permittivity and permeability tensors. It is antisymmetric in the first two (μ, ν), and the last two (σ, τ) indices. ϵ^{ijk} is the Levi-Civita symbol. The Roman indices span 1, 2, 3, while the Greek indices range over 0, 1, 2, 3. Eq. (1) is the succinct form of the two known equations: $D = \varepsilon E$ and $H = \mu^{-1} B$. For an optical medium in flat spacetime, the Maxwell's equations involving sources are given by:

$$(Z^{\mu\nu\sigma\tau} F_{\sigma\tau})_{,\nu} = 0. \tag{2}$$

The tensor Z is related to the permittivity and permeability tensors in the following way:

$$Z^{0i0j} = -Z^{0ij0} = Z^{i0j0} = -Z^{i00j} = -\frac{1}{2}\varepsilon^{ij}\,, \quad and \tag{3}$$

$$Z^{ijkl} = \frac{1}{2}\epsilon^{ijm}\epsilon^{kln}\mu_{mn}{}^{-1}. \tag{4}$$

For a curved spacetime with metric $g^{\mu\nu}$, Maxwell's equations are given by:

$$\frac{\partial}{\partial x^\nu}\left[\sqrt{-g}(g^{\mu\sigma}g^{\nu\tau} - g^{\mu\tau}g^{\nu\sigma})F_{\sigma\tau}\right] = 0. \tag{5}$$

The similarity of Eqs. (2) and (5), suggest an analogy between an inhomogeneous medium and a curved spacetime, and hence, we can write:

$$Z^{\mu\nu\sigma\tau} = \frac{\sqrt{-\bar{g}}}{\sqrt{\gamma}}\left(\bar{g}^{\mu\sigma}\bar{g}^{\nu\tau} - \bar{g}^{\mu\tau}\bar{g}^{\nu\sigma}\right). \tag{6}$$

$\bar{g}^{\mu\sigma}$ is a function of the permittivity and permeability, and is used with a bar on the top to distinguish it from the actual metric. Here, γ is the determinant of the spatial metric in the ambient curvilinear coordinate system (r, θ, ϕ), defined by $\gamma_{\alpha\beta} = -g_{\alpha\beta} + (g_{0\alpha}g_{0\beta})/g_{00}$, where g is the metric of the ambient flat spacetime. In spherical polar coordinate system on a flat spacetime, $\sqrt{\gamma} = r^2\ sin\theta$.

The field strength $F_{\sigma\tau}$ is written in terms of the four-potentials A_μ, which are expanded in the vector spherical harmonics as follows [9]:

$$\begin{aligned}
F_{\sigma\tau} &= A_{\tau,\sigma} - A_{\sigma,\tau}, \\
A_\mu &= \sum_{lm}\left[\left(0, 0, \frac{a^{lm}}{\sin\theta}\frac{\partial Y^{lm}}{\partial\phi}, -a^{lm}\sin\theta\frac{\partial Y^{lm}}{\partial\theta}\right)\right. \\
&+ \left.\left(f^{lm}Y^{lm}, h^{lm}Y^{lm}, k^{lm}\frac{\partial Y^{lm}}{\partial\theta}, k^{lm}\frac{\partial Y^{lm}}{\partial\phi}\right)\right],
\end{aligned} \tag{7}$$

where a^{lm} are odd parity functions, and f^{lm}, h^{lm}, k^{lm} are even parity functions of r and t. $Y^{lm}(\theta, \phi)$ are scalar spherical harmonics.

Using the above in Eq. (5), we get the following equation for odd parity functions:

$$\left(\bar{\mu}_{\theta\theta}^{-1} a^{lm}_{,r}\right)_{,r} - \bar{\varepsilon}^{\theta\theta}\frac{\partial^2 a^{lm}}{\partial t^2} - \frac{l(l+1)}{r^2}a^{lm} = 0, \tag{8}$$

and for even parity functions, we get:

$$\left(\bar{\mu}_{\theta\theta}^{-1} b_{,r}^{lm}\right)_{,r} - \bar{\varepsilon}^{\theta\theta} \frac{\partial^2 b^{lm}}{\partial t^2} - \frac{l(l+1)}{r^2} b^{lm} = 0. \tag{9}$$

These equations are similar to those developed for the gravitational case, by J A Wheeler and D Brill [8], and also by R Ruffini, J Tiomno and C V Vishveshwara [9] (who included the source term), which are shown below:

$$\left(g^{rr} a_{,r}^{lm}\right)_{,r} - g^{00} \frac{\partial^2 a^{lm}}{\partial t^2} - \frac{l(l+1)}{r^2} a^{lm} = 0, \tag{10}$$

and for even parity functions, it is:

$$\left(g^{rr} b_{,r}^{lm}\right)_{,r} - g^{00} \frac{\partial^2 b^{lm}}{\partial t^2} - \frac{l(l+1)}{r^2} b^{lm} = 0. \tag{11}$$

The similarity in Eqs. (9), (10) and (11) gives rise to the similarity in a number of phenomena, which depend on the form of the wave equation. One can formulate an optical spacetime with an effective optical metric, whose components are related to the permittivity and the permeability tensors as defined previously, and given as: $\bar{g}^{rr} = \bar{\mu}_{\theta\theta}^{-1} = \bar{\mu}_{\phi\phi}^{-1}$, $\bar{g}^{00} = -\bar{\varepsilon}^{\theta\theta} = -\bar{\varepsilon}^{\phi\phi}$, and $\bar{\varepsilon}^{\theta\theta}\bar{\mu}_{\theta\theta}^{-1} = 1$. Metamaterials can be used to realize such an optical spacetime. Such a system has been studied, particularly, for the Schwarzschild spacetime in [10].

3. Discussions

Choosing $\bar{\varepsilon}^{\theta\theta} = \left(1 - \frac{L}{r}\right)^{-1}$ and $\bar{\mu}_{\theta\theta}^{-1} = \left(1 - \frac{L}{r}\right)$, one gets an exact analogue of Schwarzschild black hole spacetime as:

$$\left[\left(1 - \frac{L}{r}\right) a_{,r}^{lm}\right]_{,r} - \left(1 - \frac{L}{r}\right)^{-1} \frac{\partial^2 a^{lm}}{\partial t^2} - \frac{l(l+1)}{r^2} a^{lm} = 0. \tag{12}$$

Here, $r = L$ can be considered to be the location of the event horizon. Therefore, we have the simplified equation as:

$$\frac{\partial^2 a^{lm}}{\partial r_*^2} - \frac{\partial^2 a^{lm}}{\partial t^2} - \left(1 - \frac{L}{r}\right) \frac{l(l+1)}{a^{lm}} a^{lm} = 0. \tag{13}$$

This is exactly similar to the equation obtained in the gravitational case. Here, we have used the Regge-Wheeler tortoise coordinate $r_* = r + L ln(\frac{r}{L} - 1)$. Following are the two metric components of the effective optical spacetime: $g^{00} = -(1 - \frac{L}{r})^{-1}$, and $g^{rr} = (1 - \frac{L}{r})$. The necessary conditions for a spacetime around a black hole are $g_{00} = 0$, $g^{rr} = 0$ at the event horizon. We have clearly seen that the effective metric satisfy these. This indicates that the medium can be used to simulate a Schwarzschild black hole spacetime. The analysis of the gravitational bending of light and quasi-normal modes will be exactly like that in the gravitational case.

3.1. The optical black hole of Narimanov and Khildishev

In order to simulate black hole phenomena, we must have: (i) a medium, which simulates black hole spacetime, and (ii) a perfect absorber, which acts like a black hole, and absorbs the entire incident light. We have already developed the formalism for designing the equivalent optical spacetime around a black hole. The omni-directional, broadband optical absorber developed in [7] can be used to mimic the black hole. The model developed by Narimanov and Khildishev consists of: (i) A central core (C) of complex permittivity, which absorbs the entire incident light, and (ii) An outer shell (S), with some permittivity profile, which bends the incident light, guiding it to the core, and thus, ensuring that entire incident light converges to the core, without any loss. The model with the core and the shell is placed in an external medium. Our formalism can be used to arrive at the Hamiltonian used by Narimanov and Khildishev. Starting from our wave equation and with the Eikonal approximation, we get their Hamiltonian as:

$$\frac{(\bar{\mu}_{\theta\theta})^{-1}}{\bar{\varepsilon}^{\theta\theta}}(\frac{\partial S}{\partial r})^2 + \frac{m^2}{r^2\bar{\varepsilon}^{\theta\theta}} = H. \tag{14}$$

This gives a trajectory of the light ray as:

$$\phi = \phi_0 + \int_{\frac{m}{r_1}}^{\frac{m}{r}} \frac{d\xi}{\sqrt{C_0\bar{\varepsilon}^{\theta\theta}(\frac{m}{\xi}) - \xi^2}}, \qquad \xi = \frac{m}{r}. \tag{15}$$

There is no need to introduce an optical spacetime here, as we have been considering bending of light into the absorbing core, inside the event horizon. Here, the focus is on absorbing the entire incident light rather than mimicking a spacetime. Their permittivity choice $\bar{\varepsilon}^{\theta\theta} = \bar{\varepsilon_0}^{\theta\theta}[1 + (\frac{R}{r})^n]$ serves the purpose well.

References

[1] Barcelo C, Liberati S and Visser M 2005 *Living Rev. Rel.* **8** 87
[2] DeLorenci V A, Klippert R and Obukhov Yu N 2003 *Opt. Express* **95** 061502
[3] Visser M 1998 *Class. Quant. Grav.* **15** 1767
[4] Volovik G E 2006 *J. Low Temp. Phys.* **145** 337
[5] Leonhardt U and Philbin T G 2006 *New J. Phys.* **8** 247
[6] Leonhardt U and Philbin T G 2009 *Prog. Opt.* **53** 69
[7] Narimanov E E and Khildishev A V 2009 *Appl. Phys. Lett.* **95** 041106
[8] Brill D and Wheeler J A 1957 *Rev. Mod. Phys.* **29** 3
[9] Ruffini R, Tiomno J and Vishveshwara C V 1972 *Lett. Nuovo Cimento* **3** 5
[10] Chen H, Miao R-X and Li M 2010 *Opt. Express* **18** 15183

On the comparison of results regarding the post-Newtonian approximate treatment of the dynamics of extended spinning compact binaries

S Hergt[1], J Steinhoff[2] and G Schäfer[1]
[1] Theoretisch-Physikalisches Institut, Friedrich-Schiller-Universität, Max-Wien-Platz 1, 07743 Jena, Germany
[2] Centro Multidisciplinar de Astrofísica (CENTRA), Departamento de Física, Instituto Superior Técnico (IST), Universidade Técnica de Lisboa, Avenida Rovisco Pais 1, 1049-001 Lisboa, Portugal

E-mail: steven.hergt@uni-jena.de

Abstract. A brief review has been given of all the Hamiltonians, and effective potentials calculated hitherto covering the post-Newtonian (pN) dynamics of a two-body system. A method has been presented to compare (conservative) reduced Hamiltonians with non-reduced potentials directly at least up to the next-to-leading-pN order.

1. Post-Newtonian modelling and results of the two-body dynamics

The treatment of the two-body dynamics in general relativity through higher order post-Newtonian (pN) order has to incorporate both spin- and tidal-force-induced mass multipoles of the involved bodies [1]. In our analysis, we have focused on the spin-multipole degrees of freedom, and presented some illuminating results that will be useful for the extraction of binary-system parameters from the emitted gravitational waves. One easy way for the modelling of such systems is the making use of the Tulczyjew's singular stress-energy tensor [2] $T_{\mu\nu}$, with the Greek coordinate indices running from 0 to 3, of the following form:

$$\begin{aligned}\sqrt{-g}T^{\mu\nu}(x^\sigma) &= \int d\tau\Big[u^{(\mu}p^{\nu)}\delta_{(4)} + \left(u^{(\mu}S^{\nu)\alpha}\delta_{(4)}\right)_{||\alpha} + \frac{1}{3}R^{(\mu}_{\alpha\beta\rho}J^{\nu)\rho\beta\alpha}\delta_{(4)} \\ &\quad -\frac{2}{3}\left(J^{\mu\alpha\beta\nu}\delta_{(4)}\right)_{||(\alpha\beta)} + \dots\Big], \quad u^\mu = \frac{dz^\mu}{d\tau}, \quad \delta_{(4)} = \delta(z^\sigma - x^\sigma), \quad (1)\end{aligned}$$

with the body's 4-velocity u^μ, the 4-momentum p_μ, the antisymmetric spin tensor $S^{\mu\nu}$ modelling the pole-dipole structure, and Dixon's reduced quadrupole moment tensor $J^{\mu\alpha\beta\nu}$ modelling the first order finite size effects while possessing the same symmetries as the Riemann tensor $R^{\mu\alpha\beta\nu}$. $\delta_{(4)}$ is a 4-dimensional Dirac delta function, $\int d^4x\delta_{(4)} = 1$, and $||$ denotes the 4-dimensional covariant derivative. The tensor $J^{\mu\alpha\beta\nu}$ is decomposed into stress, flow and the symmetric trace-free mass quadrupole $Q_{\mu\nu}$. The latter is given by the ansatz with a vector f_μ to which, the spin is orthogonal; i.e., $S^{\mu\nu}f_\nu = 0$ (See [3], Eq. (5.11) therein), and

$$Q_{\mu\nu} = \frac{C_Q}{m}\left(S_{\mu\rho}S_\nu{}^\rho - \frac{1}{3}P_{\mu\nu}S^{\rho\sigma}S_{\rho\sigma}\right), \quad P^{\mu\nu} = g^{\mu\nu} - \frac{1}{f_\rho f^\rho}f^\mu f^\nu, \quad (2)$$

Table 1. Post-Newtonian Hamiltonians known to date.

Order	1.0	1.5	2.0	2.5	3.0	3.5	4.0	4.5
H^N								
PM	$+\,H^{1pN}$		$+\,H^{2pN}$	$+\,H^{2.5pN}(t)$[6]	$+\,H^{3pN}$	$+\,H^{3.5pN}(t)$[7, 8]	$+\,(H^{4pN})$	$+\,\{H^{4.5pN}(t)\}$
SO		$+\,H^{LO}_{SO}$		$+\,H^{NLO}_{SO}$		$+\,H^{N^2LO}_{SO}$[9]	$+\,H^{LO,R}_{SO}(t)$	$+\,(H^{N^3LO}_{SO})$
S_1^2			$+\,H^{LO}_{S_1^2}$		$+\,H^{NLO}_{S_1^2}$		$+\,(H^{N^2LO}_{S_1^2})$	$+\,\{H^{LO,R}_{S_1^2}(t)\}$
S_1S_2			$+\,H^{LO}_{S_1S_2}$		$+\,H^{NLO}_{S_1S_2}$		$+\,H^{N^2LO}_{S_1S_2}$[10]	$+\,H^{LO,R}_{S_1S_2}(t)$
spin^3						$+\,[H^{LO}_{S^3}]$		$+\,(H^{NLO}_{S^3})$
spin^4							$+\,[H^{LO}_{S^4}]$	

{.} Eqs. of motion known [.] For black holes only (.) Not known (yet)

and is parametrized by one constant C_Q in the leading order quadratic-in-spin level, in this way fully encoding the rotational deformation. For black holes, one has $C_Q = 1$ [4], while for neutron star models, C_Q depends on the equations of state [5] and varies between 4.3 and 7.4. The next step in performing explicit pN-calculations is complex in various ways. We have compared two prominent methods: One method aims at calculating a Hamiltonian. This is achieved by a 3+1 decomposition of Einstein's field equations and the energy-momentum tensor from Eq. (1) leading to constraints, which have to be fulfilled at all instants of time on the 3-dimensional hypersurfaces of constant coordinate time. We have then used the ADM formalism as outlined in [3] to find the canonical set of variables $(\hat{\mathbf{z}}_I, \hat{\mathbf{p}}_I, \hat{\mathbf{S}}_I)$ with the body label $I = 1, 2$ fulfilling their standard canonical Poisson bracket relations $\{\hat{z}^i_I, \hat{p}_{Jj}\} = \delta_{ij}\delta_{IJ}$ and $\{\hat{S}_{I(i)}, \hat{S}_{I(j)}\} = \epsilon_{ijk}\hat{S}_{I(k)}$ with i, j, k running from 1 to 3. Letters embraced by round brackets indicate the components in local Lorentz frames and are denoted by $a, b, \dots$ from the beginning of the alphabet; so a can take the "values" $a \in \{(0), (i)\}$. The spin-tensor components S_{ab} are connected to the coordinate-frame components through a vierbein transformation $S_{ab} = e_{a\mu}e_{b\nu}S^{\mu\nu}$.

The ADM formalism also leads to a formula for calculation of the Hamiltonian in full reduced phase-space by imposing the ADMTT or transverse-traceless gauge to the 3-metric on the 3-dimensional hypersurfaces, and by choosing the canonical spin supplementary condition (SSC) to get canonical variables for the bodies. By expansion of the constraints in pN-powers of $\frac{v^2}{c^2} \sim \frac{Gm}{r^2c^2}$ (two-body system virial theorem, speed of light c is not put equal to one) one ends up with a perturbative scheme for the calculation of the various pN-Hamiltonians. As the spin is of pN order $1/c$ or $1/c^2$ depending on its strength, the formal labelling is such that we have called the first post-Newtonian spin-Hamiltonians not 1.5pN or 2.5pN according to formal counting rules but just as leading order (LO) ones and the higher corrections we have called next-to-leading (NLO), and next-to-next-to-leading order (N^{2}LO).

In Table 1, we have given a list of all known pN Hamiltonians for the case of maximally rotating objects where $|\mathbf{S}| = \frac{Gm^2a}{c}$ with the dimensionless spin a. H^N is the Newtonian Hamiltonian, PM means "point-mass", i.e., without spin, H^{nPN} with $n \in \{1, 2, \dots\}$ are the conservative pure point-mass Hamiltonians, $H^{\frac{n}{2}PN}$ are the non-autonomous, radiation-reaction (dissipative) pure point-mass Hamiltonians, SO refers to spin-orbit coupling, S_1S_2 to spin(1)-

spin(2) coupling, and S_1^2 to spin-quadrupole coupling involving the constant C_Q. LO,R in the index indicates the leading-order reaction counterpart, so $H_{S_1^2}^{LO,R}$ is the reaction counterpart to the conservative part $H_{S_1^2}^{LO}$. Obviously, the reaction part is much higher in its pN order than the conservative part, but nevertheless they are important to cover the dynamics to 4.5pN order consistently; up until now the reaction field for point masses is known to 3.5pN order [7, 8]. One other method to arrive at pN equations of motion is the derivation of effective potentials, which are subtly related to Hamiltonians by a Legendre transformation. This derivation is most effectivly achieved by sophisticated methods from Effective Field Theory (EFT) that uses full knowledge from quantum field theoretical calculations. Up until now, pN potentials have been calculated to 3pN order [11] for point masses and to NNLO for spin(1)-spin(2) coupling [12].

2. Comparison between Effective Field Theory potentials and ADM Hamiltonians

Effective potentials are part of a Lagrangian with the Newtonian kinetic energy T_N, given as:

$$L_{eff} = T_N - V_{eff} = \frac{m_1}{2} v_1^2 + \frac{m_2}{2} v_2^2 - V_{eff}. \tag{3}$$

The conservative effective potential V_{eff} for two interacting bodies is pN expanded up to next-to-leading order (NLO) spin effects in the following way:

$$V_{eff} = V_{PM} + V_{SO}^{LO} + V_{S_1^2}^{LO} + V_{S_2^2}^{LO} + V_{S_1S_2}^{LO} + V_{SO}^{NLO} + V_{S_1^2}^{NLO} + V_{S_2^2}^{NLO} + V_{S_1S_2}^{NLO}. \tag{4}$$

One key difference between EFT potentials and ADM Hamiltonians is that in most cases the potentials still depend on the $S^{(0)(i)}$-components of the spin tensor, which have to be fixed by choosing an appropriate SSC.

For a direct comparison, a formal Legendre transformation of the non-reduced potentials is conducted yielding the effective Hamiltonian H_{eff}, which is followed by a reduction process in phase-space in order to arrive at a canonical set of variables (See [13] for details). This 'canonicalization' is most transparently accomplished by reducing the following effective action:

$$S_{eff} = \int dt\, L_{eff} = \int dt \Big(p_{1i}\dot{z}_1^i + p_{2i}\dot{z}_2^i - \frac{1}{2} S_{1ab}\Omega_1^{ab} - \frac{1}{2} S_{2ab}\Omega_2^{ab} - H_{eff}\left(\mathbf{z}_I, \mathbf{p}_I, S_{Iab}\right) \Big). \tag{5}$$

Here, we have defined the angular velocity tensor $\Omega^{ab} \equiv \Lambda_A{}^a \dot{\Lambda}^{Ab}$ rendering Ω^{ab} antisymmetric and $\Lambda_{A\mu}\Lambda^A{}_\nu = g_{\mu\nu}$, $\Lambda_{Aa}\Lambda^A{}_b = \eta_{ab}$ with $(A, B, \dots) \in \{[0], [i]\}$ being the body-fixed Lorentz-frame labels. The reduced action has to read as:

$$\hat{S}_{eff} = \int dt \left(\hat{p}_{1i}\dot{\hat{z}}_1^i + \hat{p}_{2i}\dot{\hat{z}}_2^i - \frac{1}{2}\hat{S}_{1(i)(j)}\hat{\Omega}_1^{(i)(j)} - \frac{1}{2}\hat{S}_{2(i)(j)}\hat{\Omega}_2^{(i)(j)} - H_{can}\left(\hat{\mathbf{z}}_I, \hat{\mathbf{p}}_I, \hat{\mathbf{S}}_I\right)\right) \tag{6}$$

with $\hat{\Omega}^{(i)(j)} = \hat{\Lambda}_{[k]}{}^{(i)} \dot{\hat{\Lambda}}^{[k](j)}$ given by a non-linear shift of $\Lambda^{[k](i)}$ to $\hat{\Lambda}^{[k](i)}$ so that $\hat{\Lambda}^{[k](i)}\hat{\Lambda}^{[k](j)} = \delta_{ij}$.

This reduction is achieved by inserting the covariant SSC $S_{ab}u^b = 0$ as well as its conjugate condition $\Lambda^{[i]a}u_a = 0$ into $\frac{1}{2}S_{ab}\Omega^{ab}$, and performing a pN approximate variable transformation of spin and position reading as:

$$\begin{aligned} z_1^i &= \hat{z}_1^i - \left[\frac{1}{2m_1^2} p_{1k}\hat{S}_{1(i)(k)} \left(1 - \frac{\mathbf{p}_1^2}{4m_1^2}\right) - G\frac{m_2}{m_1^2}\frac{p_{1k}\hat{S}_{1(i)(k)}}{\hat{r}_{12}} + \frac{3}{2}G\frac{p_{2k}\hat{S}_{1(i)(k)}}{m_1\hat{r}_{12}} \right. \\ &\left. + \frac{G}{2}\frac{\hat{n}_{12}^k(\hat{\mathbf{n}}_{12}\cdot\mathbf{p}_2)\hat{S}_{1(i)(k)}}{m_1\hat{r}_{12}} + G\frac{m_2}{m_1^2}\frac{\hat{S}_{1(k)(l)}\hat{S}_{1(i)(l)}\hat{n}_{12}^k}{\hat{r}_{12}^2} + G\frac{\hat{n}_{12}^k\hat{S}_{1(i)(l)}\hat{S}_{2(k)(l)}}{m_1\hat{r}_{12}^2} \right], \end{aligned} \tag{7}$$

Table 2. Agreement between EFT potentials and ADM Hamiltonians.

V^{SO}_{NLO}	Levi[14]		$H^{SO}_{NLO\,ADM}$	Damour/Jaranowski/Schäfer[15, 16]
V^{SO}_{NLO}	Porto[17]			
$V^{S_1S_2}_{NLO}$	Porto/Rothstein[18, 19]	$\approx$	$H^{S_1S_2}_{NLO\,ADM}$	Steinhoff/Hergt/Schäfer[20, 16]
$V^{S_1^2}_{NLO}$	Porto/Rothstein[21, 22]		$H^{S_1^2}_{NLO\,ADM}$	Hergt/Steinhoff/Schäfer[23]

$$S_{1(i)(j)} = \hat{S}_{1(i)(j)} - \left[\frac{p_{1[i}\hat{S}_{1(j)](k)}p_{1k}}{m_1^2}\left(1-\frac{\mathbf{p}_1^2}{4m_1^2}\right) - \frac{2Gm_2}{m_1^2\hat{r}_{12}}p_{1[i}\hat{S}_{1(j)](k)}p_{1k} + \frac{3G}{m_1\hat{r}_{12}}p_{1[i}\hat{S}_{1(j)](k)}p_{2k} + \frac{G}{m_1\hat{r}_{12}}p_{1[i}\hat{S}_{1(j)](k)}\hat{n}^k_{12}(\hat{\mathbf{n}}_{12}\cdot\mathbf{p}_2) + \frac{2Gm_2}{m_1^2\hat{r}_{12}^2}p_{1[i}\hat{S}_{1(j)](l)}\hat{S}_{1(k)(l)}\hat{n}^k_{12} + \frac{2G}{m_1\hat{r}_{12}^2}p_{1[i}\hat{S}_{1(j)](l)}\hat{S}_{2(k)(l)}\hat{n}^k_{12}\right]. \tag{8}$$

Those formulas are valid to transform the potentials at least to NLO to their canonical Hamiltonian counterpart, which enabled us to obtain an overall agreement of all EFT NLO potentials with their corresponding ADM Hamiltonian as displayed in Table 2 up to canonical transformations indicated by $\approx$ (See [13] for a thorough investigation).

Acknowledgments
This work is supported by the Deutsche Forschungsgemeinschaft (DFG) through SFB/TR7 "Gravitational Wave Astronomy," project STE 2017/1-1, and GRK 1523, and by the FCT (Portugal) through PTDC project CTEAST/098034/2008.

References

[1] Binnington T and Poisson E 2009 *Phys. Rev. D* **80** 084018
[2] Tulczyjew W M 1959 *Acta Phys. Pol.* **18** 393
[3] Steinhoff J 2011 *Ann. Phys. (Berlin)* **523** 296
[4] Thorne K S 1980 *Rev. Mod. Phys.* **52** 299
[5] Laarakkers W G and Poisson E 1999 *Ap. J.* **512** 282
[6] Schäfer G 1985 *Ann. Phys. (N.Y.)* **161** 81
[7] Jaranowski P and Schäfer G 1997 *Phys. Rev. D* **55** 4712
[8] Königsdörffer C, Faye G and Schäfer G 2003 *Phys. Rev. D* **68** 044004
[9] Hartung J and Steinhoff J 2011 *Ann. Phys. (Berlin)* **523** 783
[10] Hartung J and Steinhoff J 2011 *Ann. Phys. (Berlin)* **523** 919
[11] Foffa S and Sturani R 2011 *Phys. Rev. D* **84** 044031
[12] Levi M 2011 [gr-qc[1107.4322]]
[13] Hergt S, Steinhoff J and Schäfer G 2012 *Ann. Phys. (N.Y.)* **327** 1494
[14] Levi M 2010 *Phys. Rev. D* **82** 104004
[15] Damour T, Jaranowski P and Schäfer G 2008 *Phys. Rev. D* **77** 064032
[16] Steinhoff J, Schäfer G and Hergt S 2008 *Phys. Rev. D* **77** 104018
[17] Porto R A 2010 *Class. Quant. Grav.* **27** 205001

[18] Porto R A and Rothstein I Z 2008 *Phys. Rev. D* **78** 044012
[19] Porto R A and Rothstein I Z 2010 *Phys. Rev. D* **81** 029904(E)
[20] Steinhoff J, Hergt S and Schäfer G 2008 *Phys. Rev. D* **77** 081501(R)
[21] Porto R A and Rothstein I Z 2008 *Phys. Rev. D* **78** 044013
[22] Porto R A and Rothstein I Z 2010 *Phys. Rev. D* **81** 029905(E)
[23] Hergt S, Steinhoff J and Schäfer G 2010 *Class. Quant. Grav.* **27** 135007

Confinement and focusing of geodesics in warped spacetimes

S Kar[1], A Das Gupta[2], S Ghosh[3] and H Nandan[4]

[1] Department of Physics and Centre for Theoretical Studies, Indian Institute of Technology, Kharagpur 721302, India
[2] Department of Mechanical Engineering and Centre for Theoretical Studies, Indian Institute of Technology, Kharagpur 721302, India
[3] School of Physics, Indian Institute of Science Education and Research, Thiruvananthapuram 695016, India
[4] Department of Physics, Gurukula Kangri Vishwavidyalaya, Haridwar 249404, Uttarakhand, India

E-mail: sayan@iitkgp.ac.in, anir@mech.iitkgp.ernet.in, suman.ghosh@iisertvm.ac.in, hntheory@yahoo.co.in

Abstract. We have explored certain characteristic features of test particle trajectories in five dimensional, warped bulk geometries with a single thick brane. After a brief introduction on such spacetimes, we have first discussed timelike geodesic motion. The geodesic equations, which reduce to a first order autonomous dynamical system, have been solved using analytical and numerical methods. We have demonstrated how a growing (decaying) warp factor leads to oscillatory (runaway) trajectories, suggesting confinement (deconfinement). Further, we have pointed out differences that arise when we have a cosmological brane and/or a time-dependent extra dimension. Next, we have moved on to the kinematics of geodesic congruences. The evolution of the kinematical variables (expansion, rotation and shear) along geodesic flows have been obtained using analytical and numerical approaches, with particular emphasis on the required conditions and occurence of geodesic focusing.

1. Introduction

Warped braneworld models have been a topic of active research interest for more than a decade now [1]. Such models involve extra dimensions (beyond the usual four). It is assumed that we live on an embedded timelike mem*brane*. The higher (usually five) dimensional line element has a *warp factor*, which leads, in general, to a scaled Minkowski metric on the embedded four dimensional hypersurface. The models are useful in (i) solving the hierarchy problem without introducing new hierarchies, and (ii) providing an alternative to the usual notion of compactification, via a mechanism for *localizing* fields on the brane [2]. The zero modes of the bulk fields provide the usual classical physics on the brane, while the contributions of the higher modes to various physical processes lead to corrections reflecting the presence of extra dimensions. For extensive discussions on these models, the reader is referred to well-known review articles [2, 3].

The placing of branes in this class of models is done in various ways. Introducing a derivative discontinuity (say, a $|x|$ dependence) in the metric functions (specifically, the warp factor) yields a δ-function on the right hand side of the bulk Einstein's equations, which acts as a localized

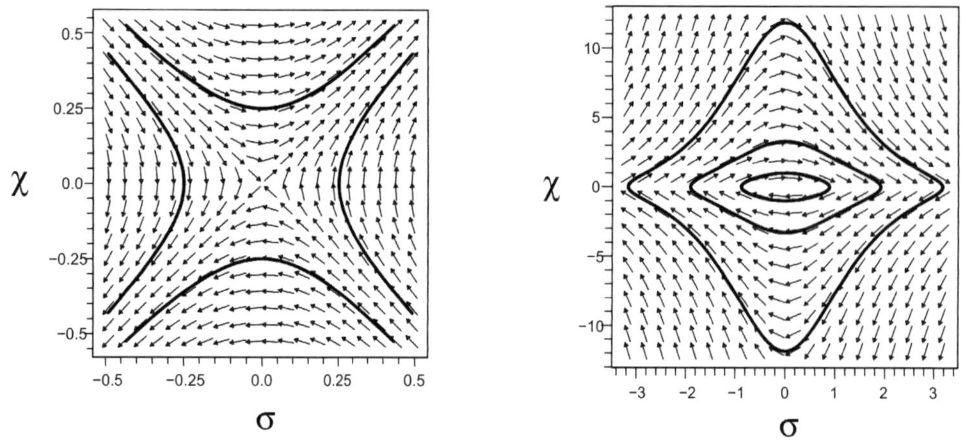

Figure 1. The figure on the left (right) are for decaying (growing) warp factors.

source (the brane). This is known as the *thin* brane model. The original Randall–Sundrum (RS) model has two such thin–branes and the bulk spacetime is orbifolded (S_1/Z_2). Alternatively, the brane can be realised as a scalar field domain wall, wherein the metric functions are smooth, resulting in the so-called *thick* brane models [2].

In this article, we have considered thick branes only. Our primary aim is to show how the nature (growing/decaying) of the warp factor is reflected in the behaviour of geodesics and geodesic congruences. We have considered a wider class of models, wherein *a cosmological on–brane metric* and a *time dependent extra dimension* have been introduced and their effects on geodesics and geodesic flows have also been studied.

2. Geodesics and confinement

The line element of interest to us here is:

$$ds^2 = e^{2f(\sigma)}\left(-dt^2 + a^2(t)|d\vec{x}|^2\right) + b^2(t)d\sigma^2, \tag{1}$$

where $e^{2f(\sigma)}$, $a(t)$ and $b(t)$ are the warp factor, cosmological scale factor and the extra dimensional scale factor respectively. The warp factor is chosen with $f(\sigma) = \pm \ln \cosh k\sigma$, where the plus corresponds to a growing one and the minus, to a decaying. The bulk timelike geodesic equations reduce to the following set of equations:

$$\begin{aligned}
\dot{\eta} &= \frac{e^{-f(\sigma)}}{a(\eta)}\sqrt{\epsilon + \frac{\chi^2}{b^2(\eta)} + \frac{\sum_{i=1}^{3} D_i^2}{e^{2f(\sigma)}a^2(\eta)}}, \\
\dot{x}_i &= \frac{D_i e^{-2f(\sigma)}}{a^2(\eta)}, \\
\dot{\sigma} &= \frac{\chi}{b^2(\eta)}, \quad \textit{and} \\
\dot{\chi} &= -f'(\sigma)\left(\epsilon + \frac{\chi^2}{b^2(\eta)}\right),
\end{aligned}$$

where we have introduced a new variable χ and also used the on–brane conformal time η. The above set of equations represent a first–order dynamical system, which can be analysed numerically.

In order to understand the role of the warp factor, we have switched off $b(t)$ (i.e., set it to 1). The σ–χ set then reduces to an independent dynamical system, which can be solved analytically (for details see [4]). In Figure 1, we have shown the phase plots for this simpler dynamical system. It is evident that for a growing (decaying) warp factor, we have oscillatory (runaway) motion. This result is further substantiated by numerically solving the dynamical system and obtaining the trajectories for specific forms of $a(t)$ and $b(t)$. The oscillatory behaviour of geodesics for the growing warp factor indicates confinement of trajectories. Varying $a(t)$ and $b(t)$ produces qualitative changes as shown in the numerous figures in [4].

3. Geodesic focusing

Having dealt with geodesics, we have now turned to geodesic congruences. It is well known that the kinematics of such congruences is described by the nature of evolution (along the geodesic flow) of the expansion θ, the shear σ_{ij} and the rotation ω_{ij}. The evolution equations are the Raychaudhuri equations, which constitute a set of first order, coupled ordinary differential equations with the affine parameter λ as the independent variable. Briefly, one defines:

$$B_{ij} = \nabla_j v_i \equiv \sigma_{ij} + \omega_{ij} + \frac{1}{3} h_{ij}\theta \quad and \quad h_{ij} = g_{ij} + v_i v_j, \tag{2}$$

and the evolution of B_{ij} is given as:

$$u^k \nabla_k B_{ij} = -B_{ik} B^k_j - R_{ikjl} u^k u^l. \tag{3}$$

One may separate the trace, symmetric traceless and antisymmetric parts of the above equation, in order to obtain individual equations for θ, σ_{ij} and ω_{ij}. The main aim is: Given *initial* values at some λ, what is the nature of θ, σ_{ij} and ω_{ij} at another future λ? To this end, we have solved the equation for B_{ij} numerically, with appropriate initial conditions, and then obtained the expansion, shear and rotation evolution. This is done for the geodesics obtained numerically in [4]. We have discussed below the behaviour of the expansion θ for various cases. It is known that if the timelike convergence condition $R_{ij}u^i u^j \geq 0$ holds, geodesics focus (meet) at a finite λ. This is the focusing theorem.

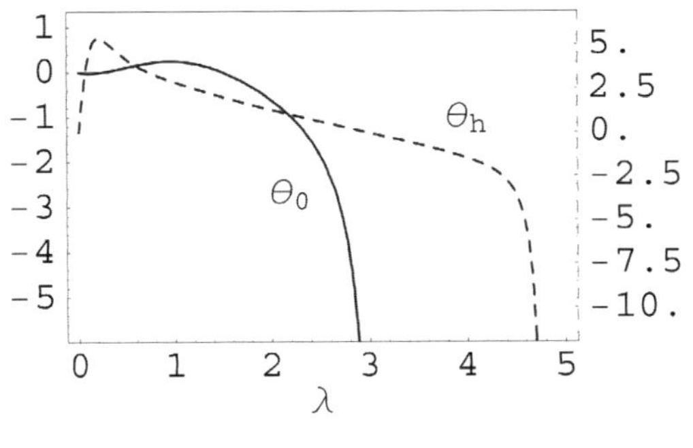

Figure 2. θ vs. λ (growing warp factor); zero (θ_0) and high (θ_h) initial rotation.

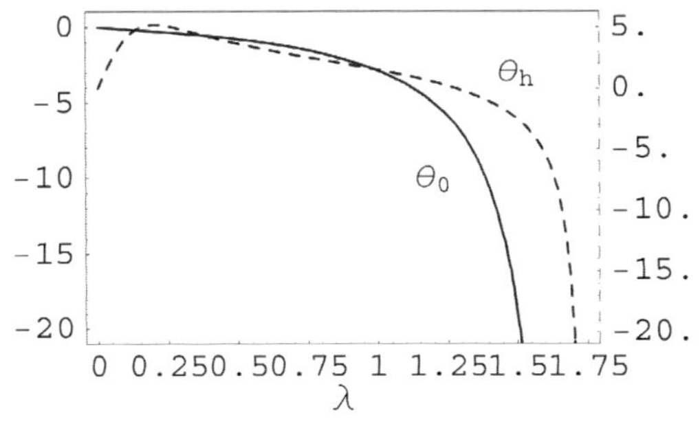

Figure 3. θ vs. λ (decaying warp factor); zero (θ_0) and high (θ_h) initial rotation.

Figure 2 shows the numerically obtained behaviour for a growing warp factor, and FRW radiation dominated $a(\eta) = 2\eta$, and $b(\eta) = 1 + \frac{1}{\eta}$. Congruences without initial rotation, expand slowly at first, but later become focused at a finite λ. For high initial rotation, expansion inflates for a very short while, but eventually the geodesics gets focused again at another finite but larger value of λ. Rotation is always dominated by shear, which grows faster. An initial large rotation delays focusing.

Figure 3 is for a decaying warp factor with $a(\eta)$, $b(\eta)$ as in Figure 2. Here, we have noted that congruences without any initial rotation, do come closer to each other monotonically, but focus only at $\sigma \to \infty$ (follows from the solution of the geodesic equation). When high initial rotation is introduced, the congruence expands initially, but eventually, geodesics tend to focus again, asymptotically.

Figure 4 is for a decaying warp factor, de Sitter $a(\eta) = \frac{1}{(1-\eta)}$, and $b(\eta) = 1 - \frac{\eta}{2}$. In the presence of a decaying warp factor, a de Sitter brane with an asymptotically static extra dimension, we have an example where geodesics are de-focused irrespective of initial rotation. Even though, it seems that at late times shear totally dominates rotation, it is the curvature term ($R_{AB}u^Au^B$) in the Raychaudhuri equation for the expansion that becomes dominant and causes the de-focusing.

Figure 5 shows the behaviour of this term and justifies the above mentioned behaviour of the expansion.

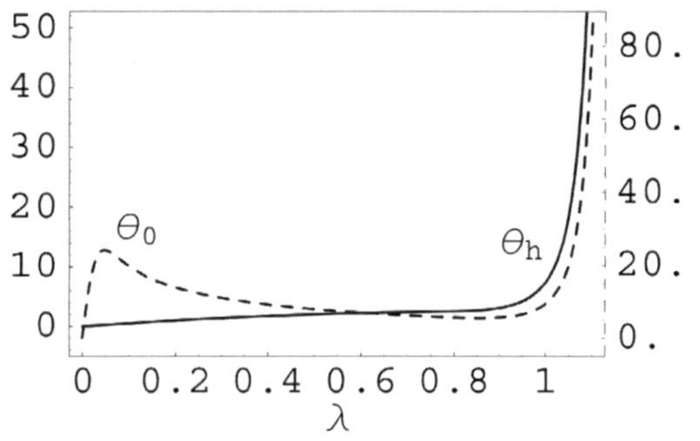

Figure 4. θ vs. λ (decaying warp factor); zero (θ_0) and high (θ_h) initial rotation.

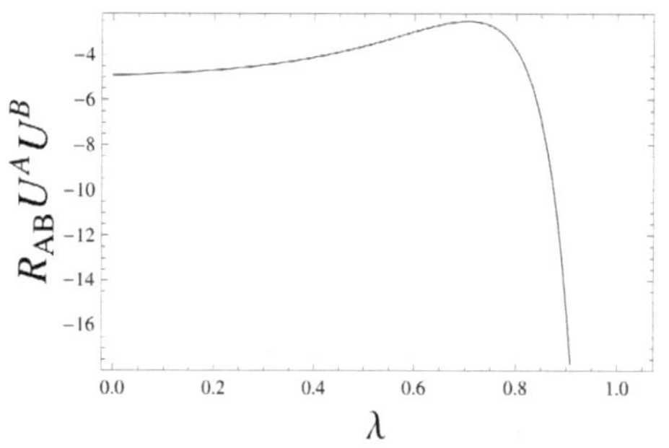

Figure 5. $R_{AB}u^Au^B$ vs. λ

4. Conclusions

In summary, we have shown how the nature of warping is reflected in the behaviour of geodesics. Confinement of trejectories is linked to a growing warp factor. We have also obtained the evolution of θ for varying initial conditions (rotation, in particular), and diferent $a(\eta)$ and $b(\eta)$. Some curious aspects of the evolution of θ have been pointed out. The interested reader is referred to [4, 5] for further details.

Acknowledgements

One of the authors (SK) thanks the organisers of ICGC-2011 for their warm hospitality during the meeting.

References

[1] Randall L and Sundrum R 1999 *Phys. Rev. Letts.* **83** 3370, *ibid.* **83** 4690
[2] Csaki C [arXiv:hep-ph/0404096]
[3] Maartens R 2010 *Living Rev. Rel.* **13** 5
[4] Ghosh S, Nandan H and Kar S 2010 *Phys. Rev. D* **82** 024040
[5] Ghosh S, Das Gupta A and Kar S 2011 *Phys. Rev. D* **83** 084001

Journal of Physics: Conference Series **484** (2014) 012020 doi:10.1088/1742-6596/484/1/012020

Galaxy rotation curves from a fourth order gravity

P Mishra and T P Singh

Tata Institute of Fundamental Research, Homi Bhabha Road, Colaba, Mumbai 400005, India

E-mail: `priti@tifr.res.in`, `tpsingh@tifr.res.in`

Abstract. While the standard and most popular explanation for the flatness of galaxy rotation curves is dark matter, one cannot at this stage rule out an explanation based on a modified law of gravitation, which agrees with Newtonian gravitation on the scale of the Solar system, but differs from it on larger length scales. Examples include Modfied Newtonian Dynamics (MOND) and Scalar-Tensor-Vector Gravity (STVG). Here, we have reported on a fourth order modification of the Poisson equation, which yields the same Yukawa type modification of Newtonian gravity as STVG, and which can explain flat galaxy rotation curves for a large sample of galaxies, once specific values for two parameters have been chosen. We have speculated on two possible origins for this modified Poisson equation: First, a possible fourth order modification of general relativity, and second, quadrupole gravitational polarization induced on a galaxy because of the pull of neighbouring galaxies.

1. Introduction

Flat galaxy rotation curves and the departure from the expected Keplerian fall-off, which they imply are most popularly attributed to the presence of cold dark matter (CDM) in the galaxy. Together with a cosmological constant Λ, CDM constitutes the highly successful ΛCDM model, which explains most of cosmological, cluster and galaxy data. Nonetheless, considering that dark matter has not yet been detected in the laboratory, one should tentatively also allow for the possibility that flat galaxy rotation curves arise, not from the presence of dark matter, but from a possible modification of the law of gravitation on scales larger than the Solar system.

Notable examples of such modifications include Modified Newtonian Dynamics (MOND) [1], where it is assumed that the gravitational force acting on a test particle of mass m is given by the relation:

$$F = ma\mu\left(\frac{a}{a_0}\right), \tag{1}$$

where a is acceleration in Newtonian mechanics, and a_0 is a new fundamental constant of nature having the value 2×10^{-10} ms^{-2}. For very small accelerations (large distances) it is assumed that:

$$\mu\left(\frac{a}{a_0}\right) = \frac{a}{a_0}, \tag{2}$$

whereas μ approaches the value one for accelerations encountered in the Solar neighbourhood.

Hence, for large distances we will have the relation:

$$\frac{GM}{r^2} = \frac{a^2}{a_0} \Rightarrow a = \frac{\sqrt{GMa_0}}{r}. \tag{3}$$

The virtue, thus, is that a falls as $1/r$ rather than $1/r^2$. Thus, equating a to the centripetal acceleration v^2/r, we get:

$$\frac{v^2}{r} = a = \frac{\sqrt{GMa_0}}{r} \quad \Rightarrow \quad v = (GMa_0)^{1/4}. \tag{4}$$

With the numerical choice of a_0 made above, one gets the desired flat value of the velocity.

Another noted example is Scalar-Tensor-Vector Gravity (STVG), where a massive vector field with a mass, denoted by μ, interacts with gravity with a coupling strength denoted by ω [2]. It can be shown that in STVG, the law of gravitation acquires a Yukawa modification, which is given by:

$$a(r) = -\frac{GM}{r^2}\left\{1 + \sqrt{\frac{M_0}{M}}\left[1 - \exp(-r/r_0)\left(1 + \frac{r}{r_0}\right)\right]\right\}. \tag{5}$$

Here, $r_0 = 1/\mu$ and M_0 is a parameter, which vanishes when $\omega = 0$. It is assumed (and justified by further considerations) that one can generalize this to the case of a mass distribution by replacing the factor GM/r^2 in Eq. (5) by $GM(r)/r^2$. The rotational velocity of a star v_c is obtained from $v_c^2(r)/r = a(r)$ and is given by:

$$v_c = \sqrt{\frac{GM(r)}{r}}\left\{1 + \sqrt{\frac{M_0}{M}}\left[1 - \exp(-r/r_0)\left(1 + \frac{r}{r_0}\right)\right]\right\}^{1/2}. \tag{6}$$

From observations of galaxies, the density profile $\mu(r)$ is known to be:

$$\mu(r) = \frac{3}{4\pi r^3}\beta M(r)\left[\frac{r_c}{r + r_c}\right], \tag{7}$$

where

$$M(r) = 4\pi\int_0^r dr' r'^2 \mu(r') = M\left(\frac{r}{r + r_c}\right)^{3\beta}, \tag{8}$$

and

$$\beta = \begin{cases} 1 & \text{for HSB galaxies,} \\ 2 & \text{for LSB and Dwarf galaxies.} \end{cases} \tag{9}$$

A good fit to a large number of galaxies has been achieved with the parameters:

$$M_0 = 9.60 \times 10^{11}\, M_\odot, \quad \textit{and} \quad r_0 = 13.92\,\text{kpc} = 4.30 \times 10^{22}\,\text{cm}. \tag{10}$$

In the fitting of the galaxy rotation curves for both LSB and HSB galaxies, using photometric data to determine the mass distribution $M(r)$, only the mass-to-light ratio $\langle M/L\rangle$ is employed, once the values of M_0 and r_0 are fixed universally for all LSB and HSB galaxies. Dwarf galaxies are also fitted with the parameters:

$$M_0 = 2.40 \times 10^{11}\, M_\odot, \quad \textit{and} \quad r_0 = 6.96\,\text{kpc} = 2.15 \times 10^{22}\,\text{cm}. \tag{11}$$

2. Fourth order gravity

Remarkably enough, it has been shown that the modified law of acceleration given in Eq. (5) of STVG, along with the consequent velocity profile in Eq. (6) is also a solution of the following modified fourth order Poisson equation [3]:

$$\nabla^4\phi - k^2\nabla^2\phi = -4\pi G k^2 \mu(r), \tag{12}$$

provided the density profile is chosen as in Eq. (7), and one identifies $k = 1/r_0$. M_0 appears as a constant of integration in the solution of Eq. (12).

Greater insight into the modified acceleration can be had by expanding the acceleration law:

$$a(r) = -\frac{GM(r)}{r^2}\left\{1 + \sqrt{\frac{M_0}{M}}\left[1 - \exp(-r/L)\left(1 + \frac{r}{L}\right)\right]\right\}, \tag{13}$$

around $r = r_0$ and writing it as a sum of two terms: A part that falls as $1/r^2$ and is independent of r_0, and a part that falls as $1/r$ and depends on r_0. This gives:

$$a(r) \approx -\frac{GM(r)}{r^2}\left[1 + \sqrt{\frac{M_0}{M}}\left\{1 - \frac{3}{e}\right\}\right] - \frac{GM(r)}{r}\left[\sqrt{\frac{M_0}{M}}\frac{1}{r_0 e}\right]. \tag{14}$$

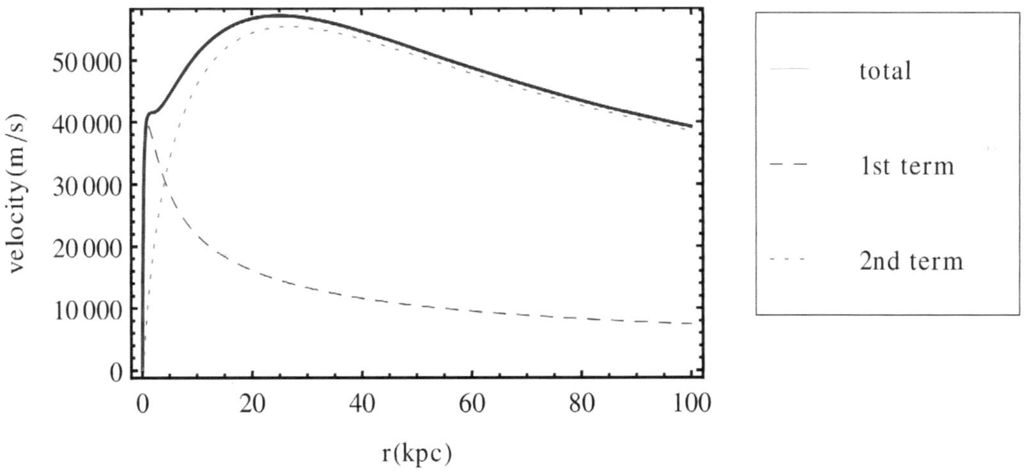

Figure 1. *Dashed curve*: Velocity due to first term, *dotted curve*: Velocity due to second term, *thick curve*: Total velocity.

The second term dominates for $r \geq r_0$. Figure 1 above plots the rotation velocity curve, and it is clear that the Keplerian fall-off due to conventional Newtonian acceleration (first term) is modified to a flat rotation curve (second term) for $r \geq r_0$. The $1/r$ fall-off brought about by the second term is MOND-like behaviour. At even larger distances, the second term is exponentially damped and the fall-off is again Keplerian, but with an effectively larger value of G.

It is significant that from our work, we have given an estimate of the theoretical value of a_0 in MOND. A simplistic guess would be to construct a quantity with the dimension of acceleration from our fundamental quantities k and M_0. This is nothing but:

$$GM_0k^2 = \frac{GM_0}{r_0^2} = \frac{6.67 \times 10^{-11} \times 9.60 \times 10^{11} \times 2 \times 10^{30}}{(13.92 \times 10^3 \times 3.08 \times 10^{16})^2}, \tag{15}$$

$$= 3 \times 10^{-10}\ \mathrm{ms}^{-2}, \tag{16}$$

which is of the same order as a_0 in MOND ($a_0 = 2 \times 10^{-10}\ ms^{-2}$). The relation is made more transparent by comparing the acceleration given by the second term of Eq. (14) with the acceleration in MOND in Eq. (3). This comparison yields the fundamental relation:

$$a_0 = \frac{GM_0}{r_0^2 e^2}. \tag{17}$$

3. Possible origin of fourth order gravity

The fourth order modified Poisson equation considered above is the weak field limit of the following modification of general relativity:

$$R^{\mu\nu} - \frac{1}{2} g^{\mu\nu} R = \frac{8\pi G}{c^4} T^{\mu\nu} + k^{-2} R^{\mu\nu\alpha\beta}{}_{;\alpha\beta}. \tag{18}$$

It is possible that such an effective modification of Einstein gravity comes into play during the late stages of the evolution of the Universe, when large scale structures form.

An intriguing alternate possibility, which has been discussed in [3] is that the fourth order modification of the Poisson equation is a consequence of induced polarization of the gravitational field of a galaxy, caused by neighbouring galaxies. Thinking of galaxies as 'molecules' made of atoms (the stars), one would like to analyze how the averaged gravitational field inside a galaxy, is modified by the polarization of the molecules, due to the external pull of other galaxies. Indeed, one has at the back of the mind, the polarization induced modification of electromagnetic fields in an electrically charged medium.

It can be shown that as a consequence of polarization, the averaged Einstein's equations are effectively modified to:

$$R^{(0)}_{\mu\nu} - \frac{1}{2} g^{(0)}_{\mu\nu} R^{(0)} = -\kappa \Big(T^{(free)}_{\mu\nu} + \frac{1}{2} c^2 Q^{;\rho\sigma}_{\mu\rho\nu\sigma} \Big), \tag{19}$$

where $T^{(free)}_{\mu\nu}$ is the energy-momentum tensor of molecules, and $Q_{\mu\rho\nu\sigma}$ is the quadrupole gravitational polarization tensor. It is modelled by assuming it be proportional to the Weyl tensor, in analogy with the electrostatic case:

$$Q_{i0j0} = \epsilon_g C_{i0j0}, \tag{20}$$

and the gravitational dielectric constant ϵ_g is modelled [4] as:

$$\epsilon_g = \frac{1}{4} \frac{mA^2c^2}{\omega_0^2} N. \tag{21}$$

Here, A is the typical linear dimension of a molecule, m is the typical mass of a molecule, ω_0 is a typical frequency of harmonically oscillating atoms in the molecule, and N is the number density of the molecules in the medium. In the weak field limit, Eq. (19) can be shown to go over to the modified Poisson Eq. (12) with $k^2 = c^2/2\pi G\epsilon_g$ [5]. Furthermore, the theoretically predicted values of k and M_0 agree well with the values phenomenologically assumed for fitting data, suggesting that gravitational quadrupole polarization is a contender for explaining flat galaxy rotation curves [3].

References

[1] Bekenstein J D 2006 *Contemp. Phys.* **47** 387
[2] Brownstein J D and Moffat J W 2006 *Ap. J.* **636** 721
[3] Mishra Priti and Singh T P 2011 [arXiv:1108.2375[astro-ph.GA]]
[4] Szekeres P 1971 *Ann. Phys.* **64** 599
[5] Montani G, Ruffini R and Zalaletdinov R 2003 *Class. Quant. Grav.* **20** 4195.

Future non-linear stability of the Einstein-Vlasov system with Bianchi II symmetry

E Nungesser

Max-Planck-Institut für Gravitationsphysik, Albert-Einstein-Institut, Am Mühlenberg 1, 14476 Potsdam, Germany

E-mail: ernesto.nungesser@aei.mpg.de

Abstract. Assuming that the spacetime is close to the Collins-Stewart solution, which will play the role of the ω-limit, and that the maximal velocity of the particles is small, we have been able to show that for Bianchi II symmetry spacetimes with collisionless matter, the asymptotic behaviour at late times is close to the special case of dust. The key was a bootstrap argument.

1. The Einstein-Vlasov system

A cosmological model represents a universe at a certain averaging scale. It is described via a Lorentzian metric $g_{\alpha\beta}$ (we will use the signature $-+++$) on a manifold M, and a family of fundamental observers. The metric is assumed to be time-orientable, which means that at each point of M, the two halves of the light cone can be labelled past and future in a way which varies continuously from point to point. This enables to distinguish between future-pointing and past-pointing timelike vectors. This is physically a reasonable assumption from both a macroscopic point of view, e.g., the increase of entropy, and also from a microscopic point of view, e.g., the kaon decay. The interaction between the geometry and the matter is described by the Einstein's field equations (we use geometrized units, i.e., the gravitational constant G, and the speed of light in vacuum c are set equal to one) as:

$$G_{\alpha\beta} = 8\pi T_{\alpha\beta} \ ,$$

where $G_{\alpha\beta}$ is the Einstein tensor, and $T_{\alpha\beta}$ is the energy-momentum tesnsor. The Einstein tensor satisfies:

$$\nabla^{\alpha} G_{\alpha\beta} = 0.$$

Thus, the energy-momentum tensor has to satisfy the same equation, which expresses the conservation of energy. For the matter model, we have taken the point of view of kinetic theory [1]. (The sign conventions of [2] are used. Also, the Einstein summation convention that repeated indices are to be summed over is used. Latin indices run from one to three and Greek ones from zero to three.)

Consider a particle with non-zero rest mass, which moves under the influence of the gravitational field. The mean field that we have mentioned in the introduction has been described now by the metric and the components of the metric connection. The worldline x^{α} of a particle is a timelike curve in spacetime. The unit future-pointing tangent vector to this curve is the

4-velocity v^α, and $p^\alpha = mv^\alpha$ is the 4-momentum of the particle. Let T_x be the tangent space at a point x^α in the spacetime M, then we have defined the *phase-space* P_m *for particles of mass* m as:

$$P_m = \{(x^\alpha,\ p^\alpha) : x^\alpha \in M,\ \ p^\alpha \in T_x,\ \ p_\alpha p^\alpha = -m^2,\ \ p^0 > 0\}.$$

We have considered from now on that all the particles have *equal* mass m. (For, how this relates to the general case of different masses, see [3].) We have chosen units such that $m = 1$, which means that a distinction between velocities and moments is not necessary. We have then that the possible values for the 4-momenta are all future-pointing unit timelike vectors. These values form the hypersurface:

$$P_1 = \{(x^\alpha,\ p^\alpha) : x^\alpha \in M,\ \ p^\alpha \in T_x,\ \ p_\alpha p^\alpha = -1,\ \ p^0 > 0\},$$

which we have called the *mass shell.* The collection of particles (galaxies or clusters of galaxies) will be described (statistically) by a non-negative real valued distribution function $f(x^\alpha,\ p^\alpha)$ on P_1. This function represents the density of particles at a given spacetime point with given 4-momentum. A free particle travels along a geodesic. Consider now a future-directed timelike geodesic parametrized by proper time s. The tangent vector is then at any time future-pointing unit timelike. Thus, the geodesic has a natural lift to a curve on P_1 by taking its position and tangent vector. The equations of motion, thus, define a flow on P_1, which is generated by a vector field L, which is called the *geodesic spray or Liouville operator.* The geodesic equations are:

$$\frac{dx^\alpha}{ds} = p^\alpha\ , \quad \textit{and} \quad \frac{dp^\alpha}{ds} = -\Gamma^\alpha_{\beta\gamma} p^\beta p^\gamma\ ,$$

where the components of the metric connection, i.e., $\Gamma_{\alpha\beta\gamma} = g(e_\alpha, \nabla_\gamma e_\beta) = g_{\alpha\delta}\Gamma^\delta_{\beta\gamma}$ can be expressed in the vector basis e_α as (1.10.3 of [4]):

$$\Gamma_{\alpha\beta\gamma} = \frac{1}{2}\Big(e_\beta(g_{\alpha\gamma}) + e_\gamma(g_{\beta\alpha}) + e_\alpha(g_{\gamma\beta}) + \eta^\delta_{\gamma\beta} g_{\alpha\delta} + \eta^\delta_{\alpha\gamma} g_{\beta\delta} - \eta^\delta_{\beta\alpha} g_{\gamma\delta}\Big)\ .$$

The commutator of the vectors e_α can be expressed with the following formula:

$$[e_\alpha,\ e_\beta] = \eta^\gamma_{\alpha\beta} e_\gamma\ ,$$

where $\eta^\gamma_{\alpha\beta}$ are called the *commutation functions.* The restriction of the Liouville operator to the mass shell is defined as:

$$L = \frac{dx^\alpha}{ds}\frac{\partial}{\partial x^\alpha} + \frac{dp^a}{ds}\frac{\partial}{\partial p^a}\ .$$

Using the geodesic equations, it has the followings form:

$$L = p^\alpha \frac{\partial}{\partial x^\alpha} - \Gamma^a_{\beta\gamma} p^\beta p^\gamma \frac{\partial}{\partial p^a}\ .$$

This operator is sometimes also called *geodesic spray.* If we denote, now, the phase-space density of collisions by $C(f)$, then the Boltzmann equation [5] in curved spacetime in our notation looks as follows:

$$L(f) = C(f)\ ,$$

which describes the evolution of the distribution function. Between collisions, the particles follow the geodesics. We have considered the collisionless case, which is described via the Vlasov equation, given as:

$$L(f) = 0\ .$$

The unknowns of our system are a 4-manifold M, a Lorentz metric $g_{\alpha\beta}$ on this manifold, and the distribution function f on the mass shell P_1 defined by the metric. We have the Vlasov equation defined by the metric for the distribution function and the Einstein's equations. It remains to define the energy-momentum tensor $T_{\alpha\beta}$ in terms of the distribution and the metric. Before that, we need a Lorentz invariant volume element on the mass shell. A point of the tangent space has the volume element $|g^{(4)}|^{\frac{1}{2}}dp^0dp^1dp^2dp^3$, (where $g^{(4)}$ is the determinant of the spacetime metric), which is Lorentz invariant. Now, considering p^0 as a dependent variable, the induced (Riemannian) volume of the mass shell considered as a hypersurface in the tangent space at that point is:

$$\varpi = |p_0|^{-1}|g^{(4)}|^{\frac{1}{2}}dp^1dp^2dp^3 \ .$$

Now, we have defined the energy-momentum tensor as:

$$T_{\alpha\beta} = \int f(x^\alpha,\ p^\alpha)p_\alpha\ p_\beta\ \varpi \ .$$

One can show that $T_{\alpha\beta}$ is divergence-free and thus, it is compatible with the Einstein's equations. For collisionless matter, all the energy conditions hold. In particular, the dominant energy condition is equivalent to the statement that in any orthonormal basis, the energy density dominates the other components of $T_{\alpha\beta}$, i.e., $T_{\alpha\beta} \leq T_{00}$ for each α, β (p. 91 of [6]). Using the mass shell relation, one can see that this holds for collisionless matter. The non-negative sum pressure condition in our case is equivalent to $g_{ab}T^{ab} \geq 0$.

The Vlasov equation in a fixed spacetime can be solved by the method of characteristic (see Chapter 3.2 of [7]):

$$\frac{dX^a}{ds} = P^a \ , \quad and \quad \frac{dP^a}{ds} = -\Gamma^a_{\beta\gamma}P^\beta P^\gamma \ .$$

Let $X^a(s,\ x^\alpha,\ p^a)$, $P^a(s,\ x^\alpha,\ p^a)$ be the unique soultion of that equation with initial conditions $X^a(t,\ x^\alpha,\ p^a) = x^a$, and $P^a(t,\ x^\alpha,\ p^a) = p^a$. Then the solution of the Vlasov equation can be written as:

$$f(x^\alpha,\ p^a) = f_0(X^a(0,\ x^\alpha,\ p^a),\ P^a(0,\ x^\alpha,\ p^a)),$$

where f_0 is the restriction of f to the hypersurface $t = 0$. It follows that, if f_0 is bounded, the same is true for f. We have assumed that f has compact support in momentum space for each fixed t (note that it is not possible in the Boltzmann case). This property holds if the initial data f_0 has compact support, and if each hypersurface $t = t_0$ is Cauchy hypersurface [8].

2. Main result

Using the $3+1$ formulation, our initial data are: $(g_{ij}(t_0),\ k_{ij}(t_0),\ f(t_0))$, i.e., a Riemannian metric, a second fundamental form, and the distribution function of the Vlasov equation respectively, on a 3-dimensional manifold $S(t_0)$. This is the initial data set at $t = t_0$ for the Einstein-Vlasov system. We have decomposed the second fundamental form, by introducing σ_{ab} as the trace-free part: $k_{ab} = \sigma_{ab} - Hg_{ab}$, where $H = -\frac{1}{3}k$ is the Hubble parameter. We have defined: $\Sigma^b_a = \frac{\sigma^b_a}{H}$, $\Sigma_+ = -\frac{1}{2}(\Sigma^2_2 + \Sigma^3_3)$ and $\Sigma_- = -\frac{1}{2\sqrt{3}}(\Sigma^2_2 - \Sigma^3_3)$, $N^j_i = \frac{R^j_i}{H^2}$, and $(N_1)^2 = -2\frac{R}{H^2}$, where R_{ij} is the Ricci tensor. We have a number (different from zero) of particles at possible different momenta, and we have defined P as the supremum of the absolute value of these momenta at a given time t as:

$$P(t) = sup\left\{|p| = \left(g^{ab}p_ap_b\right)^{\frac{1}{2}}|f(t,p) \neq 0\right\} \ .$$

We have obtained the following:

Theorem: *Consider any C^∞ solution of the Einstein-Vlasov system with Bianchi II symmetry, and with C^∞ initial data. Assume that* $|\Sigma_+(t_0)-\frac{1}{8}|$, $|\Sigma_-(t_0)|$, $|\Sigma_2^1(t_0)|$, $|\Sigma_3^1(t_0)|$, $|\Sigma_3^2(t_0)|$, $|\Sigma_2^3(t_0)|$, $|\Sigma_1^2(t_0)|$, $|\Sigma_1^3(t_0)|$, $|N_1(t_0)-\frac{3}{4}|$, $|N_2^1(t_0)|$, $|N_3^1(t_0)|$, *and* $P(t_0)$ *are sufficiently small. Then at late times, after possibly a basis change, the following estimates hold:*

$$\begin{aligned}
H(t) &= \frac{2}{3}t^{-1}(1+O(t^{-\frac{1}{2}}))\ ,\\
\Sigma_+ - \frac{1}{8} &= O(t^{-\frac{1}{2}})\ ,\\
\Sigma_- &= O(t^{-1})\ ,\\
\Sigma_2^1 &= O(t^{-\frac{1}{2}})\ ,\\
\Sigma_3^1 &= O(t^{-\frac{1}{2}})\ ,\\
\Sigma_3^2 &= O(t^{-1})\ ,\\
\Sigma_2^3 &= O(t^{-1})\ ,\\
\Sigma_1^2 &= O(t^{-1})\ ,\\
\Sigma_1^3 &= O(t^{-1})\ ,\\
N_1 - \frac{3}{4} &= O(t^{-\frac{1}{2}})\ ,\\
N_2^1 &= O(t^{-\frac{1}{2}})\ ,\\
N_3^1 &= O(t^{-\frac{1}{2}})\ , \quad \textit{and}\\
P(t) &= O(t^{-\frac{1}{2}})\ .
\end{aligned}$$

We will present the proof in a different paper.

References

[1] Stewart J M 1971 *Non-equilibrium Relativistic Kinetic Theory* (Lecture Notes in Physics, Berlin: Springer) **10**
[2] Rendall A D 2008 *Partial Differential Equations in General Relativity* (Oxford University Press)
[3] Calogero S and Heinzle J M 2010 *SIAM J. Appl. Dyna. Sys.* **9** 1244
[4] Stewart J M 1991 *Advanced General Relativity* (Cambridge University Press)
[5] Boltzmann L 1896 *Vorlesungen über Gastheorie I. Theil* (Leipzig: Verlag von Johann Ambrosius Barth)
[6] Hawking S W and Ellis G F R 1973 *The Large Scale Structure of Spacetime* (Cambridge University Press)
[7] Evans L C 2010 *Partial Differential Equations* (American Mathematical Society)
[8] Rendall A D 2004 *The Einstein Equations and the Large Scale Behaviour of Gravitational Fields* (eds Chrusciel P T and Friedrich H (Basel: Birkhauser))

Variational principle for gravity in the extended phase-space

P Sharan

Department of Physics, Jamia Millia Islamia, New Delhi 110025, India

E-mail: psharan@jmi.ac.in

Abstract. Variational formalism in the extended phase-space for fields is applied to gravity.

1. Introduction

The purpose of this paper is to show that torsion, even when zero, plays the crucial role of a mediating field between the stress-energy 3-form of matter fields and a 3-form related to the Einstein tensor, by coupling with both in the extended phase-space formalism recently given by the author. The extended phase-space formalism [1, 2] can be summarized as follows: When the evolution is along the one-dimensional time, the dynamics is determined by the stationary Poincare-Cartan or action 1-form $\Xi = p_i dq^i - Hdt$ in the extended phase-space (EPS) with coordinates (t, q, p). The EPS has a fiber bundle structure with time as the base manifold and the phase-space with coordinates (q, p) as the fiber. The evolution trajectories are *sections* from the one-dimensional base manifold into the EPS.

The Lie derivative of action is zero, i.e., $L_X \int \Xi = 0$, for a vector field X of variation of variables in the EPS. Using $L_X = i_X \circ d + d \circ i_X$, we see that we should first calculate $d\Xi$ and obtain equations of motion by contracting (i.e., applying i_X) with different independent variation fields X. The second term simply gives the Noether theorem $d(i_X \Xi) = 0$ for symmetry fields X, which satisfy $L_X \Xi = 0$.

2. Co-frames and frame-gauge invariance

We have described the gravitational field by giving a system of orthonormal co-frames. This involves four differential 1-forms $n^a = N^a_\mu dx^\mu$, $a = 0, 1, 2, 3$, such that $\langle n^a, n^b \rangle = \eta^{ab}$, where η is the matrix with $(-1, 1, 1, 1)$ on the diagonal and zero elsewhere. The co-frame n^a can be replaced by Lorentz rotated co-frame at each point without any change in the metric.

The essence of general theory of relativity is that physics is independent of the choice of local inertial frames represented by orthonormal co-frames. We have called this invariance, the "frame-gauge invariance". A local gauge group, in turn, involves its own connection forms or gauge potentials. The exterior derivatives dn^a are not frame-gauge invariant. They have to be replaced by covariant derivatives $dn^a + \omega^a{}_b \wedge n^b$, where the connection matrix $\omega^a{}_b$ of 1-forms, under the frame change by Lorentz transformation $\Lambda : n' = \Lambda n$, transforms as $\omega' = \Lambda\omega\Lambda^{-1} + \Lambda d\Lambda^{-1}$. (We have omitted the indices and the wedge symbol when there is no confusion.) The covariant derivatives of the co-frame n^a are called the *torsion* 2-forms of the

Einstein-Cartan theory, given as:

$$T^a = dn^a + \omega^a{}_b \wedge n^b. \tag{1}$$

The frame-gauge fields, or *curvature 2-forms*, are determined by:

$$\Omega^a{}_b \equiv d\omega^a{}_b + \omega^a{}_c \wedge \omega^c{}_b. \tag{2}$$

The torsion and curvature (given in Eqs. (1) and (2) respectively) forms yield the *Bianchi identities* on exterior differentiation:

$$dT + \omega T = \Omega n, \tag{3}$$

$$d\Omega + \omega\Omega = \Omega\omega. \tag{4}$$

The nature of our gauge group (Lorentz group) is reflected in the antisymmetry of the connection matrix when one index is lowered by Minkowski metric η,

$$\omega_{ab} \equiv \eta_{ac}\omega^c{}_b = -\omega_{ba}, \quad \textit{and} \quad \Omega_{ab} \equiv \eta_{ac}\Omega^c{}_b = -\Omega_{ba}. \tag{5}$$

We have used constant matrices η's for raising and lowering of indices.

3. EPS for fields

It has been suggested by the author that the fiber bundle structure of the extended phase-space in mechanics should be taken over for field systems by replacing the 1-dimensional base manifold by the 4-dimensional spacetime. The Poincare-Cartan form (or action) is now a 4-form [2]. Let ϕ be a scalar field. Then its canonical momentum is a differential 1-form p in this formalism. The Poincare-Cartan form has the structure:

$$\Xi_\phi = (*p) \wedge d\phi - H, \tag{6}$$

where the Hodge star operator [3] is used to convert a 1-form p into the 3-form $*p$ in order to obtain a 4-form of the type 'pdq'. The 'covariant Hamiltonian' $H = H(\phi, p)$, is a 4-form constructed from coordinate (0-form ϕ) and its canonical momentum (1-form p). The scalar field covariant Hamiltonian is:

$$H = \frac{1}{2}(*p) \wedge p + \frac{1}{2}m^2\phi^2(*1). \tag{7}$$

Here, we have used the notation $*1 = n^0 \wedge n^1 \wedge n^2 \wedge n^3$ for the Hodge dual of the constant function 1.

Allowed field configurations, here, are those four-dimensional sub-manifolds of the extended phase-space, which are sections on which the 4-form $i_X \circ d\Xi_\phi = 0$ for independent variational fields X. We calculate:

$$-d\Xi_\phi = \left[dp_a(*n^a) - m^2\phi(*1)\right]\left[p - d\phi\right] - \Theta^b T_b, \tag{8}$$

using formulas

$$p = p_a n^a, \quad dp = dp_a n^a + p_a dn^a, \quad d(*p) = dp_a \wedge (*n^a) + p_a d(*n^a),$$

$$d(*n^a) = *(n^a \wedge n^b) \wedge dn_b, \quad (*n^a) \wedge n^b = -\eta^{ab}(*1), \quad \textit{and} \quad d(*1) = (*n^a) \wedge dn_a,$$

replacing dn_b by the torsion 2-form $T_a = dn_a + \omega_{ab}n^b$, because the added term containing 5 factors of 1-forms n in a four dimensional space is zero. The *stress-energy 3-form* Θ is given by:

$$\begin{aligned} \Theta^b &= p_a(*n^a n^b)d\phi + \frac{1}{2}p_a p^a(*n^b) - \frac{1}{2}m^2\phi^2(*n^b), \\ &= \left[p^b p^c - \eta^{bc}\left(\frac{1}{2}p_a p^a + \frac{1}{2}m^2\phi^2\right)\right](*n_c), \qquad (9)\\ &\equiv T^{bc}(*n_c), \qquad (10) \end{aligned}$$

where T^{ab} is the familiar *stress-energy tensor*. The Hamiltonian equations for matter fields from the variation of fields ϕ and p while keeping frame fields and connection fixed are read off from the two factors of the first term:

$$p = d\phi, \qquad dp_a(*n^a) - m^2\phi(*1) = 0. \tag{11}$$

If we substitute $p = d\phi$ in the second equation, we obtain the Klein-Gordon equation for ϕ in the curved background determined by the fixed frame field n^a.

It is worth emphasizing that the exterior derivative acts on the full phase-space, and $d\phi$ is independent of n^a (or dx^μ). The momentum is a differential 1-form $p = p_a n^a$, but p_a are not functions of x. They are coordinates in the extended phase-space just as in classical mechanics, and p_i are coordinates in the cotangent bundle $T^*(Q)$. Only when we look for a section, which makes action stationary, will independent phase-space coordinates p_a become functions of x.

The term $-\Theta^a T_a$ containing torsion of the gravitational field has been seen below to combine with the gravitational part of action when the frame field and connection are varied.

4. Gravitational field in EPS

The gravitational field is jointly determined by the co-frame fields n^a and the frame-gauge connection fields ω_{ab}. The "velocities" for these are the torsion $T = dn + \omega n$ and curvature $\Omega = d\omega + \omega\omega$ respectively. Since action is gauge invariant, the true variation fields X of co-frames n and gauge potentials ω contract with gauge-covariant T and Ω and not dn or $d\omega$.

We have taken the Poincare-Cartan 4-form for gravity as the Einsten-Hilbert action:

$$\Xi_{EH} = -\frac{1}{2\kappa} * (n^a n^b)\Omega_{ba} = \frac{1}{4\kappa}n_a E^a, \qquad \kappa = 8\pi G_N, \tag{12}$$

where G_N is Newton's gravitational constant, and we have defined the *Einstein 3-form* as:

$$E^a = *(n^a \wedge n^b \wedge n^c)\Omega_{bc} \equiv -2G^{ab}(*n_b). \tag{13}$$

The components G^{ab} are defined by this equation. They reduce to the usual symmetric Einstein tensor for the Riemannian geometry.

The variation of Einstein-Hilbert action is simply:

$$d\Xi_{EH} = \frac{1}{2\kappa}ET. \tag{14}$$

The proof goes as follows: Using formulas such as: $d * (n^a n^b) = *(n^a n^b n^c)dn_c$, and the Bianchi identity,

$$\begin{aligned} d[*(n^a n^b)\Omega_{ba}] &= *(n^a n^b n^c)dn_c\Omega_{ba} + *(n^a n^b)d\Omega_{ba} \\ &= *(n^a n^b n^c)\Omega_{ba}(T_c - \omega_{cd}n^d) + *(n^a n^b)(\Omega\omega - \omega\Omega)_{ba}. \end{aligned}$$

Furthermore,

$$*(n^a n^b n^c)n^d = -\eta^{ad} * (n^b n^c) + \eta^{bd} * (n^a n^c) - \eta^{cd} * (n^a n^b),$$

we can rearrange and simplify the terms not containing T. They all add up to

$$*(n^a n^b)(\Omega\omega + \omega\Omega)_{ba} = *(n^a n^b)(\Omega_{bd}\omega^d{}_a + \Omega_{ad}\omega^d{}_b) = 0,$$

being a sum over a product of a symmetric and an antisymmetric expression.

Combining with the derivative of the matter action gievn in Eq. (8) for the scalar field, we get:

$$-d(\Xi_\phi + \Xi_{EH}) = \big[dp(*n) - m^2\phi(*1)\big]\big[p - d\phi\big] - \big[E/2\kappa + \Theta\big]T. \tag{15}$$

The allowed field configurations given by the terms above can be read off easily. For variational fields in the direction of ϕ and p, we get the Klein-Gordon equations in curved background (determined by n and ω). In the second term, varying n and keeping ω constant, gives the Einstein equation:

$$E = -2\kappa\Theta \qquad \text{or} \qquad G^{ab} = (8\pi G_N)T^{ab}, \tag{16}$$

using Eqs. (13) and (10). And lastly, if ω is varied and n is kept fixed, we get the equation $T = 0$.

References

[1] Arnold V I 1978 *Mathematical Principles of Classical Mechanics* (Springer-Verlag, New York, Section 9.C)
[2] Pankaj Sharan [arxiv:1201.4092]
[3] Pankaj Sharan 2009 *Spacetime, Geometry and Gravitation* (Hindustan Book Agency, New Delhi, Birkhauser, Basel, Section 6.6 for notation)

Journal of Physics: Conference Series **484** (2014) 012023 doi:10.1088/1742-6596/484/1/012023

Spacetime inhomogeneity and gravitational collapse

R Sharma

Department of Physics, P. D. Women's College, Jalpaiguri 735101, India

E-mail: rsharma@iucaa.ernet.in

Abstract. A model for the non-adiabatic collapse of a spherically symmetric anisotropic distribution of matter accompanied with radial heat flux has been presented. This is an inhomogeneous generalization of the model, describing collapse of a homogeneous distribution of isotropic fluid and, therefore, provides a mechanism to investigate the impacts of inhomogeneity and anisotropic stresses on the collapse of a radiating star.

1. Introduction

In cosmology and astrophysics, there are many outstanding issues relating to the collapse of a self-gravitating system which continue to bother us even today [1, 2]. In the absence of any established theory governing collapse, it has been found to be a good idea to make separate investigations on the impact of various factors such as shear, inhomogeneity, anisotropy, electromagnetic field, etc. on collapse. Through such investigations, one hopes to get a proper understanding of the gross physical behaviour of a gravitationally collapsing body. In classical GR, solutions to Einstein's field equations for a self-gravitating system and their interpretation provide valuable insight into a collapsing system. The mechanism of solving Einstein's field equations describing a physically realistic dynamical system and predicting its subsequently evolutionary stages got a tremendous impetus when Vaidya [3] derived the metric corresponding to the exterior gravitational field of a radiating star, and Santos [4] presented the junction conditions joining the interior spacetime of the collapsing body to the Vaidya exterior metric.

The objective, here, is to examine the role of spacetime inhomogeneity and anisotropic stresses on the collapse of a radiating star. We have reported a model, which describes the evolution of a spherically symmetric inhomogeneous anisotropic fluid distribution radiating away its energy in the form of radial heat flux and shrinking in size as the collapse proceeds. The model is a generalization of an earlier work presented by Banerjee *et al* [5] describing collapse of a homogeneously distributed isotropic fluid. In the present work, inhomogeneity in the background spacetime has been introduced by considering the $t = constant$ hyperspace of the 4-D manifold as having a geometry of a 3-spheroid rather than a 3-sphere. Impacts of the inhomogeneous nature of the background spacetime and anisotropic stresses on the collapse have been examined by comparing the physical parameters in this set up to the behaviour of the corresponding parameters in Banerjee *et al* [5] model, admissible as a special class in this model.

2. An inhomogeneously collapsing model

The spacetime metric formulated by Maiti [6] represents the gravitational field of a spherically symmetric conformally flat shear-free and rotation-free fluid with heat flux as source. Making use of Maiti's prescription, Banerjee *et al* [5] have presented a simple model for the collapse of a homogeneously distributed isotropic fluid. Various aspects of the model have been extensively examined by Schäfer and Goenner [7], by giving constraints on the model parameters complying with various physical plausibility requirements. To generalize the model given in [5], let us assume that the collapsing configuration is an inhomogeneous distribution of anisotropic fluid with its background spacetime described by the metric:

$$ds_-^2 = -A_0^2(r)dt^2 + R^2(t)\left[\frac{1+\lambda kr^2}{1-kr^2}dr^2 + r^2(d\theta^2 + \sin^2\theta d\phi^2)\right]. \tag{1}$$

In Eq. (1), $k \neq 0$ is a constant and λ is a parameter measuring departure from homogeneous geometry. The $t = constant$ hypersurface of the spacetime metric (1) has a geometry of a 3-spheroid representing a perturbation from that of a 3-sphere [8, 9, 10, 11, 12]. The energy-momentum tensor of the fluid with anisotropic stresses filling the interior of the collapsing body is written explicitly in the form:

$$T_{\alpha\beta} = (\rho + p_t)u_\alpha u_\beta + p_t g_{\alpha\beta} + (p_r - p_t)\chi_\alpha\chi_\beta + q_\alpha u_\beta + q_\beta u_\alpha\ , \tag{2}$$

where ρ = energy density, p_r = radial pressure, p_t = tangential pressure, χ^i = unit space-like four vector along the radial direction, u^i = 4-velocity of the fluid, and $q^\alpha = q\delta_r^\alpha$ is the radially directed heat flux vector so that $q^\alpha u_\alpha = 0$. In view of Eqs.(1) and (2), Einstein's field equations lead to the following system of four independent equations (rendering $G = c = 1$):

$$8\pi\rho = \frac{1}{R^2}\left[\frac{1}{r^2} - \frac{1}{r^2B_0^2} + \frac{2B_0'}{rB_0^3}\right] + \frac{3\dot{R}^2}{A_0^2R^2}\ , \tag{3}$$

$$8\pi p_r = \frac{1}{R^2}\left[-\frac{1}{r^2} + \frac{1}{B_0^2r^2} + \frac{2A_0'}{rA_0B_0^2}\right] - \frac{1}{A_0^2}\left(\frac{\dot{R}^2}{R^2} + 2\frac{\ddot{R}}{R}\right)\ , \tag{4}$$

$$8\pi p_t = \frac{1}{R^2}\left[\frac{A_0''}{A_0B_0^2} + \frac{A_0'}{rA_0B_0^2} - \frac{B_0'}{rB_0^3} - \frac{A_0'B_0'}{A_0B_0^3}\right] - \frac{1}{A_0^2}\left[\frac{2\ddot{R}}{R} + \frac{\dot{R}^2}{R^2}\right]\ , \quad and \tag{5}$$

$$8\pi q = -\frac{2A_0'\dot{R}}{A_0^2B_0^2R^3}\ , \tag{6}$$

where, $B_0(r) = \sqrt{(1+\lambda kr^2)/(1-kr^2)}$. Combining Eqs. (4) and (5), a time-independent differential equation of the form:

$$\frac{A_0''}{A_0B_0^2} - \frac{A_0'}{rA_0B_0^2} - \frac{B_0'}{rB_0^3} - \frac{A_0'B_0'}{A_0B_0^3} - \frac{1}{B_0^2r^2} + \frac{1}{r^2} - \delta(r) = 0\ , \tag{7}$$

may be obtained, if it is assumed that anisotropy in this model evolves as:

$$8\pi(p_t - p_r) = \Delta(r,t) = \frac{\delta(r)}{R^2(t)}. \tag{8}$$

Eq. (7) may be rewritten as:

$$(1+\lambda k - \lambda kx^2)\frac{d^2A_0}{dx^2} + \lambda kx\frac{dA_0}{dx} + \left(\lambda k(\lambda k+1) - \frac{(1+\lambda k - \lambda kx^2)^2\delta(r)}{k(1-x^2)}\right)A_0 = 0\ , \tag{9}$$

where a transformation $x^2 = 1 - kr^2$ has been introduced. Eq. (9) admits a solution for $A_0(r)$, if the function $\delta(r)$ is specified as:

$$\delta(x) = \frac{k(1-x^2)[2\lambda k(\lambda k+1)(2\lambda k+1) - (4\lambda k+7)\lambda^2k^2x^2]}{4(1+\lambda k-\lambda kx^2)^3}\ , \tag{10}$$

which is regular at all interior points of the configuration. The solution for the metric function $A_0(r)$ is then obtained as:

$$A_0(r) = (1+\lambda k^2 r^2)^{1/4}\Big(C + D\sqrt{1-kr^2}\Big), \tag{11}$$

where C and D are arbitrary constants of integration. Consequently, the spacetime of the collapsing shear-free stellar body with anisotropic stresses is described by metric

$$ds_-^2 = -(1+\lambda k^2 r^2)^{1/2}\Big(C + D\sqrt{1-kr^2}\Big)^2 dt^2 + R^2(t)\left[\frac{1+\lambda kr^2}{1-kr^2}dr^2 + r^2 d\Omega^2\right]. \tag{12}$$

The model in [5] turns out to be a sub-class of a solution of the metric in Eq. (12), which follows on setting $\lambda = 0$ and $D = -1$ [7]. Thus, without any loss of generality we have set $D = -1$ and investigated the impact of inhomogeneity ($\lambda \neq 0$) on collapse.

The evolution of the collapse in the stellar body with interior spacetime metric in Eq. (12) is governed by the function $R(t)$, which can be determined from the boundary conditions across the boundary surface Σ of the collapsing configuration. The spacetime outside the collapsing body is appropriately described by the Vaidya [3] metric:

$$ds_+^2 = -\left(1 - 2m(v)/\bar{r}\right) dv^2 - 2dv d\bar{r} + \bar{r}^2 d\Omega^2, \tag{13}$$

where v denotes the retarded time and $m(v)$ represents the total mass of the collapsing star. The matching condition $(p_r)_\Sigma = (qR(t)B_0)_\Sigma$ [4], linking smoothly the interior and the exterior spacetimes across the boundary Σ, yields a differential equation of the form:

$$\ddot{R}R + \frac{1}{2}\dot{R}^2 - \alpha\dot{R} + \beta = 0, \tag{14}$$

where, α and β are constants evaluated at the surface. A simple solution of Eq. (14) has been obtained as [7]:

$$R(t) = nt, \quad n = \alpha - \sqrt{\alpha^2 - 2\beta}\,. \tag{15}$$

For a collapsing configuration, the model parameters should be so chosen that $\dot{R} < 0$. Similarly, the rate of expansion is given by:

$$\Theta = u^\alpha_{;\alpha} = \frac{3\dot{R}}{A_0 R} = \frac{3}{t(1+\lambda k^2 r^2)^{1/4}(C - \sqrt{1-kr^2})}, \tag{16}$$

should also be negative as it describes a contracting body. Therefore, during the collapse from time $t = -\infty$ to $t = 0$, we must have $(1+\lambda k^2 r^2)^{1/4}(C - \sqrt{1-kr^2}) > 0$, implying $C > 1$. The upper bound on C can be obtained by constraining $\alpha^2 > 2\beta$ so as to ensure that n remains real.

The dynamical variables of matter density, the radial and transverse pressures, and the heat flux parameter associated with the collapsing stellar structure are obtained as:

$$8\pi\rho = \frac{k(\lambda+1)(3+\lambda kr^2)}{n^2t^2(1+\lambda kr^2)^2} + \frac{3}{t^2(1+\lambda k^2r^2)^{1/2}(C-\sqrt{1-kr^2})^2}, \tag{17}$$

$$8\pi p_r = 8\pi\rho - \frac{4}{t^2(1+\lambda k^2r^2)^{1/2}(C-\sqrt{1-kr^2})^2}, \tag{18}$$

$$8\pi p_t = 8\pi p_r + \frac{k^2r^2\left[2\lambda k(1+\lambda k)(1+2\lambda k) - \lambda^2k^2(1-kr^2)(4\lambda k+7)\right]}{4n^2t^2(1+\lambda k^2r^2)^3}, \quad \textit{and} \tag{19}$$

$$8\pi q = -\frac{r\sqrt{1-kr^2}\left[C\lambda k\sqrt{1-kr^2} + (2 - \lambda k + 3\lambda k^2r^2)\right]}{n^2t^3(1+\lambda kr^2)(C-\sqrt{1-kr^2})^2(1+\lambda k^2r^2)^{5/4}}. \tag{20}$$

It can be shown that ρ, p_r, $p_t > 0$ and ρ', p_r', $p_t' < 0$, if the conditions $k > 0$ and $\lambda \geq 0$ are satisfied simultaneously. Moreover, the weak energy condition $\rho > p_r$, pt is satisfied in this model.

The proper radius $[R(t)r]_\Sigma = ntr_\Sigma$ is infinite when collapse begins, positive at any later instant t and shrinks to zero at $t = 0$. The explicit expression for mass at any instant t within the boundary radius r_Σ is obtained as:

$$m(v) \stackrel{\Sigma}{=} m(r_\Sigma, t) = \frac{nr_\Sigma t}{2}\left[\frac{(1+\lambda)kr_\Sigma^2}{(1+\lambda kr_\Sigma^2)} + \frac{n^2 r_\Sigma^2}{\left(C - \sqrt{1-kr_\Sigma^2}\right)^2\left(1+\lambda k^2 r_\Sigma^2\right)^{1/2}}\right], \tag{21}$$

which shows that the collapse begins with an infinite mass and size of the configuration at $t \to -\infty$ and it evaporates completely as the epoch $t = 0$ approaches. The collapse does not have any impact on the mass to size ratio $(2m(r,t)/rR)_\Sigma$ as it is independent of time, and therefore, remains constant throughout the evolution. Consequently, no event horizon is formed during the collapse, since $g_{tt} = 1 - [2m(r,t)/(rR)]_\Sigma = 1 - constant$, always remains positive. As the singularity is reached at $t = 0$, $m(r,t) = 0$, which implies that the collapsing body radiates all its mass energy in the process and the collapse terminates into a naked singularity. The Ricci curvature, in this model, diverges as $1/t^2$, which suggests that it is a weak curvature singularity.

3. Conclusion

In this paper, a general framework to investigate the impacts of inhomogeneous nature of background spacetime and anisotropic stresses on a gravitationally collapsing system has been formulated by generalizing the Banerjee *et al* [5] model. The set up provides a mechanism to investigate the impact of inhomogeneity on collapse by comparing the evolution of an inhomogeneous distribution ($\lambda \neq 0$) of anisotropic matter with the evolution of a homogeneous distribution ($\lambda = 0$) of isotropic matter [7]. The parameter λ is the measure of departure from homogeneity in this construction. Though, in general, the physical parameters evolve differently for two different background spacetimes, an in-depth analysis shows that the gross features of the evolving system remain unaffected by the inhomogeneous perturbation of the background spacetime.

Acknowledgements

The author gratefully acknowledges the support from IUCAA, Pune, India, where a part of this work was carried out under its Visiting Associateship Programme. He is also thankful to Ramesh Tikekar for useful discussions.

References

[1] Joshi P S Joshi and Malafarina D 2011 *Int. J. Mod. Phys. D* **20** 2641
[2] Herrera L and Barreto W 2011 *Int J. Mod. Phys. D* **20** 1265
[3] Vaidya P C 1951 *Proc. Ind. Acad. Sci. A* **33** 264
[4] Santos N O 1985 *MNRAS* **216** 403
[5] Banerjee A, Choudhury S and Bhui B 1989 *Phys. Rev. D* **40** 670
[6] Maiti S R 1982 *Phys. Rev. D* **25** 2518
[7] Schäfer D and Goenner H F 2000 *Gen. Rel. Grav.* **42** 2119
[8] Vaidya P C and Tikekar R 1982 *J. A. & A.* **3** 325
[9] Tikekar R 1990 *J. Math. Phys.* **31** 2454
[10] Knutsen H 1988 *MNRAS* **232** 163
[11] Maharaj S D and Leach P G L 1996 *J. Math. Phys.* **37** 430
[12] Mukherjee S, Paul B C and Dadhich N K 1997 *Class. Quant. Grav.* **14** 3475

Full-analytic frequency domain gravitational wave forms from eccentric compact binaries to second post-Newtonian accuracy

M Tessmer and G Schäfer
Friedrich-Schiller-Universität, Max-Wien-Platz 1, 07743 Jena, Germany
E-mail: m.tessmer@uni-jena.de

Abstract. Full-analytic gravitational wave forms for inspiralling eccentric compact binaries of arbitary mass ratio have been provided in the frequency domain for the case of vanishing spins. Tail terms are not considered. In the given prescription, the semi-analytical property of recent descriptions, which demand inverting the higher order Kepler equation numerically, but keeping all other computations analytic is avoided.

1. Introduction

The dynamics of inspiralling compact binaries, consisting of neutron stars, black holes or a mixture of both, can be modelled accurately in the post-Newtonian (PN) approximation to general relativity. Roughly speaking, the PN approximation provides the equations of motion of a compact binary as corrections to the Newtonian equations of motion in powers of $(v/c)^2 \sim G\mathcal{M}/(c^2 R)$, where v, $\mathcal{M}$, and R are the characteristic orbital velocity, the total mass, and the typical orbital separation of the binary respectively. The PN accurate compact binary dynamics is required to construct temporally evolving gravitational wave forms associated with inspiralling compact binaries. Those wave froms are essential to make functioning gravitational wave astronomy with ground, and space based laser interferometric detectors, for which only the frequency domain of gravitational waves is crucial. We like to avoid numerical computations, because of limited computer power resources, so we have aimed to generate analytical frequency domain expressions (see also [1] for a parameter estimation concerning the second PN accurate case of decaying orbits, assuming small eccentricities, and [2] for eccentric orbits using the Newtonian equations of motion and a post-circular expansion). For the sake of simplicity, we have employed the tool of the spin-2 tensor spherical harmonics, as done in an older publication [3]. These harmonics are decomposed appropriately, where after several re-casts, the final expressions can be Fourier transformed analytically.

2. Essential ingredients

2.1. Multipole expansion of the radiation field far away

The far-zone radiation field for a source at the coordinate origin, and an observer at point $\mathbf{R}$ can be expressed in terms of symmetric trace-free (STF) Cartesian tensors with basis:

$$N_{\mathcal{A}_{l-2}} := N_{a_1} \cdots N_{a_{l-2}}, \tag{1}$$

where $\mathbf{N}:= \mathbf{R}/R$ is the direction of the line of sight (source – observer), given in [4, 5] as:

$$h^{TT}_{km}(\mathbf{R},t) = \frac{G}{c^4 R}\mathcal{P}_{kmij}(\mathbf{N})\sum_{l=2}^{\infty}\Bigg[\left(\frac{1}{c}\right)^{l-2}\left(\frac{4}{l!}\right)\overset{(l)}{\mathcal{I}}_{ij\mathcal{A}_{l-2}}\left(t-\frac{R}{c}\right)N_{\mathcal{A}_{l-2}}$$
$$+\left(\frac{1}{c}\right)^{l-1}\left(\frac{8l}{(l+1)!}\right)\epsilon_{pq(i}\overset{(l)}{\mathcal{J}}_{j)p\mathcal{A}_{l-2}}\left(t-\frac{R}{c}\right)N_{q\mathcal{A}_{l-2}}\Bigg]. \tag{2}$$

The projector $\mathcal{P}_{kmij}$ extracts the STF parts from the explicit sum of time derivatives of mass-type ($\mathcal{I}$), and current-type ($\mathcal{J}$) multipole moments. We have truncated this series beyond 2 *PN* order and take care of the *instantaneous point mass parts only.* Because the wave form depends on the direction of **N**, the frequency domain will also do so. Therefore, to deal with irreducible components, we have recast h^{TT}_{ij} into the basis of the spin-2 tensor spherical harmonics (T^{E2}_{ij} and T^{B2}_{ij}, and the scalar expansion coefficients I and S in the expression below have been processed further) as:

$$\begin{aligned}
h^{TT}_{ij} &= \frac{\eta}{c^4 R}\Bigg\{\sum_{m=-2}^{2}\overset{(2)}{I}{}^{2m}T^{E2,2m}_{ij} \\
&+ c^{-1}\left[\sum_{m=-2}^{2}\overset{(2)}{S}{}^{2m}T^{B2,2m}_{ij}+\sum_{m=-3}^{3}\overset{(3)}{I}{}^{3m}T^{E2,3m}_{ij}\right] \\
&+ c^{-2}\left[\sum_{m=-3}^{3}\overset{(3)}{S}{}^{3m}T^{B2,3m}_{ij}+\sum_{m=-4}^{4}\overset{(4)}{I}{}^{4m}T^{E2,4m}_{ij}\right] \\
&+ c^{-3}\left[\sum_{m=-4}^{4}\overset{(4)}{S}{}^{4m}T^{B2,4m}_{ij}+\sum_{m=-5}^{5}\overset{(5)}{I}{}^{5m}T^{E2,5m}_{ij}\right] \\
&+ c^{-4}\left[\sum_{m=-5}^{5}\overset{(5)}{S}{}^{5m}T^{B2,5m}_{ij}+\sum_{m=-6}^{6}\overset{(6)}{I}{}^{6m}T^{E2,6m}_{ij}\right]\Bigg\}.
\end{aligned} \tag{3}$$

The numbers in round brackets denote time derivatives, which are computed using the quasi-Keplerian parameterization, which we have introduced below. The conservative dynamics of the binary can be reduced to a one-body dynamics in a non-precessing plane. We can connect the radial separation r and the orbital phase ϕ to the elapsed time $(t - t_0)$ via the 2 *PN* Kepler equation:

$$r = a_r(1 - e_r\cos\mathcal{E})\ , \tag{4}$$

$$\frac{(\phi-\phi_0)}{\Phi/2\pi} = v + c^{-4}f_{4\phi}\sin 2v + c^{-4}g_{4\phi}\sin 3v\ , \tag{5}$$

$$\mathcal{M} := \mathcal{N}(t-t_0) = \mathcal{E} - e_t\sin\mathcal{E} + c^{-4}\Big[F_{v-\mathcal{E}}(v-\mathcal{E}) + F_v\sin v\Big]\ , \tag{6}$$

$$defining \qquad v := 2\arctan\left[\sqrt{\frac{1+e_\phi}{1-e_\phi}}\tan\frac{\mathcal{E}}{2}\right]\ , \tag{7}$$

with $\mathcal{N} := 2\pi/P$ as the mean motion ($\mathcal{P}$ being the radial period), the functions a_r (semi-major axis), $e_{\{r,\ t,\ \phi\}}$ (radial, time and phase eccentricity), $f_{4\phi}$ and $g_{4\phi}$ are the PN accurate functions of the binding energy $|E|$, and the orbital angular momentum L (see e.g., [6]). Using the above quasi-Keplerian parameterization, and PN accurate expressions for the acceleration in terms of the separation and velocity, we have been able to express the higher order time derivatives as functions of r and ϕ.

3. Expansion procedure

Our way to full-analytical Fourier-domain wave forms can be compressed symbolically as in the schedule below:

(i) Write down time derivative of I^{lm} and S^{lm} as functions of the eccentric anomaly $\mathcal{E}$ and collect for *even* and *odd* functions separately.

(ii) Convert *even terms in* $\mathcal{E}$ to $\sum_{m\,\geq\,0} c_m \cos m\mathcal{E}$, and *odd terms* to $\sum_{m\,>\,0} s_m \sin m\mathcal{E}$.

(iii) Compute coefficients of $\cos m\mathcal{E} = \sum_{j\,\geq\,0} \gamma_j \cos j\mathcal{M}$, and $\sin m\mathcal{E} = \sum_{j\,>\,0} \sigma_j \sin j\mathcal{M}$.

(iv) Perform the Fourier transformation of sines and cosines, which is straightforward.

(v) For the inclusion of radiation reaction, the method of the stationary phase will be applied for the appearing integrals.

4. Technicalities

The aim is to reduce the treatment of the scalar coefficients I and S to that of their constituents. We shall first extract the very slow periastron precession from the phase, and convert it to an overall factor. Defining $A(\mathcal{E}) := 1 - e_t \cos(\mathcal{E})$, the structure of the components reads:

$$I^{lm} = e^{-im\phi} f_{lm}(A(\mathcal{E}),\ \sin(\mathcal{E})), \tag{8}$$

where f_{lm} being PN accurate functions, and similarly for S. Extraction of the periastron shift yields:

$$I^{lm} = e^{-im\mathcal{K}\mathcal{M}} \left(f^*_{lm}(A(\mathcal{E})) + \sin(\mathcal{E}) \tilde{f}^*_{lm}(A(\varepsilon)) \right) , \tag{9}$$

where f^* and $\tilde{f}^*$ are some polynomials in $A(\mathcal{E})$. An infinite Taylor series in e_t, and a successive sorting for trigonometric functions to positive multiples of $\mathcal{E}$ gives:

$$A^n(\mathcal{E}) = \sum_{j=1}^{\infty} A_j^{(n)} \cos(j\mathcal{E}), \quad \textit{and} \quad \frac{\sin(\mathcal{E})}{A^n(\mathcal{E})} = \sum_{j=0}^{\infty} S_j^{(n)} \sin(j\mathcal{E}) \ . \tag{10}$$

These trigonometric functions of $j\mathcal{E}$ can be further expanded as:

$$\sin(n\mathcal{E}) = \sum_{j=1}^{\infty} \mathcal{S}_j^{(n)} \sin(j\mathcal{M}) \ , \quad \textit{and} \quad \cos(n\mathcal{E}) = \sum_{j=0}^{\infty} \mathcal{C}_j^{(n)} \sin(j\mathcal{M}) \ . \tag{11}$$

The coefficients $\mathcal{S}_j^{(n)}$ and $\mathcal{C}_j^{(n)}$ can be obtained by computing Fourier-Bessel integrals, like $\mathcal{S}_j^{(n)} = \int \sin(n\mathcal{E}(\mathcal{M})) sin(j\mathcal{M}) d\mathcal{M}$, inserting the 2 PN Kepler equation and expanding each term up to c^{-4}.

5. Results for the conservative case

Collecting all the intermediate results, we have obtained a time-domain representation for the non-tail part of the gravitational wave, having sines and cosines of multiples of the mean anomaly. The frequency domain in the conservative case, from here onwards, is trivial to obtain. It is to be emphasized that everything is given in the positive domain of the spectrum [7]:

$$\overset{(a)}{I}{}^{am} = \mathcal{I}_{am} e^{-mi\mathcal{K}\mathcal{M}} \sum_{j=0}^{\infty} \left\{ \sin(j\mathcal{M}) \overset{(a)}{I}{}^{am}_{Sj} + \cos(j\mathcal{M}) \overset{(a)}{I}{}^{am}_{Cj} \right\} , \quad \textit{and} \tag{12}$$

$$\overset{(b)}{S}{}^{bm} = \mathcal{S}_{bm} e^{-mi\mathcal{K}\mathcal{M}} \sum_{j=0}^{\infty} \left\{ \sin(j\mathcal{M}) \overset{(b)}{S}{}^{bm}_{Sj} + \cos(j\mathcal{M}) \overset{(b)}{S}{}^{bm}_{Cj} \right\} . \tag{13}$$

6. Including radiation reaction

Binary systems in the early part of their evolution are characterized by $|E|$ and L as slowly varying functions of time. In a stage where the variation is estimated to be small enough, the stationary phase approximation can be applied to compute the frequency domain. We have used the Peters-Matthews equations for the orbital decay of $|E|$ and e_t [8]. Under these conditions, it turns out that the Fourier domain of any term appearing in Eqs. (12) and (13), shorthand notated by $\tilde{h}(f)$, omitting the multipole indices, takes the prototype form:

$$\tilde{h}(f) = \int \mathcal{B}(t) e^{2\pi i f t - \phi(t)} dt. \tag{14}$$

Here, $\mathcal{B}(t)$ is an amplitude, slowly varying in time. Having found the stationary point in $\dot{\Phi} = 0$, where $\Phi(t) := 2\pi f t - \phi(t)$, the quantity $\tilde{h}$, consisting of complex sine and cosine contributions S and C, which are to be read off from Eq. (14), and finally written as:

$$\overset{(n)}{\tilde{h}}(f)_{n,m} = \frac{1}{c^4 R} \sum_{j=0}^{\infty} \frac{1}{2} \left(i S^*_{m(j>0)} + \Big(1 + \theta(m)\delta_{0j}\Big) C^*_{mj} \right) \times \left[\frac{e^{i\left(\Phi_{mj}(t^*_{mj}) - \pi/4\right)}}{\sqrt{\ddot{\phi}_{mj}(t^*_{mj})}} \right] e^{-im\phi_0} , \tag{15}$$

$$\ddot{\phi}_{mj}(t^*_{mj}) \equiv \mathcal{N} \, . \, (j + m\mathcal{K})|_{t=t^*_{mj}} \, . \tag{16}$$

The function $\theta(m)$ is "zero" if $m = 0$ and "one" if $m > 0$. A quantity with a star has to be evaluated at the stationary point, which in turn, depends on the number m, and the position j in the summation.

7. Outlook

For a complete picture of the frequency domain wave forms, tail effects in the far-zone amplitude are necessary to be included. The inclusion of spin interactions and higher-order radiation reaction effects are demanded too.

Acknowledgements

This project is funded by the DFG (Deutsche Forschungsgemeinschaft) through SFB TR - 7, and by the DLR (Deutsches Zentrum für Luft - und Raumfahrt). MT wishes to thank Johannes Hartung and Steven Hergt, for many useful discussions.

References

[1] Królak A, Kokkotas K D and Schäfer G 1995 *Phys. Rev. D* **52** 2089
[2] Yunes N, Arun K G, Berti E and Will C M 2009 *Phys. Rev. D* **80** 084001
[3] Junker W and Schäfer G 1992 *MNRAS* **254** 146
[4] Thorne K S 1980 *Rev. Mod. Phys.* **52** 299
[5] Blanchet L, Damour T and Iyer B R 1995 *Phys. Rev. D* **51** 5360
[6] Memmesheimer R M, Gopakumar A and Schäfer G 2004 *Phys. Rev. D* **70** 104011
[7] Tessmer M and Schäfer G 2011 *Ann. Phys. (Berlin)* **523** 813
[8] Peters P C and Matthews J 1963 *Phys. Rev.* **131** 435

Black hole binary OJ287 as a testing platform for general relativity

M J Valtonen[1,2], **A Gopakumar**[3], **S Mikkola**[4], **K Wiik**[4] **and H J Lehto**[4]

[1] Finnish Centre for Astronomy with ESO, University of Turku, 21500 Piikkiö, Finland
[2] Helsinki Institute of Physics, FIN-00014 University of Helsinki, Finland
[3] Tata Institute of Fundamental Research, Homi Bhabha Road, Colaba, Mumbai 400005, India
[4] Tuorla Observatory, Department of Physics and Astronomy, University of Turku, 21500 Piikkiö, Finland

E-mail: mvaltonen2001@yahoo.com

Abstract. The blazar OJ287 is the most promising (and the only) case for an extragalactic binary black hole system inspiralling under the action of gravitational radiation reaction. At present, though it is not possible to directly observe the binary components, it is possible to observe the jet emanating from the primary black hole. We have argued that the orbital motion of the secondary black hole is reflected in the wobble of the jet, and demonstrated that the wobble is orbital position dependent. The erratic wobble of the jet is analyzed by taking into account the binary nature of the system, and we have found that the erratic component of jet wobble is very small.

1. Introduction

OJ287 is a very special case among all quasars: Its optical light curve of 120 yrs length shows a double periodicity of 60 yrs and 12 yrs [1]. Another remarkable feature of this quasar is that while its radiation is dominated by synchrotron radiation at most times, during the major outbursts, its optical to UV spectrum is bremsstrahlung at around 3×10^5 K [2]. These main outbursts occur in unpolarized light, as is expected of bremsstrahlung [3]. Further, it is the only known quasar showing bremsstrahlung outburst peaks.

Most of these facts were not known in 1995 when a detailed model of this source was constructed [4, 5]. The model proposed that the major outbursts, which occur at roughly 12 yrs intervals are related to a 12 yrs period in a black hole binary system. The detailed mechanism for generating the outbursts, which occur in pairs separated by 1 to 2 years, was presumed to be impacts on the accretion disk of the primary by the secondary. Hot bubbles of plasma pulled off the accretion disk then generate the bremsstrahlung bursts when the bubbles become optically thin.

At present, the model has been verified in many ways [6] and subsequently, the model is invoked to constrain the spin of the primary black hole and to explore the possibility of testing black hole no-hair theorems [7, 8]. It should be noted that the binary black hole model implies typical orbital velocity $v \sim 0.1\,c$, and it explains why higher order post-Newtonian corrections are required to explain observed major outbursts and to make predictions about future major outbursts. In contrast, binary pulsars have $v/c \sim 10^{-3}$, which is roughly a factor 10 higher than orbital velocity of the Earth. However, it is desirable to probe further observational implications arising from the presence of binary components in the model, which is pursued below.

2. Dynamical modelling of the wobbling radio jet in OJ287

Supermassive black holes in active galactic nuclei typically have bi-directional jets consisting of collimated beams of matter that are expelled from the innermost regions of their accretion disks. Further, highly variable quasars like OJ287 are expected to contain relativistic jets that are beamed towards the observers. The jet from OJ287, usually observed in centimetre wavelengths, has exhibited prominent variations in its position angle (PA) in the sky, ever since the observations began in early 1980s. Our aim is to infer the presence of the orbital motion by following the temporal variations in the PA measurements of OJ287.

In Figure 1, we have shown the PA observations of the jet as a function of time. The data were collected by T Savolainen and have been reported in [9]. Also, we have plotted the expected variation in the PA of the jet if the jet is connected to the primary accretion disk and follows the wobble of the disk in a binary system [9]. Here, the wobble is modelled by a doubly periodic sinusoidal function of time t (in yrs), given by:

$$\Omega_1 = 1.5 \times \sin(2\pi \times (t - 1934.8 - d)/120), \quad (1)$$

$$\Omega_2 = 0.25 \times \cos(2\pi \times (t - 1934.8 - d)/11.2), \quad \textit{and} \quad (2)$$

$$\Delta\Omega = \Omega_1 + \Omega_2 - \Omega_0. \quad (3)$$

The angle Ω, the nodal angle of the accretion disk, is given in degrees. We have assumed that the inclination i of the disk is constant, since its variation has been found to be much smaller than the variation of Ω in numerical simulations. The actual inclination is close to 90° relative to the binary plane, and in Table 1, we have used the quantity which is actually $i - \pi/2$. The mean viewing angle of the jet is given by (Ω_0, i_0) which specifies the direction of the observer. The jet PA (JPA, in degrees) is then at any moment of time, given by:

$$\Delta\phi = -70 + \text{atan}(\tan\Delta\Omega/\sin i_0). \quad (4)$$

The mean JPA, $\Delta\phi = -70°$ is obtained by fitting to the data. In actual numerical simulations, the maxima of the $\Delta\Omega$ are not quite sinusoidal but more highly peaked.

There are three free parameters in fitting this function to the data: The two components of the mean viewing angle (Ω_0, i_0) of the jet, and the time delay d (in yrs) of the response of the jet to the variations in the disk. The results of fitting are shown in Table 1. In column 1, we have indicated the data set to be compared with: TS = cm-wave data, IA = 7 mm-wave data [10] (2 yrs averages) and CV = optical polarization data in one yr bins, calculated by C Villforth and reported in [9]. Columns 2 and 3 give the parameters i_0 and Ω_0 respectively, and column 4 gives the delay time d in years. The last column gives the rms error of the fit. These are the best fits found for the radio jet data as well as for the polarization data. In the latter case, it is assumed that the electric vector arising from radiating electrons is parallel to the jet (i.e., the magnetic field is perpendicular to the jet, presumably due to a compression in the radiating knot). The opposite case, a magnetic field parallel to the jet, has been reported in [9], and it gives a considerably worse fit to the present doubly sinusoidal model.

In Figure 2, we have shown the average PA from the 7 mm observations given in [10]. The observations at this wavelength are available only since mid 1990s, but they have a better resolution than the cm-wave maps. Thus, they refer to the jet orientation closer to the core than the cm-wave maps. An interesting feature is the major jump in the PA of the jet around 2004. The line in the figure refers to a model, where the jet responds to the changing disk, but now with a different viewing angle and time delay than in the case of the cm-wave fit.

Table 1. Model fits.

Data	i_0	Ω_0	d	rms
TS	-1.08	1.91	16.9	7.5
TS	-0.60	1.96	27.7	7.7
TS	0.99	-1.92	78.5	7.2
CV	1.36	0.48	0.05	23.3
CV	0.90	1.07	11.2	22.2
IA	0.35	0.15	11.2	16.9
IA	0.27	0.90	22.1	14.4

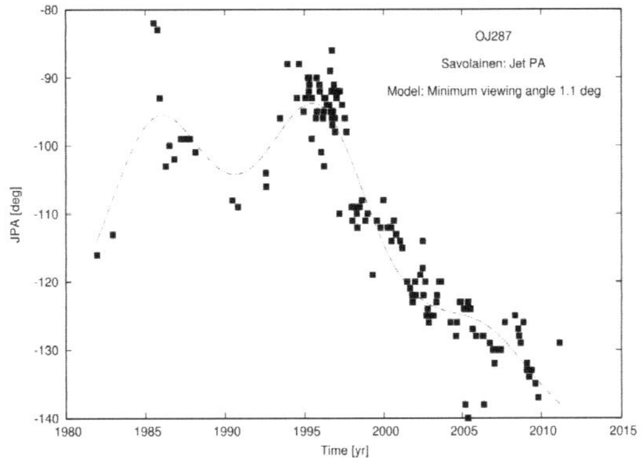

Figure 1. Observations of the jet PA in OJ287 at cm wavelengths. The line represents the model described in the text.

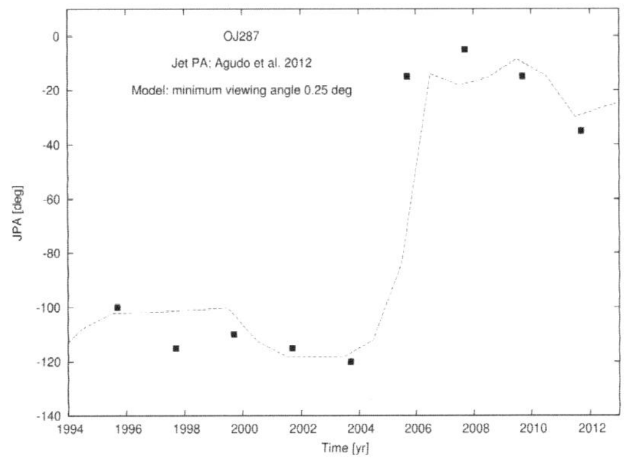

Figure 2. Observations of the JPA in OJ287 at 7 mm wavelength binned at 2 yrs intervals. The line refers to the model calculated in [9].

3. Interpretation

The variations in the orientation of the accretion disk, due to the binary influence, are transmitted to the central component in about 10 yrs [9]. We may assume that this variation is communicated to the jet, starting from the near jet, and proceeding outwards. We may now calculate the speed at which this happens; we call it *kink speed*.

In [9], it was found that the optical polarization angle responds to the disk changes with a 13 yrs time delay. The present sinusoidal model suggests a 11 yrs delay, which is not significantly different. The source of optical emission is presumably in the jet, but closer to the central black hole than the resolved radio jet. The best fit (Figure 2) to the 7 mm data is obtained if the time delay is 22 yrs, i.e., the orientation of the 7 mm jet lags behind the variations in the central component by about 11 yrs. The best solution for the cm-wave jet is provided by the 79 yrs time delay, i.e., its response is delayed by $\Delta t_{kink} \sim$ 68 years with respect to the optical core (Figure 1). A slightly worse solution is obtained with $\Delta t_{kink} \sim 6$ years.

However, since we are looking at a relativistic jet almost head-on (the viewing angle is $\sim 1°$ to $2°$), the observed time intervals Δt are compressed depending on the speed of the outflow. The knots in the jet presumably take part in a flow with close to the speed of light (Lorentz factor $\Gamma_{knot} \sim 20$ [11]). They take about a year to come from the central component to the observed cm-wave jet. The kink in the jet proceeds more slowly, and reaches the 7 mm-wave jet before the cm-wave jet. Thus, typically, the two parts of the jet are misaligned. Our fits show that the mis-alignment is about 1 degree.

Let us estimate the kink speed β_{kink} (relative to the speed of light c). The apparent speed of the kink is given by:

$$\beta_{akink} = \beta_{kink} \sin\theta/(1 - \beta_{kink}\cos\theta), \tag{5}$$

while for the knot, a corresponding equation is valid.

Taking the ratio:

$$\beta_{aknot}/\beta_{akink} = \Delta t_{kink}/\Delta t_{knot} = \beta_{knot}(1 - \beta_{kink}\cos\theta))/\beta_{kink}(1 - \beta_{knot}\cos\theta), \tag{6}$$

we have found, using the observed $\Delta t_{kink}/\Delta t_{knot} \sim 50$ and $\theta \sim 1°$, the Lorentz factor Γ_{kink}. From

$$\beta^2_{kink} = 1 - 1/\Gamma^2_{kink}, \tag{7}$$

we get $\Gamma_{kink} \sim 3$. For $\Delta t_{kink}/\Delta t_{knot} \sim 5$ we have found $\Gamma_{kink} \sim 8$. Thus, the kink speed is lower than the knot speed, perhaps because the jet has a fast spine, which carries the knots and a slower sheath responsible for the jet confinement.

References

[1] Valtonen M J, Lehto H J, Sillanpää A, Nilsson K, Mikkola S, Hudec R, Basta M, Teräsranta H, Haque S and Rampadarath H 2006 *Ap. J.* **646** 36
[2] Valtonen M and Ciprini S 2012 *Mem. Soc. Astron. Italia* **81** N.1
[3] Valtonen M J, Lehto H J, Nilsson K, Heidt J, Takalo L O, Sillanpää A, Villforth C, Kidger M, Poyner G, Pursimo T *et al* 2008 *Nature* **452** 851
[4] Lehto H J and Valtonen M J 1996 *Ap. J.* **460** 207
[5] Sundelius B, Wahde M, Lehto H J and Valtonen M J 1997 *Ap. J.* **484** 180
[6] Valtonen M and Sillanpää 2011 *Acta Polytech.* **51** 76
[7] Valtonen M J, Mikkola S, Merritt D, Gopakumar A, Lehto H J, Hyvönen T, Rampadarath H, Saunders R, Basta M and Hudec R 2010 *Ap. J.* **709** 725
[8] Valtonen M J, Mikkola S, Lehto H J, Gopakumar A, Hudec R and Polednikova J 2011 *Ap. J.* **742** 22
[9] Valtonen M J and Wiik K 2012 *MNRAS* in press
[10] Agudo I, Marscher A P, Jorstad S G, Gomez J L, Perucho M, Piner B G, Rioja M and Dodson R 2012 *Ap. J.* **747** 63
[11] Jorstad S G, Marscher A P, Lister M L, Stirling A M, Cawthorne T V, Gear W K, Gomez J L, Stevens J A, Smith P S, Forster J R and Robson E I 2005 *Astron. J.* **130** 1418

Journal of Physics: Conference Series **484** (2014) 012026 doi:10.1088/1742-6596/484/1/012026

Effect of calibration errors on Bayesian parameter estimation for gravitational wave signals from inspiral binary systems in the advanced detectors era: Further investigations

S Vitale[1], W Del Pozzo[1], T G F Li[1], C Van Den Broeck[1], B Aylott[2] and J Veitch[1]

[1] Nikhef, Science Park 105, 1098 XG Amsterdam, The Netherlands
[2]School of Physics and Astronomy, University of Birmingham, Edgbaston, Birmingham B15 2TT, UK

E-mail: salvatore.vitale@ligo.org

Abstract. By 2015, the advanced versions of the gravitational wave detectors Virgo and LIGO will be online. They will collect data in coincidence with enough sensitivity to potentially deliver multiple detections of gravitational waves from inspirals of compact-object binaries. In a previous work, we have studied the effects introduced in the estimation of the physical parameters of the source by uncertainties in the calibration of the interferometers. Our bias estimator for parameter errors introduced by calibration uncertainties consisted of two terms: A genuine bias due to the calibration errors, and a contribution coming from the limited number of samples used to explore the parameter space. In this article, we have focused on this second term, and we have shown how it is smaller than the former (about 10 times smaller), and how it decreases as the signal-to-noise ratio increases.

1. Introduction

In a recent paper [1], we have shown that the bias introduced by realistic calibration errors (CEs) in the parameter estimation process for gravitational wave (GW) signals are usually smaller than the uncertainty due to the noise in the instruments. As parameters are estimated using numerical algorithms, in our case, the Nested Sampling algorithm [2] as implemented and described in [3], the biases we have found were really made up of two contributions: A real bias, driven by the size and shape of the calibration errors, and an error coming from the finite sampling of the parameter space. In the rest of this work, we have referred to the first term as *calibration bias* and to the second as *sampling errors*. These are not related to CEs, and in fact, have been presented in any results obtained using Monte Carlo or Nested Sampling based algorithms. In this work, we have quantified the contribution of this term and we have shown that (i) it is much smaller than the calibration bias, (ii) it should not affect in any important way the parameter estimation (PE) process, and (iii) it is smaller for loud signals.

This article is organized as follows: In Section 2, we have described the method used to estimate the bias, in Section 3, we have reported the main results and discussed them, and in Section 4, we have summarized and concluded. We refer to [1] for the main methodology and results on calibration induced PE biases, to [2, 3] for the Nested Sampling methodology and implementation, and to [6] for a description of the waveform and the parameters on which it depends.

2. Method

In [1], we have estimated the bias introduced by mock, but realistic, calibration errors by running a Nested Sampling algorithm on three catalogues of simulated GW signals. The analysis was performed twice: A first time on the "exact" signals, and then on signals with added calibration errors, while keeping fixed all the relevant parameters. The runs were set up to use two chains to explore the parameter space in parallel to make optimal use of computer clustering, which were then combined to provide a single estimate for the posterior distributions of the parameters and the Bayes factor. Because of the randomness in the exploration of the parameter space, and the finite sample size, running the code on the same stretch of data but starting the exploration from a different point would result in finding slightly different posterior distributions. As mentioned before, this effect is a general one, and is not related to the presence of calibration errors.

Note, however, that in our previous study, calibration errors were the cause for this randomness to arise. In fact, adding calibration errors results in a slight change of the profile of the likelihood function, which is the real driving engine of the Nested Sampling algorithm. As a consequence, even if the two chains in the runs with CE started from the same point of the parameter space as the two chain in the run without CE (which is, indeed, the case, as we had kept *all* parameters and settings to be the same while adding calibration errors, and that includes the seed for the generation of the chain) they would ultimately follow different paths in their exploration of the parameter space. This effect has to be added to the genuine bias (the calibration bias), which would be there even in the case of perfect sampling with an infinite number of iterations, and its magnitude will depend on the particular setting of the run (number of live points and MCMC iterations) and the GW event. Questions which were left unanswered in [1] are: What can be told about the relative weight of these two biases? How does the sampling errors depend on the signal-to-noise ratio (SNR) of the injected signals? In this article, we have answered these questions by analysing in greater details the sampling errors, their typical magnitude and shape.

In particular (a) Take the set of binary systems composed of one neutron star and one black hole (BHNS) used in [1], limiting ourselves to the signals which had passed the SNR cut described therein, (b) Analyse each signal using 12 parallel chains (each run used 1700 live points and 210 MCMC iterations, see [3] for details). These chains are then combined into pairs, always using different chains, to yield 6 independent estimations of the source parameters, and (c) Calculate the differences in the estimated parameters one obtains by using a pair of chains instead of another one. To be more precise, for each event, we have calculated the quantity (somehow similar to the one used in [1]) as:

$$\Sigma_{ij}^{\alpha} \equiv \frac{\overline{\alpha}_i - \overline{\alpha}_j}{\sqrt{\frac{1}{2}\left[\sigma_n(\alpha_i)^2 + \sigma_n(\alpha_j)^2\right]}}, \qquad \text{with } 6 \geq i > j \geq 1 \tag{1}$$

where $\overline{\alpha}_k$ is the median of the parameter α as calculated by the k-th run ($k = 1, 2, .., 6$) and $\sigma_n(\alpha_k)$ is the corresponding standard deviation. The advantage of normalizing by a symmetric combination of the standard deviations is that Σ will be a dimensionless quantity of the order of 1 or less (while the various parameters on which the waveform depends have very different ranges of variation as well as different units). Nevertheless, we have found occasionally useful to work with the unweighted bias, for which we have used the symbol $\Delta_{ij}^{\alpha} \equiv \overline{\alpha}_i - \overline{\alpha}_j$.

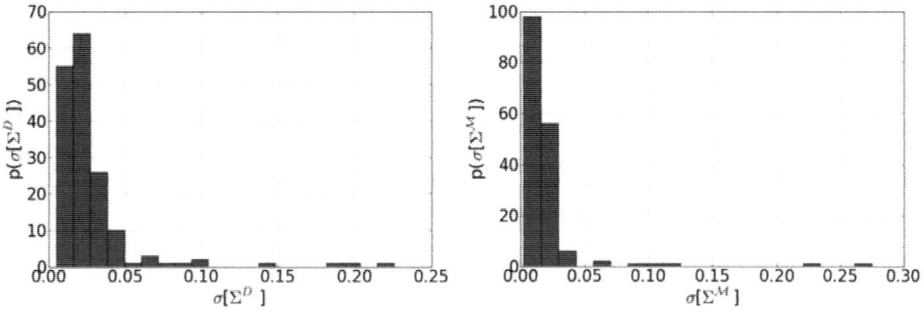

Figure 1. *Left*: The distribution of the standard deviation of the weighted bias among the signals in the catalogue for the distance estimation. *Right*: A similar plot for the chirp mass estimation

3. Results

For each event and each parameter, one can build a distribution of Σ^{α}_{ij}, with $6 \geq i > j \geq 1$. Its standard deviation can give an idea of the typical sampling errors one might get and we have labelled it $\sigma[\Sigma]$. One can then show how the width of the sampling errors distributions varies by building an histogram of $\sigma[\Sigma]$ for all the signals in the catalogue. This is shown in Figure 1 for the chirp mass and the distance (see [6] for a description of the waveform and parameters).

The results for all the parameters are summarized in Table 1, where we have given the mean and standard deviation of $\sigma[\Sigma]$ among the 167 events, as well as the 5th and 95th percentiles (we have left out the statistics for the polarization and the coalescence phase as those are, in practice, not estimable with Advanced LIGO and Virgo). In the last column, we have reported the results we had found in [1] for the standard deviation of the weighted bias due to the joint effect of calibration bias and sampling errors.

	$\overline{\sigma[\Sigma]}$	$\Delta\{\sigma[\Sigma]\}$	5th	95th	$\sigma[\Sigma]$ [CB+SE]
$\mathcal{M}$	$2.02{\cdot}10^{-2}$	$2.96{\cdot}10^{-2}$	$7.02{\cdot}10^{-3}$	$3.36{\cdot}10^{-2}$	$1.02{\cdot}10^{-1}$
η	$1.99{\cdot}10^{-2}$	$2.99{\cdot}10^{-2}$	$6.71{\cdot}10^{-3}$	$3.67{\cdot}10^{-2}$	$1.28{\cdot}10^{-1}$
RA	$5.84{\cdot}10^{-2}$	$2.22{\cdot}10^{-1}$	$2.15{\cdot}10^{-3}$	$5.32{\cdot}10^{-2}$	$3.87{\cdot}10^{-1}$
dec.	$5.54{\cdot}10^{-2}$	$1.88{\cdot}10^{-1}$	$4.60{\cdot}10^{-3}$	$1.11{\cdot}10^{-1}$	$4.49{\cdot}10^{-1}$
t_0	$5.68{\cdot}10^{-2}$	$2.07{\cdot}10^{-1}$	$4.90{\cdot}10^{-3}$	$5.74{\cdot}10^{-2}$	$4.05{\cdot}10^{-1}$
D	$2.75{\cdot}10^{-2}$	$2.98{\cdot}10^{-2}$	$8.81{\cdot}10^{-3}$	$6.88{\cdot}10^{-2}$	$2.80{\cdot}10^{-1}$
ι	$6.03{\cdot}10^{-2}$	$1.40{\cdot}10^{-1}$	$8.99{\cdot}10^{-3}$	$2.98{\cdot}10^{-1}$	$3.75{\cdot}10^{-1}$

Table 1. The mean $\overline{\sigma[\Sigma]}$, standard deviation $\Delta\{\sigma[\Sigma]\}$, 5th and 95th percentiles of the standard deviation of Σ for all the parameters. The last column shows the value of the standard deviation that was found in [1], for the bias due to the joint effect of calibration errors and sampling errors. These numbers take into account all 167 events.

On average, we can say that using a different pair of chains to explore the parameter space has little effect on the estimated parameters. From Table 1, we have seen, in fact, that the averaged $\sigma[\Sigma]$ is smaller than 6% of the noise error for all the parameters (and only $\sim 2\%$ for the intrinsic parameters). Moreover, the distributions are quite compact: For the declination and inclination angle, 90% of the times the standard deviation is contained in a region corresponding, respectively to $\sim 11\%$ and 30% of the noise error; for the other parameters, this region is even narrower (less than 7%). Moreover, comparing the first and last columns, we can say that the

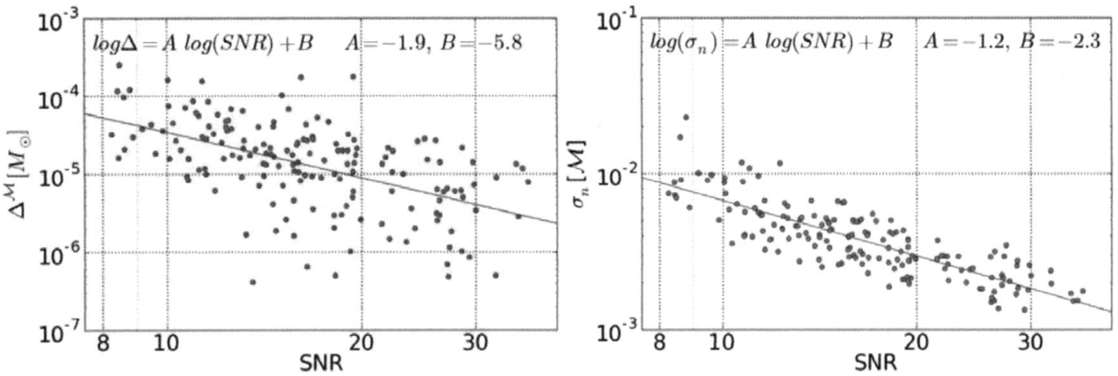

Figure 2. *Left*: The sampling errors in the estimation of the median of the chirp mass for all the signals in our catalogue together with a best fit. *Right*: The width of the posterior for the chirp mass against the SNR together with a best fit (see the text for more details).

sampling errors had a small weight in the biases we have found in [1]. Generally speaking, the results of Table 1 are good as they confirm that the results found by parameter estimation codes are robust against the particular pattern followed to explore the parameter space, at least with the set up we have used.

We have briefly investigated the dependence of the sampling errors on the SNR of the simulated signals. In the left panel of Figure 2, we have shown for each event the mean of the bias $\Delta^{\mathcal{M}} \equiv \frac{1}{15}\sum_{6\geq i>j\geq 1}\Delta^{\mathcal{M}}_{ij}$ among the 15 combined runs plotted against the SNR. If the samples used to build the posterior distributions are independent, we expect the errors to go like $\sigma_n/\sqrt{N}$, where σ_n is the width of the posterior distribution (due to the noise), and N is the number of samples in the chain. From Fisher Information studies [4, 5, 6], we expect that at least for medium-high SNRs (≥ 15), $\sigma_n \propto 1/\mathrm{SNR}$ so that the bias we measure should also depend on the inverse of SNR. The fit in the left panel of Figure 2 (blue line), shows, instead, that the bias dependence on the SNR is more close to an inverse quadratic relationship, and that there is a non-negligible scattering. On the other hand, the noise standard deviation $\sigma_n[\mathcal{M}] \equiv \frac{1}{15}\sum_{6\geq i>j\geq 1}\sqrt{\frac{1}{2}\left[\sigma_n(\mathcal{M}_i)^2 + \sigma_n(\mathcal{M}_j)^2\right]}$ for the chirp mass is quite close to an inverse relationship with the SNR, and the points are close to the best fit line. The reasons why the bias seems not to follow the expected distribution, as well as a check on the dependence of the Nested Sampling algorithm on the number of samples will be the subject of a forthcoming work.

4. Conclusions

In this work, we have investigated the typical contributions of statistical fluctuations due to limited posterior sample size to the fluctuation of the mean estimator of posterior parameter values, and we have compared it with the bias introduced by calibration errors, which we had quantified in [1]. We have built a catalogue of 167 events, injected into simulated Advanced LIGO/Virgo noise, and we have run the parameter estimation code using 12 parallel chains to explore the parameter space. For each event, the chains were combined into 6 independent pairs, and the results delivered by each pair were confronted with those of the others. We have found that the code is robust against the statistical fluctuations arising from using a pair of chain than another, the typical shifts in the estimated parameters being a tiny fraction of the random error due to the noise in the instruments: For all parameters but inclination and declination, 90% of the signals had a spread in the estimation smaller than 4% of the noise random error. We also have verified that these errors decrease as the SNR of the event increases, even though they do not seem to follow the expected inverse relationship.

References

[1] Vitale S *et al* 2012 *Phys. Rev. D* **85** 064034
[2] Skilling J 2004 in *AIP Conf. Proc.: 24th International Workshop on Bayesian Inference and Maximum Entropy Methods in Science and Engineering* **735** 395
[3] Veitch J and Vecchio A 2010 *Phys. Rev. D* **81** (6) 062003 [gr-qc/0911.3820]
[4] Zanolin M, Vitale S and Makris N 2010 *Phys. Rev. D* **81** 124048
[5] Vitale S and Zanolin M 2010 *Phys. Rev. D* **82** 124065
[6] Vitale S and Zanolin M 2011 *Phys. Rev. D* **84** 104020

Treatment of calibration uncertainty in multi-baseline cross-correlation searches for gravitational waves

J T Whelan[1], **E L Robinson**[2], **J D Romano**[3] **and E H Thrane**[4]
[1]Center for Computational Relativity and Gravitation, School of Mathematical Sciences, Rochester Institute of Technology, 85 Lomb Memorial Drive, Rochester NY 14623, USA
[2]Max-Planck-Institut für Gravitationsphysik (Albert-Einstein-Institut), D-14476 Potsdam, Germany
[3]Center for Gravitational Wave Astronomy, Department of Physics and Astronomy, The University of Texas at Brownsville, 80 Fort Brown, Brownsville TX 78520, USA
[4]Department of Physics, University of Minnesota, Minneapolis, MN, USA

E-mail: `john.whelan@astro.rit.edu`

Abstract. Uncertainty in the calibration of gravitational wave (GW) detector data leads to systematic errors, which must be accounted for in setting limits on the strength of GW signals. When cross-correlation measurements are made using data from a pair of instruments, as in searches for a stochastic GW background, the calibration uncertainties of the individual instruments can be combined into an uncertainty associated with the pair. With the advent of multi-baseline GW observation (e.g., networks consisting of multiple detectors such as the LIGO observatories and Virgo), a more sophisticated treatment is called for. We have described how the correlations between calibration factors associated with different pairs can be taken into account by marginalizing over the uncertainty associated with each instrument.

1. Calibration uncertainty with one baseline

Consider an experiment to measure a physical quantity μ (e.g., the stochastic GW background strength $\Omega_{\rm gw}(f)$ [1, 2]). An optimal combination x of cross-correlation measurements provides a point estimate of μ with error bar σ. Given a likelihood function $p(x|\mu)$ and a prior $p(\mu)$, one can use Bayes's theorem to construct the posterior:

$$p(\mu|x) = \frac{p(x|\mu)p(\mu)}{p(x)} = \frac{p(x|\mu)p(\mu)}{\int d\mu\, p(x|\mu)\, p(\mu)}\,. \tag{1}$$

Due to calibration uncertainties in each of the instruments that make up the baseline for the cross-correlation, x is an estimator not of μ, but of $\lambda\mu$, where λ is an unknown calibration factor (for the baseline) described by an uncertainty ε. Thus, the likelihood function also depends on the calibration factor λ and given by:

$$p(x|\mu,\lambda) = \frac{1}{\sigma\sqrt{2\pi}} \exp\left[-\frac{(x-\lambda\mu)^2}{2\sigma^2}\right], \tag{2}$$

so the posterior given in Eq. (1) is now constructed from the *marginalized* likelihood:

$$p(x|\mu) = \int d\lambda\, p(x|\mu,\lambda)p(\lambda). \tag{3}$$

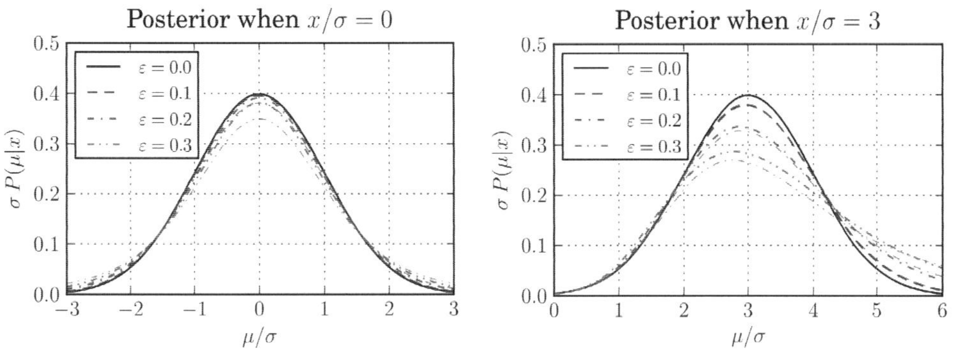

Figure 1. Effects of marginalizing analytically over a single calibration factor for two different values of the measured cross-correlation, $x = 0$ and $x = 3\sigma$. The thick line is a numerical marginalization with a log-normal prior on λ, and the thin line is an analytic marginalization with a Gaussian prior on λ.

If we assume a Gaussian distribution for λ as:

$$p(\lambda) = \frac{1}{\varepsilon\sqrt{2\pi}} \exp\left[-\frac{(\lambda-1)^2}{2\varepsilon^2}\right], \tag{4}$$

then we can do the marginalization analytically if the range of λ values is taken to be $(-\infty, \infty)$[1]. This leads to:

$$p(x|\mu) = \frac{1}{\sqrt{2\pi(\sigma^2+\varepsilon^2\mu^2)}} \exp\left[-\frac{1}{2}\frac{(x-\mu)^2}{(\sigma^2+\varepsilon^2\mu^2)}\right]. \tag{5}$$

This is the method used in stochastic GW searches with two LIGO sites, e.g., [3, 4].

A more physically-motivated prior, which explicitly takes into account that λ is multiplicative and takes on only positive values, is a log-normal distribution:

$$p(\lambda) = \frac{1}{\lambda\varepsilon\sqrt{2\pi}} \exp\left[-\frac{(\ln\lambda)^2}{2\varepsilon^2}\right] \quad \text{or} \quad p(\Lambda) = \frac{1}{\varepsilon\sqrt{2\pi}} \exp\left[-\frac{\Lambda^2}{2\varepsilon^2}\right], \quad \text{where } \Lambda = \ln\lambda. \tag{6}$$

This was the approach taken in the stochastic GW search using LIGO and ALLEGRO [5], but has the drawback of requiring numerical integration over Λ because,

$$p(x|\mu,\Lambda) = \frac{1}{\sigma\sqrt{2\pi}} \exp\left[-\frac{(x-e^{\Lambda}\mu)^2}{2\sigma^2}\right], \tag{7}$$

gives a factor which is not Gaussian in Λ. Figure 1 compares the posterior distributions for the two different choices of priors for different measured values of x and different values of ε.

2. Calibration uncertainty with multiple baselines

With more than two instruments, there are multiple baselines and multiple calibration uncertainties to marginalize over. For instance, the stochastic background search using initial

[1] Since λ is an amplitude calibration factor, it should take on only positive values. But, for small ε (e.g., of order 0.10 to 0.20), integrating over λ from 0 to ∞ is numerically the same as integrating over λ from $-\infty$ to $+\infty$.

LIGO and Virgo data [6, 7] involved 4 different instruments $I \in \{\mathrm{H1}, \mathrm{H2}, \mathrm{L1}, \mathrm{V1}\}$ and 5 different baselines $\alpha \in \{\mathrm{H1L1}, \mathrm{H1V1}, \mathrm{H2L1}, \mathrm{H2V1}, \mathrm{L1V1}\}$.

Since the cross-correlation measurements for different baselines involve different calibration factors, all of the baselines cannot be optimally combined before marginalizing over calibration. Instead, all of the measurements for a baseline α can be combined into a single point estimate x_α with error bar σ_α. Each baseline has unknown calibration factor λ_α. Since the statistical errors for the different baselines are independent [8], the likelihood is the product, given as:

$$p(\mathbf{x}|\mu, \boldsymbol{\lambda}) = \prod_\alpha \left\{ \frac{1}{\sigma_\alpha\sqrt{2\pi}} \exp\left[-\frac{(x_\alpha - \lambda_\alpha \mu)^2}{2{\sigma_\alpha}^2} \right] \right\}, \tag{8}$$

where $\mathbf{x} \equiv \{x_\alpha\}$ and $\boldsymbol{\lambda} \equiv \{\lambda_\alpha\}$. The calibration factor λ_α for each baseline is $\lambda_{IJ} = \xi_I \xi_J$, and is determined by the per-instrument calibration factors $\boldsymbol{\xi} \equiv \{\xi_I\}$. If each instrument's calibration has an underlying uncertainty δ_I, the per-baseline calibration factors $\boldsymbol{\lambda}$ have the following means, variances and covariances:

$$\langle \lambda_{IJ} \rangle = 1, \quad \langle \lambda_{IJ}\lambda_{IJ} \rangle = 1 + {\delta_I}^2 + {\delta_J}^2 + \mathcal{O}(\delta^4), \quad \textit{and} \quad \langle \lambda_{IJ}\lambda_{JK} \rangle = 1 + {\delta_J}^2 + \mathcal{O}(\delta^4), \quad \text{if } I \neq K. \tag{9}$$

2.1. Per-baseline calibration marginalization

One approach is to marginalize over the per-baseline calibration factors assuming a multivariate Gaussian prior $p(\boldsymbol{\lambda})$. This has the advantage that the marginalization integral:

$$p(\mathbf{x}|\mu) = \int d\boldsymbol{\lambda}\, p(\mathbf{x}|\mu, \boldsymbol{\lambda})\, p(\boldsymbol{\lambda}), \tag{10}$$

can be done analytically if the integrals over the per-baseline calibration factors $\boldsymbol{\lambda}$ are taken over $(-\infty, \infty)$. However, the relationship $\lambda_{IJ} = \xi_I \xi_J$ implies that:

$$\lambda_{IJ}\lambda_{KL} - \lambda_{IK}\lambda_{JL} = 0. \tag{11}$$

For a multivariate Gaussian prior on $\boldsymbol{\lambda}$, this relation is true *only as an expectation value*, not as an identity for all values of λ_{IJ}, λ_{KL}, etc.

2.2. Per-instrument calibration marginalization

An alternative approach, which enforces identities such as in Eq. (11) is to set a prior, which is the product of independent priors on each per-instrument calibration factor ξ_I or equivalently on $\Xi_I = \ln \xi_I$. Similarly, defining the log-calibration factor for a baseline $\Lambda_\alpha = \ln \lambda_\alpha$, we have:

$$\Lambda_{IJ} = \ln \lambda_{IJ} = \ln(\xi_I \xi_J) = \Xi_I + \Xi_J. \tag{12}$$

The likelihood is given by:

$$p(\mathbf{x}|\mu, \boldsymbol{\Xi}) = \prod_\alpha \left\{ \frac{1}{\sigma_\alpha\sqrt{2\pi}} \exp\left[-\frac{(x_\alpha - e^{\Lambda_\alpha} \mu)^2}{2{\sigma_\alpha}^2} \right] \right\}, \tag{13}$$

and the marginalized likelihood is:

$$p(\mathbf{x}|\mu) = \int d\boldsymbol{\Xi}\, p(\mathbf{x}|\mu, \boldsymbol{\Xi})\, p(\boldsymbol{\Xi}). \tag{14}$$

An obvious prior is log-normal on ξ_I, i.e., Gaussian on Ξ_I as:

$$p(\Xi) = \prod_I \left\{ \frac{1}{\delta_I \sqrt{2\pi}} \exp\left(\frac{{\Xi_I}^2}{2{\delta_I}^2} \right) \right\}. \tag{15}$$

The exact integral over $\boldsymbol{\Xi}$ would need to be done numerically for each μ, but if $\boldsymbol{\delta}$ are small, one can make the approximation $e^{\Lambda_{IJ}} \approx 1 + \Lambda_{IJ} = 1 + \Xi_I + \Xi_J$ to convert the likelihood to a Gaussian integral over $\boldsymbol{\Xi}$, which *can* be done analytically. The result is a likelihood of the form:

$$p(\mathbf{x}|\mu) = \sqrt{\det\left(\frac{\mathbf{M}(\mu,\boldsymbol{\sigma},\boldsymbol{\delta})}{2\pi}\right)} \exp\left[-\frac{1}{2} \sum_\alpha \sum_\beta (x_\alpha - \mu)\, M_{\alpha\beta}(\mu,\boldsymbol{\sigma},\boldsymbol{\delta})\, (x_\beta - \mu) \right]. \tag{16}$$

This is the approach which was used for the multi-baseline upper limits in [7].

For the special case of two instruments which make up a single baseline, the matrix $\mathbf{M}$ reduces to a single number with value:

$$\mathbf{M}(\mu,\boldsymbol{\sigma},\boldsymbol{\delta}) \equiv \frac{1}{\sigma_{12}^2 + \mu^2({\delta_1}^2 + {\delta_2}^2)}. \tag{17}$$

Comparing Eq. (17) with Eq. (5), we see that this approximation gives the same result as assuming a Gaussian prior in λ_{12} with ${\varepsilon_{12}}^2 = {\delta_1}^2 + {\delta_2}^2$.

3. Ongoing work

More generally, we may be using $\mathbf{x}$ to estimate multiple physical quantities, spherical harmonic modes of a non-isotropic stochastic GW background [9, 10]. These methods of analytic marginalization with either a multivariate Gaussian prior or an approximate likelihood function can be applied to the effects of calibration uncertainty in that search as well. Additionally, these calibration effects may also be considered in other cross-correlation searches, such as the modelled cross-correlation search for periodic GW signals [11].

Acknowledgements

We wish to thank the organizers of the International Conference on Gravitation and Cosmology (ICGC-2011, Goa) and our colleagues in the LIGO Scientific Collaboration and the Virgo Collaboration, especially Albert Lazzarini. JTW was supported by NSF grant PHY-0855494. ELR was supported by the Max Planck Society. JDR was supported by NSF grants PHY-0855371 and HRD-0734800. EHT was supported by NSF grant PHY-0758035. This paper has been assigned LIGO Document Number P1200051-v4.

References

[1] Christensen N L 1992 *Phys. Rev. D* **46** 5250
[2] Allen B and Romano J D 1999 *Phys. Rev. D* **59** 102001
[3] Abbott B *et al* 2007 (LIGO Scientific Collaboration) *Ap. J.* **659** 918
[4] Abbott B P *et al* 2009 (LIGO Scientific Collaboration and Virgo) *Nature* **460** 990
[5] Abbott B *et al* 2007 (LIGO Scientific Collaboration and ALLEGRO) *Phys. Rev. D* **76** 022001
[6] Cella G *et al* 2007 *Class Quant. Grav.* **24** S639
[7] Abadie J *et al* 2012 (LIGO Scientific Collaboration and Virgo) *Phys. Rev. D* **85** 122001
[8] Lazzarini A *et al* 2004 *Phys. Rev. D* **70** 062001
[9] Thrane E et al 2009 *Phys. Rev. D* **80** 122002
[10] Abadie J *et al* 2011 (LIGO Scientific Collaboration and Virgo) *Phys. Rev. D* **107** 271102
[11] Dhurandhar S *et al* 2008 *Phys. Rev. D* **77** 082001

Residual foreground contamination in the WMAP data

P Chingangbam[1] and C Park[2]

[1] Indian Institute of Astrophysics, Koramangala II Block, Bangalore 560034, India
[2] Korea Institute for Advanced Study, Hoegiro 85, Dongdaemun-Gu, Seoul 130-722, Korea

E-mail: prava@iiap.res.in, cbp@kias.re.kr

Abstract. We have studied whether there is any residual foreground contamination in the foreground reduced WMAP-7 data for the differential assemblies (DAs) Q, V and W. We have calculated the correlation between the foreground map, from which long wavelength correlations have been subtracted, and the foreground reduced map for each DA. We have found positive correlations for *all* the channels. The statistical significance of the resulting values has been tested by comparing with correlations between the cleaned CMB maps and 1000 simulated Gaussian maps to which instrumental effects have been added. We have found high statistical significance of the observed correlations, implying the presence of residual contamination in the cleaned data, and found that, for Q and V channels, a large fraction of the contamination comes from pixels, where the foreground maps have positive values larger than three times its rms value, which shows the presence of unresolved point sources that contribute significantly to the contamination.

1. Introduction

The experimentally measured CMB temperature fluctuations is composed of the true CMB signal and foreground contamination, coming from astrophysical sources that emit photons in the frequency ranges spanned by the experiments. The major part of the contamination comes from emissions in our galaxy, and a small fraction comes from point sources such as radio galaxies and dusty star-forming galaxies. It is necessary to know the contaminating component in order to get the true signal, which can be used to extract cosmological information [1]. The galactic foreground is usually estimated from a combination of theoretical modelling and observations, and then subtracted from the measured data [2]. Further, sky regions where there are possibility of residual contamination, such as the directions along the plane of our galaxy, and the locations of localized sources that have been identified are masked. The correct removal of the foreground contamination is extremely crucial, because any residual contamination will bias the extraction of cosmological information from CMB data.

In this paper, we have investigated whether there is small residual foreground contamination in the cleaned WMAP data by using a method that is simple, but has proved to be quite effective. Our method is based on calculating correlations between the foreground field, which has been processed so as to remove long wavelength correlation of the galaxy emissions, and the cleaned CMB data. Our basic premise is that if there is no residual contamination in the cleaned data, we should obtain no correlation. However, we have found statistically significant positive correlation for WMAP-7 data for the Q, V and W differential assemblies (DAs)[1]. Further, we have found that a big fraction (as big as 30% for Q channel) comes from regions where the foreground map has large positive values, which indicates unresolved point sources. This is not surprising, in light of the fact that the PLANCK early data release has identified point sources of the order of 15000 [3]. These results imply the presence of residual contamination in the cleaned data. This work has been done with the motivation to quantify the contaminant fraction, and understand the nature of non-Gaussian deviations they introduce into the CMB. The earlier studies on point sources contamination on the CMB have been included [4, 5, 6, 7].

This paper is organized as follows: In Section 2, we have outlined our procedure for calculating the correlations between the foreground and cleaned CMB maps, and then presented our correlation results and their statistical significance. We have given the concluding remarks in Section 3.

2. Quantifying residual foreground contamination

Let f_1 and f_2 be any two random fluctuation fields having zero mean values, defined on the surface of a two dimensional sphere. Let their rms values be given by σ_1 and σ_2 respectively. Re-scaling them as $\nu_1(i) \equiv f_1(i)/\sigma_1$ and $\nu_2(i) \equiv f_2(i)/\sigma_2$, where i denotes the pixel number, we can define a correlation parameter r as:

$$r \equiv \langle \nu_1(i)\, \nu_2(i) \rangle, \tag{1}$$

where the bracket denotes average over all pixels. We expect r to be zero if the two fields are uncorrelated, and non-zero otherwise. (In practice, we will always get a small, but non-zero value of r even though the two fields are known to be totally uncorrelated, due to the finite number of pixels.)

The observed WMAP data f^{obs} is a sum of the true CMB signal and foreground contamination. The foreground component is estimated using a combination of galaxy observation and theoretical modelling [2]. We have called this field the 'apparent' foreground field, denoted by f^{appfg}, keeping in mind that there may be small error in its estimation. This field is then subtracted pixel by pixel from f^{obs} to leave behind the 'cleaned' CMB signal, which we have denoted by f^{cleaned}. By definition, f^{cleaned} has zero mean. If f^{appfg} has been correctly estimated, then we expect it to have negligibly small correlation with f^{cleaned}, since they come from totally different physical processes. However, if the estimation is on the right track but not fully correct, then we should expect some residual contamination in the signal field. This should show up as non-zero correlation between f^{cleaned} and f^{appfg}. In what follows, we have described our results for the correlations.

2.1. Peak field

Our analysis is done using the 7 years data from the eight DAs of WMAP [8], namely, Q_1, Q_2, V_1, V_2, W_1, W_2, W_3, and W_4. For each DA, we have obtained f^{appfg} as:

$$f^{\text{appfg}} = f^{\text{obs}} - f^{\text{cleaned}}. \tag{2}$$

Since, we want to remove the long distance correlations of the emissions from our galaxy, we have first defined what we call the *peak* field as:

$$f^{\text{peak}} \equiv \left(f^{\text{appfg},\theta_s} - f^{\text{appfg},3\theta_s} \right) - \langle f^{\text{appfg},\theta_s} - f^{\text{appfg},3\theta_s} \rangle, \tag{3}$$

where θ_s and $3\theta_s$ are FWHM values, at which we have smoothened the field. By definition, f^{peak} has zero mean. Figure 1 shows the peak field for Q1 channel.

[1] We have done similar analysis using WMAP-5 data and obtained similar results.

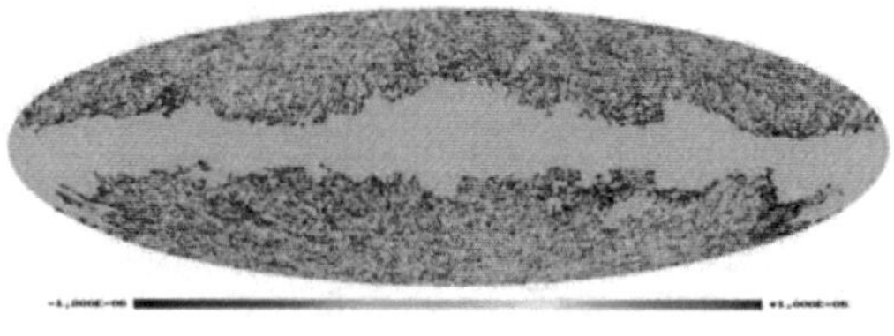

Figure 1. Peak field for Q_1 DA.

	Q_1	Q_2	V_1	V_2	W_1	W_2	W_3	W_4
r_c for $s_{mask} = 0.89$	0.026	0.025	0.020	0.019	0.010	0.008	0.009	0.006
	0.018	0.018	0.017	0.016	0.009	0.008	0.007	0.007
r_c for $s_{mask} = 0.91$	0.025	0.025	0.020	0.019	0.009	0.007	0.008	0.006
	0.018	0.017	0.016	0.016	0.008	0.008	0.006	0.006
r_c for $s_{mask} = 0.93$	0.025	0.024	0.019	0.018	0.008	0.007	0.007	0.004
	0.017	0.017	0.016	0.015	0.007	0.007	0.005	0.006

Table 1. r_c values for the eight DAs are shown. For each DA and s_{mask}, the upper value gives r_c calculated using all unmasked pixels, while the lower value has been calculated after excluding pixels having $\nu^{peak} > 3$ also. The sky fractions for the three s_{mask} values from top to bottom are roughly 62%, 60% and 58%.

We have used the KQ75y7 galaxy mask and point sources given in the 7 years data release of WMAP. Rather than using the galaxy plus point sources mask directly, we have first smoothened it with FWHM$= 3\theta_s$. The pixels near the mask boundaries have been obtained values between zero and one. To mask $f^{\rm cleaned}$ and $f^{\rm peak}$, we have chosen some threshold value, s_{mask} (between zero and one), and throw away pixels of $f^{\rm cleaned}$ and $f^{\rm peak}$, if the corresponding smoothened mask pixels have values less than s_{mask}. This method of masking makes it easy to control how far away we stay from the mask boundary of $f^{\rm cleaned}$ and $f^{\rm peak}$ in order to minimize the boundary effect. Staying 2σ (where σ is the rms of the field under consideration) away from the boundary, corresponds to choosing pixels with $s_{mask} > 0.89$. As we have chosen larger values of s_{mask}, we stay further away from the boundaries, and the sky fraction decreases.

2.2. Correlation between peak and cleaned CMB fields

Let us denote:

$$\nu^{\rm cleaned}(i) \equiv \frac{f^{\rm cleaned}(i)}{\sigma^{\rm cleaned}}, \quad and \quad \nu^{\rm peak}(i) \equiv \frac{f^{\rm peak}(i)}{\sigma^{\rm peak}}, \tag{4}$$

and define:

$$r_c \equiv \langle \nu^{\rm cleaned}\, \nu^{\rm peak} \rangle_{\theta_s}, \tag{5}$$

where the suffix θ_s is to remind us that we do the calculation for a choice of FWHM at which $f^{\rm cleaned}$ has also been smoothened. $\sigma^{\rm cleaned}$ and $\sigma^{\rm peak}$ are of the orders of 10^{-5} and 10^{-7} respectively.

Table 1 summarizes the main results for r_c. We have used $\theta_s = 35'$. Two values of r_c are shown for each DA and s_{mask}. The upper value is the case where r_c is calculated using all unmasked pixels, while the lower value is the case where pixels with $\nu^{\rm peak} > 3$ have also been excluded. The first observation we have made is that *all* r_c values are positive. For Q channels,

	Q_1	Q_2	V_1	V_2	W_1	W_2	W_3	W_4
N for $s_{\text{mask}} = 0.89$	0	0	5	5	121	156	139	224
	14	16	107	113	262	270	311	305
N for $s_{\text{mask}} = 0.91$	0	0	6	10	143	186	162	247
	15	16	110	118	276	284	332	320
N for $s_{\text{mask}} = 0.93$	0	0	11	15	169	218	196	291
	16	21	116	127	282	292	351	342

N_0 for $s_{\text{mask}} = 0.89$	0
	5
N_0 for $s_{\text{mask}} = 0.91$	0
	6
N_0 for $s_{\text{mask}} = 0.93$	0
	9

Table 2. *Left:* Number of maps N, having $r_g > r_c$ for individual DAs, out of 1000 Gaussian maps. As in Table 1, upper values are for all unmasked pixels included, while lower values are for the case when pixels with $\nu^{peak} > 3$ have also been excluded. *Right:* Number of Gaussian maps N_0, having $r_g > r_c$ simultaneously for all DAs.

we get considerably larger correlation when we keep all unmasked pixels, larger by about 30%. This indicates that there is non-trivial correlation arising from the pixels with $\nu^{\text{peak}} > 3$. For V channels, the difference is about 20%, while W channels do not seem to be affected. We have repeated these calculations for the case where ν^{cleaned} are also excluded, but not found any significant effect. The sky fractions for the three s_{mask} values are roughly, 62%, 60% and 58% respectively. For Q and V channels, as we stay further away from the mask boundaries, there is small but systematic decrease of r_c.

2.3. Statistical significance of r_c values

We have investigated how likely it is to get the observed r_c values given above by comparing with correlations between the peak field and Gaussian CMB simulations. For this purpose, we have simulated 1000 Gaussian CMB maps with WMAP-7 parameter values, and added pixel window effect, beam smearing and WMAP-7 noise characteristics. Next, we have smoothened by FWHM 35′ and mask in exactly the same way as we did when calculating r_c, and calculated the correlation with the peak field. We have denoted the correlation value by r_g. The Gaussian fields are uncorrelated with the signal field, and we should get small value of r_g. This exercise has told us what is the typical value of 'small' r_c that we can approximate it to be zero for the number of pixels under consideration, and how likely are our observed r_c values to occur by random fluctuation and not due to a true correlation.

We have counted, out of the thousand r_g values, how many are greater than r_c. The results are shown in Table 2. The left table shows the number N, of Gaussian maps having $r_g > r_c$ for each individual DA, for the three s_{mask} values used earlier for calculating r_c, and including/excluding the pixels having $\nu^{\text{peak}} > 3$. When all unmasked pixels are included, we get $N = 0$ for all s_{mask} values for Q channel, for V channels, N lies between 5 and 15, while for W channels, N lies between 121 and 291. These numbers imply that the r_c values for Q and V are statistically significant, whereas, the values for W channels have much lower significance. When pixels with $\nu^{\text{peak}} > 3$ are also excluded, we have found a reduction of N for all the channels. The table on the right side of Table 2 shows the number N_0, of Gaussian maps having $r_g > r_c$ *simultaneously for all DAs*. These values are again significant. Therefore, we have concluded that the cleaned WMAP data, particularly Q and V channels, contain small but statistically significant amount of residual foreground contamination.

3. Conclusion

We have shown that the cleaned WMAP data contains small but statistically significant amount of residual foreground contamination. The level of contamination is quantified by the level of correlation obtained between the cleaned map and the peak field. For Q and V DAs, we have found that a big fraction of the contamination could be coming from unresolved point sources. Our result has important implications for the extraction of cosmological data using the cleaned WMAP data. For example, it is known that constraints on the local primordial non-Gaussianity parameter f_{NL} using Fourier space method such as the bispectrum, and those use pixel space methods, like the Minkowski Functionals (MFs) do not agree with each other. The bispectrum gives positive best fit value of f_{NL} [9], while the MFs give negative values [10]. Our analysis in this work was motivated by an attempt to understand whether this conflict is due to residual foreground contamination. Our next goal is to understand the effect of the residual contamination on the measurement of the MFs, and we are currently pursuing this line of investigation.

Acknowledgement

We thank Korea Institute for Advanced Study for providing computing resources (KIAS Center for Advanced Computation Linux Cluster System QUEST), where a part of the computation was carried out. We acknowledge the use of the HEALPIX package [11, 12], and the Legacy Archive for Microwave Background Data Analysis (LAMBDA). Support for LAMBDA is provided by the NASA Office of Space Science.

References

[1] Komatsu E *et al* 2011 *Ap. J. Supp.* **192** 18
[2] Gold B *et al* 2011 *Ap. J. Supp.* **192** 15
[3] Planck early results VII. *The Early Release Compact Source Catalogue* 2001 *A. & A.* **536** A7
[4] Argueso F, Gonzalez-Nuevo J and Toffolatti L 2003 *Ap. J.* **598** 86
[5] Boughn S P and Partridge R B 2008 *PASP* **120** No. 865
[6] Babich D and Pierpaoli E 2008 *Phys. Rev. D* **77** 123011
[7] Lacasa F, Aghanim N, Kunz M and Frommert M [arXiv:1107.2251[astro-ph.CO]]
[8] http://lambda.gsfc.nasa.gov/
[9] Komatsu E *et al* (WMAP Collaboration) 2009 *Ap. J. Supp.* **180** 330
[10] Hikage C, Matsubara T, Coles P, Liguori M, Hansen F K and Matarrese S 2008 *MNRAS* **389** 1439
[11] Gorski K M, Hivon E, Banday A J, Wandelt B B, Hansen F K, Reinecke M and Bartelmann M 2005 *Ap. J.* **622** 759 [arXiv:astro-ph/0409513]
[12] http://healpix.jpl.nasa.gov

Dipole leakage and low CMB multipoles

S Das and T Souradeep
IUCAA, Post Bag 4, Pune University Campus, Ganeshkhind, Pune 411007, India
E-mail: santanud@iucaa.ernet.in, tarun@iucaa.ernet.in

Abstract. A number of studies of WMAP-7 have highlighted that the power at the low multipoles in CMB power spectrum is lower than their theoretically predicted value. Angular correlation between the orientations of these low multipoles have also been discovered. While these observations may have cosmological ramification, it is important to investigate possible observational artifacts that can mimic them. The CMB dipole, which is much higher than the quadrupole, can get leaked to the higher multipoles due to the non-circular beam of the CMB experiment. In this paper, an analytical method has been developed and simulations are carried out to study the effect of the non-circular beam on power leakage from the dipole. It has been shown that the small, but non-negligible power from the dipole can get transferred to the quadrupole and the higher multipoles due to the non-circular beam. Simulations have also been carried out for Planck scan strategy, and comparative results between WMAP and Planck have been presented.

1. Introduction

The standard model of cosmology emerging from recent observations is a remarkable success of theoretical physics. It can explain the cosmological observations up to an extremely high precision using a handful set of parameters. However, there are some effects that seem to be anomalous in the standard cosmological model. One of such observational facts is the power anomaly at the low multipoles of the CMB power spectrum. It has been seen that the power at the low multipoles is lower than their theoretical predictions, and there are possibly correlation between the orientations of the low multipoles. Many researchers have tried to explain the phenomenon [1, 2], but it is not satisfactorily explained. In this paper, we have analyzed the effect of the non-circular beam as origin of CMB anomalies at low multipoles.

In most of the CMB experiments such as WMAP, the beam shape of the detectors is non-circular about the pointing direction. However, in the data analysis techniques, the beam is assumed to be circularly symmetric. The CMB dipole is much stronger than the quadrupole. Therefore, due to the non-circularity of the beam, some power from dipole may leak to the quadrupole and immediate higher multipoles. However, assuming a circular beam in the data analysis technique, this leakage of the power from dipole will not be accounted for. Therefore, the contribution of this effect will contaminate the resultant map, generated by this inadequate data analysis technique. Different effects of non-circular beam have been discussed by different authors in [3, 4]. However, this particular effect of leakage from the dominant dipole has not been analyzed yet.

In this paper, analytical methods have been developed to calculate the amount of power leakage from dipole, and simulations with the actual scan pattern of WMAP have been carried out showing the order of power leakage for different WMAP beams. Our results show that the amount may not be sufficient to explain the anomalies, but the power transfer does have a measurable effect on the quadrupole. In anticipation, a similar simulation has also been carried out with Planck scan pattern. As the data for Planck beams are not available publicly, we have provided only upper limits to the beam non-circularity parameters beyond which the dipole leakage would cause detectable effect in the power spectrum.

2. Analytical description of the beam convolution

This section describes the formalism employed for measuring the power leakage from dipole to the quadrupole and higher multipoles during scanning the sky with a non-circular beam. The measured sky temperature in a CMB experiment is a convolution of the true sky temperature with the beam function. If the measured temperature along γ_i is expressed by $\tilde{T}(\gamma_i)$, whereas the sky temperature along γ is $T(\gamma)$ then:

$$\tilde{T}(\gamma_i) = \int B(\gamma_i, \gamma) T(\gamma) d\Omega_\gamma + T^N(\gamma_i)\ . \tag{1}$$

Here, T^N is the noise in the scan procedure. Since we have dealt with low multipoles, the noise is ignored in our analysis. The beam function $B(\gamma_i, \gamma)$ represents the sensitivity of the telescope around the pointing direction γ_i. This is a two point function and can be expanded in terms of spherical harmonics as:

$$B(\gamma_i, \gamma) = \sum_{l=0}^{\infty} \sum_{m=-l}^{l} b_{lm}(\gamma_i) Y_l^m(\gamma)\ . \tag{2}$$

Since we intend to measure the power leakage from dipole to the quadrupole, and it is convenient to consider only the sky dipole map and check the amount of power leakage from the dipole to higher multipole due to the non-circular beam. A dipole map $T(\gamma)$ can be written as a sum of all the spherical harmonics with $l = 1$, i.e., $T(\gamma) = \sum_{m=-1}^{1} a_{1m} Y_1^m(\gamma)$. However, its always possible to choose a coordinate system such that $a_{1,1}$ and $a_{1,-1}$ modes vanish and the sky temperature can be expressed only as $T(\gamma) = T_0 Y_1^0(\gamma)$, where $T_0 = a_{1,0}$ is a constant. In such a case, the measured sky temperature along the γ_i direction can be expressed as:

$$\begin{aligned}
\tilde{T}(\gamma_i) &= \int B(\gamma_i, \gamma) T(\gamma) d\Omega_\gamma = \int \left[\sum_{l=0}^{\infty} \sum_{m=-l}^{l} b_{lm}(\gamma_i) Y_l^m(\gamma) \right] T(\gamma) d\Omega_\gamma, \\
&= T_0 \sum_{l=0}^{\infty} \sum_{m=-l}^{l} b_{lm}(\gamma_i) \int Y_l^m(\gamma) Y_1^0(\gamma) d\Omega_\gamma = T_0 b_{10}(\gamma_i)\ .
\end{aligned} \tag{3}$$

The above expression can not be directly used for measuring the sky temperature, because it contains the beam harmonic coefficient $b_{10}(\gamma_i)$, which is a function of γ_i. It is convenient to orient the beam along some fixed direction of the sky, say, along the $\hat{z}$ direction, and consider the multipole $b_{lm}(\hat{z})$ to characterise the beam [5]. The spherical harmonic coefficients of the beam $b_{lm}(\gamma_i)$ at any particular direction γ_i can be obtained by using Wigner-D functions as:

$$\begin{aligned}
b_{10}(\gamma_i) &= \sum_{m'=-l}^{l} b_{1m'}(z) D_{0m'}^1, \\
&= b_{1,-1}(z) D_{0,-1}^1(\varphi_i, \theta_i, \rho_i) + b_{1,0}(z) D_{0,0}^1(\varphi_i, \theta_i, \rho_i) + b_{1,1}(z) D_{0,1}^1(\varphi_i, \theta_i, \rho_i), \\
&= b_{1,-1}(z) d_{0,-1}^1(\theta_i) e^{i\rho_i} + b_{1,0}(z) d_{0,0}^1(\theta_i) + b_{1,1}(z) d_{0,1}^1(\theta_i) e^{-i\rho_i}\ .
\end{aligned} \tag{4}$$

Substituting explicit expressions in terms of trigonometric functions for $d^1_{0,0}(\theta_i)$, $d^1_{0,-1}(\theta_i)$ and $d^1_{0,1}(\theta_i)$, in the above, and using Eq. (3), the expression for the scanned temperature can be written as:

$$\tilde{T}(\gamma_i) = T_0 b_{10}(z)\cos(\theta_i) + \sqrt{2}T_0\sin(\theta_i)\Big[b_r(z)\cos(\rho_i) + b_i(z)\sin(\rho_i)\Big]\,. \quad (5)$$

Here, ρ_i is the orientation of the semi-major axis of the beam at the ith scan point. The functions $b_i(z)$ and $b_r(z)$ can be defined as follows: In Eq. (4), $b_{1,1}$ and $b_{1,-1}$ are complex quantities. Since the beam is real, the beam spherical harmonic coefficients should satisfy the relation $b^*_{1,1} = -b_{1,-1}$. Here, b_r and b_i have been defined as $b_{1,1} = b_r + ib_i$, i.e., the real and the imaginary part of $b_{1,1}$.

From Eq. (5), we can calculate the power that gets leaked to the quadrupole or the higher multipoles. It can be seen that the term with $b_{10}(z)$ will not contribute to any power leakage from dipole. Therefore, the dipole to quadrupole power transfer is caused only by the terms multiplied with $b_r(z)$ or $b_i(z)$. Hence, if the experimental beam is designed in such a way that the $b_r(z)$ or $b_i(z)$ components of the beam are completely negligible, then it is possible to completely eliminate the dipole to higher multipole power transfers.

3. Dipole leakage from WMAP scan pattern

The WMAP satellite follows a unique scan pattern, in which pixels near the two poles are scanned for large number of times, whereas, those near the equator are scanned for least number of times. The satellite scans the sky temperature in five different frequency bands, named as K, Ka, Q, V and W in a differential measurements of a pair of horns. Amongst them, Q and V bands have two detectors each, and W band has four detectors. Each of these detectors has a pair of horns, both of which are about 70.5^o off the symmetry axis. It has a fast spin about the symmetry axis with the spin period of around 2.2 minutes. Along with this fast spin, the spacecraft has a slow precession, 22.5^o about the Sun-WMAP line. This precession period is about 1 hour. The earth-sun vector rotates 360° in a year.

The simulation has been carried with a dipole map, similar to the known CMB dipole. We have assumed that the shape of the two beams for a pair of horns of a detector are almost same, i.e., the parameters b_r, b_i and b_{10} are same for both the beams of a detector and the angle between them is 140 degree.

All the simulations have been carried out on Healpix map with $N_{side} = 256$ resolution. The simulation gives us three independent maps from the three independent components of Eq. (5). The figures from the three maps are shown in the Figure 1. All these maps are shown in ecliptic coordinate system. These maps can be multiplied with b_{10}, b_r and b_i, and then summed up to get the final scanned map from a detector and hence, the amount of power transfer can be calculated. The values of the beam spherical harmonic coefficients, i.e., b_r, b_i and b_{10} along with the amount of power leakage are listed in the Table 1.

The analysis shows that the temperature leaked into the quadrupole from the dipole is less than $2\mu K$ compared to the CMB quadrupole measured by WMAP of $\sim 18\mu K$. Although, the dipole power leakage is small, the amount of power leakage is sufficient to suggest a deeper analysis of the WMAP data for this effect.

4. Dipole leakage from Planck scan pattern

The Planck satellite has only one beam for each detector and thus, instead of the differential measurement, it measures the actual temperature of the sky. The Planck satellite beam is approximately 85° off symmetric axis and the precession angle for the satellite is around 7.5^o. The precession rate is taken as one revolution per six months and the spin rate as $180^o/min$.

Table 1. The dipole coefficients b_r and b_i (real and imagionary parts of $b_{1,1}$) of the beam spherical harmonics for different WMAP beams estimated from the publicly available beam maps [6]. The quadrupole and octapole temperature are calculated from the simulation considering the dipole temperature as 3.358 mK.

	b_r	b_i	T_d/T_q	$T_q(\mu K)$	T_d/T_{oc}	$T_{oc}(\mu K)$
K_1	-1.45×10^{-4}	2.37×10^{-5}	1837.9	1.82	11250.6	0.30
Ka_1	-9.42×10^{-5}	-6.03×10^{-5}	2568.3	1.31	16493.7	0.20
Q_1	8.66×10^{-5}	-1.62×10^{-4}	2748.1	1.22	14226.4	0.24
V_2	7.40×10^{-5}	8.11×10^{-5}	3256.4	1.03	19674.1	0.17
W_4	6.55×10^{-6}	2.23×10^{-5}	33048.57	0.10	133621.8	0.025

An analysis similar to that of WMAP has been carried out for Planck satellite also. Three independent maps are calculated from the Planck scan strategy, which are shown in Figure 2. As the data of the Planck beam map is not publicly available, and therefore, the amount of power leakage expected from the dipole can not be computed for the Planck scan pattern. But, the analysis shows that if the b_i and b_r are of the order of 5×10^{-4} or less, then it will cause a temperature leakage less than $1\mu K$. Therefore, that amount of leakage may be safely ignored for beams satisfying the above limits.

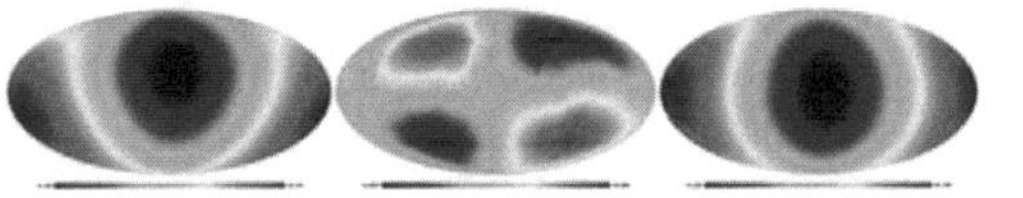

Figure 1. Simulated maps of $\cos\theta$, $\sin\theta\cos\rho$ and $\sin\theta\sin\rho$ components from WMAP scan pattern in the ecliptic coordinate system.

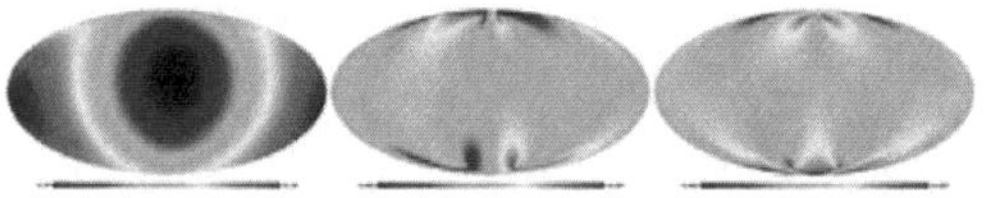

Figure 2. Simulated maps of $\cos\theta$, $\sin\theta\cos\rho$ and $\sin\theta\sin\rho$ components from Planck scan pattern in ecliptic coordinate system.

5. Conclusion

An analytical formalism has been developed to use in simulations of scan strategy to estimate the leakage of power in the dipole anisotropy to the quadrupole and higher multipoles. It has also been shown that the power leakage only depends on the two spherical harmonic coefficients of the satellite beam (b_i and b_r), and therefore, if the beam has been designed in such a way that these two parameters of the beam are small enough, then power leakage will be negligible. For WMAP, the amount of the power leakage is found to be small but not insignificant compared to the low value of quadrupole measured. The simulations also show that the amount of power leakage depends on the scan pattern. For identical beam shapes, the amount of power leakage is more in WMAP scan strategy compared to Planck scan strategy.

References

[1] Moss A, Scott D and Sigurdson K 2011 *JCAP* **1** 001
[2] Feng B and Zhang X 2003 *Phys. Lett. B* **570** 145
[3] Fosalba P, Dore O and Bouchet F R 2002 *Phys. Rev. D* **65** 063003
[4] Mitra S, Sengupta A S and Souradeep T 2004 *Phys. Rev. D* **70** 103002
[5] Souradeep T and Ratra B 2001 *Ap. J.* **560** 28
[6] The beam maps can be found on: $www.lambda.gsfc.nasa.gov/product/map/dr3/beam_profiles_get.cfm$

Journal of Physics: Conference Series **484** (2014) 012030 doi:10.1088/1742-6596/484/1/012030

Measuring the Hubble constant using gravitational waves

W Del Pozzo

Nikhef, National Institute for Subatomic Physics, Science Park 105, 1098 XG Amsterdam, The Netherlands

E-mail: walterdp@nikhef.nl

Abstract. Second-generation ground-based laser interferometers are expected to deliver a wealth of gravitational waves (GW) events from coalescing compact binaries up to a redshift of about 0.3. Being free of the systematics affecting electromagnetic measurements, GW offer the possibility of an independent measurement of H_0 . This paper presents a method based on Bayesian inference aimed at estimating the value of the cosmological parameters for any GW event. In contrast to earlier work, this framework does not require the precise identification of the putative optical counterpart, but it considers all the potential galaxy hosts consistent with the recovered sky position and distance posterior distributions. When applied to the upcoming network of second generation interferometers, 50 GW events will yield a measurement of H_0 with an uncertainty of a few per cents.

1. Introduction

The determination of the Hubble parameter is a critical issue in cosmology, with repercussions on a wide range of fields in physics ranging from the determination of the abundances of light elements in the early universe to the content of weakly interacting particles, and therefore, the nature of dark energy. While a great deal has been learnt from the original discovery of Hubble [1], the exact value of the expansion rate today is still a matter of debate. Its current accepted value lies in the range $60 - 75$ km s^{-1} Mpc^{-1}(see [2] for a review) with different methods and assumptions yielding different values. The observation of gravitational waves (GW) from coalescing compact binaries potentially offers a totally independent measurement of the Hubble (and other cosmological) parameters [3]. In GW observations, the luminosity distance can be directly measured [3, 4, 5], and if redshift information is available from other means, one could estimate the cosmological parameters from the distance-redshift relation. Obtaining a redshift measurement for a GW source is a challenge, as the typical GW error-box is $\approx 1 - 100$ deg^2. A widely explored possibility is to use short Gamma Ray Bursts (GRB). If GRB are associated to the merger of compact objects with at least a neutron star component, this would provide a redshift measurement [6, 7, 8, 9]. However, the fraction of systems producing a short GRB might be as low as 10^{-2} [10]. It is still a matter of debate, whether it is feasible to use GW binaries as a new class of standard candles when the electro-magnetic counterpart is not known. For space-based antennas, it has been shown that one can use the galactic population to obtain a set of putative redshifts for the observed source [11, 12], but, with the exception of [3], such kind of statistical inference has not been applied to ground-based instruments. In this paper, we have shown that when using wide-field sky surveys in conjunction with GW observations,

Journal of Physics: Conference Series **484** (2014) 012030 doi:10.1088/1742-6596/484/1/012030

ground-based detectors could bring a measurement of the Hubble parameter with an accuracy of a few per cents.

2. Method

Throughout the following analysis, we have assumed that the host of the GW event is *not* univocally determined. For each GW detected binary system, the world-wide network of laser interferometers will provide the 3-dimensional error box of the source location: The luminosity distance D_L and the 2-dimensional position in the sky. GW observations do not yield a direct measure of the redshift of a source, but by correlating a galaxy catalogue with the 3-dimensional GW error box, one can produce a list of possible redshifts associated with the putative source. Consider a catalogue of GW events $\mathcal{E} \equiv \epsilon_1, \ldots, \epsilon_n$. The posterior probability distribution for the cosmological parameters $\vec{\Omega}$ – the Hubble constant H_0, the density parameters Ω_m, Ω_k, and Ω_Λ, and so on, for any given event ϵ_i is given by:

$$p(\vec{\Omega}|\epsilon_i, \mathcal{H}_{\vec{\Omega}}, \mathcal{I}) = \int dD_L d\alpha d(\cos\delta) p(\vec{\Omega}, D_L, \alpha, \delta, \vec{z}|\epsilon_i, \mathcal{H}_{\vec{\Omega}}, \mathcal{I}). \tag{1}$$

The integrand in Eq. (1) can be factorized as a product of three factors: (i) $p(D_L, \alpha, \delta|\epsilon_i, \mathcal{I})$, the 3-dimensional joint probability distribution for the source luminosity distance and sky position, (ii) $p(\vec{z}|D_L, \alpha, \delta, \mathcal{H}_{\vec{\Omega}}, \mathcal{I})$, the probability distribution for the redshifts $\vec{z} \equiv z_1, \ldots, z_k$ given $D_L, \alpha, \delta, \mathcal{H}_{\vec{\Omega}}$, and (iii) $p(\vec{\Omega}|D_L, \alpha, \delta, \vec{z}, \epsilon_i, \mathcal{H}_{\vec{\Omega}}, \mathcal{I})$, the distribution for $\vec{\Omega}$ conditional on everything else. A cosmological model $\mathcal{H}_{\vec{\Omega}}$ implies that the quantities $\vec{\Omega}, D_L$ and $\vec{z} \equiv z_1, \ldots, z_k$ are not independent. For each triplet $\vec{\Omega}, D_L$ and z_j, there is a function f such that $f(\vec{\Omega}, D_L, z_j) = 0$. In particular, for all js, $f(\vec{\Omega}, D_L, z_j) = D_L - D(\vec{\Omega}, z_j) = 0$. The function $D(\vec{\Omega}, z_j)$ is the luminosity distance expression as a function of the redshift z_j, and of the cosmological parameters as dictated by $\mathcal{H}_{\vec{\Omega}}$. Hence, in full generality, $p(\vec{\Omega}|D_L, \alpha, \delta, z_j, \epsilon_i, \mathcal{H}_{\vec{\Omega}}, \mathcal{I}) \propto \delta(D_L - D(\vec{\Omega}, z_j))$. This formalism is readily extended to the full catalogue of events $\mathcal{E}$ by:

$$p(\vec{\Omega}|\mathcal{E}, \mathcal{H}_{\vec{\Omega}}, \mathcal{I}) = \left[p(\vec{\Omega}|\mathcal{H}_{\vec{\Omega}}, \mathcal{I})\right]^{1-n} \prod_{i=1}^{n} p(\vec{\Omega}|\epsilon_i, \mathcal{H}_{\vec{\Omega}}, \mathcal{I}), \tag{2}$$

where $p(\vec{\Omega}|\mathcal{H}_{\vec{\Omega}}, \mathcal{I})$ is the prior probability distribution for $\vec{\Omega}$.

3. Results

As a proof-of-principle, we have considered the case of GW observed by a network of three advanced interferometers, two Advanced LIGO [13] and Advanced Virgo [14], in conjunction with the spectroscopic galaxy sample of the Sloan Digital Sky Survey Data Release 7 (SDSS) [15]. The GW events catalogue consists of a set of 1000 compact binary coalescence events distributed uniformly on the celestial sphere coverage of SDSS. The component masses of the binary system were chosen from a uniform distribution $m_1, m_2 \in (1.0, 15.0) M_\odot$. Sources were also restricted to have $z \leq 0.1$. This redshift corresponds to the completeness limit of SDSS, and given the most plausible rates [16], should yield ~ 50 detections per year. The full event catalogue is then divided into 20 catalogues of 50 events each. Results have been presented in the form of an average over these 20 independent catalogues. The cosmological model $\mathcal{H}_{\vec{\Omega}}$ considered for the analysis is a Friedmann-Robertson-Walker-Lemaître universe depending on four parameters, $\vec{\Omega} \equiv H_0, \Omega_m, \Omega_k, \Omega_\Lambda$ with the condition $\Omega_k + \Omega_m + \Omega_\Lambda = 1$. The relevant expression for the luminosity distance $D(\vec{\Omega}, z)$ is given in [17]. If we ignore the correlations between D_L and α and δ, their joint PDF factorizes in the form $p(D_L, \alpha, \delta|\epsilon_i, \mathcal{I}) = p(D_L|\epsilon_i, \mathcal{I}) p(\alpha, \delta|\epsilon_i, \mathcal{I})$; note that, by doing so, we have taken a conservative hypothesis and over estimate the actual 3D error-box of a source. Moreover, one can assume that $p(D_L|\epsilon_i, \mathcal{I}) \propto \exp\left(-\frac{(D_L-D_0)^2}{2\sigma^2_{D_L}}\right)$ and $p(\alpha, \delta|\epsilon_i, \mathcal{I}) \propto 1$ if $\alpha^2 + \delta^2 \leq \sigma_\alpha^2 + \sigma_\delta^2$, and 0 otherwise. $\sigma_{D_L}, \sigma_\alpha$ and σ_δ are the uncertainties over the exact source 3-dimensional location. These are heuristically assigned by computing the S/N ratio corresponding to ϵ_i. The uncertainties $\sigma_{D_L}, \sigma_\alpha$ and σ_δ that characterize each event sky-box are given by [18, 19]:

$$\sigma_{D_L}/D_L = 3 \times (S/N)^{-1}, \quad and \tag{3}$$

$$\Delta\Omega \equiv \sqrt{\sigma_\alpha^2 + \sigma_\delta^2} = a \times \deg^2 \left(\frac{10}{\mathrm{S/N}}\right)^2, \tag{4}$$

where a is a factor $O(10)$ that taken to be 100. Typical values for the uncertainties are $\sim 30\%$ on D_L and ~ 50 deg^2 on the sky position. The luminosity distance - inclination degeneracy can lead to serious over estimates or under estimates of the luminosity distance. This effect is mimicked by further shifting the measured value of D_L by δD randomly drawn from the interval $\pm 3\ \sigma_{D_L}$. We have used Gaussian priors for the parameters of interest $H_0, \Omega_m, \Omega_\Lambda$ with means $\{70, 0.3, 0.7\}$ and standard deviations $\{10, 0.1, 0.1\}$ respectively. The relative uncertainty on z for SDSS spectroscopic measurements is $\sim 10^{-3}$, thus, $p(\bar{z}|D_L, \alpha, \delta, \mathcal{H}_{\vec{\Omega}}, \mathcal{I}) \propto \sum_j \delta(z_j - \bar{z}_j)$. With these assumptions, the typical number of galaxies identified as potential hosts for each GW event ranges from a *few* to ~ 20000 (see Figure 1, left panel) depending on the S/N ratio of the source. Even with such large number of potential hosts, one can obtain an estimate of $\vec{\Omega}$. This is achieved because each galaxy has the same probability of hosting a GW event, so the spacial distribution of sources follows the spacial distribution of galaxies. Galaxies are usually found in bound structures, where the typical galaxy number density is hundred times higher than the field average. It follows that sources are more likely to be observed from structures such as clusters or groups of galaxies having approximately the same redshift. For each event, the most probable value for $\vec{\Omega}$ is given by the most represented redshift in the corresponding galaxy catalogue. The remaining redshifts contribute as noise at the tails of the target posterior distribution. This effect was the base of the method in [3]. The combined PDF in Eq. (2) effectively averages out the tails of the posterior distribution for $\vec{\Omega}$.

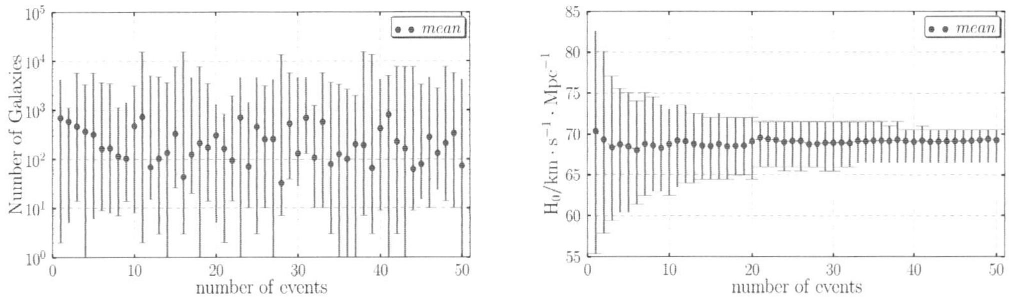

Figure 1. *Left panel*: The number of galaxies classified as potential hosts for each GW event under consideration. The dots are the mean number of galaxies per each of the events over the 20 GW catalogues realisations, while the error bars represent the minimum and maximum number per event. *Right panel*: Confidence interval evolution for H_0 as a function of the number of events considered in the analysis. The dots corresponds to the posterior mean value obtained from 20 realisations of the 50 GW sources. The error bars correspond to the mean 95% confidence interval.

Advanced interferometers will only be sensitive to H_0. The data do not offer useful information for estimating the remaining cosmological parameters and their posterior distributions coincide with their priors. For the redshift range under consideration ($z \leq 0.1$), the luminosity distance for different values of either Ω_m or Ω_Λ varies by $\sim 1\%$, that is, much smaller than the typical 30% uncertainty on D_L given the S/N ratios. H_0 is, instead, well determined already after $\sim$ 25 observations, and yields an estimate at a few per cents level. This estimate is given by $H_0 = 69^{+2}_{-4}$ km s^{-1} Mpc^{-1}after 25 observations and by $H_0 = 69^{+1.5}_{-2.5}$ km s^{-1} Mpc^{-1}after 50 observations (see Figure 1, right panel).

4. Conclusion

Second-generation GW experiments could deliver a measurement of the Hubble parameter, which is comparable with the current value derived from WMAP-7 observations [20]. Differently from most literature, the herein inference model makes use of *all* GW events, without a need for the simultaneous observation of an EM counterpart. A scenario where one considers exclusively events with a putative EM detection, neglects a significant part of the cogent information for the inference on $\vec{\Omega}$. Being a proof-of-principle, this study is characterized by some simplistic prescriptions that, in a realistic case, need to be dealt carefully. In general, the measurement of D_L cannot be dis-entangled from the inclination angle or from the sky position. This can lead to a mis-identification of the patch of the sky to be cross-correlated with a galaxy catalogue for the extraction of redshifts. In this case, the single event PDF for $\vec{\Omega}$ can be biased. The incompleteness of the survey used to obtain the redshifts will affect the PDF for $\vec{\Omega}$ in a similar fashion. Furthermore, GW are not free from systematics. The measurement of $p(D_L, \alpha, \delta | \epsilon_i, \mathcal{I})$ is known to be affected by calibration errors, uncertainties in the waveform and by spins. In the calculation of the joint PDF for multiple events, heavily biased PDFs will essentially contribute as noise, without adding any information for the inference. However, unless most of the single event PDFs are biased, the joint PDF for $\vec{\Omega}$ effectively averages out most of these effects, and it will still converge towards the true values of $\vec{\Omega}$. There are indications that considering the full inference problem, so taking into account all correlations and degeneracies among the parameters characterizing the GW signal, does not change the global inference of $\vec{\Omega}$. This will be the subject of a future study.

References

[1] Hubble E P 1929 *Proc. Nat. Acad. Sci. USA* **15** 3 168
[2] Jackson N 2007 *Living Rev. Rel.* **10** 4
[3] Schutz B F 1986 *Nature* **323** 310
[4] Chernoff D F and Finn L S 1993 *Ap. J. Lett.* **411** L5
[5] Sathyaprakash B S and Schutz B F 2009 *Living Rev. Rel.* **12** 2
[6] Holz D E and Hughes S A 2005 *Ap. J.* **629** 15
[7] Nissanke S, Holz D E, Hughes S A, Dalal N and Sievers J L 2010 *Ap. J.* **725** 496
[8] Sathyaprakash B S, Schutz B F and Van Den Broeck C 2010 *Class. Quant. Grav.* **27** 215006
[9] Zhao W, Van Den Broeck C, Baskaran D and Li T G F 2011 *Phys. Rev. D* **83** 023005
[10] Belczynski K, O'Shaughnessy R, Kalogera V, Rasio F, Taam R E and Bulik T 2008 *Ap. J. Lett.* **680** L129
[11] MacLeod C L and Hogan C J 2008 *Phys. Rev. D* **77** 043512
[12] Petiteau A, Babak S and Sesana A 2011 [arXiv:1102.0769]
[13] `http://www.ligo.caltech.edu/advLIGO/`
[14] `http://wwwcascina.virgo.infn.it/advirgo/`
[15] Abazajian K N. *et al* 2009 *Ap. J. Supp.* **182** 543
[16] Abadie J *et al* 2010 [LIGO Scientific Collaboration and Virgo Collaboration] *Phys. Rev. D* **82** 102001
[17] Hogg D W 1999 [arXiv:astro-ph/9905116]
[18] Cutler C and Flanagan É E 1994 *Phys. Rev. D* **49** 2658
[19] Wen L and Chen Y 2010 *Phys. Rev. D* **81** 082001
[20] Komatsu E *et al* [arXiv:1001.4538 [astro-ph.CO]]

Proceedings of the 7th International Conference on Gravitation and Cosmology (ICGC2011) IOP Publishing
Journal of Physics: Conference Series **484** (2014) 012031 doi:10.1088/1742-6596/484/1/012031

Evolution of spherical over-density in thawing dark energy models

N C Devi[1,2*], A A Sen[1] and T R Choudhuryi[3].

[1]Centre for Theoretical Physics, Jamia Millia Islamia, New Delhi 110025, India
[2]Departamento de Astronomia, Observatorio Nacional, 20921-400, Rio de Janeiro - RJ, Brazil
[3]National Centre for Radio Astrophysics, TIFR, Post Bag 3, Ganeshkhind, Pune 411007, India

E-mail: *chandrachani@gmail.com

Abstract. In this paper, we have studied the general evolution of spherical over-densities for thawing class of scalar field dark energy (DE) models. We have considered the scalar fields having canonical as well as non-canonical kinetic energy (particularly the Born-Infeld form of kinectic term) with various type of potentials, and also investigated the situation where DE is homogeneous as well as the case where DE virializes together with matter (inhomogeneous). Our study has shown that models with linear potential, in particular, can have significant deviation from the ΛCDM model in terms of density contrast at the time of virialization, and further study of the cluster number counts has shown that the total cluster number counts of different DE models can have substantial deviation from ΛCDM, and this deviation is most significant for all the models we have considered.

1. Introduction

One of the most significant discoveries in cosmology in recent years is the fact that our Universe is currently going through an accelerated expansion phase. This late time acceleration of the Universe can be due to the presence of an exotic fluid with large negative pressure known as *dark energy* (DE) or due to the modification of gravity itself. Inclusion of cosmological constant as DE is a minimal way to explain this late time acceleration, and also allowed by all cosmological observations. But this ΛCDM model is plagued with fine tuning and cosmic coincidence problems. Scalar field models mimicking a variable Λ can alleviate the fine tuning and coincidence problems and provide an interesting alternative to cosmological constant. These scalar field models are broadly classified into two categories depending upon the form of their potentials: Fast roll and slow roll models termed as freezing and thawing models in the literature [1]. Among these, thawing scalar field models are particularly interesting as they can naturally mimic equation of state very close to $w \sim -1$, which is preferred by all the observational data.

On the other hand, from the recent studies of the large scale structure surveys, it is found that DE, not only affects background expansion rate and the distance-redshift relation, but also affects the growth of structure in the universe. Hence, DE is expected to have an impact on observables such as cluster number counts and lensing statistics. Therefore, the studies of galaxy clusters would provide a useful tool to constrain the model parameters and would help to infer the properties of dark energy by discriminating among the different dark energy models. In order to understand accurately the effect of DE on the clustering properties of matter, one has to perform N-body simulations, but here, we try to understand it by considering the semi-analytical approach, called a spherical collapsed model. This model (first developed by Gunn and Gott [2]) is the simplest and the most fundamental tool for understanding the non-linear clustering of matter. It describes how a small spherical over-density decouples from the background evolution, slows down, then eventually turns around reaching the maximum radius and begins to collapse. After that, it virializes forming a bound system. In this series, we have calculated the matter density contrast at virialization and the cluster number counts with the thawing type of scalar field DE models [1] and compare the results with the corresponding ΛCDM model.

2. Details

2.1. Background evolution

We have considered a flat, homogeneous and isotropic background universe driven by non-relativistic matter and DE of thawing type, i.e., $\Omega_\phi + \Omega_m = 1$. These thawing type DE models are characterized by the fact that in the early universe the scalar field is frozen by very large Hubble damping, and the scalar field starts evolving slowly down its potential at the later time. So, the equation of state $w(a) = p_\phi/\rho_\phi$ initially starts with $w = -1$ and slowly departs from it in the later time. We have considered both ordinary scalar field with canonical kinetic term as well as tachyon type scalar field having Born-Infeld type kinectic term, which are minimally coupled to the gravity sector. The equations of motion for the canonical scalar field and the tachyon field are given by:

$$\ddot{\phi} + 3H\phi + dV/d\phi = 0, \quad \text{and} \quad \ddot{\phi} + 3\mathrm{H}\dot{\phi}(1-\dot{\phi}^2) + \frac{\mathrm{dV}/\mathrm{d}\phi}{\mathrm{V}}(1-\dot{\phi}^2) = 0 \tag{1}$$

respectively. (For details, see [3, 4].) H is the Hubble parameter, which describes the background expansion. Here, the energy density of DE is represented as $\rho_\phi = \rho_{\phi 0} f(a)$ with $f(a) = \exp\left[3\int_a^1 \left(\frac{1+w(u)}{u}\right) \mathrm{d}u\right]$. We have considered various types of potentials: e.g., $V = \phi$, $V = \phi^2$, $V = e^\phi$ and $V = \phi^{-2}$ as well as the Pseudo Nambu-Goldstone-Boson (PNGB) model, which is characterized by the potential $V(\phi) = m^4[\cos(\phi/f) + 1]$, with $f = 1$. The Hubble parameter is given by:

$$H^2 = \left(\frac{\dot{a}}{a}\right)^2 = \frac{8\pi G}{3}(\rho_m + \rho_\phi). \tag{2}$$

Here, $\rho_m = \rho_{m0} a^{-3}$ is the background matter density, and ρ_ϕ represents the DE density.

2.2. The spherical collapse model

The dynamics of a spherical region of radius $r(t)$ evolving in a cosmologically expanding background in the presence of DE is governed by the Raychaudury equation:

$$\frac{\ddot{r}}{r} = -4\pi G\left[\left(w(r) + \frac{1}{3}\right)\rho_{\phi c} + \frac{1}{3}\rho_{mc}\right], \tag{3}$$

where subscript "c" denotes inside the spherical over-densities. It is easier to solve the equations after normalizing at the turn-around point (denoted by subscript "t"). Therefore, with new variables $x = \frac{a}{a_t}$ and $y = \frac{r}{r_t}$, the equation of background evolution H, and that of the spherical perturbation, given in Eq. (2) become:

$$\dot{x}^2 = H_t{}^2\Omega_{m,t}[\Omega_m(x)x]^{-1}, \quad \text{and} \quad \ddot{\mathrm{y}} = -\frac{\mathrm{H_t^2}\Omega_{\mathrm{m,t}}}{2}\left[\frac{\zeta}{\mathrm{y}^2} + \frac{1-\Omega_{\mathrm{m,t}}}{\Omega_{\mathrm{m,t}}}\mathrm{yI(x,y)}\right], \tag{4}$$

where

$$I(x,y) = \begin{cases} [1 + 3w(r(y))]\,\frac{f(r(y))}{f(a_t)} & \text{Clustered DE} \\ [1 + 3w(x)]\, f(x) & \text{Homogeneous DE.} \end{cases} \tag{5}$$

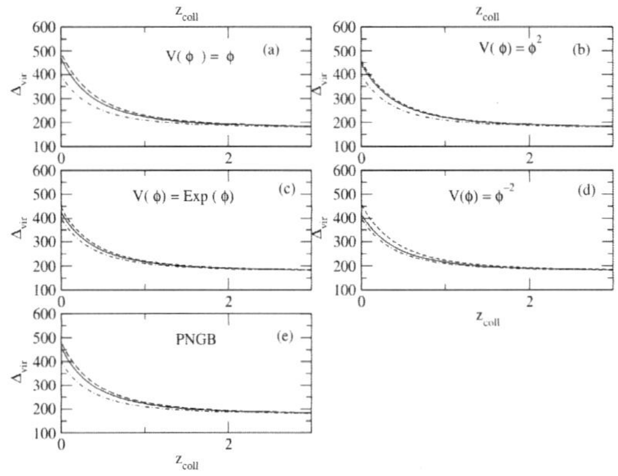

Figure 1. The non-linear density contrast at virialization vs. the collapse redshift (Z_{coll}) for $\Omega_{m0} = 0.25$. In each figure, solid, dashed and dotdashed represent inhomogeneous, homogeneous and ΛCDM models respectively. PNGB model is also compared with ΛCDM model in the last figure.

Here, ζ is the matter density contrast at turn-around, and we have calculated it from the above equations by applying the boundary conditions $(dy/dx)_{x=1} = 0$, $y_{x=0} = 0$, and $y_{x=1} = 1$ (See the details in [5]). Typically, the scalar fields have extremely small masses ($\sim H_0$ value in natural units), which is necessary for having nearly flat potentials at present day. Due to this, the scale of fluctuations for these scalar fields is extremely large, making it a smoothly distributed field within the horizon scale. So, it is safe to assume DE to be homogeneous. But it is still interesting to consider the case, where the DE clusters along with the dark matter and avoid the energy non-conservation problem examined in [6]. Hence, we have assumed both the cases.

In general, the spherical collapse formalism leads to a point singularity as the final state of the system. But physically, the objects go through a virialization process and stabilize to a finite size. Such process of virialization is not built in the spherical collpase formalism, but we have to put it by hand in order to ensure virialization. First of all, the total energy (TE) has to be conserved always, i.e., TE at the time of turn-around is equal to that at the time of virialization. At the turn-around, only the potential energy contributes, since $\dot{r} = 0$, whereas, at the virialization, the virial theorem holds. Further, we have investigated two specific cases:

- In inhomogeneous DE case, we have assumed that DE virializes together with the matter. i.e., The virial theorem holds for the total kinetic and potential energy of the system. Then, applying the energy conservation together with this, we have obtained a cubic equation of the ratio between the final (virial) r_f and the turn-around radius r_t, defined as the collapse factor $\lambda = \frac{r_f}{r_t}$ [5] as:
$$2n_1\lambda^3 - (2 + n_2)\lambda + 1 = 0, \tag{6}$$
where $n_1 = -(3w(a_f) + 1)\frac{\Omega_{\phi 0} f(a_f)}{\zeta \Omega_{m0} a_t^{-3}}$, and $n_2 = -(3w(a_t) + 1)\frac{\Omega_{\phi 0} f(a_t)}{\zeta \Omega_{m0} a_t^{-3}}$.
- In homogeneous DE case, we have assumed DE does not virialize inside the cluster. Its only effect is to contribute to the potential energy of the system. Similarly, applying the energy conservation, we have obtained λ for the homogeneous DE models.

After knowing ζ and λ, one can easily calculate the density contrast at virialization: $\Delta_{\rm vir} = \frac{\rho_{\rm mc,f}}{\rho_{\rm m,f}} = \frac{\zeta}{\lambda^3}\left(\frac{a_{\rm f}}{a_{\rm t}}\right)^3$, assuming that at the collapse point, the system has virialized fully. The relation between $a_{\rm f}$ and $a_{\rm t}$ is determined from the condition that the time needed to collapse is twice the turn-around time.

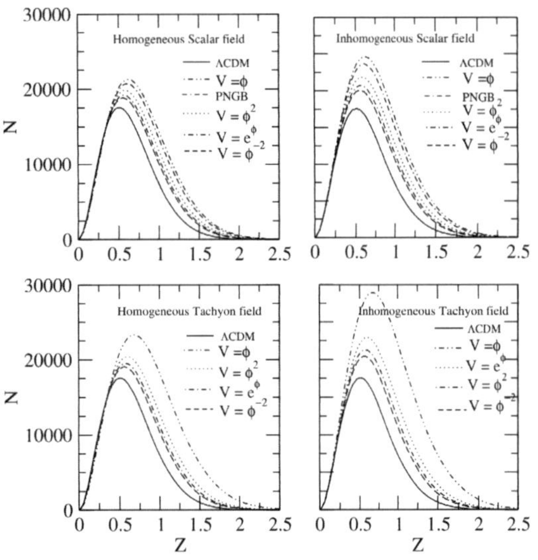

Figure 2. The total number counts N vs. Z, integrated over mass (from $M_{min} = 2 * 10^{14} M_{\odot}$) for a survey area of 10000 square degree. All the models are normalized to the same number density of haloes today. *Upper panel*: The canonical scalar field models with homogeneous (left) and inhomogeneous DE cases (right). *Lower panels*: The tachyon scalar field models with homogeneous and inhomogeneous dark energy cases in the left and right panels respectively. The different potentials correspond to the different line types. The concordance ΛCDM model (black solid line) is also plotted for comparison.

2.3. Number counts

The number of clusters in a redshift interval dz, above a given minimum (threshold) $M = M_{min}$ is obtained from $dn(M,z)/dM$ as:

$$\frac{dN}{dz}(M > M_{min}) = f_{sky}\frac{dV}{dz}\int_{M_{min}}^{\infty} dM \frac{dn}{dM}(M,z) = -f_{sky}\frac{dV}{dz}\int_{M_{min}}^{\infty} dM \frac{\rho_{m0}}{M}\frac{d\ln\sigma(M,z)}{dM} f(\sigma), \tag{7}$$

where f_{sky} is the fraction of the sky being observed and $\frac{dV}{dz}$ is the comoving volume element.

For numerical computation, the upper limit of integration in Eq. (7) is replaced by some finite mass value M_{max}. $f(\sigma(z))$ is the Sheth-Torman mass function [7]. $\sigma^2(R) = \frac{D(a)}{2\pi^2}\int_0^{\infty} k^3 P(k) W^2(kR)\frac{dk}{k}$ is the dispersion of the density field on a given comoving scale R, containing mass $M = 4\pi\rho_{m0}R(M)^3/3$. We have normalized the growth function such that $D(a) = 1$ at the present epoch. Assuming that the baryon density parameter $\Omega_{B0} \ll \Omega_{CDM,0}$, the CDM power spectrum can be approximated by $P(k) \propto kT^2(k)$, and we have used $T(k)$, Eisenstein and Hu's transfer function [8]. The effects of DE in the halo mass function come through the growth function $D(a)$ and the linear density contrast δ_c at the collapsed redshift. Using WMAP-7 data, we have normalized the power spectrum by setting $\sigma_8(\Lambda CDM) = 0.80$. We have set the other cosmological parameters as: $\Omega_{m0} = 0.25$, $\Omega_{\phi 0} = 0.75$, $h = 0.72$, $\Omega_{B0} = 0.0456$, and $n_s = 1$. (For details, see [9].)

3. Results and conclusion

In Figure 1, we have shown that although, all the models become indistinguishable for objects collapsing earlier, for objects collapsing around present time, some of the thawing models deviate significantly from the CDM model. The deviations are enhanced in the homogeneous DE case, where matter only virializes inside the cluster. This shows that inhomogeneous DE acts against the matter clustering. But, as the model is quite non-linear, it is difficult to predict the percentage of effects coming from the background and from DE clustering.

Thawing models with linear potential can have significant deviation from CDM model. One can see from Figure 2 that there exists significant difference in the number counts between different DE models considered, and the difference is most significant around $z \sim 0.5$ to 1. More specifically, seeing the top-left panel of Figure 2, the difference in number counts between ΛCDM and $V = \phi$ homogeneous scalar field model is ~ 2000 at $z \sim 0.7$, which is significantly larger than the statistical uncertainties (would be ~ 100 for the surveys like eROSITA and WFXT). Hence, the cluster number counts can be used for discriminating between different models. Although, this is a simplified approach to study the non-linear evolution of matter over-densities inside the cluster, and is not applicable to actual physical situation, it gives some interesting insight into the non-linear clustering of matter in the presence of thawing class of DE models. Given the fact that a large number of cluster surveys are currently ongoing as well as a number of future surveys are being planned, this can be a smoking gun to distinguish different dark energy models from ΛCDM. A proper analysis of how to constrain the models would involve error estimates of the parameters with a combination of different observational data sets, e.g., cluster counts, CMBR, BAO, etc.

Acknowledgements
N C Devi acknowledges the financial support provided by CSIR, Government of India.

References
[1] Caldwell R R and Linder E V 2005 *Phys. Rev. Lett.* **95** 141301
[2] Gunn J E and Gott J R 1972 *Ap. J.* **176** 1
[3] Scherrer R J and Sen A A 2008a *Phys. Rev. D* **77** 083515
[4] Scherrer R J and Sen A A 2008b *Phys. Rev. D* **78** 067303
[5] Devi N C and Sen A A 2011 *MNRAS* **413** 2371
[6] Maor I and Lahav O 2005 *JCAP* **7** 3
[7] Sheth R K and Tormen G 1999 *MNRAS* **308** 119
[8] Eisenstein D J and Hu W 1998, *Ap. J.* **496** 605
[9] Devi N C, Choudhury T R and Sen A A 2011 *MNRAS* [arXiv:1112.0728]

Characterizing the diffuse foreground for redshifted 21-cm HI signal: GMRT 153 MHz observation

A Ghosh[1*], J Prasad[2], S Bharadwaj[1], S S Ali[3], J N Chengalur[4]

[1]Department of Physics and Meteorology, and Centre for Theoretical Studies, IIT, Kharagpur 721302, India

[2] IUCAA, Post Bag 4, Pune University Campus, Ganeshkhind, Pune 411007, India

[3]Department of Physics, Jadavpur University, Kolkata 700032, India

[4]National Centre for Radio Astrophysics, TIFR, Post Bag 3, Ganeshkhind, Pune 411007, India

E-mail: *abhik.physicist@gmail.com, ghosh@astro.rug.nl

Abstract. Detailed knowledge of the foreground structure on the angular scales of $\sim 1^\circ$ to sub-arcminute will be essential for extracting the redshifted 21-cm HI signal from the observed data. We have presented results from the GMRT observations at 153 MHz, which was used to characterize the statistical properties of the diffuse radiation, using the multi-frequency angular power spectrum, across sub-degree angular scales. We have detected fluctuations in the diffuse emission on angular scales greater than $10'$ in a low galactic latitude area. The total intensity angular spectrum shows a power-law behaviour, while the detection of diffuse emission at smaller angular scales is limited by residual point sources. Also, we have estimated the level of foreground contamination.

1. Introduction

Observations of redshifted 21-cm radiation from neutral hydrogen (HI) hold the potential of tracing the large scale structure of the Universe over a large redshift range. The 21-cm radiation is expected to have an rms brightness temperature of a few mK on angular scales of a few arcminutes [1]. This signal is, however, buried in the emission from other astrophysical sources, which are collectively referred to as foregrounds. Individual sources (mostly point sources) can be identified, and removed from the image at a flux level, which depends on the sensitivity of the instrument. The rest of the foreground emission is dominated by the diffuse galactic synchrotron radiation from our galaxy [2]. Our present knowledge of the galactic diffuse radio emission at metre wavelength is not adequate. Estimates of the spatial fluctuations on arcminute scales are required for reionization experiments. In this paper, we have presented the first results from the Giant Metrewave Radio Telescope (GMRT[1], [3]) observations at 153 MHz with arcminute angular resolution, aimed at characterizing the properties of the diffuse foregrounds for reionization experiments.

Section 2 describes the observations and data analysis part, while Section 3 contains a detailed discussion on the diffuse galactic foreground component with a power spectrum analysis.

[1] http://www.gmrt.ncra.tifr.res.in

2. GMRT observations and data analysis

On January 8, 2008, we have observed a field centred on $\alpha_{2000} = 05^h30^m00^s, \delta_{2000} = +60°00'00''$ for a total time of $\sim$ 11 hours (incuding calibration). The galactic coordinates of this field are $l = 151.80°$ and $b = 13.89°$. From the 408 MHz Haslam map, the sky temperature at this location is $\sim 40\ K$ [4], and there is no structure visible at the angular resolution ($\sim 0.85°$) of the Haslam map.

In this analysis, the data reduction was carried out using a software called FLAGCAL [5], which identifies and removes bad visibilities, computes calibrated solutions using known flux and phase calibrations, and interpolates these onto the target source. The flagged and calibrated visibility data have been used to make continuum images using the standard tasks in the Astronomical Image Processing Software (AIPS).

3. Diffuse galactic foreground

We have characterized the fluctuations in the diffuse galactic foreground in our observed field. Figure 1 shows the continuum image of this field, where hundreds of point sources are visible inside the field of view of $4.0° \times 4.0°$. Several hundreds of sources were identified through a CLEAN de-convolution in a $8.0° \times 8.0°$ image, and all the point sources are removed down to a level of $\sim 20\ mJy$ from the continuum image. We have noted that below $20\ mJy$, the field is dominated by the residual imaging artifacts (see Figure 2). Figure 3 shows the resulting image, where all the point sources down to $20\ mJy$ level are removed. No diffuse structure appears in this map. However, it is not clear whether diffuse structure exist on arcminute scales. In order to investigate this point, we have made an image including only the baseline with $|\mathbf{U}| > 170$. This restriction imposes a condition such that all the angular scales $> 10'$ are absent in the resulting image (Figure 4). No diffuse structures were noticed in this map, which signify absence of diffuse power on smaller scales. The most relevant contributions left are all from residual point sources. To characterize the diffuse structures on large angular scales, we have started to taper the uv plane. We have noticed that diffuse structures begin to appear on $> 10'$ angular scales. The corresponding image, which has a synthesized beam of FWHM of $620'' \times 540''$ is shown in Figure 5, where the contribution from individual point

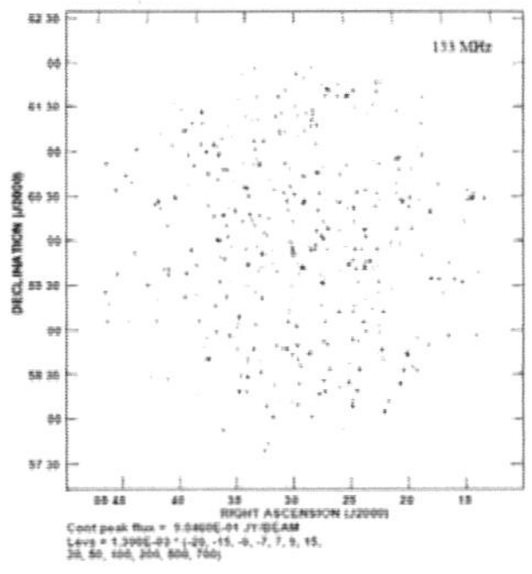

Figure 1. The figure shows our continuum image of bandwidth 3.75 MHz centred at 153 MHz. The $4° \times 4°$ field was imaged using 121 facets, which have been combined using AIPS task FLATN. The synthesized beam has a FWHM of $20'' \times 18''$.

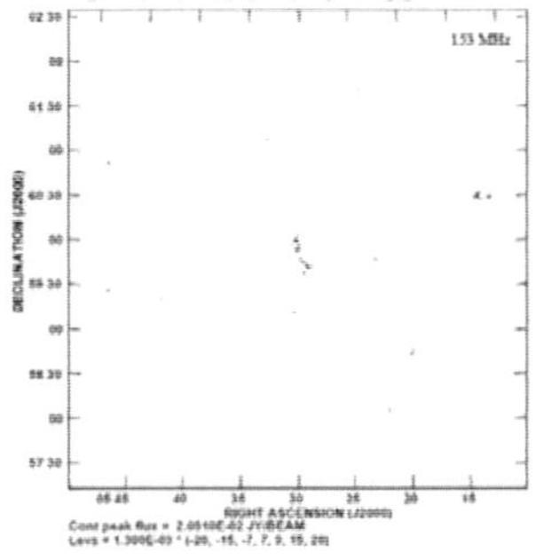

Figure 2. Pixels with flux density above $10\ mJy/beam(7\sigma)$, visually identified as genuine sources and not artifacts, have been fitted with clean components, and removed from the visibility data from which this image is made. We have noticed that the peak residual is around $20\ mJy/beam$.

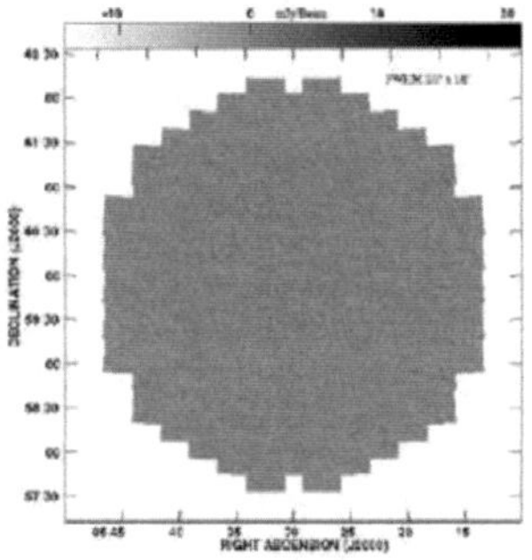

Figure 3. The image, where all the point sources down to a level of 20 mJy are removed. No structures are visible at the angular resolution ($20'' \times 18''$) of this map.

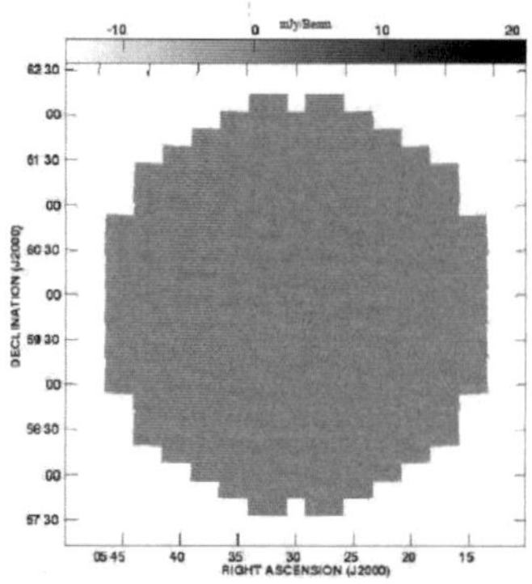

Figure 4. The image made by only including those baselines with $|\mathbf{U}| > 170$. All angular scales $> 10'$ are absent in this map.

sources have completely disappeared, and on the other hand, we have noticed that the pattern of fluctuations have started dominating the whole image. This pattern does not correlate with the

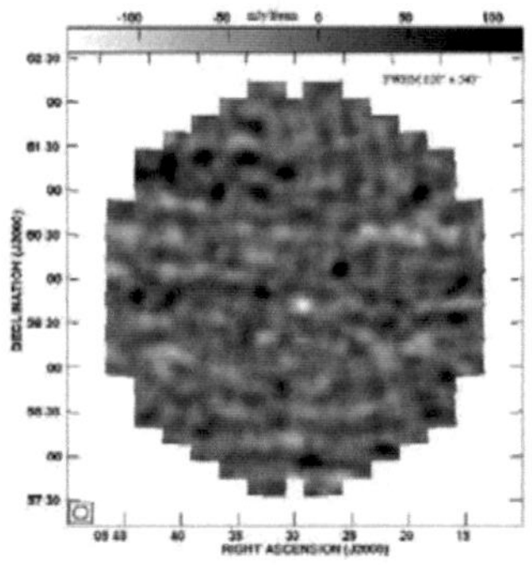

Figure 5. This shows the image, where the point sources were removed down to 20 mJy level, and the uv plane was tapered at $|\mathbf{U}| = 200$, such that the synthesized beam has a FWHM of $620'' \times 540''$. Diffuse structures begin to appear on $> 10'$ scales.

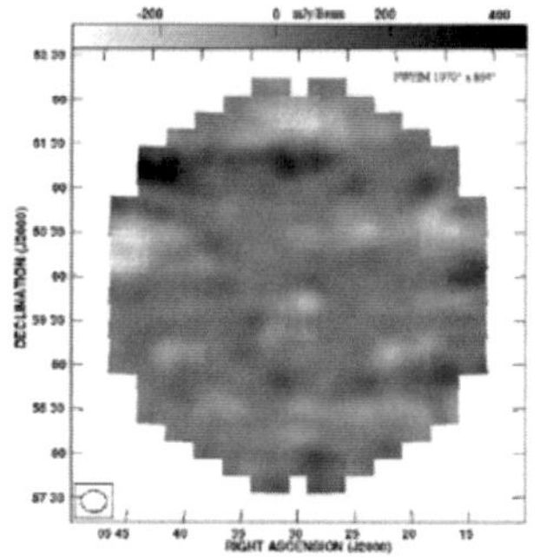

Figure 6. This shows the image, where we tapered the uv plane at $|\mathbf{U}| = 100$ such that the synthesized beam has a FWHM of $1070'' \times 864''$. All the point sources are removed down to 20 mJy level. The brighest structures in this map are at 10σ level compared to the local rms value of 35 $mJy/beam$.

distribution of point sources shown in Figure 1, indicating that possible unsubtracted sources do not contribute significantly to the diffuse emission. Then, we have continued tapering the uv plane (at $|\mathbf{U}| = 100$) to see if diffuse structures appear at even higher angular scales. Figure 6 shows the continuum image, where the synthesized beam has FWHM $1070'' \times 864''$. In this map, we have detected fluctuations in the brightest structure $> 10\sigma$ level compared to the local rms value, which is around 35 $mJy/beam$ ($\sim 2.24\ K$) in the inner region of the map. We have noticed that the rms value on the lowest resolution map accounts for diffuse emission only, whereas there is a clear indication of significant contribution from unsubtracted point sources in the highest resolution map (rms $\sim 1.3\ mJy/beam \equiv 223.34\ K$). Based on these observations, we have concluded that the observed fluctuations represent structure in the galactic foreground radio emission on scales of $10' - 20'$, but there is a clear absence of diffuse power on arcminute scales.

3.1. Power spectrum

We have used the correlation between pairs of visibilities after subtracting all the identified sources above 20 mJy to estimate the angular power spectrum $C_\ell(\Delta\nu = 0)$ (Hereafter, this will be C_ℓ) [6]. The problem of noise bias is avoided by correlating visibilities at two different baselines for which the noise is expected to be uncorrelated [7].

The total intensity angular power spectrum C_ℓ (see Figure 7) clearly shows two different scaling behaviour as a function of the angular scale. At large angular scales ($> 10'$), we have found a power law behaviour, which is typical of the galactic diffuse emission observed at higher frequencies and lower angular resolutions [8]. Above $\ell \sim 800$, the power spectrum flattens and then it remains flat down to angular scales of $\sim 1'$. Also, we have noticed similar behaviour,

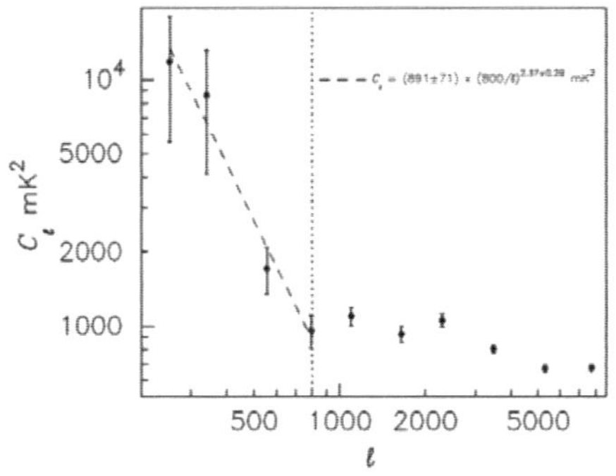

Figure 7. The total intensity power spectrum at 153 MHz, measured from the visibilities after subtracting all the identified sources above 20 mJy. The 1σ error bars have contributions from both the cosmic variance and system noise.

where the dominant contribution from the diffused emission has disppeared from the image (see Figure 4) made by discarding the short baseline ($|\mathbf{U}| > 170$). It seems that the clear break in the angular power spectrum at $\ell \sim 800$ (equivalently $\theta > 10'$) agrees well with our image plane analysis in Section 3. Based on these facts, we have concluded that the power spectrum down to $\ell \sim 800$ is representative of the diffuse galactic foreground emission, whereas, it is dominated by residual point sources at higher ℓ values. The total intensity power spectrum of the diffuse emission was fitted with a power-law $C_\ell = (891 \pm 71) \times (\frac{800}{\ell})^{2.37 \pm 0.29}$ mK^2 down to $\ell = 800$. The amplitude and the slope of the power-law also remains unchanged if we correlate two visibilities, which are sampled at different time intervals. This indicates that our estimate of the total intensity power spectrum is free from any RFI contaminations, which are mostly expected to be correlated within short time intervals. Earlier, Bernardi *et al* [9] have reported total intensity and polarization fluctuations in the galactic diffuse foreground emission at a low galactic latitude area ($b = 6.4°$), which have been measured at the frequencies and angular scales relevant for upcoming reionization experiments. We have found that the break in the total intensity angular power spectrum and the power-law slope agrees well with their results.

Acknowledgements

AG acknowledges the CSIR, Government of India, for the financial support. SSA acknowledges the support by DST, Government of India, under the project No.: SR/FTP/PS-088/2010. We thank the GMRT staff for their support.

References

[1] Zaldarriaga M, Furlanetto S R and Hernquist L 2004 *Ap. J.* **608** 622
[2] Shaver P A, Windhorst R A, Madau P and de Bruyn A G 1999 *A. & A.* **345** 380
[3] Swarup G, Ananthakrishnan S, Kapahi V K, Rao A P, Subrahmanya C R and Kulkarni V K 1991 *Current Sci.* **60** 95
[4] Haslam C G T, Salter C J, Stoffel H and Wilson W E 1982 *A. & A. Supp. Ser.* **47** 1
[5] Prasad J and Chengalur J N 2012 *Expt. Astron.* **33** (1) 157
[6] Ghosh A, Bharadwaj S, Ali S S and Chengalur J N 2011 *MNRAS* **418** 2584
[7] Begum A, Chengalur J N and Bharadwaj S 2006 *MNRAS* **372** L33
[8] Bennett C L, Hill R S, Hinshaw G *et al* 2003 *Ap. J. Supp. Ser.* **148** 97
[9] Bernardi G *et al* 2009 *A. & A.* **500** 965

Journal of Physics: Conference Series **484** (2014) 012033 doi:10.1088/1742-6596/484/1/012033

The cross-correlation of redshited 21-cm signal and Lyman-α forest: A cosmological probe

T Guha Sarkar[1]* and S Bharadwaj[2]

[1]Harish-Chandra Research Institute, Chhatnag Road, Jhusi, Allahabad 211019, India
[2] Department of Physics and Meteorology, IIT, Kharagpur 721302, India

E-mail: *tapomoy@hri.res.in

Abstract. We have investigated the cross-correlation of the Lyman-α forest and redshifted 21-cm emission as a new observational probe of the large scale structures in the post-reionization era, with a significant advantage that the problems of continuum subtraction and foreground removal, and other systematics are expected to be considerably less servere in comparison to the respective auto-correlation signals. In this paper, we have explored the possibility of detecting the baryon acoustic oscillation in the cross-correlation signal. We have developed a formalism to calculate the expected cross-correlation signal and its covariance. We have used this to predict the expected signal and noise for a range of observational parameters.

1. Introduction

The cross-correlation of the 21-cm signal with the Lyman-α forest has recently been proposed as a new probe of the post-reionization era [1]. The 21-cm emission and the Lyman-α absorption signals, both originate from neutral hydrogen (HI) at the same redshift. However, these signals emanate from distinctly different sources. The 21-cm emission comes from the damped Lyman-α systems, which harbours bulk of HI at low redshifts. On the contrary, Lyman-α forest arises from small density fluctuations in the pre-dominantly ionized IGM. On large scales, however, the fluctuations in the 21-cm signal and the Lyman-α forest transmitted flux are both excellent tracers of the underlying dark matter density field and hence, the cross-correlation between the two can be used to probe the matter power spectrum. The foregrounds and systematics in the Lyman-α forest and the redshifted 21-cm emission are expected to be uncorrelated and therefore, pose much less severe problem for the cross-correlation. The signal, if detected, shall hence, ascertain its cosmological origin conclusively. Apart from being an independent probe of the large scale matter distribution, the same cosmological and astrophysical information as the individual signals is contained in the cross-correlation signal.

The sound horizon at recombination may be used as a standard ruler to calibrate cosmological distances. The baryon acoustic oscillation (BAO) imprinted in the late time clustering of matter allows a measurement of the angular diameter distance and the Hubble parameter as functions of redshift using the transverse and the longitudinal oscillations respectively. These provide means for constraining dark energy models. We have considered the possibility of detecting BAO using the cross-correlation signal. We have quantified the cross-correlation between the Lyman-α forest and the 21-cm emission using the multi-frequency angular power spectrum

$C_\ell(\Delta z)$ [2], and presented the theoretical predictions for the expected signal and noise for a range of observational parameters.

2. The cross-correlation signal

The fluctuations in the Lyman-α forest transmitted flux $\mathcal{F}(\hat{n}, z)$ along a line of sight $\hat{n}$ to a quasar is quantified using:

$$\delta_{\mathcal{F}}(\hat{n}, z) = \mathcal{F}(\hat{n}, z)/\bar{\mathcal{F}} - 1.$$

On large scales, it is reasonable to adopt the fluctuating Gunn-Peterson approximation [3, 4] relating the flux to the matter density contrast δ. If fluctuations $\delta_{\mathcal{F}}$ are smoothened over sufficiently large length scales, then $\delta_{\mathcal{F}} \propto \delta$ [3].

We have used $\delta_T(\hat{n}, z)$ to quantify the fluctuations in the brightness temperature of redshifted 21-cm emission. In the redshift range of our interest ($z < 3.5$), it is reasonable to assume that $\delta_T(\hat{n}, z)$ is a biased tracer of δ. Simulations [5] indicate that the bias is scale independent on large scales. With these assumptions, we may express both $\delta_{\mathcal{F}}$ and δ_T in Fourier space as:

$$\Delta_\alpha(\mathbf{k}) = C_\alpha \left[1 + \beta_\alpha \mu^2\right] \Delta(\mathbf{k}),$$

where $\alpha = \mathcal{F}$ and T refer to the Lyman-α forest and 21-cm signal respectively. $\Delta(\mathbf{k})$ is the matter density contrast in Fourier space and $\mu = \hat{\mathbf{k}}.\hat{\mathbf{n}}$. We have used C_α, β_α from [6] and [1]. These are the values at our fiducial redshift $z = 2.5$.

We have used the smaller field of view $L \times L$ (L in radians) of the radio telescope to estimate the cross-correlation signal. In the flat sky approximation ($L \ll 1$), it is convenient to decompose $\delta_\alpha(\vec{\theta}, z)$ into Fourier modes $\Delta_\alpha(\mathbf{U}, z)$, where $\mathbf{U}$ is conjugate to θ. The multi-frequency angular power spectrum (MAPS) is the correlation of angular modes on two planes at comoving distances r and $r + \Delta r$, defined as:

$$\langle \Delta_\alpha(\mathbf{U}, z) \Delta^*_\gamma(\mathbf{U}', z + \Delta z) \rangle = L^2 \delta_{\mathbf{U}\mathbf{U}'} P_{\alpha\gamma}(\mathbf{U}, \Delta z),$$

where the indices α and γ both assume values T and $\mathcal{F}$. The entire three dimensional information is contained in $P_{\alpha\gamma}(\mathbf{U}, \Delta z)$ through the $\mathbf{U}$ and Δz dependence. It is related to the $3D$ matter power spectrum $P(k)$ as [2]:

$$P_{\alpha\gamma}(U, \Delta z) = \frac{1}{\pi r^2} \int_0^\infty dk_\parallel \cos(k_\parallel \Delta r)\ F_{\alpha\gamma}(\mu)\ P(k)\ ,$$

where $\Delta r = c\Delta z/H(z)$, $k = \sqrt{k_\parallel^2 + (2\pi U/r)^2}$, $\mu = k_\parallel/k$, and

$$F_{\alpha\gamma}(\mu) = C_\alpha\ C_\gamma\ \left[1 + \beta_\alpha \mu^2\right] \left[1 + \beta_\gamma \mu^2\right] .$$

The angular mode U may be identical with the angular multipole ℓ as $2\pi U = \ell$. We have referred the $2D$ transverse angular power spectrum as:

$$P_{\mathcal{F}T}(\ell) \equiv P_{\mathcal{F}T}(U, \Delta z = 0)\ .$$

This has contribution from all the $3D$ modes $\mathbf{k}$, whose projection on the transverse plane matches $\ell/r \equiv 2\pi U/r$, such that only modes $k > \ell/r$ contribute to $P_{\mathcal{F}T}(\ell)$. The amplitude of $P_{\mathcal{F}T}(\ell)$ depends on C_T and $C_{\mathcal{F}}$, whose values are highly uncertain. On the contrary, the shape of $P_{\mathcal{F}T}(\ell)$ is related to the matter power spectrum and the comoving distance r. At large ℓ (> 1000), we have $P_{\mathcal{F}T}(\ell) \propto \ell^{-1.76}$. The angular power spectrum flattens out at $\ell \sim k_{eq} r$, which corresponds to the peak in $P(k)$.

Journal of Physics: Conference Series **484** (2014) 012033 doi:10.1088/1742-6596/484/1/012033

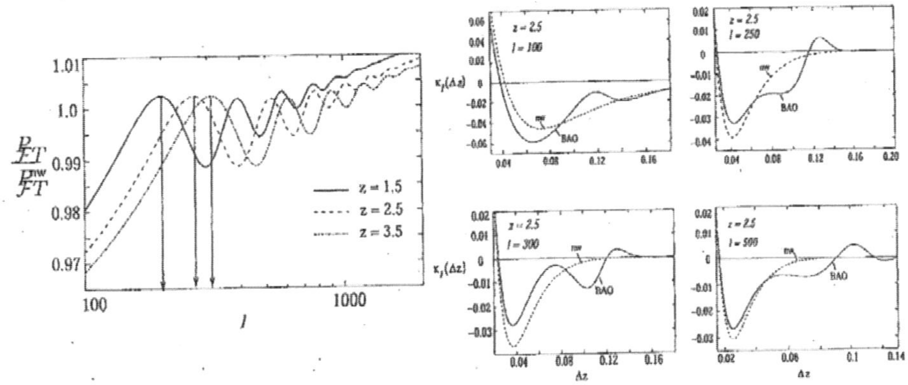

Figure 1. This shows the imprint of the BAO on the transverse angular power spectrum $P_{\mathcal{F}T}(\ell)$ and the radial de-correlation $\kappa(\Delta z)$ for the cross-correlation signal. To highlight the BAO, we have divided $P_{\mathcal{F}T}(\ell)$ by $P^{nw}_{\mathcal{F}T}(\ell)$, which corresponds to $P(k)^{nw}$ [7]. This is shown for three redshifts $z = 1.5, 2.5$, and 3.5. The $\kappa(\Delta z)$ is shown for the fiducial redshift $z = 2.5$ for several ℓ values. The BAO in κ shows ringing around its 'no wiggles' counterpart.

At fixed ℓ, $P_{\mathcal{F}T}(U, \Delta z)$ is the correlation between $\Delta_{\mathcal{F}}(\mathbf{U}, z)$ and $\Delta_T(\mathbf{U}, z + \Delta z)$ at fixed $\mathbf{U}$, but located on two planes separated by Δr. The cosine term arises due to a single $3D$ mode projecting onto planes with a phase difference of $k_{\|}\Delta r$. We have introduced the de-correlation function as:

$$\kappa_\ell(\Delta z) = P_{\mathcal{F}T}(\ell, \Delta z)/P_{\mathcal{F}T}(\ell),$$

which varies in the range $0 \leq |\kappa_\ell(\Delta z)| \leq 1$. With increasing Δz, $\kappa_\ell(\Delta z)$ falls, crosses zero and then gets anti-correlated . The de-correlation is faster for large ℓ values in the entire z range.

The BAO scale s defines an angular scale $\theta_s = s[(1+z)D_A(z)]^{-1}$ and a redshift interval $\Delta z_s = \frac{sH(z)}{c}$, where $D_A(z)$ and $H(z)$ are the angular diameter distance and Hubble parameter respectively. The MAPS allows us to measure θ_s and Δz_s separately, and thereby, $D_A(z)$ and $H(z)$. Figure 1 shows the BAO feature in $P_{\mathcal{F}T}(\ell)$. The BAO appears as a series of oscillations in $P_{\mathcal{F}T}(\ell)$, the positions of the peaks being consistent with $\ell \sim kr$. The amplitude of the first oscillation in $P_{\mathcal{F}T}(\ell)$ is around 1%, in contrast to the $\sim 10\%$ feature in $P(k)$. The reduction is caused by the projection on to plane. At $z = 2.5$, the first peak occurs at $\ell \sim 270$, and has a full width of $\Delta\ell \sim 200$. These values scale as r if the redshift is changed.

The first BAO peak imprints on $\kappa_\ell(\Delta z)$ only if $\ell < k_1 r$, where k_1 is the position of the first peak in the $3D$ matter power spectrum. Thus, for $z = 2.5$, the first peak has no impact on $\kappa_\ell(\Delta z)$ at $\ell > 500$, and has an impact at all angular modes $\ell \leq 500$. Figure 1 shows $\kappa_\ell(\Delta z)$ along with $\kappa_\ell(\Delta z)$ calculated using $P^{nw}(k)$, which does not have the BAO features. The BAO has little impact near $\Delta z = 0$, where $\kappa_\ell(\Delta z)$ is positive. The significant effect (as large as 40% to $\sim$ 100% relative to the no-wiggles model) is seen only at large Δz, typically in the range $\Delta z = 0.04$ to 0.16, where $\kappa_\ell(\Delta z)$ is negative. The BAO ringing feature around the smooth 'no wiggles' $\kappa_\ell(\Delta z)$, is distinct from the slow oscillation seen in the no-wiggles model.

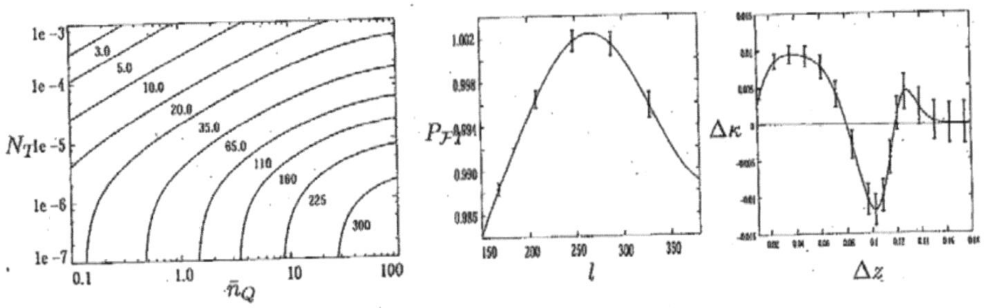

Figure 2. The SNR contours for a single observation at $z = 2.5$. The binned $P_{\mathcal{F}T}(\ell)$ and κ are shown with the error bars corresponding to a 5σ detection of the central bin. The no-wiggles [7] spectrum is divided out.

2.1. Detecting the cross-correlation signal

The observed Lyman-α flux $\delta_{\mathcal{F}0}(\vec{\theta}, n)$ can be written as:

$$\delta_{\mathcal{F}0}(\vec{\theta}, n) = \rho(\vec{\theta}) \left[\delta_{\mathcal{F}}(\vec{\theta}, z_n) + \delta_{\mathcal{F}N}(\vec{\theta}, n) \right],$$

with $z_n = z_0 + n\Delta z_c$ and $n = 1, 2, \cdots, N_c$. Here, z_0 is a reference redshift N_c, which is the total number of frequency channels, $\delta_{\mathcal{F}N}(\vec{\theta}, z)$ is the pixel noise, and $\rho(\vec{\theta}) = \sum_a \omega_a \delta_D^2(\vec{\theta} - \vec{\theta}_a) / \sum_a \omega_a$ is the sampling function. Here, $\vec{\theta}_a$ and ω_a respectively denote the positions and weights of the quasars with $a = 1, 2, \cdots, N$ and

$$\langle \delta_{\mathcal{F}N}(\vec{\theta}_a, n)\ \delta_{\mathcal{F}N}(\vec{\theta}_b, m) \rangle = \delta_{a,b}\ \delta_{n,m} \sigma^2_{\mathcal{F}N},$$

where $\sigma^2_{\mathcal{F}N}$ is the noise variance. We have used weights $\omega_a = 1$ assuming high SNR of flux for all the quasars. In Fourier space, we have:

$$\Delta_{F_0}(\mathbf{U}, z) = \tilde{\rho}(\mathbf{U}) \otimes \left[\Delta_{\mathcal{F}}(\mathbf{U}, z) + \Delta_{N\mathcal{F}}(\mathbf{U}, z) \right].$$

Radio observations directly measure $\Delta_T(\mathbf{U}, z_n)$. Each pair of antennas measure:

$$\Delta_{T_0}(\mathbf{U}, n) = \Delta_T(\mathbf{U}, z_n) + \Delta_{TN}(\mathbf{U}, n),$$

where Δ_{TN} is a noise term. We have defined an estimator $\hat{E}$ as:

$$\hat{E}(\mathbf{U}, p) = \sum_{n=1}^{N} \sum_{m=1}^{N} \frac{1}{2} \left[\Delta_{\mathcal{F}_0}(n) \Delta^*_{T_0}(m) + \Delta^*_{\mathcal{F}_0}(n) \Delta_{T_0}(m) \right] \delta_{|n-m|,p} \Big/ \sum_{n=1}^{N} \sum_{m=1}^{N} \delta_{|n-m|,p} \,,$$

with the property that:

$$\langle E(\mathbf{U}, p) \rangle = P_{\mathcal{F}T}(\mathbf{U}, p\Delta_{z_c}),$$

and a covariance matrix is given by:

$$\mathbf{Cov}(p, q) = \sum_{\alpha,\beta} \sum_{n=1}^{N_{c-p}} \sum_{m=1}^{N_{c-q}} P_{\alpha\beta 0}(p + m - n) P_{\alpha'\beta' 0}(m - n - q) / 8(N_c - p)(N_c - q) \,,$$

where the variable α' has the value of $\mathcal{F}$ when $\alpha = T$ and vive-versa, and β, β' are defined in a similar way. Here,

$$P_{\mathcal{F}\mathcal{F}_0}(\mathbf{U}, p) = P_{\mathcal{F}\mathcal{F}}(\mathbf{U}, p\Delta z_c) + \frac{1}{\bar{n}_Q}\left[\xi_{\mathcal{F}}(p\Delta_{z_c}) + \delta_{p,0}\sigma^2_{\mathcal{F}N}\right],$$

where $\xi_{\mathcal{F}}(\Delta z)$ is the $1D$ flux correlation function and $\bar{n}_Q = N/L^2$ is the quasar density on the sky. Further,

$$P_{TT_0}(\mathbf{U}, p) = P_{TT}(\mathbf{U}, p\Delta_{z_c})) + \delta_{p,0}N_T(\mathbf{U}) \; ,$$

where $N_T(\mathbf{U})$ is the 21-cm noise power spectrum.

We have considered the quasars in the redshift range $2 < z < 3$. The region 10000 $\text{km}s^{-1}$ blue-wards of the quasar's Lyman-α emission is excluded due to the quasar's proximity effect and only the pixels at least 1000 $\text{km}s^{-1}$ red-ward of the quasar's Lyman-β and O-VI lines are considered. Also, we have considered 21-cm observations of bandwidth $B = 128$ MHz centred around 406 MHz. Only a fraction (approximately 40%) of the total number of quasars will contribute the cross-correlation in the entire z range 2 to 3. We have considered a radio array with antennas of diameter $D = 2m$ ($L = 20°$) distributed such that all the baselines $\vec{d}$ within $d_{max} = 50$ m are uniformly sampled.

We have used $\bar{n}_Q$ and N_T to parametrize the respective sensitivities of the QSO survey and the 21-cm observations. We have first considered the $P_{\mathcal{F}T}(\ell) \equiv P_{\mathcal{F}T}(U, \Delta z)$ at $\Delta z = 0$. We have assumed that the baseline range has been divided into 5 equal bins and that the Lyman-α forest is smoothened to the channel width 100 KHz of the 21-cm data. Figure 2 shows SNR contours as a function of $\bar{n}_Q$ and N_T. We have seen that the SNR does not increase very significantly beyond $\bar{n}_Q > 60$ deg^{-2} or $N_T < 10^{-7}$ mK^2 with a maximum value of the SNR (= 300) at the bottom right corner. We have found that a 5σ detection of $P_{\mathcal{F}T}(\ell)$ is possible for $\bar{n}_Q \sim 0.1$ deg^{-2} and $N_T \sim 10^{-4}$ mK^2.

The first BAO peak is a 1% feature in $P_{\mathcal{F}T}(\ell)$, and a 5σ detection of the BAO peak requires an SNR of 500 for $P_{\mathcal{F}T}(\ell)$, and it is necessary to consider multiple pointings. The BOSS[1] is expected to have a QSO density of 16 deg^{-2}, which corresponds to $\bar{n}_Q = 6.4$ deg^{-2}. This survey is expected to cover ~ 10000 deg^2 of the sky, and we could ideally have $N_{poin} = 25$ independent pointings of the $20° \times 20°$ field of view. We have SNR = 100 for $N_T = 1.1 \times 10^{-5}$ mK^2, and a 5σ detection of the first BAO peak is possible with $N_{poin} = 25$ fields of view. BigBOSS [8] may achieve a QSO density of $\sim$ 64 deg^{-2}, which corresponds to $\bar{n}_Q = 25.6$ deg^{-2}. At $N_T = 3.3 \times 10^{-5}$ mK^2, 6σ detection is possible with $N_{poin} = 25$. The radial oscillations is a $\sim$ 1% deviation, which is comparable to the deviation introduced by the angular oscillations. The BAO signal is maximum in the vicinity of $\ell \approx 250$. We have collapsed the three central U bins in order to enhance the SNR for the radial oscillations. We have found that a 5σ detection is possible with BOSS if we observe 25 fields of view with $N_T = 6.25 \times 10^{-6}$ mK^2. We require $N_T = 1.7 \times 10^{-5}$ mK^2 for a similar detection with BigBOSS.

BAO observations can be used to constrain cosmological parameters and the equation of state of the dark energy. In this paper, we have estimated the range of observational parameters, for which it will be possible to detect the BAO using the cross-correlation signal. Despite being a promising tool, several observational issues including the 21-cm foreground subtraction shall, however, require serious consideration towards detecting the signal.

References

[1] Guha Sarkar T, Bharadwaj S, Choudhury T R and Datta K K 2011 *MNRAS* **410** **2** 1130
[2] Datta K K, Choudhury T R and Bharadwaj S 2007 *MNRAS* **378** 119
[3] Bi H and Davidson A F 1997 *Ap. J.* **479** 523

1 http://cosmology.lbl.gov/BOSS/

[4] Croft R A C, Weinberg D H, Katz N and Hernquist L 1998 *Ap. J.* **495** 44
[5] Guha Sarkar T, Mitra S, Majumdar S and Choudhury T R 2012 *MNRAS.tmp.2568G* [arXiv:1109.5552]
[6] McDonald P 2003 *Ap. J.* **585** 34
[7] Eisenstein D J and Hu W 1998 *Ap. J.* **496** 605
[8] Schlegel D J, Bebek C, Heetderks H, Ho S and Lampton M 2009 [arXiv: e-print]

Cosmic microwave and infrared backgrounds cross-correlation for ISW detection

S Ilić

Université Paris-Sud, Institut d'Astrophysique Spatiale, UMR8617, Orsay, F-91405, and CNRS, Orsay, F-91405, France

E-mail: stephane.ilic@ias.u-psud.fr

Abstract. We have investigated the cross-correlation between the cosmic infrared and microwave backgrounds (CIB and CMB) anisotropies through the integrated Sachs-Wolfe effect. We have first described the CIB anisotropies using a linearly biased power spectrum, then derive the theoretical angular power spectrum of the CMB-CIB cross-correlation for different instruments and frequencies. We have discussed the detectability of the ISW signal by performing a signal-to-noise (SNR) analysis with our predicted spectra. The significances obtained range from 6σ to 7σ in an ideal case, depending on the frequency; in realistic cases which account for the presence of noise including astrophysical contaminants, the results span the range $2-5\sigma$, depending strongly on the major contribution to the noise term.

1. Introduction

The discovery of the acceleration of the expansion of the Universe, made through supernovæ observations at the end of the last century, led us to hypothesize the existence of an unknown "Dark Energy", contributing 70% to the energy budget of our Universe. In this paper, we have focused on a specific probe of the dark energy, namely the integrated Sachs-Wolfe (ISW) effect: This effect of gravitational origin induces secondary anisotropies in the CMB and is due to large-scale structures. The gravitational potentials of the latter are slowly decaying in a Λ-dominated Universe, and therefore, give a net difference in energy to the CMB photons that travel across them. This effect shows in the power spectrum of the Cosmic Microwave Background (CMB) temperature anisotropies at large angular scales, but its small amplitude and the cosmic variance at those ℓ make its direct detection very challenging, if not impossible, when using only the CMB itself. To circumvent this limitation, cosmologists have devised a way to exploit the link between this imprint on the CMB and the large-scale structures causing it, by simply cross-correlating the CMB with matter density maps (galaxy maps in practice), and then comparing the results to a null hypothesis and to what is expected from theory.

During the last decade or so, a growing interest has risen in this field thanks to the development of large galaxy surveys in many wavelengths. However, this method has yet to produce a definitive and conclusive detection of the ISW effect from current surveys, with significances ranging so far from negligible up to more than 4σ, and with sometimes conflicting results throughout the literature. We have made, here, a short report of our work presented in [1]. The originality of our work is to consider the Cosmic Infrared Background (CIB), first discovered by Puget *et al* in 1996. This background, visible roughly from 10 to 1000 μm in wavelength, arises from the accumulated emission of star-forming galaxies spanning a large range of redshifts including our current dark energy dominated era. The CIB also features anisotropies that are underlined by the galaxy density field, and thus, the matter density fluctuations. It is, therefore, reasonable to expect that it has a positive correlation with the CMB through the aforementioned ISW effect.

2. Modelling the expected signal

2.1. CIB anisotropies

Ever since its discovery, many efforts have been deployed to detect the CIB and its anisotropies with increasing precision. The most recent studies on the subject make use of the halo occupation distribution and the dark matter halos properties, in order to predict the power spectrum of these anisotropies. Such models are particularly useful when describing the small, non-linear scales of the CIB. Since, we have focused, here, on the ISW effect which only concerns much larger scales, we can use a simpler model for the CIB anisotropies power spectrum, similar to the description made by Knox *et al* in [2]. We first write the CIB temperature anisotropies as the following line-of-sight integral:

$$\delta T_{CIB}(\hat{\mathbf{n}}, \nu) = \int_{\eta_{far}}^{\eta_0} dz \, \frac{d\eta}{dz} \, a(z) \, \delta j\Big((\eta_0 - \eta)\hat{\mathbf{n}}, \nu, z\Big), \tag{1}$$

with δj being the emissivity fluctuations of the CIB. Again, similar to [2], we have assumed that the CIB anisotropies are direct tracers of the matter density fluctuations up to a factor $b_j(\nu, z)$, a frequency and redshift-dependent matter-emissivity bias, given as:

$$\delta j\Big((\eta_0 - \eta)\hat{\mathbf{n}}, \nu, z\Big)/\bar{j}(\nu, z) = b_j(\nu, z)\delta\Big((\eta_0 - \eta)\hat{\mathbf{n}}, z\Big),$$

where $\bar{j}(\nu, z)$ is the mean emissivity per comoving unit volume at frequency ν as a function of redshift z, and is derived using the empirical, parametric model of [3]. We have described the matter density field δ in our analysis by a linear power spectrum valid for the scales of interest in our work ($\ell < 100$, comprising most of the ISW signal). Lastly, we have chosen the linear bias $b_j(\nu, z)$ to be constant in redshift: $b_j(\nu, z) = b_{lin}(\nu)$. As detailed in [1], to obtain it at each frequency, we have computed the value of $b_{\rm lin}$ that gives the best agreement between our predicted linear CIB power spectrum and the one obtained from the *Planck* data, and published in [4]. We get the following values for these bias factors: $b_{lin} = 1.74$ at 857 GHz, 2.09 at 545 GHz, 2.63 at 353 GHz and 2.51 at 217 GHz; these values are consistent with the results found in the literature. The linear bias increases with the wavelength; this is coherent with the fact that as we go deeper into the infrared, the galaxies probed are more luminous at higher z. They reside in more massive and rarer halos, and are, therefore, more biased.

2.2. Correlation with the ISW

In the CMB anisotropies, the temperature contribution due to the ISW effect is an integral over the conformal time of the growth rate of the gravitational potentials, given as:

$$\delta T_{ISW}(\hat{\mathbf{n}}) = \int_{\eta_r}^{\eta_0} d\eta \, e^{-\tau(\eta)} \, (\dot{\Phi} - \dot{\Psi})\Big((\eta_0 - \eta)\hat{\mathbf{n}}, \eta\Big), \tag{2}$$

where η_r is some initial time deep in the radiation era, Φ and Ψ are the Newtonian gauge gravitational potentials, $\tau(\eta)$ is the optical depth, and the dot denotes differentiation with respect to η. Using Eqs. (1) and (2), and going to Fourier space, we can express the CMB-CIB cross-power spectrum at a frequency ν as:

$$C_\ell^{\rm cr}(\nu) = 4\pi\frac{9}{25}\int\frac{dk}{k}\Delta^2_{\mathcal{R}}\times\int_{\eta_0}^{\eta_r}d\eta\; e^{-\tau(\eta)}\; j_\ell\Big(k[\eta-\eta_0]\Big)\;(c_{\Psi\Phi}\dot{\psi}-\dot{\phi})$$
$$\times c_{\delta\Psi}\int_{\eta_0}^{\eta_r}d\eta\; j_\ell\left(k[\eta-\eta_0]\right)a(\eta)\,b_{\rm lin}(\nu)\,\bar{j}(\nu,\eta)\,\tilde{\delta}(k,\eta), \quad (3)$$

where $\Delta^2_{\mathcal{R}}$ comes from the primordial curvature power spectrum $P_{\mathcal{R}}\equiv 2\pi^2\Delta^2_{\mathcal{R}}/k^3$; $j_\ell(\cdot)$ are the spherical Bessel functions, while $\tilde{\delta}$, ϕ and ψ are the time-dependent parts of (respectively) the matter density contrast δ, and the two Newtonian gravitational potentials Φ and Ψ. The two coefficients $c_{\Psi\Phi}$ and $c_{\delta\Psi}$ give the relations between δ, Φ and Ψ for adiabatic initial conditions: $c_{\delta\Psi}\equiv\delta/\Psi=-3/2$, $c_{\Psi\Phi}\equiv\Psi/\Phi=-(1+(2/5)R_\nu)$, where $R_\nu\equiv\rho_\nu/(\rho_\nu+\rho_\gamma)$, with ρ_ν and ρ_γ respectively the energy densities in relativistic neutrinos and photons.

3. Signal-to-noise analysis

To compute the aforementionned spectra, we have adapted a code named CROSS_CMBFAST [5]: for a given cosmology and emissivity function $\bar{j}(\nu,z)$, our code calculates the $C_\ell^{\rm cr}$ from Eq. (3), and at the same time the predicted power spectrum of the CIB fluctuations described by Eq. (1).

3.1. Ideal case

We have now investigated the detection level of the ISW effect using CMB-CIB cross-correlation by performing a signal-to-noise ratio analysis in the context of several past (IRAS) and present (*Herschel* SPIRE, *Planck*) experiments, and their associated frequencies. Using our computed power spectra, we can write, for each given frequency ν, the total signal-to-noise ratio of the ISW detection as:

$$\left[\frac{S}{N}\right]^2(\nu) = \sum_{\ell=2}^{\ell_{\rm max}}(2\ell+1)\frac{[C_\ell^{\rm cr}(\nu)]^2}{[C_\ell^{\rm cr}(\nu)]^2+C_\ell^{\rm CIB}(\nu)\times C_\ell^{\rm CMB}}, \quad (4)$$

where the total (or cumulative) signal-to-noise is summed over multipoles between $\ell=2$ and $\ell_{\rm max}\leqslant 100$, where the signal has its major contribution.

We have first considered the ideal situation, where the CIB and CMB maps used for cross-correlation are noiseless and cover the whole sky; with these ideal assumptions, we have obtained high levels of detection for the CIB-CMB correlation, which reach ~ 6 - 7σ, with the largest contribution to the SNR coming from multipoles lower than $\simeq 50$. Our results for the *Planck* instrument are shown on Table 1; in light of these, the optimal frequency for ISW detection appears to be around 353 GHz with a maximum SNR reaching 7σ. However, in practice, the CIB extraction at this frequency might prove challenging since the CMB becomes dominant here, and increasingly so as we go down in frequency. Therefore, the possible residuals in the extracted CIB map have to be accounted for, and other sources of noise as well, which we have done next.

3.2. More realistic SNR

We have now carried out a more realistic study by including several possible sources of contamination: First, the strong emission of the Milky Way masks the CIB signal over the galactic plane, reducing the usable fraction of the sky ($f_{\rm sky}$) by at least $\sim 25\%$, and consequently reducing the SNR by a factor $\sqrt{f_{\rm sky}}$. Furthermore, the rest of the sky is also quite polluted by dust foregrounds, which will have to be removed from our CIB maps, although some residuals might remain. There may even be a significant CMB residual in this map, due to an imperfect separation of components. Consequently, we have assessed the impact of these contaminants in our study, by adding a noise term $N_\ell^{\rm CIB}$ to the CIB power spectrum in the SNR. We have broken this noise component into several independant parts as:

$$N_\ell^{\rm CIB}(\nu) = R_\ell^{\rm CMB}(\nu) + R_\ell^{\rm fore.}(\nu) + N_\ell^{\rm instr.}(\nu) + N_\ell^{\rm correl.}(\nu),$$

where these four different terms represent the power spectra of the CMB residual, the galactic foregrounds residuals, the instrumental noise, and finally the noise due to correlation between residuals and the CIB (which appears when auto-correlating the final CIB map). The instrumental noise is fixed by the characteristics of the considered experiment. We have then chosen three parameters to govern the other three contributions: $f_{\rm sky}$, the fraction $f_{\rm CMB}$ of the total CMB map present as a residual in the CIB map, and $\mathcal{A}_{\rm fore.}$ the amplitude of the foreground residuals spectrum relative to the CIB spectrum (at $\ell = 10$). The next step would be to explore this 3D parameter space at each frequency and compute the SNR at each point. Considering the very large number of possible combinations of parameters, it would not be practical to display the complete results of this exploration here. Therefore, we have first chosen to fix $f_{\rm sky}$ to two values of interest: $f_{\rm sky} = 0.75$, corresponding to an optimistic case, where only the galactic plane is discarded, and $f_{\rm sky} = 0.15$, a pessimistic estimate of the area of the sky, where the current data allows for an efficient CIB extraction. For our other two parameters, we have limited ourselves to reasonable values, with $f_{\rm CMB} \in [0, 0.1]$ and $\mathcal{A}_{\rm fore.} \in [0, 10]$. We have also explored the combination of cross spectra at different frequencies, which allow us to increase the total signal-to-noise ratio of the ISW detection by combining their constraints, although this will be limited by the possible intrinsic correlations between the different CIB frequencies.

After a thorough exploration of these parameters, we have shown a part of our results in Table 1, and we have drawn a few conclusions: First, the gain brought by the joint correlation is small, which can be attributed to the high correlation (therefore, redundancy) between the CIB at its different observed frequencies. When focusing on the 353 GHz results (the best SNR in the ideal case), the influence of the CMB is clearly visible, quickly reducing the SNR as its residual level increases, which is even more pronounced at 217 GHz, due to the proximity to the maximum of the SED of the CMB. The presence of instrumental noise becomes significant at the two lowest frequencies (217 and 353 GHz), again reducing their value in the cross-correlation. As expected, the galactic foreground residuals also decrease the SNR, though their influence is roughly the same at all frequencies as they are defined relatively to the CIB spectrum. Lastly, the biggest influence comes from the fraction of the sky through the $f_{\rm sky}$ parameter, as the total SNR scales as $\sqrt{f_{\rm sky}}$. This makes it a crucial requirement for future applications to have the largest possible coverage to minimize this effect.

4. Conclusions

We have investigated the cross-correlation between the cosmological infrared and microwave backgrounds through the ISW effect, and studied its detectability under various observational situations. Using an advanced SNR analysis, which included the main sources of noise, both instrumental and astrophysical, and all their possible correlations, we have pointed out the most promising frequency in the ideal case of noiseless full-sky maps (353 GHz) with an expected significance as high as 7σ for the cross-correlation signal. The same frequency turned out to be less optimal with more realistic assumptions about sky coverage and possible sources of noise (here CMB, dust residuals and instrumental noise). In this case, higher frequencies such as *Planck* HFI's 545 and 857 GHz are favoured, with an expected significance ranging from 2 to more than 5σ depending on the frequency, the levels of noise and the fraction of the sky available for analysis. We have also found that a joint cross-correlation using all available frequencies is of minor interest. Nevertheless, our best results for $f_{\rm sky} = 0.75$ are higher than the significances of

Table 1. Total signal-to-noise ratio of the CIB-CMB cross-correlation for four of the CIB frequencies of *Planck* HFI and three different noise scenario.

Frequency (GHz)	**857**	**545**	**353**	**217**
Wavelength (μm)	**350**	**550**	**850**	**1380**
Perfect Single SNR	6.26	6.83	6.98	6.95
Joint SNR	7.12			
Realistic SNR 1 ($f_{\rm sky} = 0.75$, $f_{\rm CMB} = 0.01$, $\mathcal{A}_{\rm fore.} = 0.01$)	5.36	5.73	5.39	3.56
Joint SNR	5.88			
Realistic SNR 2 ($f_{\rm sky} = 0.15$, $f_{\rm CMB} = 0.01$, $\mathcal{A}_{\rm fore.} = 0.01$)	2.40	2.56	2.41	1.59
Joint SNR	2.63			

all current CMB-galaxies cross-correlation, with $\sigma > 5$, although a less optimistic estimate for the sky coverage quickly reduces our signal-to-noise ratios. This stresses once again the requirement of good component separation techniques and foreground removals for future applications, in order to have the largest fraction of common clean sky $f_{\rm sky}$ possible. We are currently applying our method and framework to the actual *Planck* data, and the results of this study will be published with the next public release of *Planck* papers and data at the beginning of 2013.

References

[1] Ilić S, Douspis M, Langer M, Pénin A and Lagache G 2011 *MNRAS* **416** 2688
[2] Knox L, Cooray A, Eisenstein D J and Haiman Z 2001 *Ap. J.* **550** 7
[3] Béthermin M, Dole H, Lagache G, Le Borgne D and Pénin A 2011 *A. & A.* **529** A4
[4] Planck Collaboration 2011 *A. & A.* **536** A18
[5] Corasaniti P, Giannantonio T and Melchiorri A 2005 *Phys. Rev. D* **71** 123521

Observational cosmology and the cosmic distance duality relation

S Jhingan[1], **D Jain**[2] **and R Nair**[1],

[1] Centre for Theoretical Physics, Jamia Millia Islamia, New Delhi 110025, India
[2] Deen Dayal Upadhyaya College, University of Delhi, New Delhi 110015, India

E-mail: `sanjay.jhingan@gmail.com`

Abstract. We have studied the validity of cosmic distance duality relation between angular diameter and luminosity distances. To test this duality relation, we have used the latest Union2 Supernovae type Ia (SN Ia) data for estimating the luminosity distance. The estimation of angular diameter distance comes from the samples of galaxy clusters (real and mock) and FRIIb radio galaxies. We have parameterized the distance duality relation as a function of redshift in four different ways, and found that the mock data set, which assumes a spherical isothermal β model for the galaxy clusters does not accommodate the distance duality relation, while the real data set which assumes elliptical β model does.

1. Introduction

The reciprocity relation relates the distances between two events (say the source S and the observer O), which are connected by null geodesics. This relation is of fundamental importance in observational cosmology, especially, measurements such as distant type Ia supernovae (SN Ia), cosmic microwave background (CMB), galaxy observations and gravitational lensing [1]. Bassett and Kunz [2] have explored whether this relation can shed some light on the presence of exotic physics, and they have ruled out non-accelerating models of universe (replenishing dust model) by more than 4σ level.

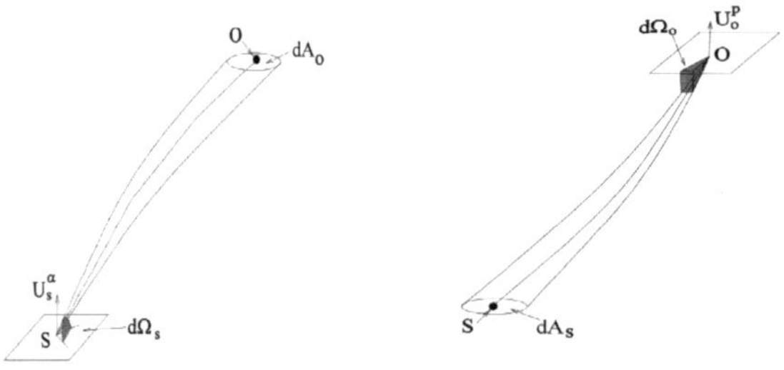

Figure 1. Consider a bundle of null geodesics emanating from the source, subtending a solid angle $d\Omega_s$. This bundle has a cross section dA, and the source angular distance $D(U_s^\alpha)$ is defined as $D(U_s^\alpha) = \left(\frac{dA}{d\Omega_s}\right)^{\frac{1}{2}}$. Similarly, the observer area distance $D(U_s^p)$ is defined as $D(U_s^p) = \left(\frac{dA_s}{d\Omega_o}\right)^{\frac{1}{2}}$.

1.1. *The reciprocity relation*

If the geodesic deviation equation holds and photon travels on null geodesics, it can be shown that (see Figure 1), $D(U_s^\alpha)$ and $D(U_s^p)$ are related as [3, 4]:

$$D(U_s^\alpha) = (1+z)D(U_s^p). \tag{1}$$

This is known as the *reciprocity relation.* Note that this relation holds regardless of the metric or the matter content of the spacetime. Using the reciprocity relation and assuming that the total number of photons are conserved on the cosmic scales (above mentioned observational methods are all based on this fundamental assumption), one can derive a relation between the angular diameter distance and the luminosity distance [4] as:

$$d_L = d_A(1+z)^2. \tag{2}$$

This equation is known as the *distance duality* relation (DD), and it is testable by astronomical observations if d_L and d_A are known for a source. To study the validity of this relation, we have analyzed the following red-shift dependence of DD, given by:

$$\eta(z) \equiv \frac{d_L}{d_A(1+z)^2}\,. \tag{3}$$

Uzan *et al* [5] have used the combined measurements of Sunyaev-Zeldovich effect and X-ray emission data of galaxy clusters and shown that if DD relation does not hold then the angular diameter distance measured from the clusters is $d_A^{cluster}(z) = d_A\,\eta^2$, and hence, the DD relation in Eq. (3) gets modified to:

$$\eta(z) \equiv \frac{d_A^{cluster}(1+z)^2}{d_L}\,. \tag{4}$$

In this paper, our aim is to re-analyze the validity of DD relation in a comprehensive manner by using different data samples and parameterizations [6].

2. Parameterizations, data and method

2.1. $\eta(z)$ parameterizations

Our parameterizations are inspired by model independent parameterizations for the dark energy equation of state [7]. We are parameterizing $\eta(z)$, whose value stays one when photon number is conserved, gravity is described by a metric theory and photons travel on null geodesics. Any significant violation from the DD relation will hint at the break down of one or more of these assumptions. To model any departure from unity, we have parameterized η with four parametric representations, which are given by:

$$\begin{aligned} \eta_I(z) &= \eta_0 + \eta_1 z, \\ \eta_{II}(z) &= \eta_2 + \eta_3\frac{z}{1+z}, \\ \eta_{III}(z) &= \eta_4 + \eta_5\frac{z}{(1+z)^2}, \quad \textit{and} \\ \eta_{IV}(z) &= \eta_6 - \eta_7\ln(1+z). \end{aligned}$$

2.2. Data

For the luminosity distances, we have used latest Union2 SNe Ia data [8], and considered that pair of galaxy cluster/radio galaxy and SN, for which $\Delta z < 0.005$ [9]. Because of this condition, the number of data points are limited to 24, 222 and 12 for the data set I, II and III respectively.

- Data set I: This sample consists of 25 galaxy clusters (isothermal elliptical β model, with concordance model for the cosmological distance-redshift relationship) [10]. The redshift interval for clusters in this sample is $0.02 < z < 0.78$.
- Data set II: This data set contains 578 angular diameter distances of mock clusters. The redshift distribution of this sample is $0.05 < z < 0.76$. (For more details, see [11, 12].)
- Data set III: In this sample, the calculation of angular diameter distance is obtained by using the physical size of extended radio galaxies [13]. This data set contains 20 radio galaxies up to redshift z = 1.8.

2.3. Method

We have performed the χ^2 analysis to fit the parameters of the assumed parameterizations as:

$$\chi^2(p) = \sum_i \frac{(\eta_{th}(z_i, p) - \eta_{obs}(z_i))^2}{\sigma(z_i)^2}, \tag{5}$$

where η_{th} is the assumed form of parameterizations, and η_{obs} is the observed value of η, which is calculated by using d_L and d_A at a particular value of redshift. The unknown parameters are denoted by the variable p.

The η_{obs} for the radio galaxies are obtained by assuming $d_A^{radio} = d_A$ (R A Daly, private communication) in Eq. (3). The angular diameter distance for radio galaxies is given in terms of the dimensionless coordinate distance y, as:

$$d_A(i) = \frac{y(z_i)}{H_0(1 + z_i)}, \tag{6}$$

where $y = a_0 r H_0$, and $a_0 r$ is the coordinate distance [13]. Treating Hubble constant H_0, as a nuisance parameter we have marginalized over it by assuming Gaussian prior. In this analysis, distances are obtained by assuming the flat ΛCDM universe.

3. Results and discussions

We have studied the validity of DD relation using different data sets, assuming four general parameterizations of $\eta(z)$, completely in analogy with the varying equation of state of dark energy $\omega(z)$. Our results are summarized as follows:

- Tables 1, 2 and 3 show the best fit values of the parameters for the four parameterizations for the first, second and third data set respectively.

Table 1. Best fit values for all parameterizations: Data set I

χ^2_ν	Parameters	Parameters
1.217	$\eta_0 = 0.999 \pm 0.144$	$\eta_1 = -0.058 \pm 0.507$
1.216	$\eta_2 = 1.007 \pm 0.170$	$\eta_3 = -0.118 \pm 0.822$
1.216	$\eta_4 = 1.013 \pm 0.205$	$\eta_5 = -0.198 \pm 1.345$
1.216	$\eta_6 = 1.003 \pm 0.157$	$\eta_7 = 0.087 \pm 0.651$

Table 2. Best fit values for all parameterizations: Data set II

χ^2_ν	Parameters	Parameters
1.076	$\eta_0 = 0.973 \pm 0.048$	$\eta_1 = -0.108 \pm 0.159$
1.078	$\eta_2 = 0.971 \pm 0.058$	$\eta_3 = -0.138 \pm 0.267$
1.080	$\eta_4 = 0.972 \pm 0.072$	$\eta_5 = -0.141 \pm 0.453$
1.077	$\eta_6 = 0.972 \pm 0.053$	$\eta_7 = 0.124 \pm 0.208$

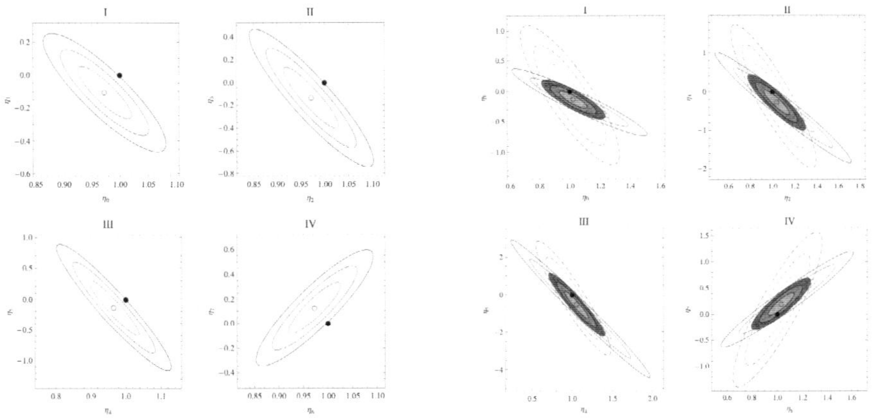

Figure 2. The left panels shows 1σ, 2σ and 3σ contours in $\eta_i - \eta_j$ plane with data set II. In the right panel, dashed, solid and filled contours correspond to data sets I, III and (I + III) respectively. In both the figures, the position of filled circle in the contours indicate the point where $\eta(z = 0) = 1$ and the position of empty circle indicates the best fit value of the parameters.

Table 3. Best fit values for all parameterizations: Data set III

χ^2_ν	Parameters	Parameters
0.944	$\eta_0 = 1.063 \pm 0.198$	$\eta_1 = -0.180 \pm 0.244$
0.971	$\eta_2 = 1.099 \pm 0.274$	$\eta_3 = -0.415 \pm 0.632$
1.021	$\eta_4 = 1.099 \pm 0.385$	$\eta_5 = -0.749 \pm 1.624$
0.958	$\eta_6 = 1.081 \pm 0.235$	$\eta_7 = 0.282 \pm 0.404$

- All the parameterizations have shown degenerate behaviour with the given data sets. The parameterizations are in complete concordance with DD relation within 1σ level for data set I, and within 2σ level for data set III. As expected, the bigger data set of mock galaxy clusters (data set II), which is generated by assuming the spherical isothermal β model for clusters gives tighter constraints on various parameterizations, but most of the parameterizations show significant deviation from DD relation (see figure 2). But the real galaxy cluster data (data set I), which is obtained by assuming isothermal elliptical β model for clusters is in good agreement with DD relation.

References

[1] Ellis G F R 2007 *Gen. Rel. Grav.* **39** 1047
[2] Bassett B A and Kunz M 2004 *Phys. Rev.* D **69** 101305
[3] Etherington I M H 1933 *Phil. Mag.* **15** 761
[4] Schneider P, Ehlers J and Falco E E 1992 *Gravitational Lenses* (Berlin: Springer-Verlag)
[5] Uzan J P, Aghanim N and Mellier Y 2004 *Phys. Rev.* D **70** 083533
[6] Nair R, Jhingan S and Jain D 2011 *JCAP* **05** 023
[7] Efstathiou G 1999 *MNRAS* **310** 842
[8] Amanullah R et al 2010 *Ap. J.* **716** 712
[9] Holanda R F L, Lima J A S and Ribeiro M B 2010 *Ap. J.* **722** L233
[10] Filippis E, Sereno M, Bautz M W and Longo G 2005 *Ap. J.* **625** 108
[11] Khedekar S and Majumdar S 2010 *Phys. Rev.* D **82** 081301
[12] Khedekar S Majumdar S and Das S 2010 *Phys. Rev.* D **82** 041301
[13] Daly R A and Djorgovski S G 2003 *Ap. J.* **597** 9

Proceedings of the 7th International Conference on Gravitation and Cosmology (ICGC2011) IOP Publishing
Journal of Physics: Conference Series **484** (2014) 012036 doi:10.1088/1742-6596/484/1/012036

Weighing neutrinos using high redshift galaxy luminosity functions

C Jose[1]*, S Samui[2], K Subramanian[1] and R Srianand[1]

[1] IUCAA, Post Bag 4, Pune University Campus, Ganeshkhind, Pune 411007, India

[2] School of Physics, UKZN, Durban 4001, South Africa

E-mail: *`charles@iucaa.ernet.in`

Abstract. We have proposed a novel way to constrain the neutrino mass using UV luminosity function (LF) of high-z Lyman break galaxies. Combining the constraints from the Wilkinson Microwave Anisotropy Probe 7 year (WMAP-7) data with the LF data at $z \sim 4$, we have got a limit on the sum of the masses of 3 degenerate neutrinos at the 95 % CL. The additional constraint of using the prior on Hubble constant strengthens this limit to at 95 % CL. As different astronomical measurements may suffer from different set of biases, the method presented here provides a complementary probe of sum of neutrino masses.

1. Introduction

A cosmic background of neutrinos is one of the key predictions of standard big bang cosmology. The absolute mass scale of neutrinos could be inferred from various β-decay experiments [1]. However, at present, stronger constraints on neutrino mass are obtained from cosmological observations [2]). The recent WMAP-7 data by itself sets an upper limit to sum of neutrino masses, $\Sigma m_\nu < 1.3$ eV [3]. It is well known that the presence of massive neutrinos leads to the suppression of density perturbations below a time dependent scale known as the free streaming scale [2, 4]. This suppression results in a decrease in the formation of dark matter halos below a characteristic mass scale. Thus, observations related to large scale structure formation in the universe can be used to probe the absolute mass scale of neutrinos. In this work, we have explored the possibility of using the luminosity functions (LF) of Lyman break galaxies (LBG) for constraining Σm_ν.

The basic idea is as follows: The reduction in the matter power spectrum in models with $m_\nu > 0$, compared to models with $m_\nu = 0$ implies a reduced abundance of galactic scale dark matter halos at high redshifts. In order to account for the observed LF of these galaxies, the light to mass ratio of each galactic halo has to be systematically higher in the models with $m_\nu > 0$. However, changing the light to mass ratio is degenerate with the unknown extinction correction, one applies to the observed LF. Nevertheless, this degeneracy can be lifted if one has a feature in the LF at some characteristic mass scale, introduced by various feedback processes, like the radiative feedback after reionization. In such cases, the shape of the predicted galaxy LF depends on the neutrino mass. We have used this idea to constrain neutrino masses. We have explained our basic model in Sections 2 and 3, presented our limits on Σm_ν in Section 3.1 and concluded in Section 4.

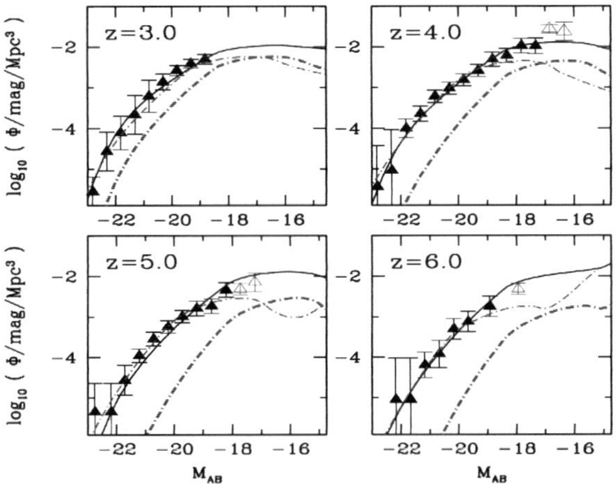

Figure 1. UV LF of LBGs at redshifts 3, 4, 5 and 6. The solid (black) line shows the predicted best fit LF for our model with $\Sigma m_\nu = 0$. The thick dashed-dotted (blue) curve shows our model predictions with $\Sigma m_\nu = 1$ eV and using the same best fit f_*/η as the $\Sigma m_\nu = 0$ eV model. This is to illustrate the suppression due to massive neutrinos. The thin dashed-dotted (blue) curves are best fits for the models with $\Sigma m_\nu = 1$ eV. The data points (filled and open triangles) for $z = 3$ are taken from [5] and for $z = 4 - 6$ are from [6].

2. High redshift galaxy luminosity functions

We have modelled high redshift galaxy LF using the semi-analytical treatment in [7], and the models are successful in explaining the observed UV LF of high redshift galaxies at $3 \leq z \leq 8$ (see also [8]). The crucial ingredient of their model is the star formation rate (SFR) of an individual dark matter halo of mass M collapsed at redshift z_c and observed at redshift z. This is assumed to rise linearly after the collapse of the halo, and decay exponentially in a dynamical time scale. A key component of the SFR is f_*, the fraction of the total baryonic mass that is converted into stars over the entire lifetime of the galaxy. The assumed star formation rate of a galaxy can be converted to a luminosity (See Eq. (6) and Figure 1 of [7]). Only a fraction $(1/\eta)$ of the total light produced by the stars comes out of the galaxy due to the absorption by dust. The LF, $\Phi(L)dL$ can be obtained by counting all galaxies collapsed at any redshift z_c, which can produce a luminosity L at the redshift of observation. To calculate the LF, one requires the formation rate of dark matter halos of mass M. The authors of [7] have modelled this formation rate as the time derivative of Sheth and Tormen (ST) mass function [9] (see also [10]).

Star formation in a given halo also depends on various feedback processes. The ionization of the IGM by UV photons increases the temperature of the gas, thereby increasing the Jean's mass for collapse. In ionized regions, we have incorporated this feedback by a complete suppression of star formation for halos with circular velocity $v_c \leq 35$ km s^{-1}, and no suppression with $v_c \geq 95$ km s^{-1} [11]. For intermediate circular velocities, a linear fit from 1 to 0 is adopted as the suppression factor [11, 7]. This is called the *radiative feedback*. (The details of other feedback mechanisms can be found [7].)

3. Effect of neutrino mass on high redshift LF

We have shown in Figure 1 that our model predictions of LF at different redshifts $z = 3 - 6$, along with the observational data. The solid line shows the predicted best fit LF at various redshifts for the fiducial cosmology and with $\Sigma m_\nu = 0$. In order to fit the observed data points, we have adjusted the free parameter f_*/η in our models, using χ^2 minimization. The flattening of the predicted LF as seen in Figure 1 at the faint end is due to the radiative feedback. The thick dashed-dotted (blue) curves show the predicted LF if we use the same f_*/η for a model with $\Sigma m_\nu = 1.0$ eV. It is clear that there is an order of magnitude suppression in the number density of galaxies at a given luminosity. This is because the presence of neutrinos suppresses the formation rate of halos at the mass and redshift scales of our interest. We can make our model predictions match with the observed data by increasing f_*/η (i.e., shifting this curve

Data	$10^2\Omega_b h^2$	$10\Omega_{DM} h^2$	τ	n_s	σ_8	H_0	Σm_ν(eV)
WMAP-7	$2.223^{+0.060}_{-0.058}$	$1.172^{+0.070}_{-0.068}$	$0.087^{+0.014}_{-0.014}$	$0.962^{+0.016}_{-0.015}$	$0.717^{+0.071}_{-0.072}$	$66.1^{+4.1}_{-4.9}$	< 1.08
WMAP-7+LF	$2.229^{+0.055}_{-0.056}$	$1.228^{+0.065}_{-0.064}$	$0.090^{+0.013}_{-0.014}$	$0.961^{+0.015}_{-0.014}$	$0.820^{+0.051}_{-0.052}$	$65.5^{+2.8}_{-2.8}$	< 0.52
WMAP-7+H_0	$2.263^{+0.050}_{-0.056}$	$1.107^{+0.051}_{-0.050}$	$0.091^{+0.015}_{-0.015}$	$0.973^{+0.013}_{-0.013}$	$0.756^{+0.049}_{-0.047}$	$70.3^{+2.5}_{-2.5}$	< 0.54
WMAP-7+H_0+LF	$2.259^{+0.054}_{-0.054}$	$1.169^{+0.056}_{-0.058}$	$0.095^{+0.015}_{-0.014}$	$0.969^{+0.014}_{-0.016}$	$0.821^{+0.042}_{-0.041}$	$68.7^{+2.1}_{-2.2}$	< 0.29

Table 1. Results of our MCMC analysis to constrain Σm_ν. The first column lists the various data used for our analysis. The last column gives limits on Σm_ν, at the 95% CL. For all other parameters, their mean values and 1σ range are given.

along the luminosity axis). These best fit LF, obtained with the new f_*/η (thin dashed-dotted curves), have a very different shape compared to the zero neutrino mass case. In particular, the predicted LF in models with $\Sigma m_\nu = 1$ eV, are suppressed at the low luminosity end compared to the zero mass case. This is basically because increasing f_*/η increases the light to mass ratio, which brings even small mass galaxies whose star formation has been suppressed due to radiative feedback, into the observable luminosity range. Therefore, strong constraints on the neutrino mass can in principle be obtained by comparing the shape of the predicted LF with observations.

3.1. Limits on neutrino mass
In order to obtain quantitative upper limits by exploring the full range of cosmological parameters consistent with both the WMAP-7 data and the observed LF of LBGs, we have performed a Markov Chain Monte Carlo (MCMC) analysis using the publically available CosmoMC code [12]. In particular, we have concentrated on the LF of LBGs at $z = 4$. We have given the constraints on Σm_ν obtained from the MCMC analysis in Table 1 and Figure 2. The constraints on neutrino mass from WMAP-7 data alone [3] is $\Sigma m_\nu < 1.08$ eV at the 95% CL. Combining the UV luminosity function data at $z = 4$ with the WMAP-7 data, significantly lowers this limit to $\Sigma m_\nu \leq 0.52$. Further, addition of the constraints from the H_0 determination of the HST SHOES (Supernova H_0 for the Equation of State) programme [13] (which we refer to as H_0) leads to a lower limit of $\Sigma m_\nu \leq 0.29$ eV.

All the above neutrino mass limits along with the other cosmological and astrophysical parameters are summarized in Table 1. The corresponding 1D marginalized distribution for Σm_ν, from various MCMC analysis, is shown in the top left panel of Figure 2. The figure also clearly shows that adding the constraint from the $z = 4$ UV luminosity function significantly improves the constraint on neutrino masses. These results are not very much sensitive to various feedback mechanisms. (For a detailed discussion, interested reader may refer to the original paper [10].)

4. Conclusions
We have proposed here a novel probe of neutrino mass using the high redshift UV LF of Lyman break galaxies. In particular, our model constructed in the framework of ΛCDM cosmology with massive neutrinos, shows that the observed shape of the UV LF of high redshift LBGs can be used to constrain the mass of the neutrinos. We have carried out an MCMC analysis to obtain quantitative upper limits on Σm_ν, by combining observed LF data with other data, and exploring the full range of cosmological parameters. Our results are summarized in Table 1 and Figure 2. We have noticed that our best limit on neutrino mass is almost a factor ~ 4 improvement compared to the limit obtained using the WMAP-7 data alone.

Journal of Physics: Conference Series **484** (2014) 012036 doi:10.1088/1742-6596/484/1/012036

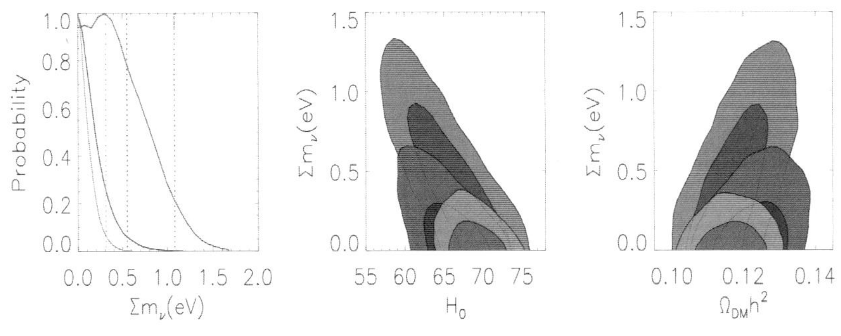

Figure 2. The 1D and 2D marginalized distributions of Σm_ν from our analysis. The left panel gives the marginalized 1D distribution for Σm_ν. The vertical dashed lines correspond to 95% confidence levels. The other two panels show the regions of 68% (dark color) and 95% (light color) confidence levels for H_0 and $\Omega_{DM}h^2$ against m_ν. The various contours corresponds to constraints obtained using WMAP-7 only (red), WMAP-7+LF (blue) and WMAP-7+HST+LF (green) data.

Some of the current limits on Σm_ν from various cosmological probes are given by [14, 15, 16, 17, 18, 19, 20]. The constraints on Σm_ν obtained here adding in the $z \sim 4$ UV LF data to the WMAP-7 and HST data, are comparable (or better in several cases) to the above limits. Our work is mainly a demonstrative first step, where we have suggested the utility of the LF of high redshift galaxies to constrain Σm_ν. Improvements in the LF data, especially at the faint end, and at higher redshifts together with a better understanding of the astrophysics of galaxy formation, will allow us to place more stringent constraints on Σm_ν.

Acknowledgements
CJ thanks CSIR, Government of India, for providing support for this work.

References
[1] Vogel P [arXiv:0807.2457]
[2] Lesgourgues J and Pastor S 2006 *Phys. Report* **429** 307
[3] Komatsu E, Smith K M, Dunkley J *et al* 2010 *Ap. J. Supp.* **192** 18 [arXiv:astro-ph/1001.4538]
[4] Bond J R and Szalay A S 1983 *Ap. J.* **274** 443
[5] Reddy N A, Steidel C C, Pettini M *et al* 2007 *Ap. J.* **175** 48
[6] Bouwens R J, Illingworth G D, Franx M and Ford H 2007 *Ap. J.* **670** 928
[7] Samui S, Srianand R and Subramanian K 2007 *MNRAS* **377** 285
[8] Samui S, Subramanian K and Srianand R 2009 *New Astron.* **14** 591
[9] Sheth R K and Tormen G 1999 *MNRAS* **308** 119
[10] Jose C, Samui S, Subramanian K and Srianand R 2011 *Phys. Rev. D* **83** 123518
[11] Bromm V and Loeb A 2002 *Ap. J.* **575** 111
[12] Lewis A and Bridle S 2002 *Phys. Rev. D* **66** 103511
[13] Riess A G, Macri L, Casetano S *et al* 2009 *Ap. J.* **699** 539
[14] Thomas S A, Abdalla F B and Lahav O 2010 *Phys. Rev. Lett.* **105** 031301
[15] Tereno I, Schimd C, Uzan J P, Kilbinger M, Vincent F H and Fu F 2009 *A. & A.* **500** 657
[16] Reid B A, Verde L, Jimenez R and Mena O 2010 *JCAP* **01** 3
[17] Veil M, Haehnelt M G and Springel V 2010 *JCAP* **06** 015
[18] Mantz A, Allen S W and Rapetti D 2009 *MNRAS* **406** 1805
[19] Swanson M E C, Percival W J and Lahav O 2010 [arXiv:1006.2825]
[20] Ichiki K, Takada M and Takahashi T 2009 *Phys. Rev. D* **79** 023520

Non-Gaussianity from extragalactic point sources

F Lacasa
Institut d'Astrophysique Spatiale (IAS), Orsay, France
E-mail: fabien.lacasa@ias.u-psud.fr

Abstract. We have described briefly the population of compact sources appearing at CMB frequencies, and studied their non-Gaussianity using publicly available full-sky simulations. Introducing a parametrization permitting to visualise efficiently the bispectrum, we have described the configuration and scale dependence of the bispectrum of radio and IR point sources, as well as its frequency dependence, and shown that it is well fitted by an analytical prescription. We have shown further that the clustering of IR sources increases their non-Gaussianity by several orders of magnitude, and that their bispectrum peaks in the squeezed triangles. Examining the impact of these sources on primordial non-Gaussianity estimation, we have found that the radio sources yield an important positive bias to local $f_{\rm NL}$ estimation at low frequencies, but this bias is efficiently reduced by masking detectable sources. On the other hand, IR sources produce a negative bias at high frequencies, which is not dimmed by the masking, as their clustering is dominated by faint sources.

1. Introduction

The Cosmic Microwave Background is a powerful probe of the early and late universe, it is the dominant signal on the sky around 100 GHz, and its observation has contributed to the establishment of the standard model of cosmology, the latest surveys being WMAP ACT and SPT. However, in the 30 − 350 GHz frequency window, other signals are present, and they become important at small angular scales, where the CMB plummets due to Silk damping, and in non-Gaussian part of correlation function, as the CMB is close to Gaussian. This is of special importance, e.g., for Planck, which covers a large frequency range at high angular resolution. There are principally three populations of extragalactic sources:

(i) Galaxy clusters, in which the electrons of the hot ionised gas scatter off the CMB photons leaving a distinct spectral signature (Sunyaev-Zeldovich effect).

(ii) Radio loud galaxies with an Active Galactic Nucleus and emitting through synchrotron and free-free processes.

(iii) Dusty star forming galaxies, where the UV emission from stars heats the dust which consequently re-emits in the infrared domain.

We have concentrated on the spatial distribution of this two former populations. Radio sources can be considered randomly distributed on the sky and can, hence, be modelled as a white-noise, entirely characterized by the number counts of the sources; the harmonic correlation function (power spectrum, bispectrum, etc.) are then flat and related to the corresponding moment of the 1-point distribution. On the other hand, IR sources are highly clustered in dark matter halos and have, hence, non-trivial correlation functions; there are models and observations of their number counts and power spectrum, but all higher order moments are theoretically necessary to describe the field.

The study of higher order moments also has an interest for the CMB, as it can break the degeneracy between the primordial processes generating the cosmological perturbations, e.g., the inflation models. While most models predict similar power spectra, they may be distinguished at the 3-point or higher level; as an example, standard single-field models predict undetectably small non-Gaussianity, while multifield models, non-canonical kinetic terms or vacuum initial condition, etc. produce potentially detectable non-Gaussianity with distinct shapes. As of today, no definite deviation from Gaussianity has been found in the CMB, even if there is a hint of NG of the 'local' shape in WMAP-7 data [1]. The local shape of non-Gaussianity is produced generically by several inflation models and takes the form:

$$\Phi(\mathbf{x}) = \Phi_G(\mathbf{x}) + f_{\rm NL}\left(\Phi_G(\mathbf{x})^2 - \langle\Phi_G(\mathbf{x})^2\rangle\right), \tag{1}$$

where Φ is the Bardeen potential, and the subscript G denotes the Gaussian linear part.

To study the distribution of radio and IR galaxies, we have made use of publicly available full-sky simulated maps of these sources [2] between 30 and 350 GHz at high angular resolution made through N-body dark matter simulations. The results presented in this paper are explained in greater details in [3].

2. Bispectrum

A Gaussian field on the sphere is entirely characterised by its mean and 2-point correlation function (or power spectrum in harmonic space), and the 3-point correlation function is, thus, the lowest order indicator of non-Gaussianity, and possesses a priori more signal to noise ratio than higher order moments. The bispectrum is the 3-point correlation function in harmonic space given as:

$$\langle a_{\ell_1 m_1} a_{\ell_2 m_2} a_{\ell_3 m_3}\rangle = G^{m_1 m_2 m_3}_{\ell_1 \ell_2 \ell_3} \times b_{\ell_1 \ell_2 \ell_3} \quad \text{with} \quad G^{m_1 m_2 m_3}_{\ell_1 \ell_2 \ell_3} = \int \mathrm{d}^2 n\, Y_{\ell_1 m_1}(n) Y_{\ell_2 m_2}(n) Y_{\ell_3 m_3}(n). \tag{2}$$

A full calculation of the bispectrum at WMAP or Planck resolution is computationally very intensive, and so one usually resort to binning multipoles. The bispectrum is invariant under permutation of (ℓ_1, ℓ_2, ℓ_3), and it is a function of the triangle shape only. Three triangle shapes are of particular importance: squeezed (one side is much smaller than the other two sides), equilateral, and folded (flat isosceles).

To plot a bispectrum in an efficient way, e.g., without redundancy of information, we have introduced a parametrization accounting for this permutation invariance, and defined three parameters (P, F, S) based on the elementary symmetric polynomials $\sigma_1 = \ell_1 + \ell_2 + \ell_3$, $\sigma_2 = \ell_1\ell_2 + \ell_1\ell_3 + \ell_2\ell_3$, and $\sigma_3 = \ell_1\ell_2\ell_3$. The locations of the different triangles of constant perimeter P in the (F, S) plan is shown in Figure 1. One can then plot a bispectrum by making slices of perimeters and colour-coding the value of the bispectrum.

The local type primordial non-Gaussianity defined in Eq. (1) leaves a characteristic imprint on the CMB, which can be detected at the 3-point level. The bispectrum, thus, generated peaks in the squeezed triangles, and because it has a separable form, it has been shown [4] that a fast estimator for its amplitude factor $f_{\rm NL}$ can be built, which does not need the prohibitive computation of the full bispectrum. This, commonly called *KSW estimator* is optimal in the

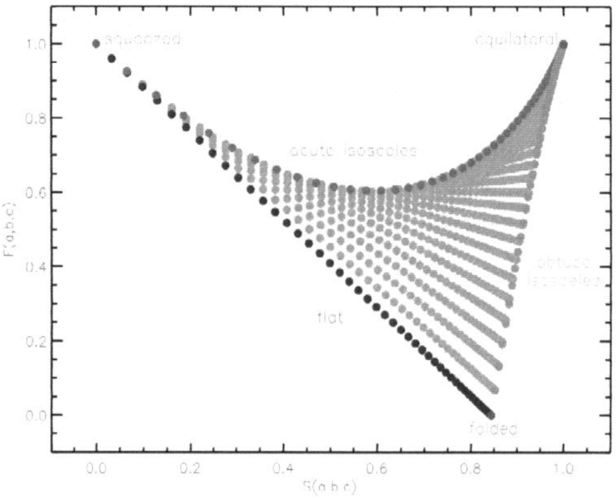

Figure 1. Positions of the triangles of a given perimeter in the (F, S) plan of the parametrization

sense that it minimizes the χ^2 of the fit of the observed bispectrum to the local bispectrum by:

$$\hat{f}_{\rm NL} = \sigma^2(f_{\rm NL}) \sum_{\ell_1 \leq \ell_2 \leq \ell_3} N_{\ell_1\ell_2\ell_3} \frac{b^{\rm obs}_{\ell_1\ell_2\ell_3}\, b^{\rm loc}_{\ell_1\ell_2\ell_3}}{C_{\ell_1} C_{\ell_2} C_{\ell_3}}, \tag{3}$$

$$\text{with} \quad \sigma^2(f_{\rm NL}) = \sum_{\ell_1 \leq \ell_2 \leq \ell_3} N_{\ell_1\ell_2\ell_3} \frac{\left(b^{\rm loc}_{\ell_1\ell_2\ell_3}\right)^2}{C_{\ell_1} C_{\ell_2} C_{\ell_3}}, \tag{4}$$

where $N_{\ell_1\ell_2\ell_3} = \frac{(2\ell_1+1)(2\ell_2+1)(2\ell_3+1)}{4\pi} \begin{pmatrix} \ell_1 & \ell_2 & \ell_3 \\ 0 & 0 & 0 \end{pmatrix}^2$ is the number of such configurations on the sky.

3. Point sources

As stated in the introduction, radio sources are randomly distributed on the sky, and so they produce a constant bispectrum $b_{\rm ps} \propto \ \ S^3 \frac{{\rm d}N}{{\rm d}S}{\rm d}S$, but as higher frequencies are surveyed, e.g., with Planck, the IR population becomes of growing importance, and one must account for the superposition of these signals. We have used a prescription, originally proposed by [5], and developped it to a full sky multi-population and analytical prescription; its assumptions are that radio and IR populations are independent, and that IR distribution can be modelled as a white-noise convolved by the known power spectrum. Analytically, at the bispectrum level, it yields:

$$b^{\rm IR}_{\ell_1\ell_2\ell_3} = \alpha\sqrt{C^{\rm IR}_{\ell_1} C^{\rm IR}_{\ell_2} C^{\rm IR}_{\ell_3}}, \quad \text{with} \quad \alpha = \frac{S^3 \frac{{\rm d}N}{{\rm d}S}{\rm d}S}{\left(\int S^2 \frac{{\rm d}N}{{\rm d}S}{\rm d}S\right)^{3/2}}. \tag{5}$$

We have used publicly available simulations of these two populations between 30 and 350 GHz by [2] to compare with this prescription. Figure 2 shows the observed and predicted bispectra at 350 GHz in the case of superposition of IR and radio sources.

We have found a good agreement, within cosmic variance error bars, between the prescription and the observed bispectrum at all frequencies and in all cases: radio sources alone, IR alone, and radio and IR together. It is also remarkable that the clustering of IR sources increases its non-Gaussianity, compared to un-clustered case, by several order of magnitudes at large angular scales. While the radio bispectrum is flat, the IR one exhibits a configuration dependence due to clustering, which can be visualized with the parametrization described previously (see Figure 3).

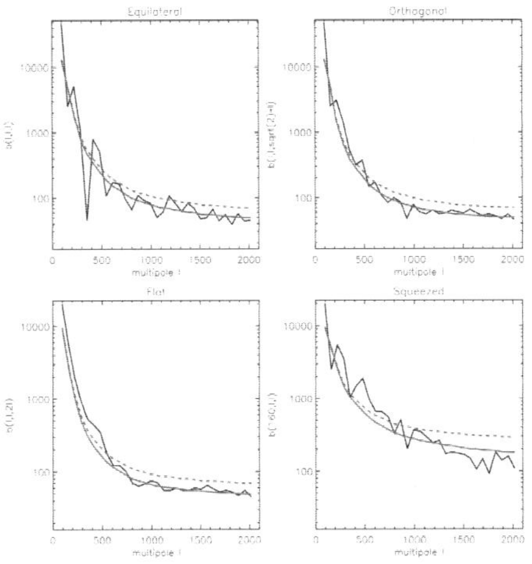

Figure 2. IR+radio bispectrum at 350 GHz in some configurations. Black line is the observed bispectrum of Sehgal *et al* simulations, red line is the prescription, blue dashed line is the case, where radio and IR populations are assumed totally correlated.

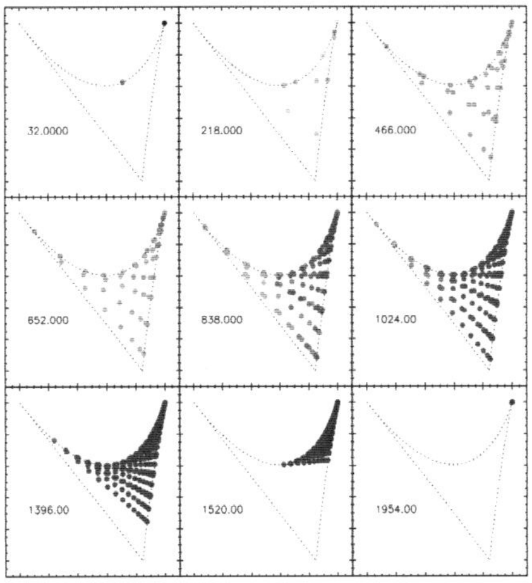

Figure 3. IR bispectrum at 148 GHz in the parametrization

We have found that the IR bispectrum peaks in the squeezed triangles, as predicted by the prescription, with a slight flattening at high multipoles due to shot-noise, especially at the higher frequencies.

4. Contamination of primordial non-Gaussianity estimate

We have used the KSW estimator [4] described previously by Eq. (3) on the simulated maps, to estimate how much foreground signals would bias the estimation of the primordial local NG parameter $f_{\rm NL}$, and computed this bias for several maximum angular resolution (WMAP-like, Planck-like), with the raw maps and after masking sources with flux above the flux cut of the Planck ERCSC catalogue [6]. Figure 1 summarizes these results respectively for radio and IR sources.

without flux cut						
ν (GHz)	30	90	148	219	277	350
$\ell_{\max} = 50$	-4.2	-0.0025	-0.00037	-0.00021	-0.00027	-0.00068
$\ell_{\max} = 700$	3850	2.5	0.38	0.21	0.27	0.65
$\ell_{\max} = 2048$	177000	117	18	9.7	12	30
with flux cut						
ν (GHz)	30	90	148	219	277	350
$\ell_{\max} = 700$	108	0.17	0.0071	0.0031	0.0035	0.0064
$\ell_{\max} = 2048$	4930	7.5	0.31	0.14	0.16	0.29

without flux cut						
ν (GHz)	30	90	148	219	277	350
$\ell_{\max} = 50$	$-3.5 \cdot 10^{-8}$	$-5.9 \cdot 10^{-6}$	$-9.2 \cdot 10^{-5}$	-0.0027	-0.023	-1.0
$\ell_{\max} = 700$	$-1.3 \cdot 10^{-6}$	-0.00019	-0.0033	-0.063	-0.55	-9.0
$\ell_{\max} = 2048$	$-1.8 \cdot 10^{-5}$	-0.0026	-0.039	-0.68	-4.8	-67
with flux cut						
ν (GHz)	30	90	148	219	277	350
$\ell_{\max} = 700$	$-1.3 \cdot 10^{-6}$	-0.00019	-0.0033	-0.078	-0.74	-11
$\ell_{\max} = 2048$	$-1.8 \cdot 10^{-5}$	-0.0026	-0.039	-0.67	-6.3	-66

Table 1. Bias from radio (*left panel*) and IR (*right panel*) to local NG parameter $f_{\rm NL}$, depending on frequency, maximum angular resolution, and masking or not detectable sources

The relative error bars of these biases are of 1.5 – 2.5% for radio sources and 3 to 7% for IR sources. We have found that for both population, the bias increases at higher maximum multipoles, as the CMB signal plummets due to Silk damping. The radio bias is positive and maximal at low frequencies, while the IR bias is negative and peaks at the higher frequencies. At a WMAP-like resolution, except at 30 GHz, these biases can be neglected compared to the expected error bars on $f_{\rm NL}$ due to cosmic variance. However, at Planck-like resolution, these biases become non-negligible; while the radio bias is very efficiently reduced by masking detectable sources, it is not the case for IR sources, where the bias is unaffected. Indeed, the IR population, and especially the clustered galaxies, are dominated by faint sources much below the detection limit.

References

[1] Komatsu E, Smith K M, Dunkley J, Bennett C L, Gold B, Hinshaw G, Jarosik N, Larson D *et al* 2011 *Ap. J. Supp. Ser.* **192** 18 [ISSN 0067-0049]
[2] Sehgal N, Bode P, Das S, Hernandez-Monteagudo C, Huffenberger K, Lin Y T, Ostriker J P and Trac H 2010 *Ap. J.* **709** 920 [ISSN 0004-637X] [arViv:0908.0540]
[3] Lacasa F, Aghanim N, Kunz M and Frommert M 2011 *To be published in MNRAS* [arXiv:1107.2251]
[4] Komatsu E, Spergel D N and Wandelt B D 2005 *Ap. J.* **634** 14 [ISSN 0004-637X]
[5] Argueso F, GonzalezNuevo J and Toffolatti L 2003 *Ap. J.* **598** 86 [ISSN 0004-637X]
[6] Planck Collaboration 2011 *A. & A.* **536** A7 [arXiv:1101.2041]

Proceedings of the 7th International Conference on Gravitation and Cosmology (ICGC2011) IOP Publishing
Journal of Physics: Conference Series **484** (2014) 012038 doi:10.1088/1742-6596/484/1/012038

High-order clustering of WiggleZ galaxies

F A Marín and The WiggleZ Team[1]

Centre for Astrophysics and Supercomputing, Swinburne University of Technology, P.O. Box 218, Hawthorn, VIC 3122, Australia

E-mail: `fmarin@astro.swin.edu.au`

Abstract. High-order statistics are a useful and complementary tool for measuring the clustering of galaxies, containing information on the non-Gaussian evolution and morphology of the large structures in the Universe. In this work, we have presented results of the 3-point correlation function in the WiggleZ spectroscopic galaxy survey, at three different epochs. Using N-body simulations to predict the clustering of matter, we have constrained the linear and non-linear bias parameters of WiggleZ galaxies with respect to total matter, and marginalized over them to obtain constraints on $\sigma_8(z)$, the variance of perturbations, and its evolution with redshift. These measurements are consistent with the predictions of the ΛCDM concordance cosmology and test this model in a new way.

1. Introduction

In the current structure formation paradigm [1], galaxies are formed inside dark matter halos, which evolved from small perturbations in the early universe, allowing us to connect the galaxy field with the cosmological models of structure formation. This connection, however, is not a perfect one, since galaxy observables such as luminosity, colour, etc. are also shaped by baryonic physics, leading to systematics and degeneracies when using clustering statistics of galaxies to extract information on cosmological parameters, and therefore, it is necessary to resort to other observables or objects to break these degeneracies. One way to use the *same data sample* to separate the galaxy formation imprints in the large-scale clustering from the cosmological background are the higher-order correlations functions [2].

In this work, we have presented results of the measurement of the 3-point correlation function (3PCF) for a sample of 187,000 spectroscopy-selected galaxies from the WiggleZ galaxy survey [3], with galaxies in the range $0.1 < z < 1.0$ at a median redshift of $z \sim 0.6$. Using N-body simulations to study dark matter statistics, we have estimated the WiggleZ galaxy bias to put constraints on the parameter σ_8, the rms over-density of spheres at a radius 8 h^{-1}Mpc, which provides a normalization of the primordial power spectrum. Also, we have splitted our galaxy sample in redshift slices, which allow us to measure σ_8 as a function of redshift, and in this way get constraints on the structure growth history.

2. Data and simulations

The WiggleZ Dark Energy Survey has obtained spectroscopic redshifts of galaxies over a cosmic volume of approximately 1 Gpc^3 in the range $0 < z < 1$, measured on 3.9 m Anglo-Australian

[1] http://wigglez.swin.edu.au/site/team.html

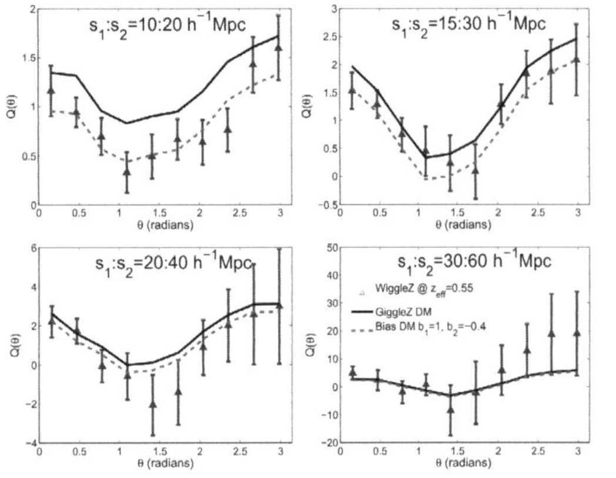

Figure 1. The redshift-space reduced 3PCF, $Q(s_1, s_2, \theta)$ of WiggleZ galaxies in the $z_{\mathrm{eff}} = 0.55$ slice (blue triangles), and of total matter from the GiggleZ simulation (black solid line), and the best bias model (red dashed line) for different sets of scales s_1.

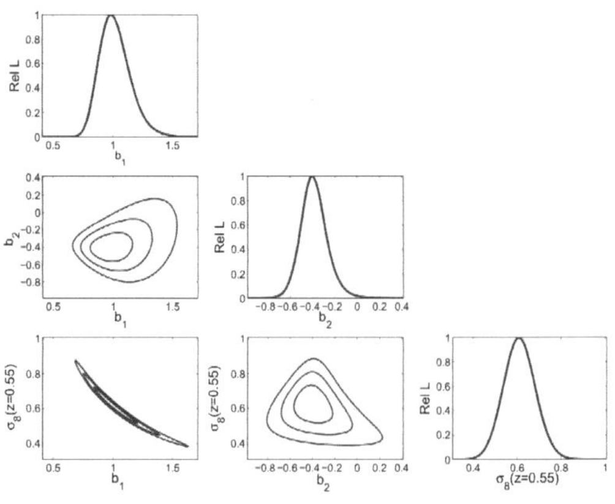

Figure 2. Constraints in the bias parameters and σ_8 from the redshift-space 2PCF and 3PCF for WiggleZ galaxies in the $z_{\mathrm{eff}} = 0.55$ slice. The contours represent 1σ, 2σ and 3σ joint confidence regions for a two-parameter fit.

Telescope. Target galaxies in seven different regions were chosen using UV photometric data from the *GALEX* survey [4], matched with optical photometry from the Sloan Digital Sky Survey [5], and the Red Sequence Cluster Survey 2 [6], applying magnitude and colour cuts to select star-forming galaxies with bright emission lines in a redshift distribution centred around $z \sim 0.6$. To study the evolution of clustering with redshift, we have used three overlapping slices of redshifts $[0.1, 0.5]$, $[0.4, 0.8]$, and $[0.6, 1.0]$ with effective (correlation-function averaged) redshifts of $z_{\mathrm{eff}} = [0.35, 0.55, 0.68]$ respectively.

The dark matter correlation functions have been measured on the GiggleZ simulations [7]. In a 1 $(h^{-1}\mathrm{Gpc})^3$ periodic cube, 2160^3 dark matter particles with individual masses of $m_p = 7.5 \times 10^9$ $h^{-1}M_\odot$ are evolved on a WMAP-5 cosmology [8]. In order to compare the correlation functions of WiggleZ galaxies and dark matter, we have measured them in snapshots of the simulation at the effective redshift of the WiggleZ sub-samples.

3. The WiggleZ 3-point correlation function

The galaxy n-point correlation functions are the average of galaxy over-density δ_g measured at n different points [2]. Whereas, the 2-point correlation function $\xi(r)$ (2PCF) depends only on the scale of clustering r, but the 3-point correlation function $\zeta(r_1, r_2, r_3)$ (3PCF), is sensitive as well to the shapes of spatial structures [9]. We have calculated the 2PCF and 3PCF using estimators, which count pairs and triplets in the galaxy data and random catalogues, optimally correcting for edge effects and incompleteness of the survey [10, 11]. The random catalogues were built taking into account the radial selection function and angular completeness [12], and the covariance matrix is calculated from jack-knife re-sampling of the different regions. To count the pairs and triplets, we use the *ntropy-npoint*, which is an exact n-point calculator, which uses a kd-tree framework with enhanced routine performance [13, 14].

We have shown our measurements in terms of the redshift-space reduced 3PCF, $Q(s_1, s_2, \theta) = \zeta(s_1, s_2, s_3)/[\xi(s_1)\xi(s_2) + \mathrm{perm}]$, where θ is the angle between the first two sides s_1 and s_2, which determines s_3. Figure 1 shows $Q(s_1, s_2, \theta)$ of the WiggleZ galaxies (from optimally combining the seven independent regions) in the intermediate redshift slice, $z_{\mathrm{eff}} = 0.55$ for a range of scales,

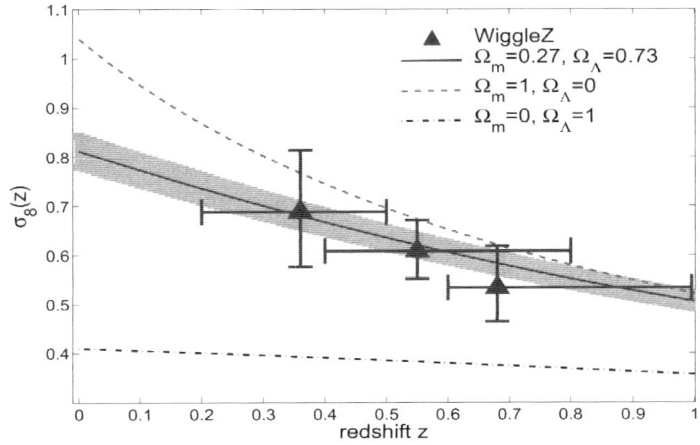

Figure 3. Evolution of $\sigma_8(z)$. Blue symbols correspond to the estimates of $\sigma_8(z_{\rm eff})$ from the 2PCF and 3PCF of WiggleZ galaxies at different redshifts, marginalizing over the linear and non-linear bias parameters. Blue solid line corresponds to the evolution of $\sigma_8(z)$ in a flat ΛCDM universe, with $\Omega_{\rm m} = 0.27$ and $\sigma_8(z = 0) = 0.812$; blue shaded region corresponds to combined WMAP-5 errors. Red dashed line corresponds to the evolution of $\sigma_8(z)$ in a flat $\Omega_m = 1$ CDM universe. Black dash-dotted line shows the evolution of a flat $\Omega_\Lambda = 1$ universe. All models are normalized at the epoch of recombination.

where the $s_2 = 2s_1$ as a function of θ. It can be seen that we recover the shape dependence of the 3PCF, observed on smaller redshifts [14, 15], where there is a bigger 3PCF amplitude at very small or large θ, i.e., the collapsed and elongated configurations. This 'V-shape' is more prominent on large scales; it is a consequence of the morphology of galaxy structures going from spherically-shaped clusters and groups on small $\sim 1\ h^{-1}$Mpc scales, to filaments on the largest scales. The dark matter reduced 3PCF follows the same characteristics of the WiggleZ 3PCF, and moreover, the difference in amplitude in $Q(\theta)$ is small, contrary to what is found in other types of galaxies (e.g., LRGs [15]), which is indicative of differences of the dark matter halos that host different types of galaxies [16].

4. Constraints on galaxy bias and $\sigma_8(z)$

Since different types of galaxies form inside different dark matter halos, they are an imperfect tracer of dark matter [16], and their n-point correlations will differ as well. To model these differences, we have adopted a deterministic and local bias formalism to relate galaxy over-density δ_g with the underlying dark matter density δ_{dm} [17, 18], where $\delta_{gal} = b_1\delta_{dm} + (b_2/2)\delta_{dm}^2$. As a consequence of adopting this model, the 2PCF and the 3PCF of dark matter and galaxies are related by these bias parameters b_1, b_2. In addition to this, the total amplitude of the correlation functions depends on the amplitude of the initial perturbations, in the form of the parameter σ_8. At leading order, for the 2PCF, $\xi \propto \sigma_8^2$ and for the 3PCF, $\zeta \propto \sigma_8^4$ [19], we have then,

$$\xi_{gal} = b_1^2 \left(\frac{\sigma_8}{\sigma_8^{fid}}\right)^2 \xi_{dm,\sigma_8^{fid}}, \quad and \quad Q_{gal} = \frac{1}{b_1}\left(Q_{dm} + \frac{b_2}{b_1}\right), \tag{1}$$

where $\sigma_8^{fid}(z = 0) = 0.812$, the value used in our GiggleZ simulations. Though these relations are valid only on large scales (where perturbations are mildy non-linear) and in real space, with the level of the statistical errors of our measurements, it is adequate to use them in redshift space [20, 21, 15]. Therefore, the reduced 3PCF Q breaks the degeneracy between the linear bias and σ_8 in the 2PCF.

Comparing both the 2PCF and 3PCF for configurations with $s_1 > 10\ h^{-1}$Mpc, from WiggleZ galaxies and the biased dark matter model from the GiggleZ simulation, we have used a χ^2-minimization procedure to estimate the bias parameters b_1, b_2 and σ_8. In Figure 2, we have shown the constraints in the $z_{\rm eff} = 0.55$ redshift slice. Even though the confidence intervals are bigger than other measurements at lower redshifts [21, 22, 15], the value of the linear bias $b_1 \sim 1$ indicates that WiggleZ galaxies are hosted by low-to-intermediate mass dark matter halos. In addition, we have detected a non-zero value for the non-linear bias $b_2 \sim -0.4$. For σ_8, the 1σ values are inside of what is expected at that redshift from the WMAP-5 cosmology (where $\sigma_8^{fid} = 0.62$ at z=0.55). It is important to note that these estimates are independent of any other observable than the galaxy clustering itself.

We have repeated the measurement and analysis of the correlation functions for the other two redshift slices, in order to get constraints on σ_8 as a function of redshift. The evolution of this parameter gives us an indication on how the linear growth factor $D(z)$ evolves with redshift, since $\sigma_8(z) = [D(z)/D(0)]\sigma_8(0)$. In Figure 3, we have plotted σ_8 as a function of redshift for our measured values in the WiggleZ survey, and though the measurements of the different redshift slices are correlated, given that they overlap, we have found a good agreement with the measurements of WMAP-5 in a concordance cosmology. In the near future, with improved measurement techniques and with the arrival of bigger surveys, we will be able to use this method to get constraints not only on σ_8 in the context of the ΛCDM model, but also to discriminate it against modified gravity models [23].

References

[1] Press W H and Schechter P 1974 *Ap. J.* **187** 425
[2] Peebles P J E 1980 *The Large-Scale Structure of the Universe* (Princeton University Press)
[3] Drinkwater M J *et al* 2010 *MNRAS* **401** 1429 [arXiv:0911.4246]
[4] Martin D C *et al* 2005 *Ap. J.* **619** L1 [arXiv:astro-ph/0411302]
[5] Adelman-McCarthy J K *et al* 2006 *Ap. J. Supp.* **162** 38
[6] Gilbank D G, Gladders M D, Yee H K C and Hsieh B C 2011 *Astron. J.* **141** 94 [arXiv:1012.3470]
[7] Poole G *et al* 2012 *In preparation*
[8] Komatsu E *et al* 2009 *Ap. J. Supp.* **180** 330 [arXiv:0803.0547]
[9] Peebles P J E and Groth E J 1975 *Ap. J.* **196** 1
[10] Landy S D and Szalay A S 1993 *Ap. J.* **412** 64
[11] Szapudi S and Szalay A S 1998 *Ap. J. Lett.* **494** L41
[12] Blake C *et al* 2010 *MNRAS* **406** 803 [arXiv:1003.5721]
[13] Gardner J P, Connolly A and McBride C 2007 *Astronomical Data Analysis Software and Systems XVI* (*ASP Conf. Ser.* **376** ed. R A Shaw, F Hill, and D J Bell 69
[14] McBride C K, Connolly A J, Gardner J P, Scranton R, Newman J A, Scoccimarro R, Zehavi I and Schneider D P 2011 *Ap. J.* **726** 13 [arXiv:1007.2414]
[15] Marín F 2011 *Ap. J.* **737** 97 [arXiv:1011.4530]
[16] Berlind A A and Weinberg D H 2002 *Ap. J.* **575** 587
[17] Fry J N and Gaztanaga E 1993 *Ap. J.* **413** 447
[18] Frieman J A and Gaztanaga E 1994 *Ap. J.* **425** 392
[19] Pan J and Szapudi I 2005 *MNRAS* **362** 1363
[20] Scoccimarro R, Couchman H M P and Frieman J A 1999 *Ap. J.* **517** 531
[21] Gaztañaga E, Norberg P, Baugh C M and Croton D J 2005 *MNRAS* **364** 620
[22] McBride C K, Connolly A J, Gardner J P, Scranton R, Scoccimarro R, Berlind A A, Marín F and Schneider D P 2011 *Ap. J.* **739** 85 [arXiv:1012.3462]
[23] Linder E V and Cahn R N 2007 *Astroparticle Phys.* **28** 481 [arXiv:astro-ph/0701317]

Joint QSO - CMB constraints on reionization history

S Mitra

Harish-Chandra Research Institute, Chhatnag Road, Jhusi, Allahabad 211019, India

E-mail: smitra@hri.res.in

Abstract. We have tried to give an overview of model-independent semi-analytical approach to study the observational constraints on reionization. We have implemented and investigated a method to do a detailed statistical analysis using principal component analysis (PCA) technique. We have also discussed different observations related to reionization and shown how to use PCA for constraining the reionization history. Using Markov Chain Monte Carlo methods, we have found that all the quantities related to reionization can be severely constrained at $z < 6$, whereas a broad range of reionization histories at $z > 6$ are still permitted by the current data sets. We have shown that with the forthcoming PLANCK data on large-scale polarization, the $z > 6$ constraints will be improved considerably.

1. Introduction

In the past few years, the understanding of reionization process has become increasingly sophisticated in both the observational and theoretical communities, thanks to the availability of good quality data related to reionization. Mainly, the observations by the Wilkinson Microwave Anisotropy Probe (WMAP) satellite of cosmic microwave background (CMB) and highest redshift quasars put very tight constraints on the reionization history of the universe. The physical processes relevant to reionization are so complex that neither the analytical nor the numerical simulations alone can capture the overall picture. That is why, it is often studied using semi-analytical models of reionization, with limited computational resources. The semi-analytical model used in this work is based on Choudhury and Ferrara [1, 2]. This model treats the inter-galactic medium (IGM) as multi-phase medium, and takes into account the inhomogeneities in the IGM. The sources of ionizing radiation taken, here, are stars and quasars. The model computes *radiative feedback* (suppressing star formation in low-mass haloes using a Jeans mass prescription) self-consistently from the evolution of the thermal properties of the IGM. The major uncertainty in modelling any semi-analytical reionization scenario is to model the parameter $N_{\rm ion}$, the number of photons entering the IGM per baryon in collapsed objects.

2. Principal component analysis

Most likely, $N_{\rm ion}$ is a function of halo mass and the redshift, but the dependences are not well understood, and that is why it is usually taken to be a constant. A method which is ideally suited to tackle this problem is to use the principal component analysis (PCA); this is a technique to compute the most meaningful basis to re-express the unknown parameter set, and the hope is that this new basis will reveal hidden detailed statistical structure [3, 4].

In this paper, we have assumed $N_{\rm ion}(z)$ to be an unknown function of z and decompose it into principal components. We have included only one stellar population and hence, there should not be any explicit chemical feedback in our model. A detailed likelihood analysis is done using three different data sets: The photoionization rates $\Gamma_{\rm PI}$, obtained using Ly-α forest Gunn-Peterson optical depth observations and a large set of hydrodynamical simulations, the redshift distribution of LLS $\mathrm{d}N_{\rm LL}/\mathrm{d}z$ at $0.36 < z < 6$, and the angular power spectra C_l for TT, TE and EE modes using WMAP-7 and forecasted PLANCK data. The moment we have included the C_l's (TT+TE+EE) in our analysis, we have realized that parameters related to reionization may have strong degeneracies with (some of) the cosmological parameters, mainly with the normalization of the matter power spectrum σ_8 and the slope n_s, so we have carried out our analysis by varying only these two parameters (in addition to the parameters related to reionization model) and keeping all the other cosmological parameters fixed to their best-fit WMAP-7 value.

2.1. *Formalism*

We have represented the unknown function $N_{\rm ion}(z)$ by a set of $n_{\rm bin}$ discrete free parameters with redshift bins. Then we have constructed the Fisher matrix:

$$F_{ij} = \sum_{\alpha=1}^{n_{\rm obs}} \frac{1}{\sigma_\alpha^2} \frac{\partial \mathcal{G}_\alpha^{\rm th}}{\partial N_{\rm ion}^{\rm fid}(z_i)} \frac{\partial \mathcal{G}_\alpha^{\rm th}}{\partial N_{\rm ion}^{\rm fid}(z_j)}, \tag{1}$$

where $\mathcal{G}_\alpha$, $\alpha = 1, 2, \ldots, n_{\rm obs}$ represent the observational data points, σ_α are their corresponding error, $\mathcal{G}_\alpha^{\rm th}$ is theoretical value of $\mathcal{G}_\alpha$ and $N_{\rm ion}^{\rm fid}$ is the fiducial model, which is, in principle, close to the underlying "true" model. Here, we have taken the fiducial model $N_{\rm ion}^{\rm fid}$ to be the one, which matches the current data points up to an acceptable accuracy, and also which is characterized by a higher $N_{\rm ion}$ at higher redshifts. Once the Fisher matrix is constructed, we can determine its eigenvalues and corresponding eigenvectors. Because of the ortho-normality and completeness of the eigenfunctions, we can expand the deviation of $N_{\rm ion}$ from its fiducial model as $\delta N_i = N_{\rm ion}(z_i) - N_{\rm ion}^{\rm fid}(z_i) = \delta N_i = \sum_{k=1}^{n_{\rm bin}} m_k S_k(z_i)$, where $S_k(z_i)$ are the principal components of $N_{\rm ion}(z_i)$, and m_k are the expansion coefficients. The advantage is that, unlike $N_{\rm ion}(z_i)$, the coefficients m_k are uncorrelated with variances. It can be shown that the largest eigenvalues correspond to minimum variance and vice versa. Hence, most of the information relevant for the observed data points is contained in the first few modes with larger eigenvalues. We can then reconstruct the function δN_i using only the first $M \leq n_{\rm bin}$ modes. So, the important step in this analysis is to decide on how many modes M to be used.

3. Results: WMAP-7 vs PLANCK

3.1. *The principal components of $N_{\rm ion}(z)$*

We have found that the components of the Fisher matrix vanish for $z < 2$, because there are no data points considered at these redshifts. For $2 < z < 6$, the values of F_{ij} are considerably higher, because it is determined by the sensitivity of $\Gamma_{\rm PI}$ and $\mathrm{d}N_{\rm LL}/\mathrm{d}z$ on $N_{\rm ion}(z)$. On the other hand, the information at $z > 6$ is determined by the sensitivity of $C_l^{\rm EE}$ on $N_{\rm ion}$, which is relatively weak. We have found that F_{ij} is negligible for $z > 14$, and this is expected because the collapsed fraction of haloes is negligible at those redshifts, and hence, there exist no free electrons to contribute to electron scattering optical depth $\tau_{\rm el}$ or $C_l^{\rm EE}$ [5, 6]. In Figure 1, we have shown the inverse of the first few eigenvalues (λ_i), i.e., the variances of the corresponding modes. For modes $i > 8$, the eigenvalues are almost zero, and the variances are extremely large. This implies that the errors on $N_{\rm ion}$ would increase dramatically if we include modes $i > 8$. The first 8 eigenmodes (i.e., those which have the lowest variances) are plotted in Figure 2. We have found that all the eigenmodes tend to vanish at $z < 2$ and $z > 15$, which is obvious because of F_{ij} being negligible at these redshifts. We can see a number of spikes and troughs in the first four modes, whose positions correspond to the presence of data points for $\Gamma_{\rm PI}$ and $\mathrm{d}N_{\rm LL}/\mathrm{d}z$ at $2 < z < 6$.

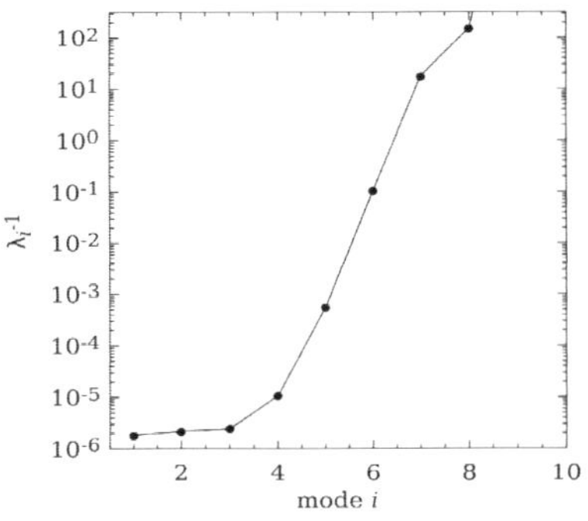

Figure 1. The inverse of eigenvalues of the Fisher matrix F_{ij}, which essentially measures the variance on the corresponding coefficient.

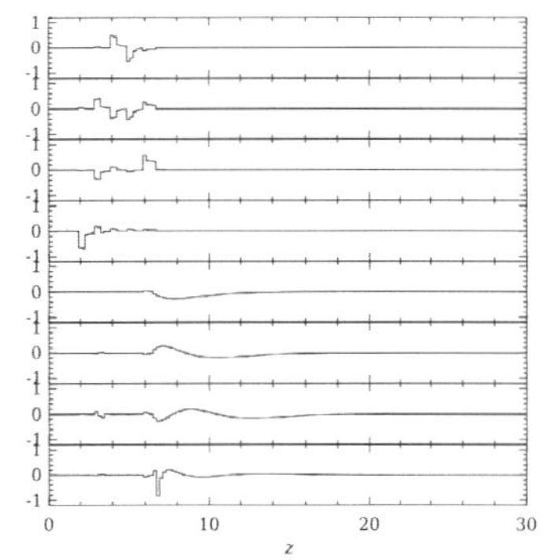

Figure 2. The first 8 eigenmodes of the Fisher matrix.

The last four modes contain the information about the sensitivity of $C_l^{\rm EE}$. The modes (> 8) with smaller eigenvalues, i.e., large variances introduce huge uncertainties in the determination of $N_{\rm ion}$, and hence, do not contain any meaningful information about the reionization history.

3.2. Choice of the number of modes

The next step in our analysis is to decide on how many modes M to use. A sophisticated and model-independent prescription is to use Akaike information criterion (AIC $= \chi^2_{\rm min} + 2M$), where smaller values of AIC are assumed to imply a more favoured model. Note that there is no reason to select one particular reconstruction, the minimum of AIC can be accompanied by an increased chance of getting the reconstructed parameters wrong. One successful strategy for this purpose is to select different M, which are near the minimum value of AIC and amalgamate them equally at the Monte Carlo stage when we compute the errors [7]. In this way, we can reduce the inherent bias, which exists in any particular choice of M. We have examined that, in our case, the family of different M reconstructions, starting from $M = 2$, which satisfy AIC $<$ AIC$_{\rm min}+\kappa$, where $\kappa = 10$ (which corresponds to $M = 8$) produces very solid results. For alternative data sets, the value of κ can be adjusted.

3.3. Markov Chain Monte Carlo constraints

The constraints on reionization are obtained by performing a Monte Carlo Markov Chain (MCMC) analysis over the parameter space of the optimum number of PCA amplitudes and other free parameters of the model. To avoid the confusion about the correct choice of number of modes, we have performed the MCMC analysis for PCA amplitudes taking from $M = 2$ to $M = 8$, all of which obey the AIC criterion.

It can be seen from the plot of $N_{\rm ion}(z)$ (*top-left* panel of Figure 3) that such quantity must necessarily increase from its constant value at $z < 6$. This rules out the possibility of reionization with a single stellar population having non-evolving IMF and/or star-forming efficiency. From the plot of $\Gamma_{\rm PI}(z)$ (*top-middle* panel) and $dN_{\rm LL}/dz$ (*top-right* panel), we have found that the mean model is consistent with the observational data at $z < 6$, as expected. The errors corresponding to 95% confidence limits are also smaller at this epoch. However, the model described by the mean values of the parameters shows a sharper and prominent peak around $z \sim 6.5$ for both cases. From the plot of the volume filling factor of ionized region $Q_{\rm HII}(z)$ (*bottom-left* panel), we have seen that the growth of this quantity for the mean model is much faster than that of fiducial model at initial stages, though the completion of reionization takes

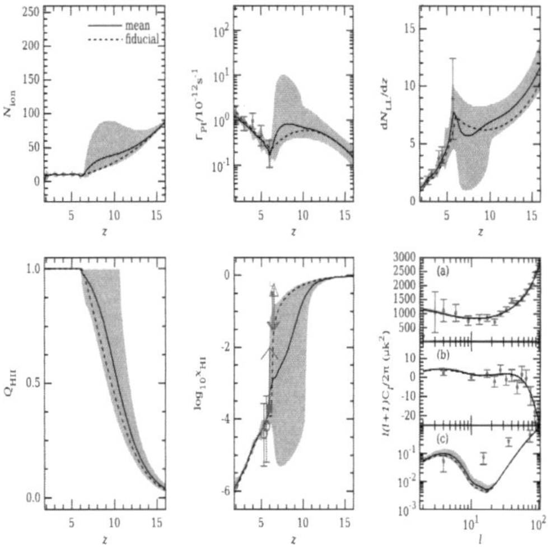

Figure 3. The marginalized posteriori distribution of various quantities related to reionization history obtained from the PCA using the AIC criterion with first 8 eigenmodes for WMAP-7 data.

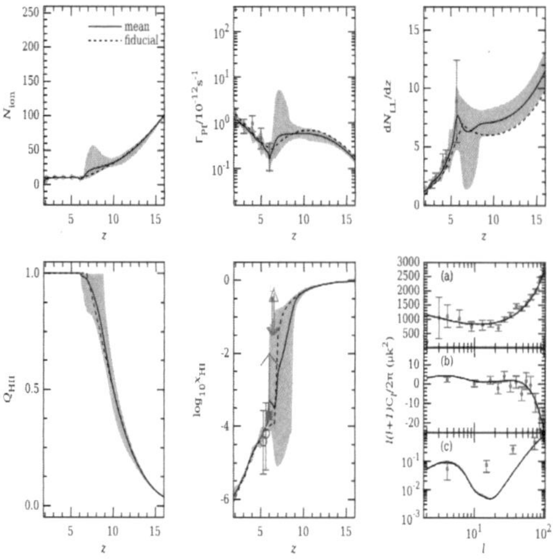

Figure 4. The marginalized posteriori distribution of various quantities related to reionization history obtained from the PCA using the AIC criterion with first 8 eigenmodes for PLANCK likelihood.

place only at $z \approx 6$. Similarly, the neutral hydrogen fraction $x_{\rm HI}(z)$ (*bottom-middle* panel) decreases much faster than the fiducial one at $6 < z < 12$ and then smoothly matches the Ly-α forest data. Finally, we have shown the values of (a) $C_l^{\rm TT}$, (b) $C_l^{\rm TE}$ and (c) $C_l^{\rm EE}$ for the mean model in the *bottom-right* panel of this figure, which is almost the same as the fiducial model. Also, here, we have got the value of $\tau_{\rm el}$ higher than the current WMAP value. So, a wide range of reionization histories is still allowed by the data we have used.

To check how better the PLANCK data can constrain the reionization scenario, we have first generated the simulated PLANCK data of CMB power spectra for our fiducial model up to $l \leq 2000$ using the exact full-sky likelihood function at PLANCK-like sensitivity, and then repeat the same MCMC analysis as we have done for WMAP-7 data. In Figure 4, we have illustrated the recovery of the same quantities as mentioned in the earlier case. We have found that our main results are in quite reasonable agreement with those obtained from the WMAP data, except that all the 2σ (95 %) limits are reduced remarkably for all redshift range, especially the 2σ limits for $Q_{\rm HII}$ reduces significantly for this case. Thus, with the forthcoming PLANCK data on large scale polarization (ignoring the effect of foregrounds), the $z > 6$ constraints will be improved considerably. In particular, we have found that using the PLANCK data, the 2σ error on $\tau_{\rm el}$ will be reduced to 0.009 and the uncertainties on the redshift at which reionization is 50% completed, i.e., $z(Q_{\rm HII} = 0.5)$ and almost or 99% completed, $z(Q_{\rm HII} = 0.99)$, would be ~ 1 and 3 (95% CL) respectively, which are much smaller than the case for current WMAP data. However, for more stringent constraints on reionization at $z > 6$, one has to rely on the data sets other than CMB.

Acknowledgements
The author would like to thank all his collaborators, who have contributed to the research work described in this article.

References

[1] Choudhury T R and Ferrara A 2005 *MNRAS* **361** 577
[2] Choudhury T R and Ferrara A 2006 *MNRAS* **371** L55
[3] Mortonson M J and Hu W 2007 *Ap. J.* **657** 1
[4] Guha Sarkar T, Mitra S, Majumdar S and Choudhury T R 2012 *MNRAS* **421** 3570
[5] Mitra S, Choudhury T R and Ferrara A 2011 *MNRAS* **413** 1569
[6] Mitra S, Choudhury T R and Ferrara A 2012 *MNRAS* **419** 1480
[7] Clarkson C and Zunckel C 2010 *Phy. Rev. Lett.* **104** 211301

Probing the anisotropic expansion history of the universe using CMBR

R K Mohapatra, P S Saumia*, and A M Srivastava

Institute of Physics, Sachivalaya Marg, Bhubaneswar 751005, India

E-mail: *saumia@iopb.res.in

Abstract. We have proposed a technique to detect any anisotropic expansion in the universe from the beginning of inflation to the last scattering. Any anisotropic expansion in the universe would deform the shape of the primordial density perturbations in the universe, and a shape analysis of the super-horizon fluctuations in CMBR will detect this shape deformation. Using this analysis, we have constrainted any anisotropic expansion in the universe to be less than 35%.

1. Introduction

Observations have shown that the present universe is homogeneous and isotropic. One of the most important observational evidences for the homogeneity and isotropy of the universe is the highly smooth and uniform Cosmic Microwave Background (CMB) radiation. This does not rule out the possibility of an initial anisotropic expansion of the homogeneous universe, which tend to isotropic expansion later [1]. The question we would like to ask, here, is: Whether we can detect any such transient anisotropic expansion in the history of the universe in a model independent way? We have proposed that an analysis of the shapes of the super-horizon sized fluctuations in the CMB can detect any anisotropic expansion in the universe, starting from inflation to the surface of last scattering. Our technique is very general. The only assumption, here, is that there is some mechanism in the early universe, which generates the density fluctuations of super-horizon scales. However, for definiteness, we have used the picture of an early inflationary stage of the universe, which is responsible for the generation of density fluctuations. Any early anisotropic expansion in the universe will deform the fluctuations.

We have discussed two different ways of identifying shape deformations of density fluctuation patches: (i) We have generated patches by considering some threshold value for over-density (or under-density) of CMBR fluctuations in a given region on the surface of last scattering (excursion sets). We then have studied shape deformation of these patches in this region using either the Fourier transform, or detected it directly in the spatial patch. Both these methods can detect the shape deformations. However, the shape analysis with spatial region has specific advantages, which we have discussed later. (ii) We have also analyzed simulated density fluctuations of specific geometrical shapes, such as spheres, and ellipsoids. This helps us in analyzing the strengths of these two techniques in differentiating various scenarios, such as anisotropies arising from initial conditions, or anisotropies arising from the anisotropic expansion, etc.

2. The analysis

The main idea underlying our analysis is the following: If the density perturbations generated initially (say during inflation) were statistically isotropic, then they would continue to have an average spherical shape (statistically), if the expansion of the universe was always isotropic. But if the universe ever went through an anisotropic expansion, then these perturbations would get deformed [2]. For example, if the universe ever expanded differently in one direction from the other two, then the *average* shape of these perturbations will become ellipsoidal. This average deformation will survive the later isotropic expansion till the time the size of the perturbation remains super-horizon. This means that any anisotropic expansion from the beginning of inflation to the surface of last scattering will leave their signature in the shapes of super-horizon perturbations in the CMBR.

We have used WMAP-7 data (ILC map) for the analysis, and chosen thin strips of the sky along the great circles so as to avoid the galaxy contamination along the equator, and also chosen the convenient pixel arrangement of HEALPix. We have done this by rotating the sky along z-axis at different angles. A strip of width $\pm 10^\circ$ along this rotated equator has been chosen, and we have chosen CMBR fluctuations above/below a particular value, thereby, forming the excursion sets. Figure 1 (a) shows a small $20^\circ \times 20^\circ$ region in this equatorial belt. Filled patches correspond to temperature anisotropies of magnitude (0.02 - 1)$\times(\Delta T)_{max}$, where $(\Delta T)_{max}$ is the maximum magnitude of CMBR temperature anisotropy in this patch. As one can see, the patches appear randomly shaped and sized. Such patches are projected on a plane to calculate the X and Y extents.

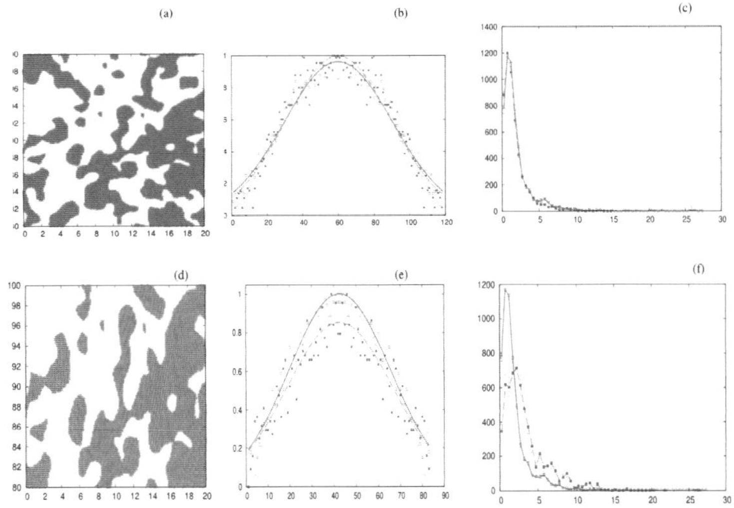

Figure 1. (a) X-axis corresponds to $(\sin\theta)\ \phi$ (degrees) and the Y-axis shows θ (degrees). The figure shows the excursion sets in a $20^\circ \times 20^\circ$ region in the equatorial belt of CMBR sky. (b) Shows the histograms in the two directions for the 2D Fourier transform. Solid and dashed curves are best fits to these histograms. (c) Shows plots of these histograms of the widths of filled patches with corresponding $(\sqrt{N})$ error bars. Distribution for width along X-axis (angle ϕ) is shown by solid plot and distribution for width along Y-axis (angle θ) is shown by the dashed line. (d), (e) and (f) correspond to (a), (b) and (c) in the case $\alpha = 2.0$

2.1. *Fourier transform method*

Detection of shape anisotropies using Fourier transforms can be done in different ways depending on what criterion one adopts to characterize the anisotropy. We have followed the approach used in [3] for analyzing anisotropic deformations in metallography. Here, one uses a digitized image of the material and then calculates the 2D Fourier transform of the image, which is thresholded to levels 0 and 1. Anisotropy in the Fourier space is then determined by the ratio of the widths of the histograms in the two directions. (See [4] for more details). Figure 1 (b) shows the plots of the histograms in the two directions corresponding to the 2D Fourier transform. The two plots completely overlap showing the statistical isotropy of the excursion sets in Figure 1 (a).

2.2. *Analysis in the physical space*

We have now described our technique for detecting shape deformations directly in the physical space. What we want to find out is the average widths of the fluctuations in X-direction, and compare it with the average widths in the Y-direction for the sky as given in Figure 1 (a). For this, we have proceeded as follows: We have divided the entire $20°$ wide equatorial belt into thin slices (varying from $0.05°$ - $1°$) in X- and Y-directions (to increase statistics). Using these slices, we have determined the X and Y extents of various filled patches. We have then plotted the frequency distributions (histograms) of X and Y widths of the intersections of all the patches with these slices in this equatorial belt. For the isotropic case, we expect the X and Y histograms to almost overlap. Any relative shift, or difference, between the X and the Y histograms will imply the presence of an anisotropy, such as an anisotropic expansion. We mention that this fine slicing of the equatorial strip is done to simplify the analysis and improve the statistics. The solid and dashed curves in Figure 1 (c) shows the X and Y histograms, and they overlap showing an isotropic expansion as also shown by the Fourier transform plots.

3. Constraining the anisotropy

To determine the level of isotropy, which is implied by the overlap of the two histograms in Figures 1 (b) and 1 (c), we have introduced artificial stretching in the CMBR patches as follows and repeated the above analysis for this *stretched* data. To *simulate* stretching by a factor α, we have simply multiplied the Y-coordinate (i.e., θ) for each point in the equatorial strip used above by a factor $\alpha \times \frac{|y|}{10°}$. This stretching represents an anisotropic expansion of the universe along the polar axis compared to the expansion in the equatorial plane by a factor α. Note again that this simple scaling expression works approximately fine for a relatively thin strip along the equator. For strips having larger widths along the longitudes, the Y-coordinates of different patches will be scaled by a more complicated factor.

Figure 1 (d) shows these artificially stretched patches corresponding to the patches shown in Figure 1 (a). The stretching factor is $\alpha = 2$ for Figure 1 (d). We have now repeated the analyses as described above for these patches. Figure 1 (e) shows the histograms calculated for the 2D Fourier transform for the stretched patches of Figure 1 (d). We have seen that these histograms do not overlap showing the anisotropy arising from the stretching in Figure 1 (d). Similarly, Figure 1 (f) shows the histograms resulting from the analysis in physical space applied to Figure 1 (d). We have seen, here, also that the two histograms do not overlap and the difference is significant. Note incidentally that the peak in the dashed curve (θ histogram) has shifted to larger widths by almost a factor of 2, which is the factor of stretching of patches for Figure 1 (d). We have tried to put stronger constraint on the anisotropic expansion factor α by repeating the analysis with different values of α (including values of $\alpha < 1$). Figures 2 (a) and 2 (b) show the cases of $\alpha = 1.35$ and 1.3. We have seen that the two histograms corresponding to X and Y widths of patches are clearly separated for $\alpha = 1.35$. However, for $\alpha \leq 1.3$, the differences in the two histograms are insignificant. It is important to note, here, that the most important, qualitative, signature of anisotropic expansion in our technique is the relative lateral

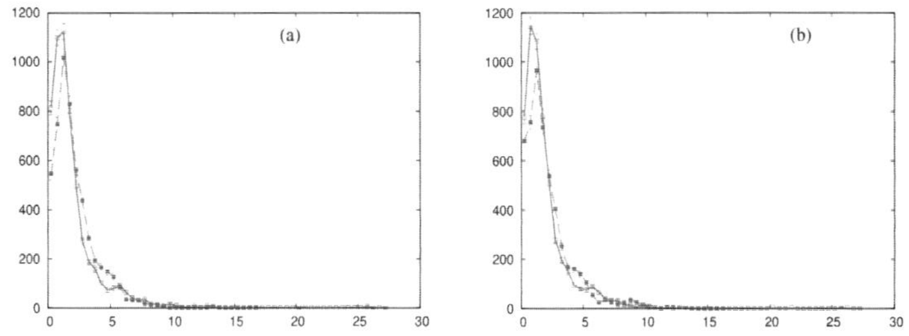

Figure 2. Plots (a) and (b) are the frequency distributions for CMBR data with stretch factor $\alpha = 1.35$, and 1.3 respectively. ϕ and θ width distributions are shown by solid and dashed curves.

shift of the curves of the X and Y frequency distributions. This is clearly seen in Figure 2 (a), which represents larger length scales in one direction compared to the other direction. This automatically is correlated with change in relative heights of the peaks. Thus, in Figure 2 (a), though peak height are very different, our focus on detecting anisotropic expansion (here, with $\alpha = 1.35$) is in the shift of the overall curve towards right. When we use smaller values of α, this lateral shift is not significant. With this, it seems reasonable to conclude that with our analysis technique, and with the present CMBR data, one can put a conservative upper bound of $\alpha < 1.35$ on the anisotropic expansion in the entire history of the universe. We have also repeated the analysis by using under-density patches in CMBR sky and the results are the same.

To demonstrate the strength of the technique, we have repeated the analysis using simulated patches with well defined geometric shapes. The isotropic case is represented by circles, whereas anisotropic case by ellipses. For this, we have created a 3 dimensional cubic region in which over-densities of constant magnitude and specific geometric shapes are created at different locations. This represents a part of the universe enclosing the CMBR sky. We have then determined the shapes of the over-density patches by embedding a surface of two-sphere (representing the CMBR sky) in this cubic region and recording the patches, which are intersected by this S^2. We have repeated the analyses, and noticed that for isotropic case, the histograms in both the analyses overlap, but for anisotropic case, they do not, and here also, for the analysis in physical space, there is a peak shift which quantifies the stretch factor or the anisotropy factor. (See [4] for the figures). We can also show with simulated patches that our technique can also distinguish fluctuations, which are created deformed with the ones, which are deformed due to anisotropic expansion. It can also be shown that the analysis in physical space has more advantages, since it can quantify the shape deformation of fluctuations, but is not sensitive to any other type of anisotropy, say anisotropic distribution of patches, whereas the Fourier transform method detects all anisotropies and cannot distinguish them in a qualitative manner.

4. Conclusions

We have emphasized the most important part of our results, that a simple technique of shape analysis is able to answer an important question in an almost model independent manner. That is whether the universe ever expanded anisotropically almost from the beginning of inflation near $t \simeq 10^{-35}$ sec. up to the stage of last scattering when the universe was 300,000 years old. Even with our qualitative approach of comparing the histograms in the two directions, we can conclude that our technique can rule out any anisotropic expansion of the universe in the past to less than 35%, apart from any sufficiently early stages of inflationary universe, which are followed by very long, isotropic inflationary stage. In particular, any anisotropic expansion stage after the end of inflation is certainly restricted to $\alpha < 1.35$. With PLANCK data, one should be able to do shape analysis of patches with high resolution. A major drawback of our present analysis is its restriction to relatively small angular scale. Thus, claims of anisotropy at quadrapole level in the literature are not examined by our analysis at present. We are working on improvement of our techniques for larger angular scales, and also trying to get better control on statistical fluctuations in our plots.

Acknowledgements

We acknowledge the use of the Legacy Archive for Microwave Background Data Analysis (LAMBDA), and HEALPix package.

References

[1] See Chapter XI *General Relativity - An Einstein Centenary Survey* 1979 eds S W Hawking and W Israel (Cambridge University Press)
[2] Buniy R V, Berera A and Kephart T W 2006 *Phys. Rev. D* **73** 063529
[3] Holota and Nemecek 2002 *Applied Electronics* p 88
[4] Mohapatra Ranjita K, Saumia P S and Srivastava Ajit M [arXiv:1111.2722] *and references therein.*

Journal of Physics: Conference Series **484** (2014) 012041 doi:10.1088/1742-6596/484/1/012041

The CFHTLS-Strong Lensing Legacy Survey (SL2S): Investigating the group-scale lenses with the SARCS sample

A More[1,2*], R Cabanac[3,4], S More[1], C Alard[5], M Limousin[2], J-P Kneib[2], R Gavazzi[5] and V Motta[6]

[1] Kavli Institute for Cosmological Physics, University of Chicago, 5640 S. Ellis Ave., Chicago IL-60637, USA
[2] Laboratoire d'Astrophysique de Marseille, 38 rue Frederic Joddress 13013 Marseille, France
[3] Université de Toulouse, UPS-OMP, IRAP, Tarbes, France
[4] CNRS, IRAP, 57, Ave. d'Azereix, 65000 Tarbes, France
[5] Institute d'Astrophysique de Paris, France
[6] Universidad de Valparaiso, Departamento de Fisica y Astronomia, Avenida Gran Bretana 1111, Valparaiso, Chile

E-mail: *anupreeta@kicp.uchicago.edu

Abstract. We have presented the Strong Lensing Legacy Survey - ARCS (SARCS) sample compiled from the T0006 data release of the Canada-France-Hawaii Telescope Legacy Survey (CFHTLS) covering a total non-overlapping area of 159 deg^2. We have adopted a semi-automatic method to find gravitational arcs in the survey that makes use of an arc-finding algorithm. The SARCS sample consists of 127 lens candidates, out of which, 54 systems are promising lenses. From our sample, we have detected a systematic alignment of the giant arcs with the major axis of the baryonic component of the putative lens in concordance with previous studies. This alignment is also observed for all arcs in the sample and does not vary significantly with increasing arc radius. Owing to the large area and depth of the CFHTLS, we have found the largest sample of lenses probing mass scales that are intermediate to cluster and galaxy lenses for the first time. We have compared the observed image separation distribution (ISD) of our arcs with theoretical models. A two-component density profile for the lenses, which accounts for both the central galaxy and the dark matter component is required by the data to explain the observed ISD. Unfortunately, current levels of uncertainties and degeneracies accommodate models both with and without adiabatic contraction. We have also shown the effects of changing parameters of the model that predict the ISD and that a larger lens sample might constrain relations such as the concentration-mass relation, mass-luminosity relation and the faint-end slope of the luminosity function.

1. Introduction

Gravitational lensing is the deflection of light coming from distant sources in the Universe, due to the gravitational potential of intervening structures [1, 2]. The last decade has seen the rise of a wide variety of applications of strong lensing such as the study of distant lensed galaxies with unprecedented magnification [3, 4], the constraints on sub-structure within lensing halos [5, 6], accurate measurements of the Hubble constant [7], constraints on the stellar initial mass function [8], and constraints on the slope of the inner density profile of the lensing halos

[9, 10, 11]. Although strong lensing is a rare event, several surveys covering a wide sky area and deep enough imaging across different wavelengths, have resulted in the discovery of over 200 strong lens systems at galaxy scales from surveys such as the Cosmic Lens All Sky Survey [12], the Sloan Lens ACS Survey [13], and about a few dozen lens systems at cluster scales such as, the Extended Medium Sensitivity Survey [14], the MAssive Cluster Survey (MACS) [15], the Las Campanas Distant Cluster Survey [16], and the Red sequence Cluster Survey [17].

Majority of the surveys in the past have primarily focused on studying galaxy-scale or cluster-scale structures. As a result, matter distribution in galaxies and galaxy clusters is relatively well-studied via both strong and weak lensing. A further improvement in our understanding has come from the use of complementary methods to lensing such as stellar kinematics, satellite kinematics and X-ray scaling relations. In contrast, little is known about galaxy groups which are intermediate to galaxies and galaxy clusters, typically corresponding to masses of $10^{12}-10^{14}\ M_{\odot}$. Relatively fewer investigations have been carried out with galaxy groups e.g., study of mass-to-light ratios with the Canadian Network for Observational Cosmology 2 sample [18], study of faint end of the luminosity function of nearby compact groups [19], study of concentration-mass (c-M) relation from the SDSS [20] via weak lensing, study of colours and star formation [21], and study of scaling relations of X-ray selected groups [22]. Since studies of groups are limited, and we still do not have a detailed understanding of matter distribution, formation and evolution of galaxy groups. Being one of the important components in the hierarchical assembly of structures in the Universe, galaxy groups are much more massive than galaxy-scale halos and are concentrated enough to act as lenses. Furthermore, since galaxy groups are quite abundant compared to massive structures like galaxy clusters, the probability to find group scale lenses is also large. Hence, lensing can be successfully used to study group-scale halos.

The Strong Lensing Legacy Survey (SL2S) [23] is a survey from the Canada-France-Hawaii Telescope Legacy Survey (CFHTLS). The design of the CFHTLS allows us to find large sample of group-scale lenses, which can be studied in detail upto high redshifts, for the first time. The SL2S is as a precursor to wide field imaging surveys such as the Dark Energy Survey, Large Synoptic Survey Telescope and Euclid. In this paper, we have presented the SL2S-ARCS sample from the final T0006 release of the CFHTLS. In Section 2, we have given an overview of the survey and procedure of sample selection and present the final sample. In Section 3, we have discussed some statistical results using the final sample, and in Section 4, we have summarized the survey and our main findings.

2. The CFHTLS-SL2S ARCS sample

In this section, we have given a brief overview of the survey from which we have derived the lens sample.

2.1. Survey overview

CFHTLS is a photometric survey made with the Canada-France-Hawaii Telescope (CFHT) in five optical bands ($u^*g'r'i'z'$) using the wide-field imager MegaPrime with a field-of-view of 1 deg^2 on the sky and a pixel size of $0.186''$. The WIDE and DEEP components of the CFHTLS are designed to carry out extragalactic research. These components are ideal for searching strong lens systems. The SL2S sample is compiled from the CFHTLS-WIDE encompassing a combined area of 171 deg^2 and CFHTLS-DEEP encompassing a combined area of 4 deg^2. However, taking into account the masked and overlapping areas, the effective area of the survey is 150.4 deg^2 (146.9 deg^2 for WIDE and 3.5 deg^2 for DEEP). The WIDE and DEEP each consist of four fields[1]. Each of the deep fields covers an area of 1 deg^2. Among the WIDE fields,

[1] These numbers are estimated from [http://terapix.iap.fr/cplt/table_syn_T0006.html] [http://terapix.iap.fr/cplt/table_syn_T0006.html]

g band imaging is the deepest of all bands with a limiting magnitude of 25.47 and a mean seeing of 0.78″ whereas the g band imaging of DEEP fields has nearly 10 times deeper exposures than the WIDE fields, and the median seeing is $\sim 0.7''$. Further details of the T0006 release, which is the first complete release of the WIDE and DEEP, can be found on Terapix website[2].

2.2. *Sample selection and output*

We have defined the SARCS sample such that lens systems with arc radius ($R_{\rm A}$) $\geq 2''$ belong to the sample. The radius of the arc is defined as the distance of the lensed image from the putative lens galaxy which is roughly the Einstein radius. Typically, lensing halos with Einstein radius larger than 2″ are very massive lenses with significant contribution from the environment of the primary lensing galaxy. Thus, the SARCS sample, predominantly, consists of group to cluster scale lenses.

The SARCS sample from the CFHTLS is compiled in a three-step process: (i) Run the arc-detection algorithm called ARCFINDER [24]. We have chosen to run the ARCFINDER on the g-band image since most of the lensed images correspond to galaxies with high star formation that have little emission at redder wavelengths. At the end of the first step, we have produced a list of arc candidates with various parameters. (ii) Apply a cut-off on arc properties such as the area, the peak flux count, the radius of curvature, and reject candidates within masked area. These cuts allow a significant reduction in false detections at the cost of losing some real arc candidates. (iii) Visual inspection, and classification are carried out to grade the quality of the candidates. Note that the final SARCS sample consists of candidates that are detected by the ARCFINDER and/or by visual inspection.

After the automatic detection and screening, about 1000 candidates/deg^2 are visually inspected. This is reduced to a total sample of 413 candidates, which is further considered for ranking by three people. Finally, we have accepted 127 lens candidates in the SARCS sample[3], which are ranked 2 or higher (average of the ranks by three people), and have $R_A > 2''$.

3. Results and discussion

In the following subsections, we have described the main findings from the SARCS sample and constraints on average properties of the lens population using statistical properties of the arcs.

3.1. *Galaxy-arcs orientation*

Triaxial dark matter halos, which appear elliptical in projection, may form lip caustics and giant arcs (high length-to-width ratio) are likely to occur in such cases (e.g., [25]). Using numerical simulations, authors of [26] have shown that the giant arcs are oriented very close to the major axis of the dark matter halo. If a central galaxy is further added to the halo, then it appeared to isotropize the angular distribution of arcs to a small extent. They have measured the angular distribution of giant arcs from the EMSS cluster sample, and found that most of the giant arcs had an orientation of < 45 deg, consistent with their predictions.

We have used the same definition as that of [26], that is, the angle between the major axis of the lens galaxy and the line connecting the centre of the lens galaxy to the midpoint of the arc. The midpoint of the arc and its orientation from the major axis of the lens is calculated manually. When there is a single dominant lensing galaxy, we have measured the position angle (PA) of the major axis of the lens galaxy with SEXTRACTOR [27], otherwise we have adopted the following strategy: If there are two or three (nearly collinear) lens galaxies with similar colours and brightness, then the PA of the ellipticity is assumed to be along the line joining the

[2] [http://terapix.iap.fr/cplt/T0006-doc.pdf] [http://terapix.iap.fr/cplt/T0006-doc.pdf]
[3] High resolution images of these systems are made available at http://kicp.uchicago.edu/~anupreeta/sarcs sample.

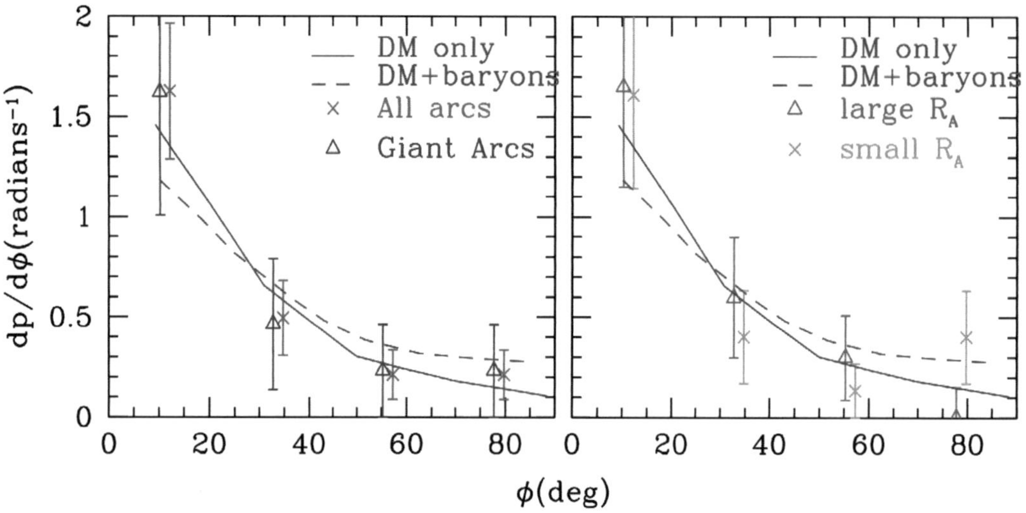

Figure 1. *Left*: Angular distribution of arcs with respect to the lens galaxy. The data points are from the SARCS sample. *Right*: The "all arcs" sample is divided into small and large R_{A}. The expected curves for dark matter (DM) only (solid) and DM+$3 \times 10^{12}h^{-1}M_{\odot}$ (dashed) are taken from [26].

lens galaxies. If a circle going through the arc encloses multiple lens galaxies with comparable brightness but no obvious elongation, then such system is rejected.

After applying the above selection cuts, we are left with 36 all arc candidates and 11 giant arcs ($l/w \geq 8$) with an average ranking > 2.5. The angular distribution of the giant arcs from the SARCS sample is shown in the left panel of Figure 1. Our measurements are compared with predictions from [26], namely, (a) for dark matter only lensing halo (solid curve), and (b) a dark matter halo to which a galaxy of $3 \times 10^{12}h^{-1}M_{\odot}$ is added (dashed curve). The distribution of giant arcs from our sample appears to follow the expected trend although it is not possible to distinguish between the two curves. We have also found that the distribution of orientation of all arc candidates (see Figure 1) is consistent with expectations. This extends the result obtained in [26] to groups-scale lens systems.

We have further split the all arcs sample into a small R_{A} ($< 5''$) and large R_{A} ($\geq 5''$) samples to see the dependence of the angular distribution on the R_{A} with an arbitrary dividing value of $5''$. The baryonic matter tends to make the matter distribution in halos more spherical at the centre. Therefore, the angular distribution of arcs for the sample with smaller R_{A} is expected to be more isotropic compared to the sample with large R_{A}. However, we do not find any clear differences in the angular distribution of small and large R_{A} for the SARCS sample within the uncertainties (see the right panel of Figure 1). We have noticed that these measurements may have some systematic errors due to the orientation dependence of the selection function. Ambiguities also exist in the definition of the orientation in few cases, especially when multiple candidate lens galaxies are involved. In addition, the current analysis consists of lens candidates as opposed to confirmed lens systems. Therefore, our conclusions should be treated more of a qualitative nature.

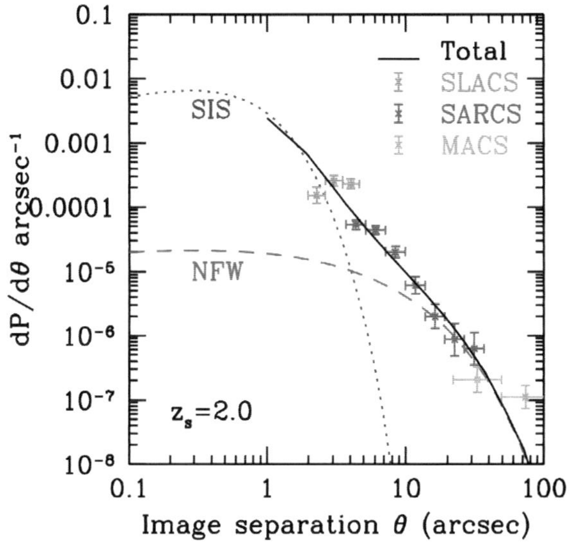

Figure 2. Image separation distribution. Theoretically calculated image separation distribution curves for SIS profile (dotted), NFW profile (dashed), and total profile (solid) following the model in [28]. The data points from the SARCS sample (magenta).

3.2. Image separation distribution

The image separation distribution (ISD) is sensitive to the halo mass, abundance of the lens population, the mass distribution in the lens, and the source redshift. Therefore, the ISD measured from galaxy to cluster scales contains information about the cosmological parameters and various scaling relations between galaxy properties and halo mass. Hitherto, the ISD has been measured either at small image separations (θ) primarily, with lens samples such as CASTLES (e.g., [29]) and CLASS (e.g., [28]) or at large θ with cluster-scale lenses, for example, the MACS sample [30]. With the SARCS sample, we can probe the intermediate mass regime corresponding to group scale lenses.

3.2.1. Model for the expected distribution: We have followed the framework developed in [28] to calculate the expected ISD in order to compare it with the observed ISD from the SARCS sample, and also adopted the cosmology, that is, $\Omega_m = 0.3$, $\Omega_\Lambda = 0.7$, and $\sigma_8 = 0.9$. The probability for a source at redshift z_s to get lensed with image separation greater than θ is $P(>\theta; z_s)$. The differential probability is then given by:

$$\left|\frac{\mathrm{d}P}{\mathrm{d}\theta}\right| = \int_0^{z_s} \mathrm{d}z_l \frac{\mathrm{d}\chi}{\mathrm{d}z_l} n(\tilde{M}[\theta], z_l)\, \sigma_{\rm lens} \Theta(M - \tilde{M}), \tag{1}$$

which can be directly related to the ISD of the observed lens sample. Here, χ is the comoving distance and z_l is the redshift of the lens, M corresponds to the halo mass, $n(M, z_l)$ is the halo mass function, $\tilde{M}$ is the minimum halo mass that causes an image separation equal to θ, and Θ is the Heaviside step function. The biased lensing cross section, $\sigma_{\rm lens}$ is measured in comoving units and includes $B(z_s)$, the magnification bias which causes sources, fainter than the limiting magnitude of the survey, to be detected in the sample.

In [28], the Eq. (1) is used to predict the ISD resulting from different components in a given halo. For the mass distribution inside a halo, the author in [28] has considered three types of density profiles: (i) isothermal (SIS) density profile, which is found to fit well galaxy-scale halos, (ii) NFW density profile, which is a good fit for dark matter halos in simulations and also in the observations of clusters, and (iii) A more complex "Total" profile, consisting of Hernquist [31] profile for the stellar distribution in the central regions and NFW profile for the underlying dark matter distribution along with adiabatic contraction model [32]. The author of [28] has assumed

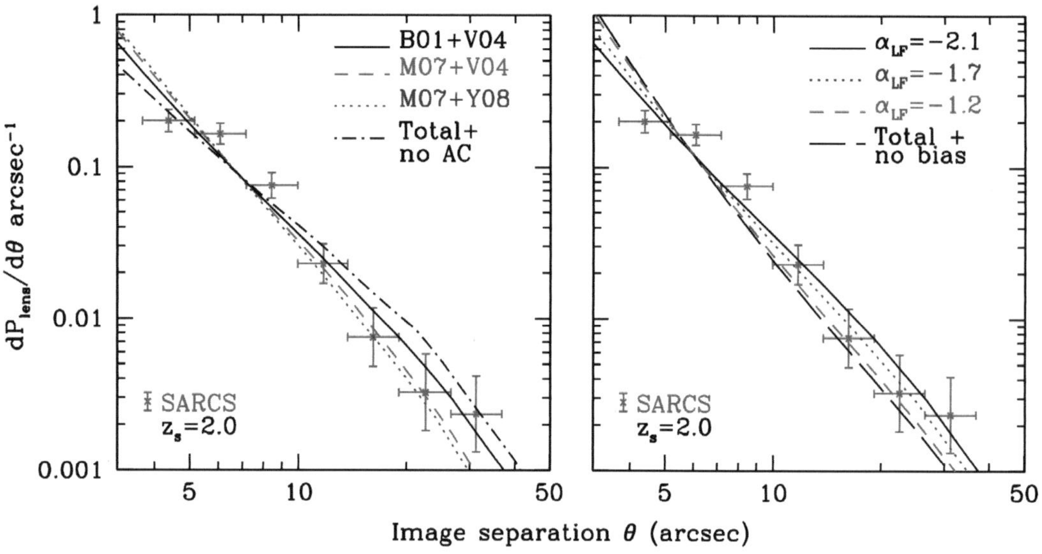

Figure 3. Image separation distribution. Theoretical curves for the total profile, shown with solid black line, are the same as in Figure 2. *Left:* Adopting the $c-M$ relation of [33] (dashed) steepens the ISD. The dotted curve, with $M-L$ relation of [35] and $c-M$ relation of [33], steepens it further but negligibly. The ISD without the AC model (dashed-dotted) is also shown. *Right:* The effect of varying slope of the source luminosity function on the ISD is shown for the total profile. The total profile without any magnification bias (long dashed) is independent of $\alpha_{\rm LF}$. Changes in various model parameters have degenerate effects on the expected ISD.

the background source population to lie at a redshift of $z_s = 2$, and the source luminosity function $\Phi(z_s, L) \propto L^{-2.1}$ appropriate for the radio survey CLASS [12][4]. We are able to reproduce the expected ISDs from [28], given by Eq. (1), for all the three density distributions mentioned above. The expected ISDs corresponding to these three density distributions are shown in Figure 2.

3.2.2. Observed distribution: The observed ISD is calculated by logarithmically binning the image separations of 125 SARCS candidates with $\theta_{\rm cut} \geq 20$ pixels (that is, $> 3.7''$) and an average ranking of 2 and above. The image separation for each lens candidate is taken as twice the Einstein radius or roughly the arc radius, which is the distance between the candidate lensed image and the centre of respective lens galaxy. Let $\theta - d\theta = \theta_{\rm l}$ and $\theta + d\theta = \theta_{\rm h}$, then the observed ISD is given by:

$$\left|\frac{dP_{\rm lens}}{d\theta}\right|_{\rm obs} = \frac{N(>\theta_{\rm h}) - N(>\theta_{\rm l})}{N(>\theta_{\rm cut})2d\theta}, \quad (2)$$

where the total number of observed lenses is $N(>\theta_{\rm cut})$. The SARCS data points demonstrate that the average density profile of the halos, giving rise to the intermediate θ values ($\sim 3'' - 12''$), is best represented by a combined profile for the main galaxy, and the dark matter halo as opposed to a pure SIS or pure NFW profile. The vertical error-bars are calculated assuming Poisson number statistics and the horizontal bars show the bin width. We have noticed that effects due to purity or incompleteness of the SARCS sample are not accounted for in this plot.

[4] The luminosity density for such a steep faint end slope ($\alpha_{\rm LF} < -2$) diverges as $L \to 0$, and necessarily requires a cutoff below some value of $L_{\rm min}$.

3.2.3. Tests with varying models: Here, we have tested the effects of varying the different components of the model in [28], and compared against the observed ISD. First, we have tested the influence of excluding the AC while computing the total profile. This has the effect of making the expected ISD shallower as shown by the dashed-dotted curve in the left panel of Figure 3. Prima facie, the AC model fits the data better than the model without AC. However, as we have shown below, there are other degeneracies in the model which prevent us from ruling out the "no AC" model at high significance. Next, we have tested the effect of using different $c - M$ relations on the ISD. For example, we have used the $c - M$ relation by [33] instead of that given by [34]. The $c - M$ relation of [33] is roughly $15 - 20\%$ lower than that of [34]. The combined profile using the $c - M$ relation of [33] is shown by the dashed curve in the left panel of Figure 3. Within the current statistical limits on the data, both the $c - M$ relations appear plausible, although the data appears to slightly favour the $c - M$ relation of [33]. We have also tested the effect of using a more recent determination of the $M - L$ relation obtained by [35] from a sample of SDSS groups along with the $c - M$ relation of [33] for the combined total profile. The $M - L$ relation of [35] differ by $\sim$ 0.2 dex from that of [36] at the intermediate mass regime which is the regime of interest. This appears to cause a very negligible change in the predicted ISD.

We have tried to quantify the effect of varying the slope ($\alpha_{\rm LF}$) of the source luminosity function ($\Phi(L) \propto L^{\alpha_{\rm LF}}$) at the faint-end. This influences the lens cross-section via the magnification bias. The solid curve in Figure 3 shows the expected ISD using the fiducial value of $\alpha_{\rm LF} = -2.1$, while the dotted and short dashed curves show the ISD in the right panel, corresponding to $\alpha_{\rm LF}$ equal to -1.7 and -1.2 respectively. It is evident from the figure that the observed ISD can be used to constrain the slope of the luminosity function, if the statistical error bars could be reduced. We have noted that the magnification bias factor in the biased lens cross-section is calculated for point sources. However, this may not be true for the sources of the lensed arcs, and their magnification bias could be negligible (e.g. [37]). Therefore, we have calculated the ISD assuming no bias. The long dashed curve in the right panel of Figure 3 shows the ISD without the bias. For all predictions, we have assumed $z_s = 2$ for the source redshift. We have checked the effect of adopting different source redshifts $z_s = 1.5$ and $z_s = 3.0$ on the predicted ISD. The predicted $\mathrm{d}P_{\rm lens}/\mathrm{d}\theta$ is not affected, because the corresponding increase (decrease) in $P(>\theta_{\rm cut})$ almost perfectly cancels out the increase (decrease) in $\mathrm{d}P/\mathrm{d}\theta$. This implies that the expected ISD would not be drastically different had we accounted for the distribution of source redshifts instead of assuming a single value for the source redshift.

4. Summary

We have presented the SARCS sample from the completed CFHTLS-WIDE and CFHTLS-DEEP, covering a combined unmasked area of $\sim$150 deg^2 in the sky. The lens sample is compiled through a semi-automatic technique consisting of using ARCFINDER algorithm, followed by visual inspection and ranking of the candidates. We have compiled a total of 127 candidates in the SARCS sample. From the complete sample, 54 are promising lens systems.

We have measured the azimuthal distribution of (giant) arcs from the SARCS sample and found consistency with the predictions by the authors of [26]. However, the simulations used for the predictions need to be improved by implementing realistic physics, whereas larger samples with better understanding of the selection function will allow us to test different physical models of gas cooling, and feedback that are used as inputs in the simulations. The SARCS sample allowed, for the first time, to probe the intermediate mass regime corresponding to group-scale halos via the ISD. We have shown that the density profile of the halos is well-reproduced by a combined profile (NFW and Hernquist) at the group-scales, which is consistent with the predictions. We have found that varying the model parameters have degenerate effects on the ISD, however, if priors from independent methods are used on one of these relations, the observed ISD could be used to constrain, for example, the $c - M$ relation.

We have shown the possibility to use arcs statistics in constraining some of the scaling relations of the models tested here. However, the models may be simplistic and may need refinement as the lens samples become larger with upcoming surveys. On the observational side, understanding the selection effects will become crucial to constrain model parameters with better accuracy. We hope to address some of these important issues in future studies.

Acknowledgements

This work is based on observations obtained with MegaPrime/MegaCam, a joint project of CFHT and CEA/DAPNIA, at the Canada-France-Hawaii Telescope (CFHT), which is operated by the National Research Council (NRC) of Canada, the Institut National des Science de l'Univers of the Centre National de la Recherche Scientifique (CNRS) of France, and the University of Hawaii. Also, this work is based in part on data products produced at TERAPIX and the Canadian Astronomy Data Centre as part of the Canada-France-Hawaii Telescope Legacy Survey, a collaborative project of NRC and CNRS.

References

[1] Blandford R D and Narayan R 1992 *Ann. Rev. A. & A.* **30** 311
[2] Kochanek C S 2006 *Saas-Fee Advanced Course 33: Gravitational Lensing: Strong, Weak and Micro*
[3] Swinbank A M, Webb T M, Richard J *et al* 2009 *MNRAS* **400** 1121
[4] Zitrin A and Broadhurst T 2009 *Ap. J. Lett.* **703** L132
[5] More A, McKean J P, More S *et al* 2009 *MNRAS* **394** 174
[6] Vegetti S, Czoske O and Koopmans L V E 2010 *MNRAS* **407** 225
[7] Suyu S H *et al* 2010 *Ap. J.* **711** 201
[8] Ferreras I, Saha P, Leier D, Courbin F and Falco E E 2010 *MNRAS* **409** L30
[9] Treu T and Koopmans L V E 2002 *Ap. J.* **575** 87
[10] Koopmans L V E, Treu T, Bolton A S, Burles S and Moustakas L A 2006 *Ap. J.* **649** 599
[11] More A, McKean J P, Muxlow *et al* 2008 *MNRAS* **384** 1701
[12] Myers S T *et al* 2003 *MNRAS* **341** 1
[13] Bolton A S, Burles S, Koopmans L V E, Treu T and Moustakas L A 2006 *Ap. J.* **638** 703
[14] Luppino G A, Gioia I M, Hammer F, Le Fèvre O and Annis J A 1999 *A. & A. Supp. Ser.* **136** 117
[15] Ebeling H, Edge A C and Henry J P 2001 *Ap. J.* **553** 668
[16] Zaritsky D and Gonzalez A H 2003 *Ap. J.* **584** 691
[17] Gladders M D, Hoekstra H, Yee H K C, Hall P B and Barrientos L F 2003 *Ap. J.* **593** 48
[18] Parker L C, Hudson M J, Carlberg R G and Hoekstra H 2005 *Ap. J.* **634** 806
[19] Krusch E, Rosenbaum D, Dettmar R J *et al* 2006 *A. & A.* **459** 759
[20] Mandelbaum R, Seljak U and Hirata C M 2008 *JCAP* **8** 6
[21] Balogh M L *et al* 2009 *MNRAS* **398** 754
[22] Rines K and Diaferio A 2010 *Astron. J.* **139** 580
[23] Cabanac R A *et al* 2007 *A. & A.* **461** 813
[24] More A, Cabanac R, More S *et al* 2012 *Ap. J.* **749** 38
[25] Hattori M, Kneib J and Makino N 1999 *Prog. Theor. Phys. Supp.* **133** 1
[26] Dalal N, Holder G and Hennawi J F 2004 *Ap. J.* **609** 50
[27] Bertin E and Arnouts S 1996 *A. & A. Supp. Ser.* **117** 393
[28] Oguri M 2006 *MNRAS* **367** 1241
[29] Keeton C R, Christlein D and Zabludoff A I 2000 *Ap. J.* **545** 129
[30] Zitrin A, Broadhurst T, Barkana R, Rephaeli Y and Benítez N 2011 *MNRAS* **410** 1939
[31] Hernquist L 1990 *Ap. J.* **356** 359
[32] Gnedin O Y, Kravtsov A V, Klypin A A and Nagai D 2004 *Ap. J.* **616** 16
[33] Macciò *et al* 2007 *MNRAS* **378** 55
[34] Bullock J S *et al* 2001 *MNRAS* **321** 559
[35] Yang X, Mo H J and van den Bosch F C 2008 *Ap. J.* **676** 248
[36] Vale A and Ostriker J P 2004 *MNRAS* **353** 189
[37] Narayan R and Wallington S 1993 *Liege Int. Ap. Coll.* **31** 217

The galaxy-dark matter connection: A cosmological perspective

S More[1]*, F van den Bosch[2], M Cacciato[3], A More[1], H Mo[4] and X Yang[5]

[1]Kavli Institute for Cosmological Physics, University of Chicago, 933 East 56th Street, Chicago, IL 60637, USA
[2]Department of Astronomy, Yale University, Box 208101, New Haven, CT 06520-8101, USA
[3]Racah Institute of Physics, The Hebrew University, Jerusalem 91904, Israel
[4]Department of Astronomy, University of Massachusetts, Amherst MA 01003-9305, USA
[5]Key Laboratory for Research in Galaxies and Cosmology, Shanghai Astronomical Observatory, The Partner Group of MPA, Nandan Road 80, Shanghai 200030, China

E-mail: *surhud@kicp.uchicago.edu

Abstract. We have presented a method that uses observations of galaxies to simultaneously constrain cosmological parameters, and the galaxy-dark matter connection (aka halo occupation statistics). The latter describes how galaxies are distributed over dark matter haloes, and is an imprint of the poorly understood physics of galaxy formation. A generic problem of using galaxies to constrain cosmology is that galaxies are a biased tracer of the mass distribution, and this bias is generally unknown. The great advantage of simultaneously constraining cosmology and halo occupation statistics is that this effectively allows cosmological constraints marginalized over the uncertainties regarding galaxy bias. We have used a combination of the analytical halo model and the conditional luminosity function to describe the galaxy-dark matter connection, which we have used to model the abundance, clustering and galaxy-galaxy lensing properties of the galaxy population. We have used a Fisher matrix analysis to gauge the complementarity of these different observables, and presented some preliminary results from an analysis based on data from the Sloan Digital Sky Survey. Our results are complementary to and perfectly consistent with the results from the WMAP mission, strengthening the case for a true 'concordance' cosmology.

1. Introduction

In the concordance cosmological picture, dark matter and dark energy dominate the energy density of the Universe, while stars in galaxies form a negligible component. However, unlike dark matter or dark energy, we can observe galaxies directly, and use them as tracers of the underlying matter density field to investigate the properties of the Universe. Unfortunately, this connection between galaxies and matter is complicated by the fact that galaxies are biased tracers of the mass distribution. Although this 'galaxy bias' is generally considered a nuisance when trying to use galaxies to constrain cosmology, it also contains a wealth of information regarding galaxy formation. After all, it is the physics of galaxy formation that determines where, how and with what efficiency galaxies form within the dark matter density field. Therefore, ideally one would like to *simultaneously* solve for cosmology and galaxy bias.

It is crucial to understand why and how galaxies are biased with respect to the matter distribution in order to break the degeneracy between galaxy bias and the cosmological parameters. Since dark matter halos, in which galaxies reside, form preferentially at the peaks of matter density field, they are biased tracers of the underlying matter density field. Galaxies inherit the bias of their parent halos. Observations of the abundance of galaxies [1, 2], the clustering of galaxies on small scales [3], the gravitational lensing signal due to the dark matter around galaxies [4], and the kinematics of satellite galaxies around halos [5, 6, 7] can all provide important clues regarding this "galaxy-dark matter connection" (i.e., what galaxies resides in what halo). Using this information, one can predict the galaxy bias, both as function of scale and as function of galaxy properties (e.g., luminosity). This allows one to break (some of) the degeneracies between galaxy bias and cosmology, and thus, to use the observed distribution of galaxies to constrain cosmology.

In this paper, we have demonstrated the strength and complementarity of a variety of galaxy observations from the SDSS to constrain cosmological parameters. In particular, we have shown how observations of galaxy abundances [1], galaxy clustering [3] and galaxy-galaxy lensing [4], can be used to constrain cosmological parameters such as the matter density in units of the critical density $\Omega_{\rm m}$, and the amplitude of the power spectrum of matter fluctuations, as characterized by σ_8. We rely on the framework of the halo model to analytically predict these observations. The halo model assumes that all the dark matter in the Universe is partitioned over dark matter halos of different sizes and masses [8]. The abundance and clustering of these halos of dark matter is set by the underlying cosmological parameters, and this dependence has been well calibrated with the use of numerical simulations [9, 10]. A parametric form of how galaxies populate halos, called the *halo occupation distribution function*, can then be used to predict the abundance and clustering of galaxies using the abundance and clustering of halos [11]. In this paper, we have used a Fisher matrix analysis to highlight the complementarity of using these different data sets, and presented some preliminary results from an analysis based on existing data.

We have used the conditional luminosity function (CLF) to specify the halo occupation distribution of galaxies [12]. The CLF describes the average number of galaxies with a given luminosity that reside in a halo of mass M. We have used a total of 9 parameters to describe the CLF. Given the parameters of the CLF, and the cosmological parameters which set the abundance and clustering of halos, we can predict all the observables that we wish to model. Because of the page-limits of these proceedings, we cannot provide the detailed expressions that we have used to calculate the observables for a given model (i.e., cosmology plus CLF). These have been presented in [13]. We have emphasized, though, that our implementation of the 'halo model' properly accounts for (i) the scale dependence of halo bias, (ii) halo exclusion, and (iii) residual redshift space distortions that can affect the determinations of galaxy bias [14].

Throughout, we have adopted a 'standard' flat ΛCDM cosmology (i.e., gravity is described by the standard general relativity, neutrino mass is neglected, initial power spectrum is a single power-law, and dark energy is modelled as Einstein's cosmological constant), which is described by 5 cosmological parameters: The matter density parameter $\Omega_{\rm m}$, the baryon density parameter $\Omega_{\rm b}$, the hubble parameter h, the power law index $n_{\rm s}$, and the parameter σ_8. Our goal is to constrain (subsets) of these cosmological parameters, fully marginalizing over the galaxy-dark matter connection as parameterized by our 9-parameter CLF model.

2. Results

2.1. Fisher information analysis

In this section, we have used the Fisher information matrix in order to gauge the accuracy with which constraints on the cosmological parameters $\Omega_{\rm m}$ and σ_8 can be obtained, given the current accuracy of the observables that we wish to model. Since, we have three different

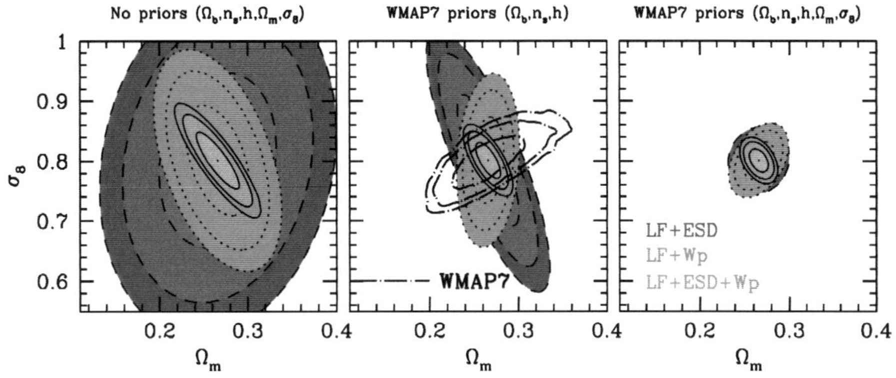

Figure 1. Fisher forecasts of 68, 95 and 99 per cent confidence constraints on the cosmological parameters $\Omega_{\rm m}$ and σ_8 when different combinations of the luminosity function (LF), galaxy clustering (Wp), and the galaxy-galaxy lensing (ESD) data are analysed. Different panels show the effect of varying priors on cosmological parameters from analysis of the cosmic microwave background data. (For a definitive version, see [15].)

observables: the luminosity function, galaxy-galaxy clustering, and galaxy-galaxy lensing, we have started by investigating how each of these different data sets contributes to our constraining power. The different panels of Figure 1 show the 68, 95 and 99 per cent confidence intervals that can be placed on the cosmological parameters $\Omega_{\rm m}$ and σ_8 under varying assumptions of prior information from the analysis of the WMAP-7 mission [16]. The left-hand panel assumes uninformative priors on all of the cosmological parameters in our model. The dashed contours are used to indicate the confidence levels when we perform a joint analysis of the abundance of galaxies and the galaxy-galaxy lensing signal around them. The constraints are fairly weak, in particular, because the galaxy-galaxy lensing signal has only been measured on fairly small scales ($r_{\rm p} \lesssim 2h^{-1}$Mpc). This results in a number of degeneracies between the cosmological parameters, and the CLF parameters such that $\Omega_{\rm m}$ and σ_8 are only weakly constrained. The dotted contours show the confidence contours obtained by combining the luminosity function with the galaxy-galaxy clustering data. The constraints are significantly tighter, and the improvement is largely due to the addition of information on intermediate scales ($2\ h^{-1}$ Mpc $\lesssim r_{\rm p} \lesssim 40\ h^{-1}$ Mpc). Finally, the solid contours show the result of a joint analysis of all three observables. Even in the absence of prior information, this joint analysis breaks a number of degeneracies that are present in our model. The resulting cosmological constraints are competitive with the existing constraints on these parameters, demonstrating the potential power of this method. The middle panel of Figure 1 shows the effect of adding prior information on the secondary cosmological parameters Ω_b, n_s and h from the WMAP-7 results and the right panel shows the effect of adding all the prior information from WMAP-7.

2.2. Cosmological constraints

We have carried out a joint analysis of all the three observables, and obtained the posterior distribution of our model parameters given these data. We have used a Monte Carlo Markov Chain to sample from the posterior probability distribution of the parameters. For our fiducial analysis, we have imposed priors on the secondary cosmological parameters $\Omega_{\rm b}$, $n_{\rm s}$ and h, and completely uninformative priors on the parameters $\Omega_{\rm m}$ and σ_8. Our model is able to fit the data sufficiently well with χ^2 per degree of freedom of the order of 2. The fits to the data have been presented in [17]. Preliminary results of our analysis, in the form of 68, 95 and 99 per cent

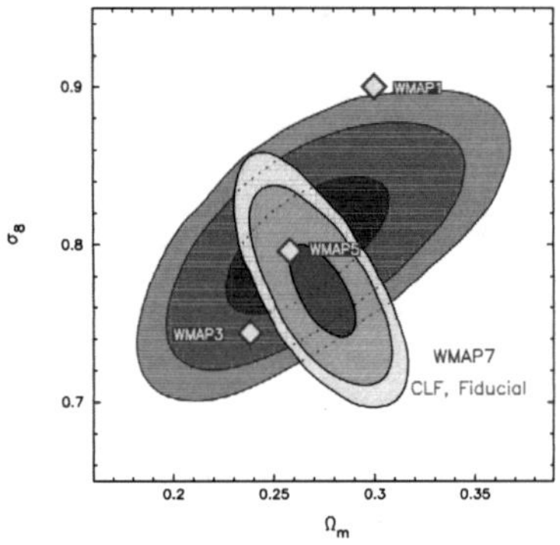

Figure 2. 68, 95 and 99 per cent confidence limits on the cosmological parameters Ω_m and σ_8 from our analysis (shown in chrome yellow) compared with the confidence limits obtained by the analysis of the WMAP-7 experiment (shown in green). (For a more definitive version, see [17].)

confidence contours, are shown in Figure 2 and compared to the WMAP-7 results.

There are two points worth making: First of all, the constraints obtained from our analysis are in remarkably good agreement with the WMAP-7 results, even though we have used no prior information on $\Omega_{\rm m}$ and σ_8. The WMAP-7 results are based on observations of the microwave background at a very early time in the Universe ($z \sim 1080$), and primarily rest on the physics of perturbations that can be treated with the help of linear perturbation theory. The results from our analysis derived from galaxy observations at redshift $z \sim 0.1$ and are obtained by modelling extremely non-linear scales, properly marginalizing over the uncertainties related to galaxy bias (i.e., the galaxy-dark matter connection). The agreement in cosmological constraints obtained from these two completely disjunct analyses is extremely striking, and provides strong support for the notion of a true 'concordance' cosmology: Clearly ΛCDM provides an excellent description of data over a large range of scales and cosmic epochs. Secondly, the constraints obtained from our analysis are both competitive with, and complementary to those obtained by the WMAP-7 analysis. This is also in agreement with the complementarity expected from the Fisher analysis presented in the previous subsection.

3. Summary

Observations of galaxies are an excellent way of probing the underlying matter distribution in the Universe and thereby, obtaining precise constraints on the cosmological model. We have shown that a joint analysis of the abundance of galaxies, the clustering of galaxies, and the galaxy-galaxy lensing signal in the framework of the halo model can be a useful way to constrain the cosmological parameters.

Using a Fisher matrix analysis, we have shown that the cosmological information contained in the three observables described above is complementary to each other and a joint analysis of these datasets is able to break a number of degeneracies between the CLF parameters and the cosmological parameters. We have followed up the Fisher analysis, by constraining our model parameters using the actual data. We have shown that the resulting constraints on the cosmological parameters $\Omega_{\rm m}$ and σ_8 are in remarkable agreement with constraints from the analysis of the WMAP-7 data. This is yet another jewel in the crown of the ΛCDM model, which continues to reign king.

References

[1] Blanton M R *et al* 2003 *Ap. J.* **592** **2** 819
[2] Vale A and Ostriker J P 2006 *MNRAS* **371** 1173
[3] Zehavi I *et al* 2011 *Ap. J.* **736** 59
[4] Mandelbaum R *et al* 2006 *MNRAS* **368** 715
[5] More S *et al* 2009 *MNRAS* **392** 801
[6] More S, van den Bosch F C and Cacciato M 2009 *MNRAS* **392** 917
[7] More S *et al* 2011 *MNRAS* **410** 210
[8] Cooray A and Sheth R 2002 *Phys. Rev.* **372** 1

[9] Tinker J *et al* 2008 *Ap. J.* **688** 709
[10] Tinker J L *et al* 2010 *Ap. J.* **724** 878
[11] Cacciato M *et al* 2009 *MNRAS* **394** 929
[12] Yang X, Mo H J and van den Bosch F C 2003 *MNRAS* **339** 1057
[13] van den Bosch F *et al* 2013 *MNRAS* **430** 725
[14] More S 2011 *Ap. J.* **741** 19
[15] More S *et al* 2013 *MNRAS* **430** 747
[16] Komatsu E *et al* 2011 *Ap. J. Supp. Ser.* **192** 18
[17] Cacciato M *et al* 2013 *MNRAS* **430** 767

Modelling non-linear large scale structure using Lagrangian perturbation theory re-expansions

S Nadkarni-Ghosh[1,2] **and D F Chernoff**[3]

[1]Department of Theoretical Sciences, S. N. Bose NCBS, Kolkata 700098, India
[2]Department of Physics, Indian Institute of Technology, Kanpur 208016, India
[3]Department of Astronomy, Cornell University, Ithaca, NY 14850 USA

E-mail: [1,2]`nsharvari@gmail.com`, [3]`chernoff@astro.cornell.edu`

Abstract. Lagrangian perturbation theory (LPT) has been widely used to model the non-linear growth of large scale structure analytically. However, the LPT series converges only for a finite time. In recent work, the authors have examined this issue in great detail, and proposed a new algorithm (called LPT re-expansions), to extend the domain of validity of the series. This article outlines the main ideas of the algorithm, first developed for the spherical top-hat system, and later generalized to deal with inhomogeneous initial conditions arising from random fields.

1. Introduction

We have presented a new algorithm to model large scale structure in the mildly non-linear regime. It is based on Lagrangian perturbation theory (LPT), a technique introduced by Zeldovich in the 1970s, and developed further by various authors. In the Lagrangian formalism, the position of a particle is the fundamental variable, and is expressed as a function of the initial position and time. The density is then reconstructed from the position by imposing mass conservation. This definition involves the Jacobian of the transformation, relating the initial (Lagrangian) coordinate space and the final (Eulerian) coordinate space, and hence, is intrinsically non-linear. A fundamental assumption in this framework is that the matter is 'cold'. Once particle trajectories cross the initial space to final space, mapping becomes many-to-one and the density formally diverges.

However, LPT has another major drawback. The series is not guaranteed to converge at all times before orbit crossing. In a recent paper [1], this issue has been investigated in detail for the spherical top-hat system. Our analysis has estimated the exact 'time of validity' of the series, and we have proposed an algorithm (dubbed LPT re-expansions) to extend the domain of validity of the series beyond the initial time interval. We have shown that this technique has worked, and is computationally feasible for the top-hat system. Convergence could be achieved by increasing the re-expansion frequency and/or the Lagrangian order.

In a subsequent paper [2], this idea of the algorithm has been applied to inhomogeneous initial conditions arising from random fields. One important additional feature emerges. The solution outlined by the traditional framework in [3] is unique only up to a purely time-dependent transformation. This time dependent function needs to be added to get a complete solution that has the correct convergence properties. This paper summarizes the main results of our work.

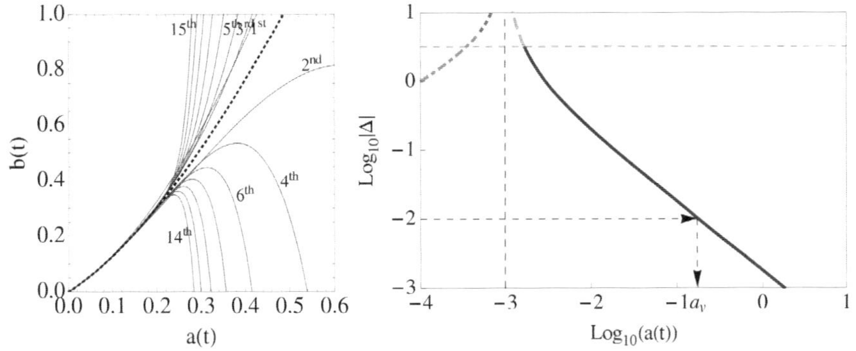

Figure 1. Lack of convergence in LPT. Initial $\Delta = 0.01$. The left panel is the scale factor $b(t)$ vs. the scale factor of the background $a(t)$. The thick black dotted curve is the exact solution, and the blue curves are approximate solutions for the scale factor. The system converges only up to $a \sim 0.2$. The right panel is the corresponding 'root plot'. Red and pink are different types of complex roots, and blue and cyan are different types of real roots. The 'time of validity' (a_ν) can be simply read off from the root plot, and it matches with that observed in the left panel.

2. Spherical top-hat system

The spherical top-hat perturbation consists of a inner over-dense or under-dense sphere surrounded by a mass compensating region. The perturbation is described by two parameters: fractional over-density δ and fractional Hubble parameter δ_v. In a $\delta - \delta_v$ phase-space, the perturbation can be represented as a point with polar coordinates (Δ, θ). Δ is taken as the expansion parameter for the series. The scale factors $a(t)$ and $b(t)$ describe the time evolution of the background and perturbed spheres respectively. The exact evolution for $b(t)$ is known as a function of (Δ, θ). By treating the variable Δ to be complex, one can estimate when the solution $b(t)$ is singular and the roots to the singularity condition give the radius of convergence $R(\Delta, t)$. This is represented on a 'root plot'. It is generally different for different values of θ. From the root plot, the time of validity can be read off for the Δ corresponding to the initial value. Figure 1 shows a specific example of the problem.

To remedy this problem, we have proposed the algorithm of LPT re-expansions. The main idea is somewhat analogous to that of 'analytic continuation'. The series is expanded up to some final time, less than the time of validity given by the initial parameters. At that time, the system is re-initialized by defining a new Δ and θ, and the series is propagated forward until a new final time, less than the new time of validity. This is done successively until the desired end point is reached. A detailed examination of the dynamics in the $\delta - \delta_v$ phase-space proves that this procedure, indeed works, and implementation is realistically feasible. Figure 2 shows the implementation of this algorithm on the same problem considered in Figure 1. The upper panel is $b(t)$ vs. $a(t)$ and the lower panels show the log of the error between the various components. It is clear that taking more steps decreases the error. Comparing the second and third plots, it is also seen that higher orders converge faster. Compare, for example, the red line (order 1) with the black line (order 9), i.e., the convergence rates improve as the order of the approximation improves.

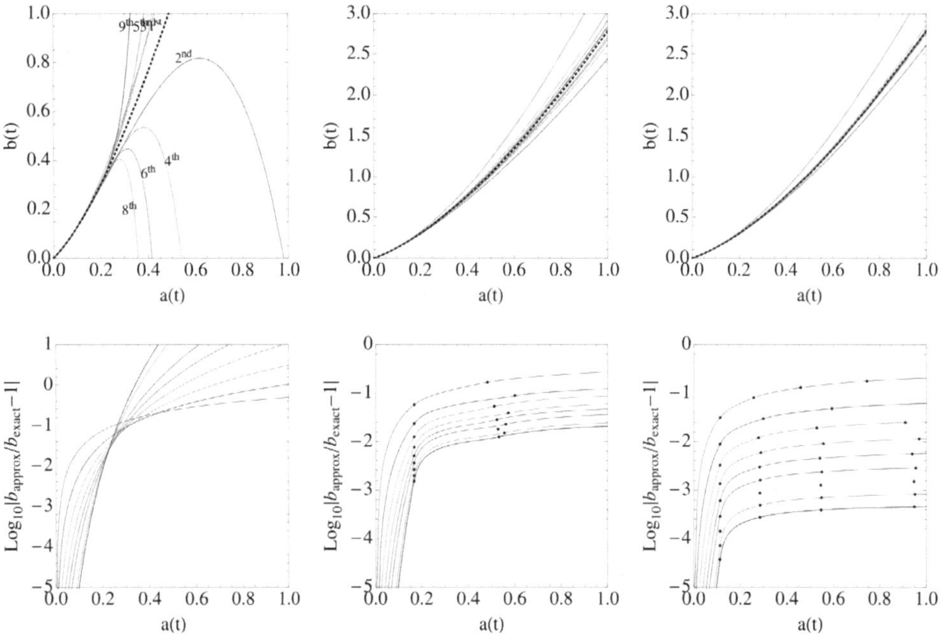

Figure 2. The LPT re-expansion algorithm applied to the example in Figure 1. The three plots correspond to one step (left), three steps (middle), five steps (right).

3. General initial conditions

In the general case, the equation of motion for the particle position $\mathbf{r}$ is given by the geodesic equation:

$$\ddot{\mathbf{r}} - \frac{\ddot{a}}{a}\mathbf{r} = -\nabla_r \psi(\mathbf{r}, t), \quad \textit{and} \tag{1}$$

$$\nabla_r^2 \psi(\mathbf{r}, t) = 4\pi G \delta\rho_m(\mathbf{r}, t). \tag{2}$$

The formal expansion of Ehlers and Buchert [3] uses a derivative form:

$$\nabla_r \cdot \ddot{\mathbf{r}} = -4\pi G \rho_m(\mathbf{r}, t), \quad \textit{and} \tag{3}$$

$$\nabla_r \times \ddot{\mathbf{r}} = 0, \tag{4}$$

where $\psi(\mathbf{r}, t)$ is the gravitational potential, $\rho_m(\mathbf{r}, t)$ and $\delta\rho_m(\mathbf{r}, t)$ are the total and perturbed matter densities. Generally, one substitutes in Eqs. (3) and (4) the ansatz $\mathbf{r} = a\mathbf{X} + \mathbf{p}(\mathbf{X}, t)$, where $\mathbf{X}$ is the Lagrangian coordinate, and solves for the displacement vector $\mathbf{p}(\mathbf{X}, t)$. The solutions to Eqs. (1) and (2) differ from those to Eqs. (3) and (4) by a time dependent transformation (called *frame shift*). Satisfying the geodesic equation implies that the total momentum in the box decays as $1/a(t)$ for the expanding universe. Ignoring the shifts violate this law and break convergence. Figure 3 shows the effect on convergence with (left panels) and without (right panels) the shifts for generic initial conditions (where the velocity can have a non-zero transverse part).

4. Conclusion

We have outlined the main idea of the LPT re-expansion scheme, and demonstrated that it works for the spherical top-hat as well as generic perturbations. The algorithm, developed from the top-hat analysis, consists of re-expanding the solution in overlapping time domains. The errors can

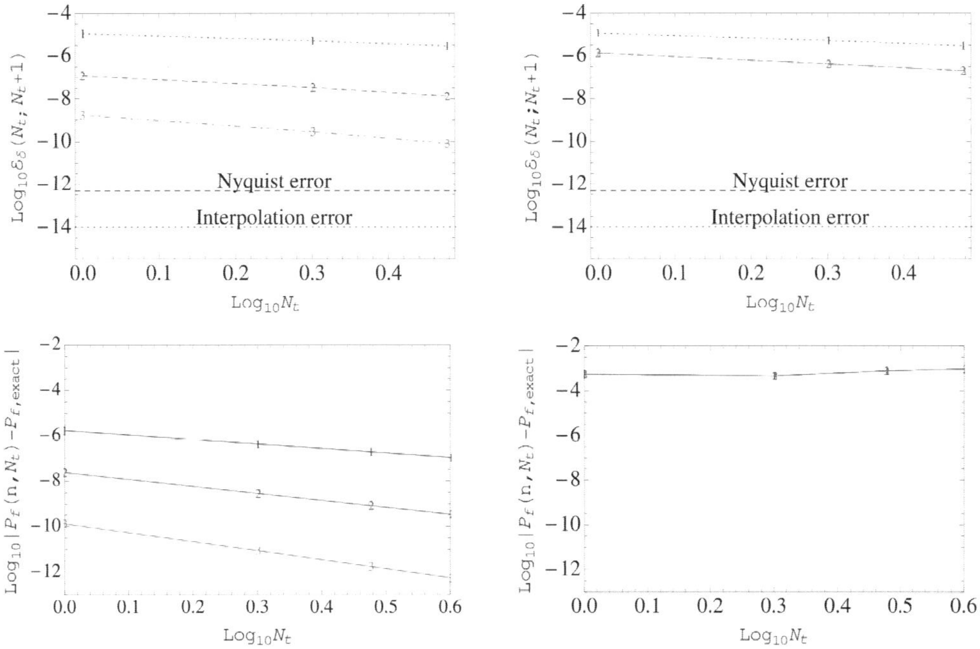

Figure 3. Two runs with (left) and without (right) the frame shift. N_t is number of steps, numbers '1, 2, 3' denote Lagrangian orders. The upper panel shows the final Cauchy errors in the density field, and the lower panel shows the log errors between the expected and observed final momentum. When the shifts are included, the convergence rates are similar to the top-hat. When they are not, the errors do not decrease.

be fine tuned by changing the order of the series or the frequency of re-expansion. Higher orders converge faster. The algorithm is then generalized to inhomogeneous initial conditions. In this case, the series solution to the differential form of the geodesic equation must be supplemented by time dependent gauge functions called frame shifts. Omission of these functions violates momentum conservation, and breaks the expected convergence.

The new technique has three major advantages: (i) The displacement and hence, the density and velocity are treated as smooth fields, making the results free from particle induced shot-noise, (ii) The errors can be precisely controlled by changing the Lagrangian order and/or the number of steps, and (iii) It is capable of handling arbitrary velocity data including cases with a initial rotational component.

Nothing in this work deals with what happens once orbits cross. Some sort of effective pressure needs to be introduced to handle the onset of orbit crossing.

References

[1] Nadkarni-Ghosh S and Chernoff D F 2011 *MNRAS* **410** 1454
[2] Nadkarni-Ghosh S and Chernoff D F 2013 *MNRAS* **431** 799
[3] Ehlers J and Buchert T 1997 *Gen. Rel. Grav.* **29** 733

Interplay between cosmological expansion and massive objects

R Nandra[1,2], A N Lasenby[1,2] and M P Hobson[1]

[1] Astrophysics Group, Cavendish Laboratory, JJ Thomson Avenue, Cambridge CB3 0HE, UK
[2] Kavli Institute for Cosmology, Institute of Astronomy, Madingley Road, Cambridge CB3 0HA, UK

E-mail: roshina.nandra@googlemail.com, a.n.lasenby@mrao.cam.ac.uk, mph@mrao.cam.ac.uk

Abstract. We have derived a metric for a point mass in an expanding universe. In the spatially flat case, a simple coordinate transformation relates our metric to that derived by McVittie. Nonetheless, our use of non-comoving (physical) coordinates greatly facilitates physical interpretation. We have also derived a coordinate-free expression for the force required to keep a particle at rest in this spacetime. For redshift $z > 0.67$, we have identified two important time-dependent physical radii; the largest possible circular orbit $r_F(t)$, and the largest stable circular orbit $r_S(t)$, which lies inside $r_F(t)$, either of which could be interpreted the egde of an object of mass m. In the case of a dynamical background dominated by a phantom fluid, we can use our force expression to predict the time of the 'Big Rip' when the universe ends. For open and closed universes, our metrics describe different spacetimes to McVittie's metrics; we believe the latter to be incorrect. In the closed case, our metric possesses an image mass at the antipodal point of the universe.

1. Introduction

Galaxies and galaxy clusters are not usually considered to have definite edges. Indeed, their densities $\rho(r)$ are often modelled using a Navarro-Frenk-White profile, which falls off gradually with radial distance from the centre. From a Newtonian perspective, the force on a constituent particle falls off as $1/r^2$, and tends to zero as $r \to \infty$ rather than at some finite value of r.

There is recent observational evidence, however, for a definite cut-off within objects such as galaxies. Since this separates bound and unbound material, one could interpret it as the object's edge. One piece of evidence has come from an analysis conducted by Peirani and de Freitas Pacheco [1]. By plotting the radial velocity profiles of constituent stars of nearby galaxies, they have observed a zero velocity surface in each case. For example, for M81 plotted in Figure 1, this appears to be at ~ 0.8 Mpc, around 3 times the extent of a typical ~ 0.26 Mpc galaxy. We have now aimed to explain this phenomenon using models for a massive object in an expanding universe.

2. Newtonian verses general relativistic analysis with Λ

The Newtonian force on a particle near some object, including a cosmological constant Λ, is given by:

$$F(r) \propto \frac{v^2(r)}{r} = -\frac{M(r)}{r^2} + \frac{1}{3}\Lambda r, \tag{1}$$

where $v(r)$ is the circular velocity and the object's mass is $M(r) = \int_0^r 4\pi\bar{r}^2\rho(\bar{r})\,d\bar{r}$. The first term points inwards and is gravitational, and the second term points outwards due to the universe's *de Sitter* expansion. For any object, $F(r) = v(r) = 0$ at a finite r. Indeed, even for a point mass with $\Lambda = 10^{-35}$ s^{-2}, one obtains results in close agreement with those observed [2].

Here, we have shown the general relativistic circular velocity function to be given by:

$$v(r) = \frac{1}{g_{rr}}\sqrt{\frac{M(r)}{r} - \frac{\Lambda}{3} + 4\pi p(r)}, \tag{2}$$

where g_{rr} represents the radial part of the spacetime metric with pressure $p(r)$. The edge of the object occurs at $v = 0$, and the pressure also vanishes there; therefore, the edge is located as the solution to $M(r)/r - \Lambda/3 = 0$. This is equivalent to where $F(r) = 0$ in the Newtonian expression given in Eq. (1). Remarkably, no general relativistic correction is required. The circular velocity profile for a typical galaxy is shown in Figure 2. The cut-off is found to lie at ~ 1.06 Mpc, just over 4 times the typical extent of a galaxy.

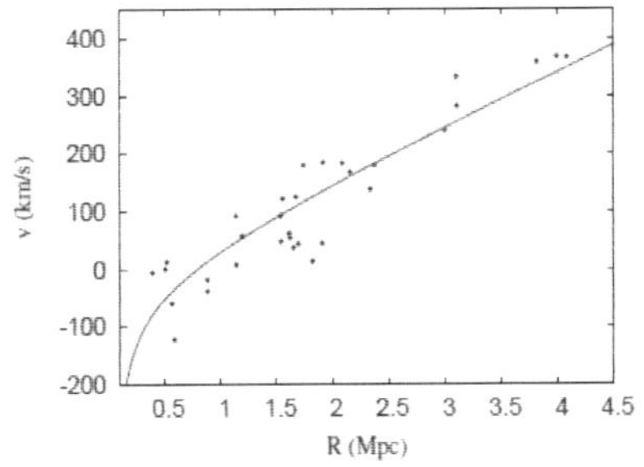

Figure 1. Radial velocity vs. distance of stars in galaxy M81 with mass 9.2×10^{11} M$_\odot$, and the best-fit line (reproduced from [1])

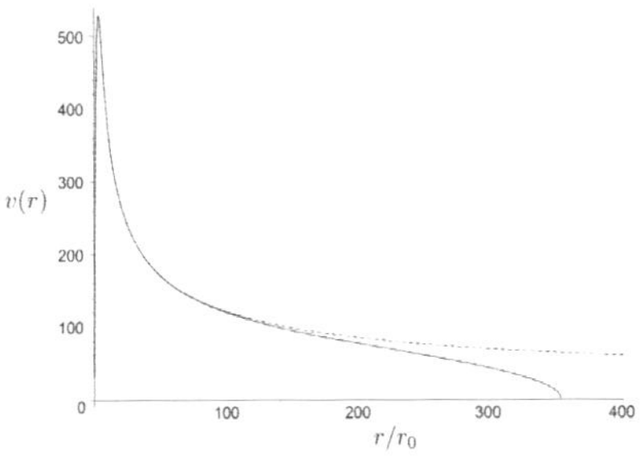

Figure 2. Circular velocity $v(r)$ (km/s) for a galaxy of mass 1×10^{12} M$_\odot$ with exponentially-decreasing radial density profile, and $r_0 = 3$ kpc. The dotted line is for $\Lambda = 0$.

3. A mass particle in a spatially flat expanding universe

In reality, the universe is not de Sitter; rather it expands at the Hubble rate $H(t) = R'(t)/R(t)$, where $R(t)$ is the scale factor for expansion. We have now considered a mass embedded in a Friedmann-Robertson-Walker (FRW) background with density $\rho(t)$, albeit at the expense of modelling the mass as pointlike. Working in 'physical' (non-comoving) coordinates, the energy contained within a radius r is $M(r,t) = m + (4/3)\pi r^3\rho(t)$. Omitting the details of the tetrad-based derivation conducted in [3], this leads to the spatially flat spacetime metric:

$$ds^2 = \left[1 - \frac{2m}{r} - r^2H^2(t)\right]dt^2 + 2rH(t)\left(1 - \frac{2m}{r}\right)^{-\frac{1}{2}}dr\,dt - \left(1 - \frac{2m}{r}\right)^{-1}dr^2 - r^2\,d\Omega^2, \tag{3}$$

where $d\Omega^2 = d\theta^2 + \sin^2\theta d\phi^2$. This describes the same spacetime as McVittie's metric via the coordinate transformation $r \to rR(t)(1 + m/(2rR(t)))^2$. Therefore, McVittie's metric *does* describe a mass particle in an FRW background, an issue which has been previously under much debate.

3.1. *Dynamical apparent horizons*

We have shown in [4] that there are two apparent horizons associated with this metric: The 'black hole horizon' and the 'cosmological horizon'. Since, the metric is time-dependent, these horizons are dynamical. In fact, they did not always exist. Their evolution is depicted in Figure 3.

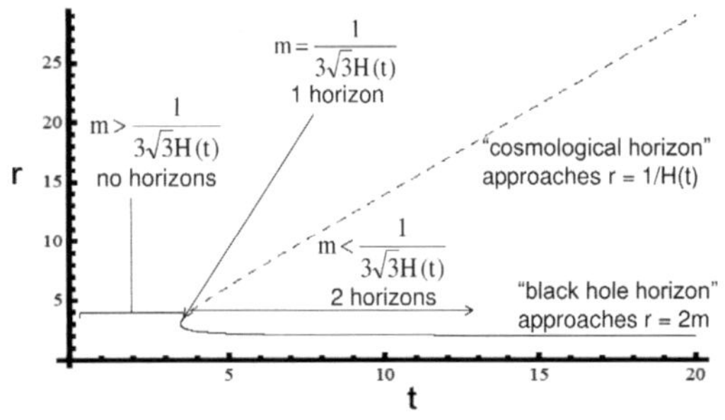

Figure 3. The existence and locations of the apparent horizons around m.

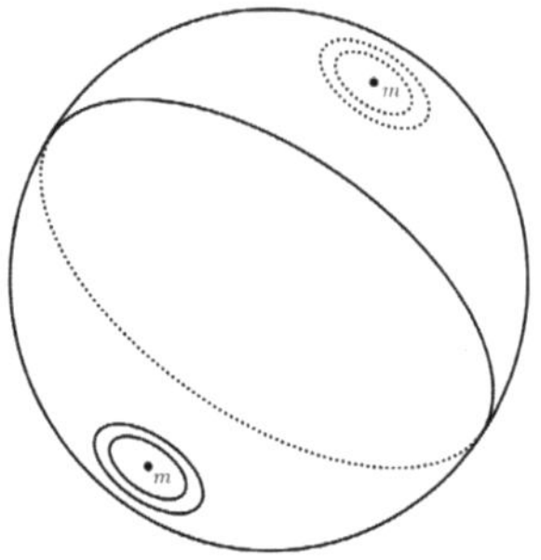

Figure 4. The spacetime geometry for $k = 1$, showing both the original and image masses.

3.2. *Force required to keep a test particle at rest*

In [3], we have derived a new, coordinate-free, fully general relativistic expression for the force required to keep a particle at rest in this spacetime as:

$$f = \frac{\frac{m}{r^2} - rH^2(t)}{\left(1 - \frac{2m}{r} - r^2H^2(t)\right)^{1/2}} + \frac{rH^2(t)(q(t)+1)\sqrt{1-\frac{2m}{r}}}{\left(1 - \frac{2m}{r} - r^2H^2(t)\right)^{3/2}} \overset{m \ll r \ll R(t)}{\approx} \underbrace{\frac{m}{r^2}}_{gravity} + \underbrace{q(t)H^2(t)r}_{cosmology}, \tag{4}$$

where $q(t) = -H'(t)/H^2(t) - 1$ is the deceleration parameter. Notice that the cosmological component of the Newtonian expression is directed outwards (inwards) when the expansion of the universe is accelerating (decelerating). This highlights the common misconception that there is a force associated simply with the expansion of space; instead it is better associated with the *acceleration/deceleration of the expansion.* The Newtonian expression does appear in the GADGET cosmological simulation codes. However, the cosmological force is thought to have reversed in the direction for redshift $z < 0.67$. It would, therefore, be interesting to study structure formation in the early universe using the full general relativistic expression for the force.

3.3. *Physical radii around point mass*

The largest possible $r_F(t)$ and largest stable $r_S(t)$ orbits around the point mass m may be calculated as the minimum in the effective potential of an orbiting particle, equivalent to the zero of its physical acceleration d^2r/dt^2. In [2], we have shown, using the Lagrangian method, that:

$$r_F(t) \approx \left[-\frac{m}{q(t)H^2(t)}\right]^{\frac{1}{3}}, \quad and \quad r_S(t) \approx \left[-\frac{m}{4q(t)H^2(t)}\right]^{\frac{1}{3}}. \tag{5}$$

Galaxy	Mass ($10^{11}M_\odot$)	P&P	$r_F(t_0)$	$r_S(t_0)$
M83	21	1.10	1.47	0.93
M81	9.2	0.75	1.12	0.70
IC 342	2.0	0.51	0.67	0.42
NGC 253	1.3	0.40	0.58	0.36

Table 1. Comparison between our theoretical physical radii and those cut-offs found from a best-fit numerical analysis for galaxies (P&P).

Time	Event
15 Gyrs	Today
$t_{rip} - 70$ Myr	Destroy Milky Way
$t_{rip} - 4$ months	Unbind Solar System
$t_{rip} - 25$ mins	Earth explodes
$t_{rip} = 35$ Gyr	Big Rip

Table 2. The history and future of the Universe with $w = -3/2$.

Physically, $r_F(t)$ corresponds to where $f = 0$ in Eq. (4), so it separates bound and unbound material. Since, orbits here may not be stable, it might make more sense to interpret $r_S(t)$ as the object's edge. These cut-offs only exist when the universe's expansion is accelerating ($q(t) < 0$). Despite both being time-dependent, $r_S(t)$ lies a constant factor of ~ 1.6 inside $r_F(t)$ at these times. Table 1 compares some $r_F(t)$ and $r_S(t)$ values with those obtained by Peirani and de Freitas Pacheco, for $H_0 = 70$ km s^{-1} Mpc^{-1}, and $q(t) = -0.55$. All three cut-offs are in close agreement.

3.4. The Big Rip

There has been recent renewed interest in a phantom background fluid, with equation of state $w = p/\rho < -1$. This has led to the prediction of a Big Rip at a finite time t_{rip} in the future [5]. If $w = -\frac{3}{2}$, the time remaining for the universe before the Big Rip, $\tilde{t}$, is calculated as ~ 22 Gyrs.

We may use our approximate force expression in Eq. (4) to calculate the time before t_{rip} at which a particle at the edge of an object becomes gravitationally unbound from it (when $f = 0$). This is interpreted as the time at which specific objects of mass m and size r are ripped apart. We have shown in [2] that at late times $H(t) \to 2/(3|1+w|\tilde{t})$, and $q(t) \to -(1/2)|1+3w|$. Therefore,

$$\tilde{t} \approx \frac{\sqrt{2|1+3w|}}{3|1+w|}\sqrt{\frac{r^3}{m}}. \tag{6}$$

Although approximate, this result matches Caldwell's precisely, despite having been derived using a different method. Actual values of $\tilde{t}$ for specific objects are given in Table 2.

4. A mass particle in a spatially curved expanding universe

The metrics for spatially curved expanding universes are found to involve elliptic integrals. To avoid these, we have instead, obtained series solutions for $m \ll r \ll R(t)$. Unlike in the spatially open case ($k = -1$), there is no notion of spatial infinity in a closed universe ($k = 1$), so we have, instead, specified a boundary condition at the outer radius of pressure divergence $r_2^p(t)$. This is determined to be merely a coordinate singularity, marking a point of reflection for the entire spacetime geometry. Thus, for $k = 1$, there is a *real* image mass at the anti-potal point of the universe, as depicted in Figure 4. To leading order, the force required to hold a particle at rest in both cases is the same as in the spatially flat case, but an extra term appears at the next order as:

$$f \approx \frac{m}{r^2} + \frac{m^2}{r^3} + \frac{3m^3}{2r^4} + q(t)H^2(t)r + \left(2q(t) + \frac{3}{2}\right)H^2(t)m + \frac{3k}{2}\frac{m}{R^2(t)}. \tag{7}$$

For $k = -1$ ($k = 1$), this can be interpreted as an 'anti-gravitational' (gravitational) force due to the virtual (real) 'image' mass being dragged out to the curvature scale of the universe.

5. Further remarks

In reality, objects are neither pointlike nor fixed. In our subsequent paper, we will investigate the model for a massive object with finite size, and constant spatial density embedded in an otherwise FRW universe. This accounts for mass accretion by construction. We will look at two special cases, which offer elegant analytical solutions: The case that the boundary of the object is fixed, and the case that its mass is fixed. Further numerical results which emerge are left for future research.

References

[1] Peirani S and de Freitas Pacheco J A 2008 *A. & A.* **488** (3) 845
[2] Nandra R, Lasenby A N and Hobson M P 2012 *Accepted at MNRAS* [arXiv:1104.4458]
[3] Nandra R, Lasenby A N and Hobson M P 2012 *Accepted at MNRAS* [arXiv:1104.4447]
[4] Faraoni V, Zambrano Moreno A F and Nandra R 2012 *Submitted to Phys. Rev. D* [arXiv:1202.0719]
[5] Caldwell R R, Kamionkowski M and Weinberg N N 2003 *Phys. Rev. Lett.* **91** 071301

Higher dimensional cosmological models: An alternative explanation for late time cosmic acceleration

I Pahwa

Department of Physics and Astrophysics, University of Delhi, Delhi 11007, India

E-mail: ipahwa@physics.du.ac.in

Abstract. Einstein's equations have been solved for a homogeneous, anisotropic $1+3+D$ dimensional universe. We have studied the late time acceleration of the universe in this model. The content of the universe is assumed to be such that it exerts zero pressure in visible dimensions and in extra dimensions, the equation of state is given. The parameters of the model are fixed by comparing with observational data on supernovae. We have shown that late time acceleration occurs in the universe with q_0 and z_{tr} being approximately 0.46 and 0.76 respectively.

1. Introduction

Observations of high redshift supernovae (SN) type 1a, large scale structures and cosmic microwave background (CMB) have shown that our universe is in a phase of accelerated expansion. Several theories have been suggested to explain this feature, but a convincing theoretical framework is still lacking. Various models are constructed in extra dimensional context, also including (but not limited to) brane world models [1, 2].

Here, we have considered a simple model, which is used in [3] to solve the horizon problem through extra dimensions in early universe, using an anisotropic fluid residing in $1+D_1+D_2$ dimensions, we have adapted the formalism to produce a late time acceleration instead. We have assumed a uniform density matter in the universe, but the pressure exerted by matter in normal dimensions is different from that in extra dimensions.

2. The framework

A spacetime, which has one temporal, three normal spatial dimensions and D extra spatial dimensions, is described by the metric:

$$ds^2 = -dt^2 + a^2(t)\left(\frac{dr^2}{1-k_1r^2} + r^2 d\Omega\right) + b^2(t)\left(\frac{dR^2}{1-k_2R^2} + R^2 d\Omega_{D-1}\right). \tag{1}$$

We have assumed spatial curvature of both sub-spaces to be zero, i.e., $k_1 = 0$ and $k_2 = 0$. The parameter D takes integral values, and is to be determined by comparing with observations. Here, $a(t)$ represents the scale factor in the normal 3-dimensions and $b(t)$ denotes the scale factor in the extra dimensions. The content in the universe is assumed to exert pressure, which is different in the normal and extra dimensions.

The energy-momentum tensor is assumed to be of the form: $T^{\mu}_{\nu} = diag\,(-\rho, P_a, P_a, P_a, P_b, \ldots, P_b)$, where ρ is the energy density of the fluid and P_a (P_b) is the pressure exerted in normal (extra) dimensions. For the line element given in Eq. (1) and energy-momentum tensor given above, Einstein's equations can be written as:

$$\frac{\dot a^2}{a^2} + D\frac{\dot a}{a}\frac{\dot b}{b} + \frac{D(D-1)}{6}\frac{\dot b^2}{b^2} = \frac{8\pi G\rho}{3}, \tag{2}$$

$$2\frac{\ddot a}{a} + D\frac{\ddot b}{b} + 2D\frac{\dot a}{a}\frac{\dot b}{b} + \frac{\dot a^2}{a^2} + \frac{D(D-1)}{2}\frac{\dot b^2}{b^2} = -\frac{8\pi G P_a}{3}, \quad and \tag{3}$$

$$(D-1)\frac{\ddot b}{b} + 3\frac{\ddot a}{a} + 3(D-1)\frac{\dot a}{a}\frac{\dot b}{b} + (D-1)(\frac{D}{2}-1)\frac{\dot b^2}{b^2} + 3\frac{\dot a^2}{a^2} = -\frac{8\pi G P_b}{3}. \tag{4}$$

The conservation equation for energy-momentum tensor implies that $T^{\mu}_{\nu;\mu} = 0$, and

$$\frac{d}{dt}(\rho a^3 b^D) + P_a b^D \frac{d}{dt}a^3 + P_b a^3 \frac{d}{dt} b^D = 0. \tag{5}$$

In 3-dimensional standard cosmology, the universe today consists of dark energy, dark matter and baryons. Both dark matter and baryons are very well approximated by pressureless matter. In extra dimensional sub-space, we have assumed a simple form for the equation of state, which is given as $P_b = w_b\rho$ with $w_b = \frac{w}{\bar\rho^{\gamma}}$, where w and γ are parameters of the model, which are determined by comparing with the observational data and $\bar\rho = \frac{\rho}{\rho_c}$, $\rho_c = (3H_0^2)/8\pi G$.

We have now solved Eq. (2) for $\dot a/a$ and bringing it to a form close to the familiar form of the FRW equations, we have:

$$\frac{\dot a^2}{a^2} = \frac{8\,\pi\,G\,\rho}{3} + \frac{D\,(2D+1)}{6}\frac{\dot b^2}{b^2} \mp \frac{D}{2}\frac{\dot b}{b}\sqrt{\frac{D\,(D+2)}{3}\frac{\dot b^2}{b^2} + \frac{32\,\pi\,G\rho}{3}}. \tag{6}$$

This equation would reduce to the standard form in the absence of the last two terms on the right hand side. In other words, the $\dot b/b$ dependent terms act as an effective dark energy source.

3. Cosmological solutions and observational constraints

It will be convenient to express the variables in terms of dimensionless quantities, namely:

$$t \equiv \tau/H_0, \qquad A' \equiv a'/a = \dot a/(a\,H_0), \quad and \quad B' \equiv b'/b = \dot b/(b\,H_0).$$

Here, primes denote derivative with respect to τ. In terms of these variables, and after eliminating $\ddot b$, the equations of motion now read:

$$\begin{aligned} 0 &= (D+2)\,A'' + 3(D+1)\,A'^2 + \frac{D\,(1-D)}{2}\,B'^2 + D\,(D-1)\,A'\,B' + 3\,D\,\Omega_0\,\bar\rho\,w_b, \\ \bar\rho' &= -\bar\rho\,[3\,A' + D\,(1+w_b)\,B'], \quad and \\ B' &= (D-1)^{-1}\left[-3\,A' \pm \sqrt{3\,D^{-1}\left\{(D+2)\,A'^2 + 6\frac{(D-1)}{D}\bar\rho\Omega_0\right\}}\right]. \end{aligned} \tag{7}$$

We have evolved the equations backward with the 'initial' conditions (at the present epoch ($\tau = 0$)) given by:

$$(\dot a/a)|_{\tau=0}\,H_0 \Rightarrow A'|_{\tau=0} = 1, \quad and \quad \rho/\rho_c|_{\tau=0} = 1 \Rightarrow \bar\rho|_{\tau=0} = 1.$$

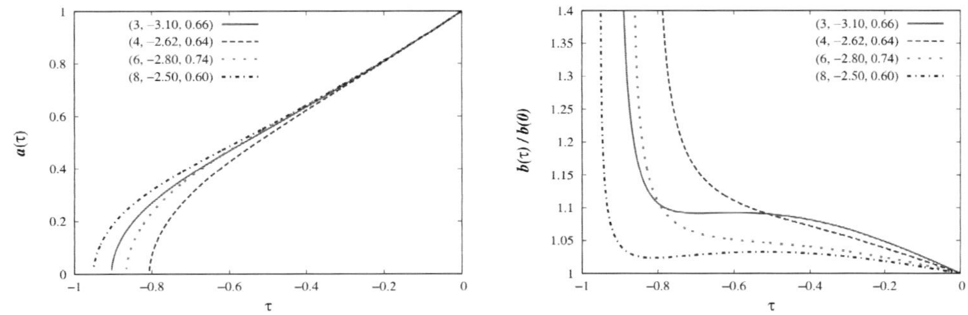

Figure 1. *Left panel*: Behaviour of the scale factors $a(\tau)$, and *Right panel*: $b(\tau)$ with the re-scaled time τ. The numbers in the parentheses refer to (D, w, γ).

Since, Einstein's equations can only determine $A'(\tau)$ and $B'(\tau)$, and not the scale factors themselves, we re-scale the solutions as:

$$a|_{\tau=0} = 1 \quad \Rightarrow \quad A|_{\tau=0} = 0, \quad and \quad b|_{\tau=0} = 1 \quad \Rightarrow \quad B|_{\tau=0} = 0.$$

There exists two solutions, one for each sign in the last of Eq. (7). We have rejected the minus sign branch, as it leads to an accelerated expansion for all times rather than a transition from a decelerated phase to an accelerated one.

Figure 1 represents the solutions for the two scale factors for some particular values of parameters (D, w, γ). (For details of calculation, see [4].) The scale factor $a(\tau)$ for our universe starts from 1 at $\tau = 0$ and decreases monotonically for negative values of τ. And these equations are valid only as long as the universe is matter dominated. As it can be easily seen, the relative evolution in $b(\tau)$ is small.

The epoch of matter-radiation equality $\tau_{eq} \equiv \tau(z = z_{eq})$, has a considerable dependence on the parameter choice. For a given value of D and w, a smaller γ shifts τ_{eq} further into the past, thus, increasing the present-day age of the universe. Similarly, for a given value of D and γ, τ_{eq} shifts further into the past with decrease in w. As we have seen above, the model

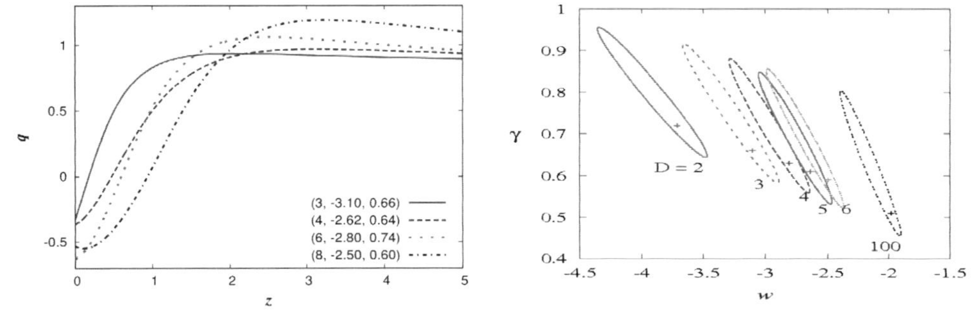

Figure 2. *Left panel*: Evolution of the deceleration parameter q with the redshift z. The numbers in the parentheses refer to (D, w, γ), and *Right panel*: 95% C.L. contours in the $\gamma - w$ plane for different values of D. The points represent the best fit values for each D.

leads to correct late time acceleration for some choice of parameters. We have now looked for it to compare with SN type 1a observational data [5], which lists the distance modulus μ

$\equiv 5\,\log d_L + 25$ as well as the redshift for 557 such SN. The luminosity distance $d_L(z)$ is defined through: $d_L(z) = (1+z)\int_0^z \frac{dz'}{H(z')}$.

We have now calculated $d_L(z)$, given our determination of $H(z)$ for a particular choice of parameters. We have defined χ^2-test through $\chi^2(D,\gamma,w) = \sum_{i=0}^{i=n} \frac{[\mu_{obs}(z_i)-\mu_{th}(D,\gamma,w;z_i)]^2}{\sigma_i^2}$, where μ_{th} defines the value expected in our model for a particular choice of parameters, whereas μ_{obs} and σ are the observational value and the associated root-mean-squared error. We have found the best-fit values of the parameters by minimizing the χ^2, which are listed in Table 1. As we have seen, with an increase in D, both $|w|$ and γ decrease. This is based on the fact that the resulting smaller extra-dimensional pressures would now have an enhanced effect in the normal world through the larger effective coupling between $\dot{a}(t)$ and $\dot{b}(t)$.

D	χ^2_{min}	best fit w	best fit γ	q_0	z_{tr}
2	538.92	$-3.71^{+0.25}_{-0.63}$	$0.72^{+0.23}_{-0.08}$	$-0.49^{+0.10}_{-0.26}$	$0.75^{+0.23}_{-0.27}$
3	539.18	$-3.10^{+0.18}_{-0.55}$	$0.66^{+0.25}_{-0.07}$	$-0.47^{+0.09}_{-0.27}$	$0.78^{+0.22}_{-0.32}$
10	539.42	$-2.31^{+0.12}_{-0.38}$	$0.57^{+0.14}_{-0.06}$	$-0.46^{+0.08}_{-0.36}$	$0.79^{+0.19}_{-0.33}$
100	539.74	$-1.98^{+0.07}_{-0.38}$	$0.51^{+0.28}_{-0.05}$	$-0.43^{+0.06}_{-0.30}$	$0.80^{+0.19}_{-0.34}$

Table 1. The error bars correspond to the projections of the 95% C.L. ellipses on the two axes.

Figure 2 (right panel) shows 95% C.L. contours in the $\gamma - w$ plane for a given D. It is easily seen that there is a strong negative correlation between the two parameters. Integral values of γ and positive w are also ruled out. As deduced from Table 1, the position of minima is getting increasingly closer with increase in D arbitrarily, and also shown are the values of q_0 and z_{tr} as determined within our model.

4. Conclusions and discussion

In this work, the effect of the extra dimensions is indirectly seen as the accelerated expansion of the scale factor in normal dimensions. To produce this effect, we have considered matter content as a pressureless gas in the normal dimensions and a monomial equation of state in the extra dimensions. This phenomenon is just produced by a very moderate contraction of the extra dimensions, and in the time interval that $a(\tau)$ has increased by nearly a factor of 3000, $b(\tau)$ has decreased by $\sim 29\%$. The model shows a good agreement with the observational data on SN type 1a for a significantly wide range of parameters as indicated by both a low χ^2 per degree of freedom, but we still need a negative pressure which is limited to the extra dimensions.

Acknowledgements

The author acknowledges the CSIR, Government of India, for assistance under grant 09/045(0908)/2009-EMR-I. Facilities at IUCAA Resource Centre are acknowledged.

References

[1] Qiang L, Ma Y, Han M and Yu D 2005 *Phys. Rev. D* **71** 061501
[2] Setare M R and Saridakis E N [arXiv:0811.4253]
[3] Sahdev D 1984 *Phys. Rev. D* **30** 2495
[4] Pahwa I, Choudhury D and Seshadri T R 2011 *JCAP* **09** 015
[5] Supernova Cosmology Project *http://supernova.lbl.gov/Union/*

Statistics of statistical anisotropy measures

N Pant, A Rotti and T Souradeep
IUCAA, Post Bag 4, Pune University Campus, Ganeshkhind, Pune 411007, India
E-mail: nidhip@iucaa.ernet.in, aditya@iucaa.ernet.in, tarun@iucaa.ernet.in

Abstract. Cosmic Microwave Background (CMB) is a Gaussian random field to a sufficient approximation, and its statistics is completely specified by the 2-point correlation function, which, most generally, can be expanded in Bipolar Spherical Harmonic (BipoSH) basis. Statistical Isotropy (SI) of 2-point correlation function is a common assumption in cosmology, which needs to be tested. Any SI violating signal can be searched in the expansion BipoSH coefficients. We have analytically evaluated the moments and the distribution of these coefficients using characteristic function approach. We have found that coefficients with $M = 0$ have an exact analytical form for any order moment. For the remaining BipoSH coefficients with $M \neq 0$, the moments have to be supplemented with a correction term. We have verified our results with measurements of BipoSH coefficients on numerically simulated statistically isotropic CMB maps.

1. Introduction

Assuming Gaussianity of CMB, BipoSH forms an orthonormal basis on $\mathbf{S}^2 \times \mathbf{S}^2$, making them a natural choice of basis to expand 2-point correlation function [1, 2, 4, 3, 5, 6]. These expansion coefficients were adopted by the WMAP team [7] to search for violations of isotropy in the WMAP data. We have found expressions for the moments of the distribution of the BipoSH coefficients analytically using the characteristic function approach. BipoSH coefficients can be expressed as linear combination of off-diagonal elements of the harmonic space covariance matrix [1]. Every term in the linear combination of BipoSH coefficients with $M = 0$, is found to be independent of every other term in summation, which allows us to use characteristic function method to find exact expressions for moments. BipoSH coefficients with $M \neq 0$ can have non-linear correlations among terms due to which we have corrected the expressions for moments derived using characteristic function method upto kurtosis with an extra additive term. We have tested these analytical results against simulations, and simulated the CMB maps using the widely used HEALPix [8] package.

In Section 2, we have briefly discussed the characteristic function approach, and in Section 3, we have discussed the BipoSH formalism and the statistics of the BipoSH coefficients. We have concluded with a discussion of our results in Section 4.

2. Characteristic function method

The characteristic function of any random variable is defined in the following manner [9]:

$$\varphi_X(t) = E[e^{itX}], \qquad t \in \Re\,. \tag{1}$$

Consider a random variable given by:

$$Z_n = \sum_{i=1}^{n} a_i X_i \ , \tag{2}$$

where a_i's are constants and X_i's are independent random variables, which are not necessarily identically distributed. Statistical independence of two random variables necessarily requires any form of correlations between them to vanish. The characteristic function of Z_n will be the product of the characteristic function of the individual terms contributing to the linear sum given as:

$$\varphi_{Z_n}(t) = \varphi_{X_1}(a_1 t)\varphi_{X_2}(a_2 t)...\varphi_{X_n}(a_n t)\, . \tag{3}$$

Cumulant generating function is defined as the logarithm of the characteristic function as:

$$g_Z(t) = \log[\varphi_Z(t)]\, . \tag{4}$$

Cumulants of the random variable Z can be obtained by taking the derivative of the cumulant generating function, and evaluating them at zero as:

$$K_n = i^n g_Z^n(t)|_{t=0}\, . \tag{5}$$

The explicit relationships between cumulants and central moments, till the fourth central moments are:

$$\begin{aligned} \mu_1 &= K_1 && \text{Mean,} \\ \mu_2 &= K_2 && \text{Variance,} \\ \mu_3 &= K_3 && \text{Skewness,} \quad \textit{and} \\ \mu_4 &= K_4 + 3K_2^2 && \text{Kurtosis.} \end{aligned} \tag{6}$$

Note that in Figures 1 and 2, we have plotted the moments normalized with standard deviation.

3. Statistics of BipoSH coefficients

BipoSH expansion of 2-point correlation function is given as:

$$C(\hat{n}_1, \hat{n}_2) = \sum_{l_1, l_2, L, M} A_{l_1 l_2}^{LM} \{Y_{l_1}(\hat{n}_1) \otimes Y_{l_2}(\hat{n}_2)\}_{LM}, \tag{7}$$

where $A_{l_1 l_2}^{LM}$ are BipoSH coefficients and $\{Y_{l_1}(\hat{n}_1) \otimes Y_{l_2}(\hat{n}_2)\}_{LM}$ are bipolar spherical harmonics [10]. BipoSH functions can be expressed as:

$$\{Y_{l_1}(\hat{n}_1) \otimes Y_{l_2}(\hat{n}_2)\}_{LM} = \sum_{m_1 m_2} C_{l_1 m_1 l_2 m_2}^{LM} Y_{l_1 m_1}(\hat{n}_1)\, Y_{l_2 m_2}(\hat{n}_2), \tag{8}$$

where $C_{l_1 m_1 l_2 m_2}^{LM}$ are Clebsch-Gordon coefficients. The indices of these coefficients satisfy the triangularity conditions $|l_1 - l_2| \leq L \leq l_1 + l_2$, and $m_1 + m_2 = M$. BipoSH coefficients are a measure of off-diagonal elements of covariance matrix:

$$A_{l_1 l_2}^{LM} = \sum_{m_1 m_2} a_{l_1 m_1} a_{l_2 m_2} C_{l_1 m_1 l_2 m_2}^{LM}\ , \tag{9}$$

where a_{lm}'s are the spherical harmonic coefficients of the CMB maps. SI of CMB implies that the 2-point correlation function is rotationally invariant $C(\hat{n}_1, \hat{n}_2) \equiv C(\hat{n}_1 \cdot \hat{n}_2) = \sum_l \frac{(2l+1)}{4\pi} C_l P_l(\hat{n}_1 \cdot \hat{n}_2)$, where $\hat{n} = (\theta, \phi)$ is a unit vector on the sphere, and C_l is well known angular power spectrum. Under this symmetry, only non-vanishing coefficients of expansion are $A_{ll}^{00} = (-1)^l C_l \sqrt{2l+1}$ [1].

BipoSH coefficients are linear combination of random variables given in Eq. (9). Therefore, to arrive at the distribution of a given BipoSH coefficient, we have begun with finding out the characteristic function of individual terms involved in the linear combination. Assuming the terms in linear sum to be independent of each other, the characteristic function of the BipoSH coefficient can then be derived from Eq. (3).

Case (a): Bipolar coefficient with $M = 0$
SI violating signal detection has been seen in these coefficients in WMAP-7 [7]. Assumption of independence among terms holds true in this case. We have found out that these coefficients have an asymmetric distribution for $l_1 = l_2$, and symmetric for $l_1 \neq l_2$ [11], as seen in Figure 1. Subsets of this case are the coefficients with $L = 0$, which are nothing but angular power spectrum $A_{ll}^{00} = (-1)^l C_l \sqrt{2l+1}$, known to have χ^2 distribution.

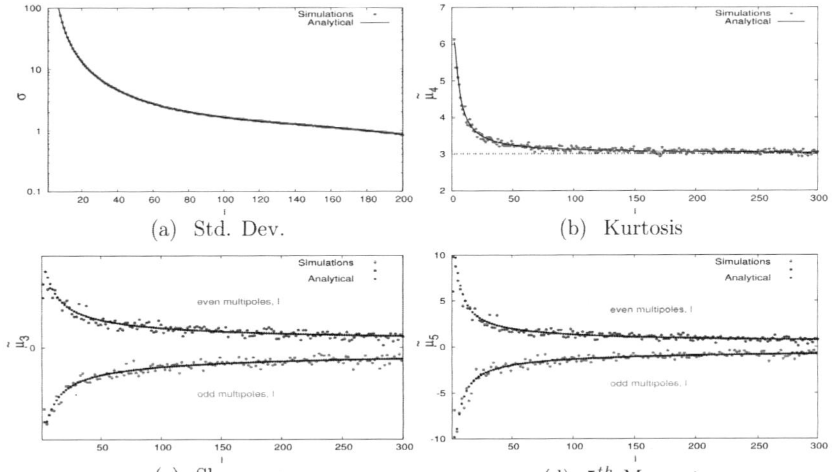

(a) Std. Dev. (b) Kurtosis (c) Skewness (d) 5th Moment

Figure 1. Standard deviation (σ), kurtosis ($\tilde{\mu}_4$), skewness ($\tilde{\mu}_3$), and 5th moment ($\tilde{\mu}_5$) of A_{ll}^{20}, from 15000 simulations.

Case (b): Bipolar coefficient with $M \neq 0$
We have first calculated the moments of distribution for these coefficients using the characteristic function method, assuming that all terms in the linear combination are independent of each other. It is observed that only even ordered cumulants exist, implying that the distribution of these coefficients is symmetric as seen in Figure 2. The mismatch in simulations and analytically derived moments at low multipoles (l) are due to the assumed underlying independence of the terms contributing to the linear combination as seen in Figure 2. The moments calculated using the characteristic function method need to be supplemented with correction terms, which account for the higher order correlations among terms. We have found that variance does not

Journal of Physics: Conference Series **484** (2014) 012046 doi:10.1088/1742-6596/484/1/012046

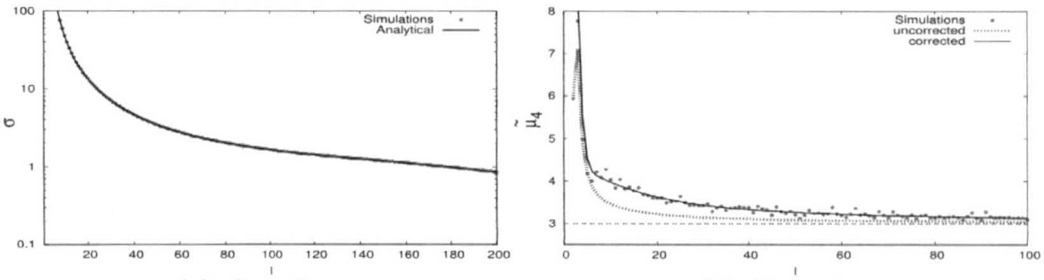

(a) Std. Dev. (b) Kurtosis

Figure 2. Standard deviation (σ) and kurtosis ($\tilde{\mu}_4$) of A^{43}_{ll} derived from 15,000 simulations. The difference between corrected and uncorrected analytical moments is prominent at low values of multipole (l).

have any corrections due to the fact that the terms are linearly uncorrelated, and given by:

$$\tilde{\mu}_4 = \bar{\mu}_4 + \frac{3\Big[\sum_i^N (K_2^i)^2 + 2\sum_{i\neq j} E[X_i^2 X_j^2]\Big]}{\Big(\sum_i^N K_2^i\Big)^2}, \tag{10}$$

where second term is the correction term. In the above expression, K_i is the cumulant of the ith term, and X_i and X_j are the ith and j^{th} terms in the summation. The calculation for correction for higher order moments becomes very tedious, hence, we have restricted ourselves to calculating corrections for moments only upto kurtosis.

4. Conclusions

Statistics of BipoSH coefficients is important to understand, as signals of isotropy violation are being searched for in CMB data using these coefficients. Detection of SI violation has been claimed by WMAP-7 team in V-band and W-band maps. Due to the difference in significance of detection in the two frequency bands and its alignment with the ecliptic, these detections were not suspected to have cosmological origin. However, more recent work has attempted at explaining these observations by accounting for gravitational lensing modifications to the BipoSH coefficients [12]. Interestingly, in our study, we have found that the BipoSH coefficients showing SI violation detection have an asymmetric PDF. The WMAP team has used band power averaged BipoSH coefficients with the large bin sizes to reduce noise, however, with experiments like PLANCK, it might be possible to achieve similar signal to noise ratio for smaller bin sizes, with which the skewness of these coefficients might become considerable, and it will then become important to account for the non-Gaussian PDF of the BipoSH coefficients.

References

[1] Hajian A and Souradeep T 2003 *Ap. J. Lett.* **597** L5
[2] Souradeep T and Hajian A 2004 *Pramana: J. Phys.* **62** 793
[3] Hajian A and Souradeep T 2004 [arXiv:astro-ph/0501001]
[4] Hajian A, Souradeep T and Cornish N 2004 *Ap. J.* **618** L63
[5] Basak S, Hajian A and Souradeep T 2006 *Phys. Rev. D* **74** 021301
[6] Hajian A and Souradeep T 2006 *Phys. Rev. D* **74** 123521
[7] Bennet C L *et al* 2011 *Ap. J. Supp. Ser.* **192** 19
[8] Gorski K M *et al* 2005 *Ap. J.* **622** 759
[9] Melvin D Springer 1979 *The Algebra of Random Variables* (John Wiley Sons Inc.)
[10] Varshalovich D A, Moskalev A N and Khersonskii V K 1988 *Quantum Theory of Angular Momentum* (Singapore: World Scientific)
[11] Joshi N, Rotti A and Souradeep T 2012 *Phys. Rev. D* **85** 043004
[12] Aditya Rotti, Moumita Aich and Tarun Souradeep 2011 [arXiv:1111.3357v1]

Cosmological parameter estimation using Particle Swarm Optimization

J Prasad and T Souradeep

IUCAA, Post Bag 4, Pune University Campus, Ganeshkhind, Pune 411007, India

E-mail: jayanti@iucaa.ernet.in, tarun@iucaa.ernet.in

Abstract. Constraining parameters of a theoretical model from observational data is an important exercise in cosmology. There are many theoretically motivated models, which demand greater number of cosmological parameters than the standard model of cosmology uses, and make the problem of parameter estimation challenging. It is a common practice to employ Bayesian formalism for parameter estimation for which, in general, likelihood surface is probed. For the standard cosmological model with six parameters, likelihood surface is quite smooth and does not have local maxima, and sampling based methods like Markov Chain Monte Carlo (MCMC) method are quite successful. However, when there are a large number of parameters or the likelihood surface is not smooth, other methods may be more effective. In this paper, we have demonstrated application of another method inspired from artificial intelligence, called Particle Swarm Optimization (PSO) for estimating cosmological parameters from Cosmic Microwave Background (CMB) data taken from the WMAP satellite.

1. Introduction

Cosmic Microwave Background (CMB) observations have entered into a new era with space based telescopes like Wilkinson Microwave Background Anisotropy Probe (WMAP) and Planck, and it has become possible to rule out many theoretical models on the basis of data of very high quality. In the framework of Bayesian formalism, this is done by asking the question that which of the theoretical models, as is represented by a set of parameters, is best fit to the observational data or how the probability distributions of various parameters look like [1, 3, 2]. Bayesian formalism of parameter estimation also needs the likelihood function $\mathcal{L}(\theta)$ or the probability $P(d|\theta)$ of realizing the data d, given a set of parameters θ, and is directly related to the probability distribution $P(\theta|d)$ (posterior) of parameters, given the data using Bayes theorem [4, 5], we have:

$$P(\theta|d) \propto P(d|\theta)P(\theta). \tag{1}$$

When we are mainly interested in finding the set of parameters for which the likelihood function is maximum, the problem of parameter estimation can be considered an optimization problem, where we have tried to find the vector θ_0 for which the function $\mathcal{L}(\theta)$ has the highest value. The simplest way to find the global maximum of the optimization function would be to set up a grid in the multi-dimensional parameter space and compare the value of the function at each grid point. Grid based search is a prohibitively expensive exercise when dimensionality of the search space is large, i.e., computational cost grows exponentially with the number of search dimensions. There are many classic methods, which use the value of the function and its derivative at some initial guess point, and then iteratively find the maximum point. These methods are quite effective in lower dimensions but become intractable in higher dimensions. Apart from that, they are good mostly for finding the local maxima. To overcome the limitation of finding only the local maxima, stochastic methods are generally employed, many of which are inspired from artificial intelligence techniques like Artificial Neural Network (ANN), Genetic Algorithms (GA), and Swam Intelligence (SI). Here, we have demonstrated the application of Particle Swarm Optimization (PSO) for cosmological parameter estimation from WMAP-7 data.

The plan of this paper is as follows: In Section 2, we have given a short introduction of PSO, and the details of which can be found in our earlier paper [6]. In Section 3, we have presented our results, and in Section 4 we have summarized and concluded.

2. Particle Swarm Optimization

Proposed by Kennedy and Eberhart [7], PSO is a population based search procedure inspired from group search, in which the members of a group use their own learning experience and that of other members of the group for exploration, as is observed in bird flocks and fish schools. Although, applications of PSO are quite common in engineering and artificial intelligence problems, it has been applied successfully in astrophysics problems relatively recently [8].

In PSO, a team of particles (computational agents) is launched in the multi-dimensional search space following certain rules for updating positions and velocities. The velocity $V^i(t)$ and position $X^i(t)$ of a particle labelled by the index i at time t is updated in the following way:

$$V^i(t+1) = wV^i(t) + c_1\xi_1(X^i_{\texttt{Pbest}} - X^i(t)) + c_2\xi_2(X_{\texttt{Gbest}} - X^i(t)), \tag{2}$$

and

$$X^i(t+1) = X^i(t) + V^i(t+1), \tag{3}$$

where c_1 and c_2 are two acceleration coefficients, w is the inertia weight, and ξ_1 and ξ_2 are uniform random members in the range $[0, 1]$ (c_1, c_2 and w are also called *design parameters*). The n-dimensional vector $X^i_{\texttt{Pbest}}$ (personal best) and $X_{\texttt{Gbest}}$ (global best) are the points in the search space, at which the particle i has found the highest value so far, and the point at which the highest value has been found by any of the particles respectively, i.e., $X_{\texttt{Gbest}}$ is one of the $X_{\texttt{Pbest}}$. Apart from setting the values of the design parameters in any PSO implementation, we need to decide on schemes for Eq. (1), setting up the initial boundary conditions given in Eq. (2), and topping criteria given in Eq. (3). In order to stop particles from leaving the search region, we also need to fix the maximum velocity of particles. In our exercise, we have set the initial values of the positions and velocities of particles randomly in the allowed range and use reflecting boundary conditions. We have stopped the search when we have found that the improvement in the value of the fitness function remains smaller than a given number for some fixed number of steps. The details of our PSO implementation can be found in [6].

3. Results

We have considered a six parameter cosmological model ($\Omega_b h^2, \Omega_c h^2, \Omega_\Lambda, A_s, n_s, \tau$), where $\Omega_b, \Omega_c, \Omega_\Lambda$ are the energy densities of baryons, cold dark matter and cosmological constants respectively, h is the Hubble parameter, and A_s, n_s and τ are the amplitude, spectral index of primordial power spectrum and reionization optical depths respectively. The range over which we have carried out the search and the best fit parameters we have obtained are given in Table 1.

The procedure we have followed in PSO has three main steps: (i) We have computed theoretical angular power spectrum C_l for a set of cosmological parameters using publicly available code CAMB [10], (ii) We have computed the likelihood function using theoretical angular power spectrum and WMAP-7 data. We have used the likelihood code provided by

Cosmological parameters from PSO						
Variable		PSO best fit			WMAP best fit[9]	
	Range	Gbest	Mean	std. dev.	ML	Mean
$\Omega_b h^2$	(0.01,0.04)	0.022044	0.022043	0.000490	0.02227	$0.02249^{0.00056}_{-0.00057}$
$\Omega_c h^2$	(0.01,0.20)	0.112111	0.112129	0.005489	0.1116	0.1120 ± 0.0056
Ω_Λ	(0.50,0.75)	0.722980	0.720416	0.030279	0.729	$0.727^{+0.030}_{-0.029}$
n_s	(0.50,1.50)	0.963800	0.963794	0.012470	0.966	0.967 ± 0.014
$A_s/10^{-9}$	(1.0,4.0)	2.445123	2.446698	0.111354	2.42	2.43 ± 0.11
τ	(0.01,0.11)	0.084124	0.084124	0.013444	0.0865	0.088 ± 0.015

Table 1. The first and the second columns in this table show the PSO fitting parameters and their range respectively, and third, fourth and fifth columns show the best fit (Gbest), average values and the errors (which are computed by fitting the sampled function) on fitted parameters respectively. In the last two columns, we have given the maximum likelihood (ML) and the average values of the cosmological parameters as reported by the WMAP-7 team respectively. We can see that the values of parameters which we get using PSO are not very different from the values reported by the WMAP-7 team.

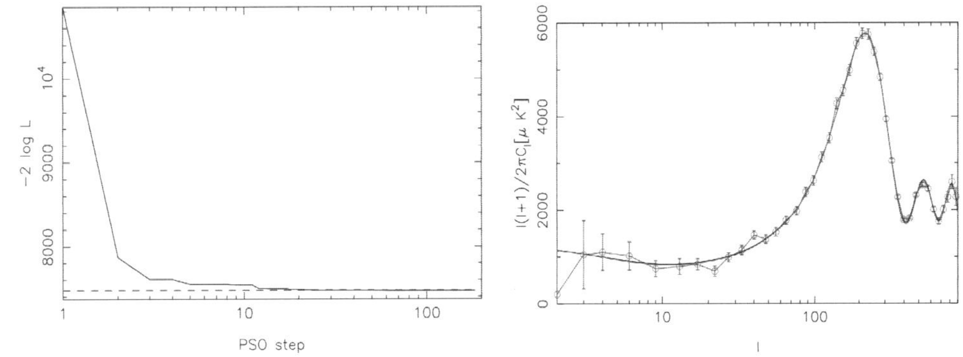

Figure 1. The left and right panel in this figure show the change in the fitness function ($\chi^2_{\text{eff}} = -2\log\mathcal{L}$) and the best fit angular power spectrum we have obtained from PSO. We can see that in the beginning, the fitness function (solid line in the left panel) changes rapidly but later on when the Gbest approaches close to the global maximum, and there is hardly any change. For reference, we have shown the value of $-2\log\mathcal{L}$ for WMAP-7 data using dashed line. The red black and red lines in the right panel show the angular power spectrum for the best fit cosmological parameter as reported by the WMAP-7 team, and for the parameters we have obtained using PSO respectively. We have also shown the binned power spectrum with error bars from the WMAP-7 data. Note that the small difference (see Table 1) between the PSO best fit parameters and the WMAP-7 best fit parameters leads to a difference $\Delta\chi^2_{\text{eff}} \approx 2$ (it is smaller for PSO)

WMAP-7 team for computing the likelihood function [11], and (iii) We have computed the new set of points in the parameter space following the PSO algorithm (see [6]).

The main results of our work are summarized in Table 1, in the form of the best fit (Gbest) values of the cosmological parameters. We have reported the value of the Gbest as well as the average location of the PSO particles at the last step, which matches well with the Gbest value.

Since PSO is an optimization method, it is designed to find the point at which the fitness (likelihood) function has the maximum value rather than sampling the fitness function, and therefore, sampling done by PSO is not good enough to do statistics as in the case of MCMC. In order to get some idea of the errors on various estimated parameters, we have fitted a multi-dimensional paraboloid to the sampled points close to the `Gbest` and identified the fitting coefficients with Hessian matrix, and got error from the inverse of it, i.e., covariance matrix (for details see [6]).

Left panel in Figure 1 shows the change in the $\chi^2_{\rm eff} = -2\log\mathcal{L}$ as PSO progresses. It is clear that in the beginning, the improvement in the value of the fitness function is large, and it decreases as `Gbest` reaches close to the global maximum. Once `Gbest` remains close to the global maximum for a given number of steps, we can stop the search. In the right panel of the Figure 1, we have shown the best fit angular power spectrum for the cosmological parameters we have obtained using PSO and for their standard values as reported by the WMAP-7 team. We have also shown the binned power spectrum from the WMAP-7 data with error bars in the right panel.

4. Conclusions

Sampling based methods like MCMC are quite common in parameter estimation from observational data, but their use is mostly limited to situations when dimensionality of the search space is low and the likelihood surface is smooth. Here, we have demonstrated that PSO, which is a population based search procedure may have advantages in problems, in which the dimensionality of the search space is large and/or the fitness function has multiple local maxima. Based on a very simple algorithm and having a few design parameters, PSO is quite easy to programme and parallelize on a multi-processor system.

References

[1] Verde L 2010 *Statistical Methods in Cosmology* (Lecture Notes in Physics, Berlin Springer Verlag) ed G Wolschin **800** 147 [0911.3105]
[2] Verde L *et al* 2003 *Ap. J. Supp. Ser.* **148** 195 [arXiv:astro-ph/0302218]
[3] Lewis A and Bridle S 2002 *Phys. Rev. D* **66** 103511 [arXiv:astro-ph/0205436]
[4] Hamimeche S and Lewis A 2008 *Phys. Rev. D* **77** 103013 [0801.0554]
[5] Hamimeche S and Lewis A 2009 *Phys. Rev. D* **79** 083012 [0902.0674]
[6] Prasad J and Souradeep T 2011 [arXiv:1108.5600v1]
[7] Kennedy J and Eberhart R C 1992 *IEEE International Conference on Neural Networks* **4**
[8] Wang Y and Mohanty S D 2010 *Phys. Rev. D* **81** 063002 [1001.0923]
[9] Larson D *et al* 2011 *Ap. J. Supp. Ser.* **192** 16 [1001.4635]
[10] http://camb.info/
[11] http://lambda.gsfc.nasa.gov/

Journal of Physics: Conference Series **484** (2014) 012048 doi:10.1088/1742-6596/484/1/012048

Effect of dark energy sound speed and equation of state on CDM power spectrum

H A Rizwan ul and S Unnikrishnan,

IUCAA, Post Bag 4, Pune University Campus, Ganeshkhind, Pune 411007, India

E-mail: rizwan@iucaa.ernet.in

Abstract. We have studied the influence of equation of state w, and effective sound speed c_e of the dark energy perturbations on the cold dark matter (CDM) power spectrum. We have considered different cases of the equation of state and the effective sound speed, and the CDM power spectrum is found to be generically suppressed in these cases as compared to the ΛCDM model. The suppression at different length scales depends on the value of w and c_e, and the effect of different w is profoundly seen at all length scales. The influence of sound speed is significantly seen only at the intermediate length scales, and is negligible at scales very much larger and smaller than the Hubble scale.

1. Introduction

Recent supernova (SN) type 1a supernova observations have confirmed that the universe is expanding at an accelerated rate at the present epoch [1]. Understanding the nature of the component which drives this accelerated expansion is one of the major challenges of present day cosmology. One of the possible candidates could be a dynamical scalar field, with a negative pressure, known as dark energy (DE). Several models of scalar field dark energy like the quintessence, k-essence, Chaplygin gas, etc. have been studied extensively [2, 3, 4]. In general, these models can lead to degenerate evolution of the scale factor [5]. The scalar field models of dark energy are described by the two parameters equation of state (EOS) w and speed of sound (ESS) c_e. Since, DE is a scalar field, it not only contributes in the background, but also through the perturbations, so we need to take into account perturbations to the DE [6, 7]. The ESS, which relates the density perturbations to pressure perturbation is useful in studying the effects of perturbations in DE. So considering the DE, perturbations can provide vital information in understanding these models, hence, distinguish and possibly ruling out some of the models. Our aim here is to study how perturbation in such models of DE will effect the CDM power spectrum. We have focused on evolving EOS and ESS, whose evolution is not fixed by a particular model of DE, and seen its effects on observable quantities such as CDM power spectrum.

We have considered a scalar field model of DE, whose action is given by:

$$S[\phi] = \int \sqrt{-g}\, \mathcal{L}(X,\phi) d^4x \ ,$$

where $\mathcal{L}(X,\phi)$ is the Lagrangian density of the scalar field. Also, we have assumed the universe to be spatially flat consisting of CDM and DE. The EOS parameter for the DE with above

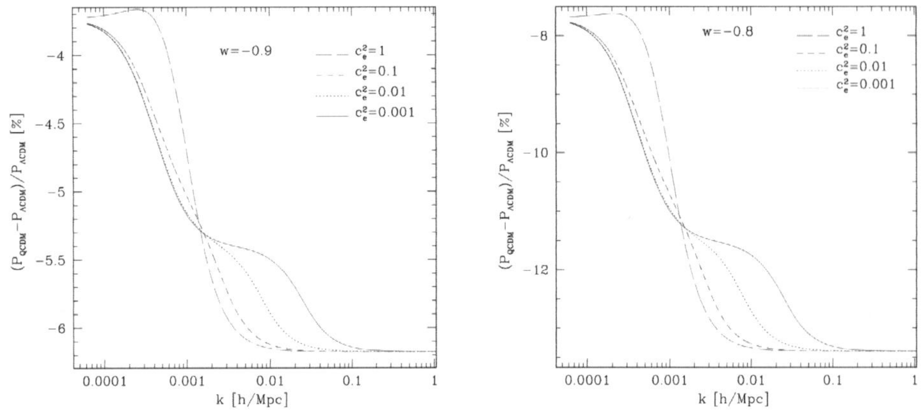

Figure 1. The percentage suppression in matter power spectrum in DE models in comparison to ΛCDM model is plotted for a constant w and for different values of c_e^2. It can be seen that suppression is more for DE models with $w = -0.8$ than with $w = -0.9$. The power spectrums shown in this figure correspond to its value at the present epoch.

mentioned action is given by: $w = \frac{\mathcal{L}(X,\phi)}{2(\frac{\partial \mathcal{L}}{\partial X})X - \mathcal{L}(X,\phi)}$, the corresponding effective ESS of DE is $c_e^2 \equiv \frac{\left(\frac{\partial \mathcal{L}}{\partial X}\right)}{\left(\frac{\partial \mathcal{L}}{\partial X}\right) + 2X\left(\frac{\partial^2 \mathcal{L}}{\partial X^2}\right)}$. From these definitions of EOS and ESS, it is obvious that their evolution is governed by the Lagrangian density of the scalar field DE. However, it is also possible to obtain the form of Lagrangian density from a given evolution of w parameter and c_e^2 (see [8]). Here, we have used this fact, and chosen a parameterized form of w and c_e^2 to study the influence of perturbations in DE on CDM power spectrum. We have started by presenting the necessary equations to study the evolution of perturbations, and the perturbed Friedmann-Robertson-Walker (FRW) metric in the longitudinal gauge is given by:

$$ds^2 = (1 + 2\Phi)\, dt^2 - a^2(t)\,(1 - 2\Phi)\left[dx^2 + dy^2 + dz^2\right], \quad (1)$$

where Φ is the gravitational potential describing the scalar metric perturbations. In a system consisting of CDM and DE, using the linearized Einstein's equations and the covariant conservation equations for DE, we can obtain the following set of equations for the evolution of perturbations:

$$\Phi'' = -\frac{\Phi'}{a}\left(\frac{\ddot{a}}{aH^2} + 4\right) - \frac{\Phi}{a^2}\left(2\frac{\ddot{a}}{aH^2} + 1\right) + \frac{4\pi G\bar{\rho}_d}{a^2H^2}\left(c_e^2\delta_d - \left(c_e^2 - c_a^2\right)\sigma_d\right), \quad (2)$$

$$\sigma_d' = \left(\left(\frac{\ddot{a}}{aH^2}\right) + 3\left(c_e^2 - c_a^2\right) + (3w - 1)\right)\frac{\sigma_d}{a} - \frac{3c_e^2\delta_d}{a} - \frac{3\Phi(1+w)}{a}, \quad and \quad (3)$$

$$\delta_d' = \left(\frac{3\,\delta_d}{a}\right)\left(w - c_e^2\right) + \frac{3\,\sigma_d}{a}\left(c_e^2 - c_a^2\right) + \frac{k^2\,\Phi}{3a^3H^2} + 3\,(1+w)\,\Phi'. \quad (4)$$

The quantities appearing in the above equations are as follows: (i) δ_d is the density contrast given by $\delta_d = \frac{\delta\rho_d}{\bar{\rho}_d}$, (ii) σ_d is a dimensionless variable defined via the equation: $\delta T^0_{\ i} = \left(\rho_d/3Ha^2\right)\sigma_{d,i}$, and (iii) $c_a^2 \equiv \dot{p}_d/\dot{\rho}_d$ is the adiabatic ESS of the DE. Here, dot and prime denote the derivatives

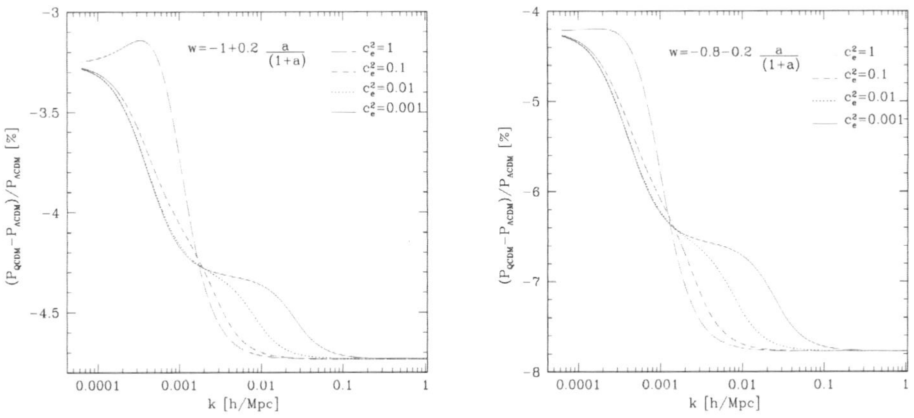

Figure 2. Figure shows the suppression in matter power spectrum compared to ΛCDM model for the case when w is evolving, but c_e^2 is constant. In the left panel, w is initially -1 and asymptotically approaches -0.8, whereas in the right panel, w is initially -0.8 and asymptotically approaches -1. It can also be verified that in both the panels w at the present epoch is -0.9.

with respect to time and scale factor respectively. Now, solving the Eqs. (2), (3) and (4), we have obtained expressions for $\Phi(a,\, k)$ and $\delta_d(a,\, k)$. Then, the density contrast δ_m, describing the evolution of perturbations in the CDM can be evaluated from the temporal component of the linearized Einstein's equation, and is given by:

$$\delta_m(a,\, k) = -\left(\frac{2}{\Omega_m(a)}\right)\left(\Phi + a\,\Phi' + \left(\frac{k^2}{3H^2a^2}\right)\Phi\right) - \left(\frac{\Omega_d(a)}{\Omega_m(a)}\right)\delta_d(a,\, k), \tag{5}$$

where Ω_m and Ω_d are the dimensionless density paramcters. It is evident from the Eq. (5) that the matter power spectrum is dependent on the effective ESS and EOS. The corresponding expression for matter perturbation in the ΛCDM is independent of c_e^2 and δ_d, as there are no perturbations to the cosmological constant. To see how the perturbations in DE effect the CDM power spectrum, we have compared the growth of CDM perturbation in DE models with respect to ΛCDM model.

2. Effect of EOS and ESS parameters on CDM power spectrum

We have now investigated, how varying the EOS w and ESS c_e^2 parameters of DE will influence CDM power spectrum. The most commonly studied scalar field DE models are with a constant value of w and c_e^2. We have considered a class of models, in which c_e^2 is constant, and allowed for evolving EOS. The EOS can have a parametric form, which depends on the scale factor, and can evolve from a initial value to asymptotic value, spanning the entire evolution of the universe. Then, there are classes of models, where c_e^2 is epoch dependent and w is constant. Again, we can use a similar parameterizations for c_e^2, which allow for all possible values of c_e^2. Finally, we can have the most general case, where both w and c_e^2 are epoch dependent (for details of parameterization used, see [8]). Next, we have quantitatively seen how the ESS and EOS influence the behaviour of power spectrum, by taking the difference in the growth of CDM perturbations in the scalar field models of DE with respect to the ΛCDM. The percentage difference can be found by numerically evaluating the following quantity:

$$\Delta\% \equiv \frac{P_{QCDM}(k) - P_{\Lambda CDM}(k)}{P_{\Lambda CDM}(k)} \times 100,$$

where $P_{QCDM}(k)$ is the CDM power spectrum at the present epoch in DE models, and $P_{\Lambda CDM}(k)$ is the corresponding power spectrum in ΛCDM model. Figure 1 shows Δ for a constant value of c_e^2 and w, it can be seen that at all length scales of perturbations percentage difference is negative. It can be inferred that the CDM power spectrum is suppressed in DE models in comparison to ΛCDM model. Figure 2 shows the percentage difference in models, where w is epoch dependent and c_e^2 is constant, again there is a suppression in CDM power spectrum at all length scales. Similarly, in the cases where c_e^2 is evolving or where both c_e^2 and w are evolving with time, the CDM power spectrum is found to be suppressed compared to ΛCDM model [8].

There are certain key points to be noted from the Figures, the suppression depends on the values of c_e^2 and w. For instance, a bigger value of $w = -0.8$, gives more suppression than a smaller value $w = -0.9$, implying that the more EOS deviates from the value ΛCDM (w = -1) the suppression in CDM power spectrum increases. Also, the effect of different values of c_e^2 (as seen in Figures 1 and 2) is profound only at the intermediate length scales of k, whereas for scales much higher and smaller than the Hubble scale, the value of c_e^2 is irrelevant. The reason why we see the effect of c_e^2 at intermediate scales is that the perturbation in DE grows only at scales greater than the sound horizon (which, for the case of $c_e^2 = 0.001$, it is $k = 0.07\ h/Mpc$, for other cases of c_e^2 considered here, sound horizon will be greater than this value). For scales much smaller than Hubble scale, say ($k \gg 0.001\ h/Mpc$), the effect of c_e^2 is inappreciable, as the perturbation in DE is negligibly smaller in comparison to that in dark matter. The DE can be approximated to be homogeneously distributed at these scales. On the other hand, at scales much larger than the Hubble radius ($k \ll 0.001\ h/Mpc$), it is the adiabatic perturbations of DE, which contribute to the CDM power spectrum, hence, the effective ESS of DE does not influence the CDM perturbations. Also, it turns out that the variation of an evolving c_e^2 does not have any considerable effect on CDM power spectrum, whereas the different evolution of w will have much more significant influence [8]. Hence, exact information about the EOS, and ESS will help in better understanding the nature of DE.

References

[1] Riess A G *et al* 1998 *Astron. J.* **116** 1009, Perlmutter S *et al* 1999 *Ap. J.* **517** 565
[2] Sahni V 2004 *Lecture Notes in Physics* (Springer Verlag, Berlin, Germany)
[3] Copeland E J, Sami M and Tsujikawa S 2006 *Int. J. Mod. Phy. D* **15** 1753
[4] Li M, Li X, Wang S and Wang Y 2011 *Commun. Theor. Phys.* **56** 525
[5] Padmanabhan T 2002 *Phys. Rev. D* **66** 021301, Feinstein A 2002 *Phys. Rev. D* **66** 063511
[6] de Putter R, Huterer D and Linder E V 2010 *Phys. Rev. D* **81** 103513
[7] DeDeo S, Caldwell R R and Steinhardt P J 2003 *Phys. Rev. D* **67** 103509
[8] Rizwan ul H A and Unnikrishnan S [arXiv:1104.4609]

Halo shapes, initial shear field, and cosmic web

G Rossi

School of Physics, Korea Institute for Advanced Study, 85 Hoegiro, Dongdaemun-Gu, Seoul 130 – 722, Korea
CEA, Centre de Saclay, Irfu/SPP, F-91191 Gif-sur-Yvette, France, and
Paris Center for Cosmological Physics (PCCP) and Laboratoire APC, Université Paris 7, 75205 Paris, France

E-mail: graziano@kias.re.kr, graziano.rossi@cea.fr

Abstract. The ellipsoidal collapse model, combined with the excursion set theory, allows one to estimate the shapes of dark matter halos as seen in high-resolution numerical simulations. The same theoretical framework predicts a quasi-universal behaviour for the conditional axis ratio distributions at later times, set by initial conditions and unaltered by non-linear evolution. The formalism for halo shapes is also useful in making the connection with the initial shear field of the cosmic web, which plays a crucial role in the formation of large-scale structures. The author has briefly discussed the basic aspects of the modelling, as well as the implications of a new formula for the constrained eigenvalues of the initial shear field, given the fact that positions are peaks or dips in the corresponding density field – and not random locations. This formula leads to a new generalized excursion set algorithm for peaks in Gaussian random fields. The results highlighted, here, are relevant for a number of applications, especially for weak lensing studies and for devising algorithms to find and classify structures in the cosmic web.

1. Introduction

The gravitational clustering of matter, as seen in large-volume N-body simulations such as the new Horizon Runs [1], show a characteristic anisotropic web-like structure known as the *cosmic web*. Figure 1 exemplifies a snapshot of the matter density field through the Horizon Run 2 at the present epoch, which clearly reveals the typical pattern of halos, filaments and sheets, connected in a network, with voids encompassing the volume in between. A zoom into a smaller region of the large survey volume shows that individual dark matter halos are, indeed, triaxial in shape. This is not surprising, as the statistics of Gaussian random fields implies that spherically symmetric configurations are highly unlikely [2]. However, defining a gravitationally bound object in a simulation is still a matter of debate, and different methods for defining virialized halos give rise to very different statistical shape distributions. This is probably, why most analytic models of structure formation still make the assumption that gravitationally bound objects are spherical, despite dark matter halos are rather elongated, close to prolate in the central parts and rounder in the outskirts (see, for example [3]). In the first part of this paper (Section 2), we have summarized the main results of a new model presented in [4, 5], which allows one to describe the triaxial distribution of halo shapes at any given time, starting from first principles. In particular, a remarkable feature of the model is the ability to explain the universality of the conditional axis ratio distributions at later times. In the second part (Section 3), we have aimed at making the connection with the statistics of the cosmic web. This is

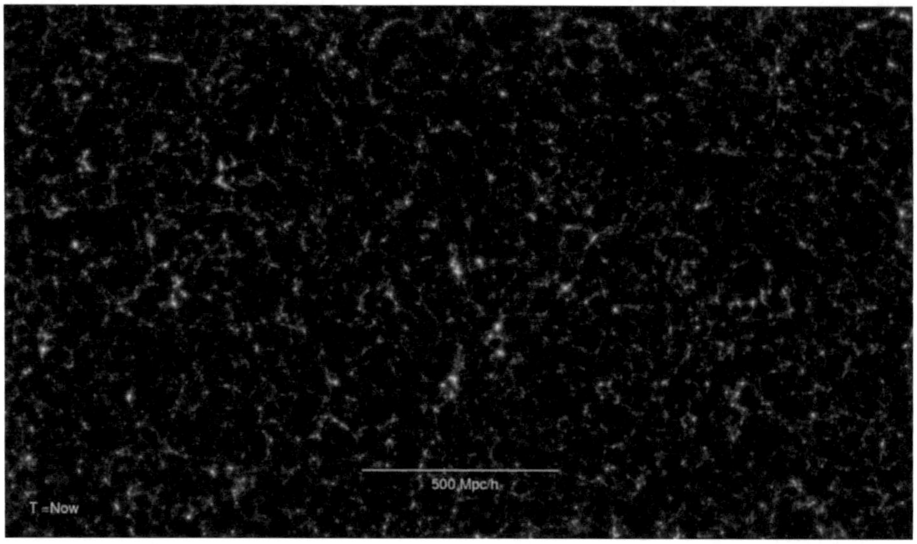

Figure 1. A snapshot of the matter density field from the Horizon Run 2 at the present time [1], which reveals the characteristic *cosmic web* pattern. (Figure courtesy of Juhan Kim.)

achieved by discussing the importance of a new formula for the constrained eigenvalues of the initial shear field presented in [6], which leads to a new generalized excursion set algorithm for peaks [7]. Section 4 concludes with a short highlight of the main findings, and with a description of ongoing and future applications.

2. Halo shapes

The model for triaxial halo shapes developed in [4, 5] has two independent parts: A scheme for how an initially spherical patch evolves and virializes, and a procedure to assign the correct initial shapes to halos of different masses. While the second part is simply given by the statistics of Gaussian random fields, the first one is based on the ellipsoidal collapse framework proposed by [8]. In the ellipsoidal model, the time required to collapse depends not only on the initial over-density $\delta_{\rm i}$ and size $R_{\rm i}$ (or variance $\sigma_{\rm i}^2$) of the patch, but also on the surrounding shear field described by two shape parameters: ellipticity and prolateness (e, p). The essence of the model has been summarized in Figure 2, where 10000 halos of fixed mass $M = 10^{13.5} M_{\odot}$ are evolved from $z = 39$ (left panel) till $z = 0$ (central panel), via a numerical implementation of the ellipsoidal collapse scheme. Comparison with high-resolution simulations, such as those of [3] has shown that results are accurate within the selected mass range (right panel), but significant differences remain at lower and higher masses; reasons for these discrepancies are discussed in detail in [5].

A remarkable feature of the proposed theory is the ability to describe the universality of the conditional distributions of halo shapes, for which simulations also have shown to be quasi-universal (i.e., independent of mass and epoch). This is mainly the reflection of initial conditions, given by the statistics of Gaussian random fields. In particular, if A_1, A_2, A_3 are the scale factors for the principal axes of the ellipsoidal perturbation (ordered as $A_1 \leq A_2 \leq A_3$), and $\delta_{\rm i}$ and e are fixed, then it is easy to show that A_1/A_2 is only a function of prolateness. Hence, for a fixed

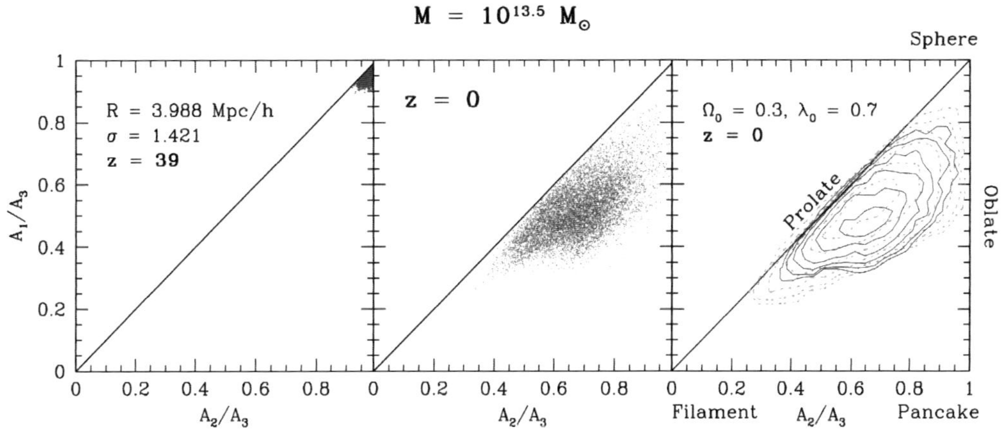

Figure 2. Essence of the halo shape model presented in [4, 5]. Halos of fixed mass $M = 10^{13.5} M_\odot$ and axis ratios A_1/A_3, A_2/A_3 (i.e., $A_1 \leq A_2 \leq A_3$) are evolved from $z = 39$ (left panel) till $z = 0$ (central panel) with a numerical implementation of the ellipsoidal collapse. The right panel compares the model (solid line) with measurements from numerical simulations [3].

mass, if one defines $A_{12} \equiv A_1/A_2$ and $A_{13} \equiv A_1/A_3$, then,

$$p(A_{12}|A_{13}, \delta, \sigma) = \frac{3}{2(1 - A_{13})} \left[1 - \frac{(2A_{12} - 1 - A_{13})^2}{(1 - A_{13})^2} \right] \exp\Big\{ - \frac{5}{8\sigma^2}(2A_{12} - 1 - A_{13})^2 \Big\}. \quad (1)$$

At small masses, the exponential term $\rightarrow 1$, making this distribution almost universal. This remarkable feature is extremely useful for interpreting various results on halo shapes.

3. Initial shear field

The previous formalism for halo shapes allows one to make the connection with the initial shear field of the cosmic web [6], and with the theory of peaks in Gaussian random fields developed in [9]. In fact, directly related morphological quantities such as e and p are crucial in characterizing the structural properties of halos, voids, sheets and filaments. A central result presented in [6] is a formula which extends Doroshkevich's unconditional relations [2] for the eigenvalues of the linear tidal field, by including correlations between the density and tidal shear tensors. The strength of their correlation is en-capsulated in the coefficient γ, the same factor, which is fundamental in peak theory [9]. In particular, if ζ_i (with $i = 1, 2, 3$) are the constrained eigenvalues of the density matrix, then,

$$p(\zeta_1, \zeta_2, \zeta_3|\gamma) = \frac{15^3}{8\sqrt{5}\pi} \frac{1}{(1 - \gamma^2)^3} \exp\left[- \frac{3}{2(1 - \gamma^2)}(2K_1^2 - 5K_2) \right] |(\zeta_1 - \zeta_2)(\zeta_1 - \zeta_3)(\zeta_2 - \zeta_3)|,$$

where

$$K_1 = \zeta_1 + \zeta_2 + \zeta_3, \quad and \quad (2)$$

$$K_2 = \zeta_1\zeta_2 + \zeta_1\zeta_3 + \zeta_2\zeta_3, \quad (3)$$

and $\zeta_i = \lambda_i + \gamma\xi_i$, with λ_i's and ξ_i's the corresponding unconstrained eigenvalues of the shear tensor and Hessian matrix respectively (see also [6] for more details).

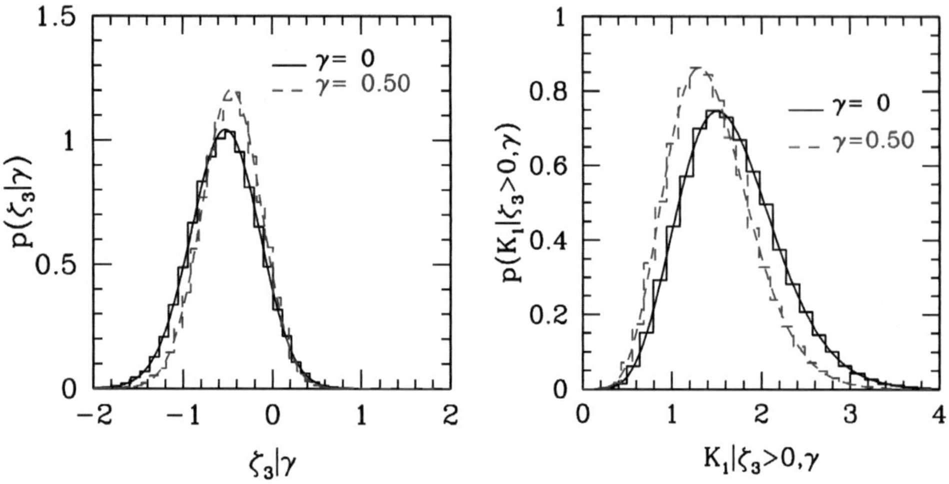

Figure 3. Conditional distributions $p(\zeta_3|\gamma)$ (left) and $p(K_1|\zeta_3 > 0, \gamma)$ (right), from 50000 mock realizations, when $\gamma = 0, 0.5$ (histograms). Curves are theoretical predictions from [6].

Eq. (2) permits to extend several results obtained in [9]; in addition, it naturally suggests a new generalized excursion set algorithm for peaks, which allows one to test the various relations derived in [6] against mock data (this analysis is presented in [7]). An example is shown in Figure 3, where 50000 realizations are used to successfully confirm the analytic expressions for the conditional distributions $p(\zeta_3|\gamma)$ and $p(K_1|\zeta_3 > 0, \gamma)$.

4. Conclusions

The theoretical model for predicting halo shapes described in Section 2 and summarized in Figure 2 has a broad spectrum of applications in cosmology, from weak lensing to the study of early-type galaxies (see [5], for more details). A related aspect is the connection of this theoretical formalism with the initial shear field of the cosmic web, and with the theory of peaks in Gaussian random fields [6, 7, 9]. To this end, in Section 3 we have discussed a new formula for the constrained eigenvalues of the initial shear field presented in [6], which accounts for the fact that halos (voids) may correspond to maxima (minima) of the density field; this leads to a new generalized excursion set algorithm for peaks [7], an example of which is provided in Figure 3. Ongoing and future research directions include applications to the skeleton of the cosmic web, extensions to non-Gaussian fields, structure reconstructions on the basis of tessellations, and the analysis of the properties of halos, voids and filaments within this framework. In addition, the outlined formalism is useful for developing algorithms to find and classify structures in the cosmic web, for shape finders, and to study the environmental dependence of galaxy formation and the role of the peculiar gravity field. Current and future weak lensing and galaxy surveys such as the SDSS-III [10] or Euclid [11] will provide useful data to test these models.

Acknowledgements

Many thanks to the organizers of the ICGC-2011 in Goa for the kind invitation, and for a very pleasant meeting, and their hospitality.

References

[1] Kim J, Park C, Rossi G, Lee S M and Gott J R III 2011 *JKAS* **44** 217

Journal of Physics: Conference Series **484** (2014) 012049 doi:10.1088/1742-6596/484/1/012049

[2] Doroshkevich A G 1970 *Astrofizika* **6** 581
[3] Jing Y P and Suto Y 2002 *Ap. J.* **574** 538
[4] Rossi G 2008 *Ph.D. Thesis, University of Pennsylvania, Publication Number: AAT 3328641*
[5] Rossi G, Sheth R K and Tormen G 2011 *MNRAS* **416** 248
[6] Rossi G 2012 *MNRAS* **421** 296
[7] Rossi G 2013 *MNRAS* **430** 1486
[8] Bond J R and Myers S T 1996 *Ap. J. Supp. Ser.* **103** 1
[9] Bardeen J M, Bond J R, Kaiser N and Szalay A S 1986 *Ap. J.* **304** 15
[10] York D G, Adelman J, Anderson J E. *et al* 2000 *Astron. J.* **120** 1579
[11] Cimatti A, Cassata P, Pozzetti L *et al* 2008 *A. & A.* **482** 21

The luminosity, colour and morphology dependence of galaxy structures in the Sloan Digital Sky Survey

P Sarkar[1], B Pandey[2,4] and S Bharadwaj[3]

[1] IUCAA, Post Bag 4, Ganeshkhind, Pune University Campus, Pune 411007, India
[2]Max-Planck Institute for Astrophysics, Karl-Schwarszchild Str. 1, D85748, Garching, Germany
[3]Department of Physics and Meteorology, and Centre for Theoretical Studies, IIT, Kharagpur, 721302, India
[4]Department of Physics, Visva-Bharati University, Santiniketan, Birbhum, 731235, India

E-mail: [1]`sarkar@iucaa.ernet.in`

Abstract. It is possible to visualize the Cosmic Web as an inter-connected network of one dimensional filaments, two dimensional sheets and three dimensional volume filling structures, which we refer to as clusters. We have considered the Local Dimension D, as a tool to locally quantify the shape in the neigbourhood of different galaxies along the Cosmic Web. We expect $D \sim 1$, 2 and 3 for a galaxy located in a filament, sheet and cluster respectively. We have analysed the Cosmic Web using Local Dimension in a LCDM N-body simulations and to a three dimensional volume limited galaxy sample from the Sloan Digital Sky Survey (SDSS). We now want to test for luminosity, colour and morphology dependence of the galaxy structures (filaments, sheets and clusters) in three dimensional volume of SDSS data.

1. Introduction

Galaxy redshift surveys map the distribution of galaxies in the universe, and tell us directly how the present universe looks like. All galaxy redshift surveys like the Center for Astrophysics (CfA) Survey [1], Las Campanas Redshift Survey (LCRS) [2], Two-Degree Field Galaxy Redshift Survey (2dFGRS) [3], and Sloan Digital Sky Survey (SDSS) [4] clearly demonstrate that galaxies are distributed in an inter-connected complex network of filaments, sheets and clusters encircling nearly empty voids. This network of galaxies is often referred as the "Cosmic Web" [5]. Quantifying the Cosmic Web and understanding its origin is one of the most interesting and challenging issues in cosmology.

It is now quite well accepted that galaxies with different physical properties are differently distributed in space. Studies over several decades have established that ellipticals and spirals are not distributed in the same way. The ellipticals are found predominantly inside regular, rich clusters, whereas the field galaxies are mostly spirals. This effect is referred to as "morphological segregation", which has been studied extensively in the literature. The dependence of clustering on galaxy properties like the luminosity, colour and morphology provides very important inputs for theories of galaxy formation.

In this paper, we have studied the luminosity, colour and morphology dependence of galaxy filaments in SDSS DR4 and morphology dependence of filaments in the Sloan Digital Sky Survey

Data Release Four (SDSS DR4). This allows us to study the distribution of different types of galaxies on the largest length-scales possible.

2. The Local Dimension

The Cosmic Web, as defined above, is an inter-connected network of filaments, sheets and clusters. We have used a simple but effective quantifier, the Local Dimension (introduced in [6] and used in [7]) to quantify these different structural elements in the Cosmic Web. The Local Dimension is based on the following argument: If we consider, for example, a particular galaxy "G" as centre of a sphere of radius R, then the count $N(< R)$ is the number of galaxies inside that sphere. It is expected that $N(< R) \propto R$. Now, if we consider a galaxy located in a sheet, as centre, we put sphere of radius R, and the count $N(< R)$ is the number of galaxies. We expect that $N(< R) \propto R^2$. Similarly, we can say for a cluster $N(< R) \propto R^3$. If we generalize this concept, then we can say that $N(< R) \propto R^D$, where D is termed as the Local Dimension. The values of $D = 1$, 2 and 3 will represent a galaxy located in a filament, sheet and cluster respectively. It is expected that the Cosmic Web looks different when viewed at different length-scales. This scaling behaviour can be easily studied using the Local Dimension over different length-scales $R_1 \leq R \leq R_2$. It is important to study the luminosity, colour, and morphology dependence with these different structural elements estimated by the Local Dimension at a particular length-scale $R_1 \leq R \leq R_2$.

3. Data and method of analysis

3.1. SDSS DR6 data

The analysis presented, here, is based on galaxy redshift data from the Main Galaxy Sample of SDSS DR6 [8]. The data set contains photometric information of ~ 290 million objects over 9583 square degrees and spectroscopic information of about $790,860$ galaxies over 7425 square degrees. We have used the Main Galaxy Sample, for which the target selection algorithm is detailed in [9]. The Main Galaxy Sample comprises of galaxies brighter than a limiting r band Petrosian magnitude 17.77. We have identified a contiguous region in the Northern Galactic Cap, which spans $50° < \lambda < 30°$ and $6° < \eta < 35°$, where λ and η are survey coordinates defined in [10]. A volume limited galaxy sub-sample was constructed in this region by restricting the extinction corrected Petrosian r band apparent magnitude to the range $14.5 \leq m_r \leq 17.77$, and restricting the absolute magnitude to the range $20 \leq M_r \leq -19$. This gives us 26108 galaxies in the redshift range $0.035 \leq z \leq 0.076$, which corresponds to the comoving radial distance range $104 \leq r \leq 223\ h^{-1}\ Mpc$.

We have studied the luminosity, colour and morphology dependence of the above volume limited sub-sample. When testing for luminosity dependence, all the galaxies are to be classified as either bright or faint. We have determined a value $M_{r,c} = -19.458$ for the luminosity, which divides the sample into equal number of faint (i.e., $M_r \leq M_{r,c}$), and bright (i.e., $M_r > M_{r,c}$) galaxies. The galaxy $u - r$ colour is known to have bimodal distribution [11]. In our analysis, we have determined a value $(u - r)_c = 2.08$ for the colour into equal number of red (i.e., $u - r > (u - r)_c$), and blue (i.e., $u - r \leq (u - r)_c$) galaxies. The morphological classification was carried out using the concentration index defined as $C_i = r_{90}/r_{50}$, where r_{90} and r_{50} are the radii containing 90% and 50% of the Petrosian flux respectively. We have chosen a cutoff $C_{i,c} = 2.50$, which partitions the galaxies into spirals (i.e., $C_i \leq C_{i,c}$) and ellipticals (i.e., $C_i > C_{i,c}$).

3.2. Method of analysis

We have chosen a particular galaxy as centre, and determined the number of other galaxies $N(< R)$ inside a sphere of comoving radius R. This counting process is repeated by varying R. We have assigned Poisson error $\Delta N(< R) = \sqrt{N(< R)}$ to each measured value of $N(< R)$. For each centre, we have determined the power law $N(R) = AR^D$ that provides the best fit to the

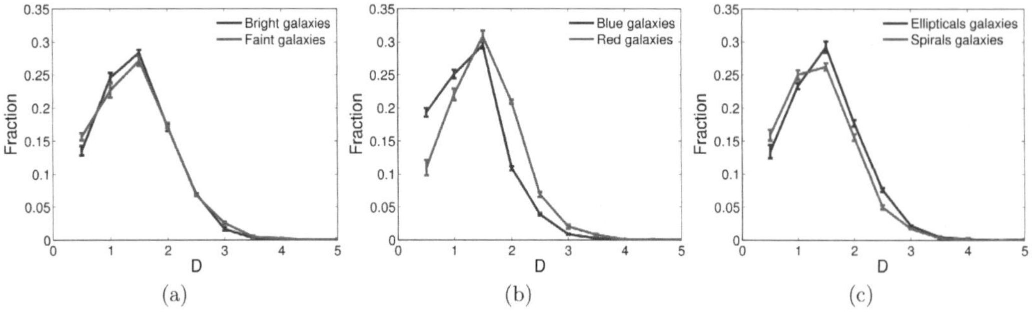

Figure 1: This shows the fraction of galaxies with a particular D values correspond to different samples based on luminosity, colour and morphology. The bins in D have size ± 0.25.

data. The fitting procedure was restricted to values within the range $0.5 \le R \le 5\ h^{-1}\,\mathrm{Mpc}$ in length-scale. The value of D is accepted as the Local Dimension corresponding to the particular centre if the chi-square per degree of freedom of the power law fit satisfies $\chi^2/\nu \le 1$. We have applied the above analysis to different sub-samples constructed by using the above cuts. The cosmic variance of the data has been estimated using bootstrap re-sampling of the data, and we have used ten bootstrap samples.

4. Results

We have first studied the comparison of different samples based on luminosity. Figure 1 (a) shows the fraction of centre with different D values. The D values in the interval 1 ± 0.5, 2 ± 0.5, and 3 ± 0.5 have respectively been binned as $D \sim 1, 2$ and 3. The centres, for which a D value could not be determined were discarded. The red curve shows the distribution for faint galaxies, while the blue curve shows for the bright galaxies. We have not found any luminosity dependence with galaxy structures.

Figure 1 (b) shows the fraction of centres with different D values. The blue curve represents blue galaxies, while the red curve represents the distribution of red galaxies. We have found that there is a statistically significant colour dependence. Red galaxies dominates at higher dimension as compared to blue galaxies, which are mostly distributed in $D = 1$.

Figure 1 (c) shows the fraction of centres with different D values. The blue curve shows the distribution for elliptical galaxies, while the red shows the same for spiral galaxies. We have found significant morphology dependence on the galaxy structures. Elliptical galaxies dominate more on the filamentary region as compared to the spirals galaxies. The spiral galaxies are more dominant on the filamentary region as compared to the elliptical galaxies.

5. Conclusion

We have noticed that the filamentary pattern seen in the galaxy distribution exhibits statistically significant colour and morphology dependence, but not luminosity dependence. We have found that the blue and the spiral galaxy dominates at filamentary region as compared to red and the elliptical galaxies. The colour and morphology are very strongly correlated galaxy properties, the red galaxies begin predominantly elliptical galaxies, and the blue ones are spirals. It is well known that the ellipticals are found in clusters, where the spirals are distributed over the filaments [12]. Our finding is consistent with the above picture of galaxy distribution in the Cosmic Web. We can interpret in terms of pressure and density of the galaxy environments. It is well known that cluster of galaxies has relatively high density ($\delta \sim 1000$) compared to field galaxies. At this high density region and pressure, the cluster galaxies strip each other of their interstellar gas and lose their ability to form new stars (elliptical galaxies). However, a better description of the relation between galaxy properties and the galaxy formation is an important issue, which we plan to address in future.

References

[1] Geller M J and Huchra J P 1989 *Science* **246** 897
[2] Shectman S A, Landy S D, Oemler A *el al* 1996 *Ap. J.* **470** 172 [arXiv:astro-ph/9604167]
[3] Colless M, Dalton G, Maddox S *et al* 2001 *MNRAS* **328** 1039 [arXiv:astro-ph/0106498]
[4] York D G, Adelman J, Anderson Jr J E *et al* 2000 *Astron. J.* **120** 1579 [arXiv:astro-ph/0006396]
[5] Bond J R, Kofman L and Pogosyan D 1996 *Nature* **380** 603 [arXiv:astro-ph/9512141]
[6] Sarkar P and Bharadwaj S 2009 *MNRAS* **394** L66 [arXiv:0812.1800]
[7] Sarkar P, Pandey B and Bharadwaj S 2011 [arXiv:1105.4469]
[8] Adelman-McCarthy J K, Agüeros M A, Allam S S *et al* 2008 *Ap. J. Supp. Ser.* **175** 297 [arXiv:0707.3413]
[9] Strauss M A, Weinberg D H, Lupton R H *et al* 2002 *Astron. J.* **124** 1810 [arXiv:astro-ph/0206225]
[10] Stoughton C, Lupton R H, Bernardi M *et al* 2002 *Astron. J.* **123** 485
[11] Strateva I, Ivezić Ž, Knapp G R *et al* 2001 *Astron. J.* **122** 1861 [arXiv:astro-ph/0107201]
[12] Dressler A 1980 *Ap. J.* **236** 351

Rydberg atom in gravity

A Agrawal[1]
Indian Institute of Technology, New Delhi 110016, India
E-mail: aniket.agrawal@physik.lmu.de

Abstract. The local position invariance of a physical system is examined using a Rydberg atom and the universality of free fall is found to be invalid for a quantum system. A Rydberg atom is analysed in Newtonian gravity and curved space. The energy is found to vary as n^2 for very large values of the principal quantum number n. The change in energy is calculated using this formalism and compared to a similar calculation by Chiao. The value that we have got from our calculation is found to be 6 orders higher in magnitude than Chiao's value. These results can be of significance in gravitational redshift experiements proposed by Muller *et al* and Wolf *et al*

1. Introduction

In the 16th century, Galileo performed his famous experiment of dropping two iron balls of different weights from the leaning tower of Pisa. Scientists have, since then, time and again used this classic experiment to demonstrate the equality of objects under gravitational attraction. This equality has come to be known as *The Equivalence Principle*. Since the inception of general relativity in 1917, the equivalence principle has become one of the best tested principles in physics and also a subject of much debate. Modern physics recognizes three major forms of this principle:

(i) **Weak Equivalence Principle**: Which states that two point objects which have the same mass, have equal accelerations under gravity.
(ii) **Einsteinian Equivalence Principle**: The outcome of any local non-gravitational experiment in a freely falling laboratory is independent of the velocity of the laboratory and its location in spacetime.
(iii) **Strong Equivalence Principle**: Which requires both the above conditions to hold for any theory of gravity.

The equivalence principle has been tested to accuracies of about 10^{-13} [1].

Gravitational redshift experiments measure the frequency of an oscillating system at two different locations in a gravitational field and use the change in frequencies to predict the redshift. However, these measurements are assumed to be independent of the size of the system, thus, being independent of the tidal gravitational forces. For quantum systems, we have shown that these measurements also depend on the tidal gravitational forces, which appear as a shift in the frequency of transitions. In this paper, we have shown that if such experiments are carried out

[1] Present Address: Sektion Physik, Ludwig Maximilians University, Schellingstr. 4, Munich, Germany

in highly elliptical orbits in space, tidal gravitational effects can measurably affect the results of these experiments. Therefore, highly accurate atom interferometry experiments must take these effects into account. We have examined the equivalence principle in this paper and concentrated especially on the implications of size on the equivalence principle. We have presented below two approaches to the problem: (i) An approach used by Chiao [2] to show that universality of free fall may not be true for quantum systems, and (ii) A Newtonian and curved space analysis, which shows that tidal gravity changes the frequency of transitions in Rydberg atoms and the effect is more significant than claimed by Chiao.

2. Quantum incompressibility of a falling Rydberg atom

If we leave two point-like objects for a free fall, they will converge towards the centre of the earth. But an extended quantum object such as Hydrogen atom prepared in a high principal quantum number state is predicted to fall more slowly than a classical point like object. This can be proved quantitatively as follows: Let us first consider the single electron of a circular Rydberg atom (ignoring electron spin), which is prepared in the state:

$$|n, l = n-1, m = n-1\rangle,$$

where n is a very large principal quantum number, $l = n - 1$ is the maximum possible orbital angular momentum quantum number for a given n and $m = l$ is the maximum possible azimuthal quantum number for a given l, i.e., the 'stretch' state. We have now considered what happens when such an atom falls under Earth's gravity.

2.1. Using the DeWitt gravitational vector potential [2]

The DeWitt Hamiltonian for a freely falling hydrogen atom in the presence of weak electromagnetic and gravitational field is given by:

$$\hat{H} = \frac{1}{2m}(\mathbf{p} - m\mathbf{h} - e\mathbf{A})^2 + \frac{e^2}{4\pi\epsilon_0 r}, \tag{1}$$

where $\mathbf{h}$ is the DeWitt gravitational vector potential and $\mathbf{A}$ is the electromagnetic vector potential. Let us consider the case when weak tidal gravitational fields are present without any electromagnetic fields, that is, when $\mathbf{A}= 0$ and $\mathbf{h}\neq 0$. The horizontal tidal gravitational field experienced by the atom, as observed by a distant inertial observer is given by:

$$\mathbf{h}(x, y, t) = \mathbf{v}(x, y, t) = \mathbf{g}'t = \frac{gt}{R_E}(\mathbf{e}_x x + \mathbf{e}_y y), \tag{2}$$

where $\mathbf{v}(x,\ y,\ t)$ is the velocity of a freely falling, point-like test particle located at (x, y) and observed at time t by the distant inertial observer, $\mathbf{g}'$ is the horizontal component of the Earth's gravitational acceleration arising from the radial convergence of free-fall trajectories towards the centre of the Earth, and $\mathbf{e}_x$ and $\mathbf{e}_y$ are respectively the unit vectors pointing along the x and y axes, in this observer's coordinate system.

Therefore, the shift in energy of the atom in the circular Rydberg state due to the Earth's tidal gravitational fields is given in first-order perturbation theory [3] by:

$$\Delta E_{\mathbf{h}.\mathbf{h}} = \frac{mg^2t^2}{2R_E^2}\langle \Psi_{n,n-1,n-1}|x^2 + y^2|\Psi_{n,n-1,n-1}\rangle \approx \frac{ma_n^2}{2R_E^2}g^2t^2. \tag{3}$$

It is easy to calculate the value of t required for an object to fall from a height of (say) 10 km. The value of t is found to be ≈ 45.18 s, from which the change in energy ΔE for an electron with $n = 100$ is found to be $\approx 10^{-56}$ J. The energy shift arising from the tidal gravitational perturbations leads to a force on the atom given by:

$$\nabla(\Delta E_{\mathbf{h.h}}) \approx -\frac{1}{2} m a_n^2 t^2 \nabla \left(\frac{g^2}{R_E^2} \right),$$

which causes the atom to fall more slowly compared to a point object.

2.2. Using the Newtonian gravitational potential

Now, consider a hydrogen atom at a height h from the Earth's centre, under Earth's gravitation. The electron of the atom experiences a gravitational potential, given as:

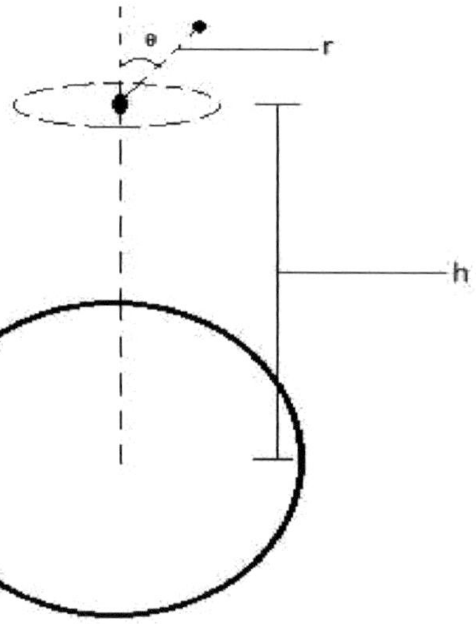

Figure 1. Hydrogen atom at height h from the Earth's centre, and electron at radius r

$$V = \frac{GM_E \cdot m_e}{h + r\cos\theta}, \tag{4}$$

where G is Newton's gravitational constant, M_E is the mass of the Earth and m_e is the mass of electron. Assuming $h \gg r\cos\theta$, we can expand the above potential to quadrupole order as:

$$V = \frac{GM_E \cdot m_e}{h} \left[1 - \frac{r\cos\theta}{h} + \left(\frac{r\cos\theta}{h} \right)^2 \right]. \tag{5}$$

Considering the effects of this potential as a perturbation to the electron, we have found that the first term gives a constant contribution to all levels and the second term has a zero contribution. We have now looked at the contribution from the third term which represents the quadrupole action of gravity. The change in energy of the electron in the nth 'stretch' orbit due to this term is given by:

$$\Delta E_n \propto \frac{1}{h^2} \langle \Psi_{n,n-1,n-1} | (r\cos\theta)^2 | \Psi_{n,n-1,n-1} \rangle. \tag{6}$$

The wavefunction $\Psi_{n,n-1,n-1}$ for the hydrogen atom is given by:

$$\Psi_{n,n-1,n-1} = R_{nl}(r) \cdot Y_l^m(\theta, \phi),$$

where

$$R_{nl}(r) = \frac{1}{r} c_0 \left(\frac{r}{na}\right)^{l+1} e^{-(r/na)}, \quad and$$

$$Y_l^m(\theta, \phi) = \sqrt{\frac{(2l+1)}{4\pi} \frac{(l-|m|)!}{(l+|m|)!}} P_l^m(\cos\theta) e^{im\phi}.$$

Thus, we can write the change in energy as:

$$\Delta E_n = \langle \Psi_{n,n-1,n-1} | V | \Psi_{n,n-1,n-1} \rangle = \frac{GM_E \cdot m_e}{h} \left[1 - \frac{\alpha}{h^2}\right], \tag{7}$$

where

$$\alpha = 2\pi \int_0^{\pi} \int_0^{\infty} c_0^2 \left(\frac{r}{na}\right)^{2(l+1)} \frac{e^{-2(r/na)}}{r^2} r^2 dr \sin\theta d\theta \, r^2 \cos^2\theta \left[P_l^m(\cos\theta)\right]^2 \frac{(2l+1)}{4\pi} \frac{(l-|m|)!}{(l+|m|)!}. \tag{8}$$

For $l = m > 0$, we have:

$$\alpha = \frac{(2l+1)}{8} (na)^2 (2l+4)(2l+3) \left[\frac{4l}{(2l+1)^2(2l-1)} + \frac{2}{(2l+1)(2l+3)}\right], \tag{9}$$

where a is the Bohr radius $= 0.529$ angstrom.

Considering an atom at the same height, i.e., 10 km, and $n = 100$, we get $\Delta E_n \approx 10^{-50}$ J, which is 6 orders higher in magnitude compared to Chiao's result calculated above.

2.3. Hydrogen atom in curved space [4]

We have now considered a hydrogen atom in a curved space, and a positive curvature space for which the metric is given by:

$$ds^2 = R^2 d\alpha^2 + R^2 \sin^2\alpha (d\theta^2 + \sin^2\theta d\phi^2).$$

We have considered that the electron moves in this space under the action of an electrostatic potential, which must satisfy the Laplace's equation:

$$\Delta V = g^{ij} V_{,ij} = 0, \tag{10}$$

where

$$\Delta = \frac{\partial}{\partial x^\alpha} \sqrt{g} g^{\alpha\beta} \frac{\partial}{\partial x^\beta}, \quad and \quad g = |g_{ij}| = R^6 \sin^4\alpha \sin^2\theta. \tag{11}$$

For the above metric, the Laplace's equation takes the form:

$$\frac{\partial}{\partial\alpha} \sqrt{g} \frac{\partial V}{\partial\alpha} + \frac{\partial}{\partial\theta} \sqrt{g} \mathrm{cosec}^2\alpha \frac{\partial V}{\partial\theta} + \frac{\partial}{\partial\phi} \sqrt{g} \mathrm{cosec}^2\alpha \mathrm{cosec}^2\theta \frac{\partial V}{\partial\phi} = 0. \tag{12}$$

This simplifies to:

$$\left[\frac{\partial^2 V}{\partial\alpha^2} + 2\cot\alpha \frac{\partial V}{\partial\alpha}\right] + \mathrm{cosec}^2\alpha \left[\frac{\partial^2 V}{\partial\theta^2} + \cot\theta \frac{\partial V}{\partial\theta}\right] + \mathrm{cosec}^2\alpha \mathrm{cosec}^2\theta \frac{\partial^2 V}{\partial\phi^2} = 0. \tag{13}$$

Now, we have considered a spherically symmetric potential, which is independent of θ and ϕ. Thus, we are left with:

$$\frac{\partial^2 V}{\partial\alpha^2} + 2\cot\alpha \frac{\partial V}{\partial\alpha} = 0, \tag{14}$$

from which, the potential is found to be:

$$V = C_1 \cot \alpha + C_2. \tag{15}$$

We have solved the Schrödinger's equation:

$$\frac{-\hbar^2}{2m_e}\frac{\partial}{\partial x^\alpha}\sqrt{g}g^{\alpha\beta}\frac{\partial \psi}{\partial x^\beta} + V(x^\alpha)\psi = E\psi \tag{16}$$

to get [4]:

$$E_n = C + A\left(-\frac{1}{n^2} + (n-1)(n+1)\frac{a^2}{R^2}\right). \tag{17}$$

3. Results

We have concluded that the energy of the electron in the hydrogen atom under gravity, depends on its principal quantum number, which is also seen from the curved space analysis. Also, our analysis predicts that the change will be about 6 orders higher in magnitude compared to the result predicted by Chiao. Therefore, the effect predicted by him will be masked to a very large extent by the effect calculated above.

We have also concluded that the frequency of transitions in a hydrogen atom will be changed to a measurable extent for atoms used in lab experiments such as that of Muller *et al* [1] and of Wolf *et al* [5] falling onto airless planets/moons, atoms falling into white dwarfs and atoms falling into black holes.

References

[1] Muller *et al* 2010 *Nature* **463** 926
[2] Chiao R Y [arXiv:1003.1149.0]
[3] Griffiths D 2004 *An Introduction to Quantum Mechanics* 2nd Ed (New Jersey: Pearson)
[4] Schrödinger 1940/1941 *Proc. Roy. Iri. Acad., Sec. A: Math. Phys. Sci.* **46** 9
[5] Wolf *et al* 2010 *Nature* **467** E1

Cosmological magnetogenesis

K Atmjeet

Department of Physics and Astrophysics, University of Delhi, Delhi 110007, India

E-mail: katmjeet@physics.du.ac.in

Abstract. Observations indicate the presence of a magnetic field at galactic and cosmological scales. However, the origin of these magnetic fields is not well understood. There is enough motivation to look into the primordial origin of magnetic field, which essentially requires the breaking of conformal invariance of Maxwell's theory. Several mechanisms to generate primordial magnetic field have been proposed. A brief overview of those models has been presented. Central problem of the models within inflationary paradigm has been addressed. Possibilities to generate primordial magnetic field beyond inflationary framework are mentioned. A toy model for bouncing cosmology has been presented to understand the idea of magnetogenesis in such models.

1. Introduction

Universe is magnetized at all scales. Magnetic field of the order of nano G has been observed in intergalactic medium (IGM). This has been confirmed by various measurements, like Faraday rotation measurements and CMBR polarization. The origin of magnetic fields at cosmological scales is an unsolved problem. Inflation becomes the natural choice to produce effects at scales larger than the Hubble horizon. The required strength of the cosmic magnetic field at the epoch of structure formation, adiabatically re-scaled to the present time, should be 10^{-10} to 10^{-7} G. On the other hand, if the galactic dynamo mechanism is at play, then the seed fields of strength 10^{-22} to 10^{-16} G is required at the end of inflation [1, 2]. (We have used the metric signature $(-,+,+,+)$. Natural units have been used for all calculations, i.e., $\hbar = c = G = 1$.)

2. Magnetogenesis in FRW-universe

Consider a homogeneous, isotropic and flat universe described by Friedman-Robertson-Walker (FRW) metric:

$$ds^2 = a^2(\eta)(d\eta^2 - \delta_{ik}dx^i dx^j), \quad where\ \eta = \int \frac{dt}{a(t)}. \tag{1}$$

Here, x, y, z are comoving coordinates and η is the conformal time. The general form of the standard electromagnetic field action is given by:

$$S = \frac{-1}{16\pi}\int F_{\mu\nu}F^{\mu\nu}\sqrt{-g}d^4x, \tag{2}$$

where $F_{\mu\nu} = \partial_\mu A_\nu - \partial_\nu A_\mu$. This action is conformally invariant. It turns out that electromagnetic wave fluctuations can not be amplified in FRW universe, as $B \propto 1/a^2(t)$. Therefore, mechanisms for magnetic field generation require the breaking of conformal invariance of the electromagnetic action. If one can construct a model in which $B \propto 1/a^{\epsilon}$, with typically $\epsilon << 1$, one would have achieved a way to obtain strong magnetic fields.

3. Models within inflationary paradigm

Inflation becomes a natural choice to bring the effects at cosmological scales. There have been several mechanisms proposed to break conformal invariance of electromagnetic action [3, 4, 8]. A wide variety of models can be incorporated together into a general action of the form [5]:

$$S = -\int \left[\frac{1}{16\pi} f^2(\phi) F_{\mu\nu} F^{\mu\nu} + \frac{1}{2} g^{\mu\nu} \partial_\mu \phi \partial_\nu \phi + V(\phi) + M^2 A_\mu A^\mu \right] \sqrt{-g} d^4x, \tag{3}$$

where

$$f(\phi) \to f(\phi, R, R^2, R_{\mu\nu} R^{\mu\nu} ...),$$

M is the mass associated with the massive vector field, and ϕ is the scalar field. Conformal invariance is broken by the presence of a time dependent coupling of electromagnetic field with scalar fields, curvature, and other such terms. This can be achieved also by the presence of a massive vector field. Here, it is important to mention that the mass term is rather put in by hand and not been generated through Higgs mechanism.

3.1. Evolution of magnetic field

For the gauge $A_0 = 0$, $\nabla . A = 0$, we have found that the Fourier coefficients $\mathcal{A} \equiv a(\eta) f(\eta) A(\eta, k)$ satisfy:

$$\mathcal{A}''(\eta, k) + (k^2 - \frac{f''}{f}) \mathcal{A}(\eta, k) = 0. \tag{4}$$

From the above equation, one can see that if f is a constant in time, one gets the usual harmonic oscillator solution, and particularly $f = 1$ gives the results of usual Maxwell's equations. For a growing or decaying mode solution, we require, $f''/f > k^2$. The electric and magnetic fields can be obtained from:

$$E_i = \dot{A}_i, \quad \textit{and} \quad B_i = \frac{1}{a} \epsilon_{ijk} \partial^j A^k. \tag{5}$$

3.2. Solution for power law inflation

We have worked with the form of scale factor as [6, 7]:

$$a(\eta) = a_0 \left[\frac{\eta}{\eta_0} \right]^{1+\beta}, \tag{6}$$

and the coupling function is taken as $f \propto a^{\alpha}$, where α and β are parameters and $\beta = -2$ corresponds to the de Sitter solution. So, the equation for vector field takes the form:

$$\mathcal{A}''(k, \eta) + \left[k^2 - \frac{\gamma(\gamma - 1)}{\eta^2} \right] \mathcal{A}(k, \eta) = 0, \tag{7}$$

where $\gamma = \alpha(1 + \beta)$. At much smaller length scale $(-k\eta >> 1)$, one can neglect the effect of spacetime curvature and the solution will be similar to that found in Minkowski space time, i.e., a plane wave solution. Thus, we have obtained the initial conditions as:

$$\mathcal{A}(k, \eta) \to \frac{1}{\sqrt{2k}} exp(-ik\eta), \quad \textit{and} \quad \mathcal{A}'(k, \eta) \to -\iota \sqrt{\frac{k}{2}} exp(-ik\eta). \tag{8}$$

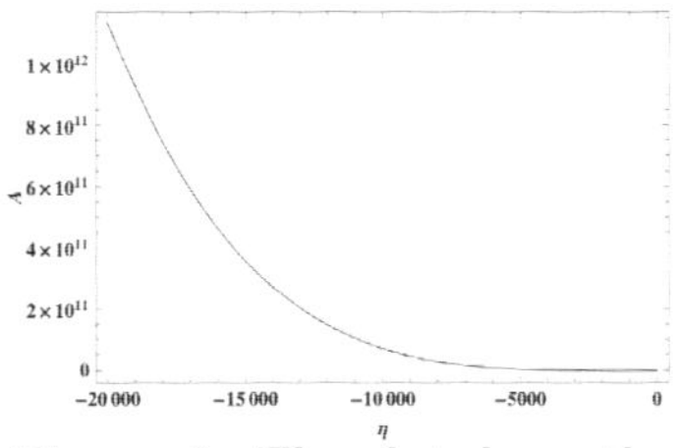

Figure 1. The plot shows the evolution of vector potential vs η for $k = 10^{-2}$

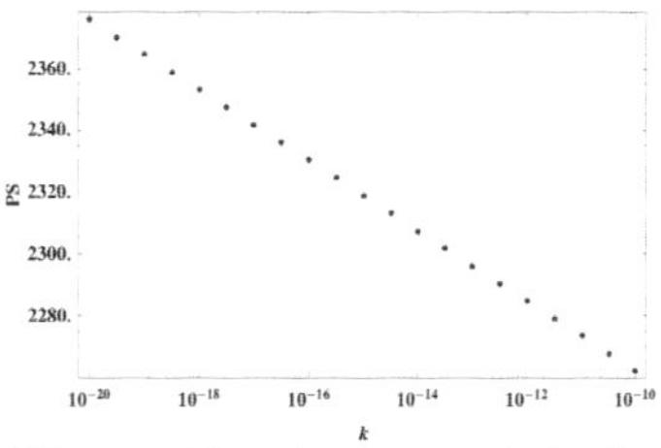

Figure 2. Log-log plot shows nearly scale invariant PSM, for $\gamma = -2$

3.3. Energy-momentum tensor for EM fields

The energy-momentum tensor can be calculated by varying the EM action with respect to background metric. This gives:

$$T_{\mu\nu} = -\frac{2}{\sqrt{-g}}\frac{\delta[\sqrt{-g}L_{EM}]}{\delta g^{\mu\nu}} = \frac{f^2}{4\pi}\left[g^{\gamma\beta}F_{\mu\gamma}F_{\nu\beta} - g_{\mu\nu}\frac{F_{\alpha\beta}F^{\alpha\beta}}{4}\right]. \quad (9)$$

Vacuum expectation values of energy densities for electromagnetic field can be calculated by:

$$\rho_B = \langle 0|T^B_{\mu\nu}u^\mu u^\nu|0\rangle, \quad and \quad \rho_E = \langle 0|T^E_{\mu\nu}u^\mu u^\nu|0\rangle, \quad (10)$$

where, $T^B_{\mu\nu}u^\mu u^\nu = \frac{f^2 B_i B^i}{8\pi}$ and $T^E_{\mu\nu}u^\mu u^\nu = \frac{f^2 E_i E^i}{8\pi}$.

3.4. Scale invariant power spectrum

The power spectrum is defined as: $d\rho/dlnk$. Power spectra for electric (PSE) and magnetic (PSM) fields are obtained from:

$$\frac{d\rho_B}{dlnk} = \frac{1}{2\pi^2}\left(\frac{k}{a}\right)^4 k\,|\mathcal{A}(k,\eta)|^2, \quad and \quad \frac{d\rho_E}{dlnk} = \frac{f^2}{2\pi^2}\frac{k^3}{a^4}\left|\left[\frac{\mathcal{A}(k,\eta)}{f}\right]'\right|^2. \quad (11)$$

It is possible to obtain scale invariant spectrum for magnetic field for $\gamma = -2$ and $\gamma = 3$. In fact, $\gamma = -2$ seems a plausible solution, as it can be shown that electric field decays for this solution. For the other case $\gamma = 3$, the electric field will grow rapidly, which is not an acceptable solution [6, 7].

4. Problems and possibilities

Getting a scale invariant power spectrum is not enough. One has to check whether magnetic field can survive through the expansion. It is shown that in such models, inflation cannot generate a seed magnetic field of sufficient strength. The first reason is the back-reaction. Once we start with small coupling regime, inflation will halt before the universe has expanded to the required number of e-foldings. On the other hand, if we start with large coupling regime, it is possible to generate sufficient amount of magnetic field. But, this is not a very acceptable scenario, as large couplings theories are not trusted at all. The case for massive vector field has also been ruled out [5].

4.1. Magnetogenesis beyond inflationary paradigm

There are possibilities to look within inflationary paradigm, as well as there are enough motivations to look beyond the standard inflationary models. As inflation is not the only outcome of possible cosmological theories, possibilities of generating primordial magnetic fields are being probed into models based upon $f(R)$ (modified gravity theories), higher dimensional cosmology and bouncing cosmological theories [8].

Here, we have presented a toy model for bouncing cosmology with the idea of magnetogenesis.

4.2. *A toy model*

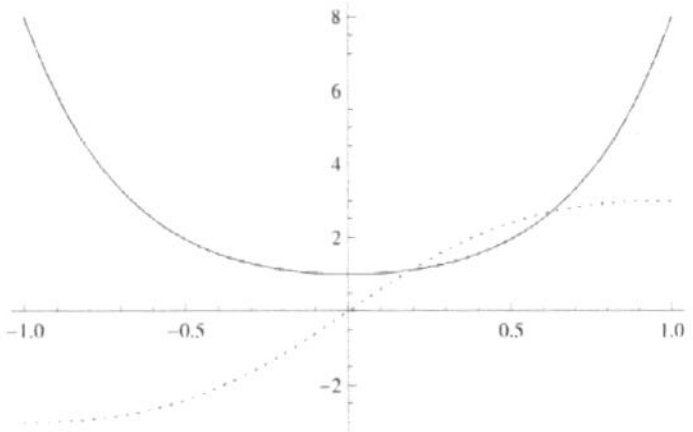

Figure 3. The solid curve shows the evolution of scale factor and the dotted curve correspondingly to Hubble parameter. This toy model represents a symmetric bounce and has no singularity.

A toy model for symmetric bouncing universe is taken as:

$$a(t) = a_0 \left[1 + \left(\frac{t}{t_0}\right)^2\right]^p.$$

Problem of big bang singularity gets resolved automatically in bouncing cosmological models. Universe goes through a contracting phase upto finite size, and then bounces back expanding. This also solves the horizon problem as the modes which are seen outside horizon could be easily brought back inside during the contracting phase. For magnetogenesis, the idea is such that magnetic field can grow rapidly during contracting phase (through flux conservation) and evolve through the bounce. It again decays during the expansion phase as in FRW universe. A more plausible scenario should be asymmetric bounce.

5. Conclusion

Conformal invariance needs to be broken for electromagnetic fields for the generation of magnetic fields. Array of models based upon inflation can produce a seed magnetic field. Most of the models are unable to generate seed of desired magnitude. Back reaction problem is there in almost all models based upon inflation. Conformal invariance can be broken through non-inflationary models also. There are motivations to look beyond the standard cosmological models like bouncing cosmology, f(R), and brane-world cosmology.

Aknowledgement

The author would like to thank UGC, Government of India, for financial support through SRF. He is also thankful to IUCAA Reference Centre, University of Delhi, for providing technical and computational facilities.

References

[1] Brandenburg A and Subramanian K 2005 *Phys. Report* **417** 1
[2] Grasso D and Rubinstein H R 2001 *Phys. Report* **348** 163
[3] Turner M S and Widrow L M 1988 *Phys. Rev. D* **37** 2743
[4] Ratra B 1992 *Ap. J.* **391** L1
[5] Demozzi V, Mukhanov V and Rubinstein H 2009 *JCAP* **08** 025
[6] Subramanian, K 2010 *Astron. Nachr.* **331** 110
[7] Martin J and Yokoyama J 2008 *JCAP* **01** 025
[8] Barrow J D, Maartens R and Tsagas C G 2007 *Phys. Report* **449** 131

Baryon inhomogeneities due to CP violating QCD Z(3) walls

A Atreya[1], A Sarkar[2] and A M Srivastava[1]

1 Institute of Physics, Sachivalaya Marg, Bhubaneswar 751005, India
2 Physical Research Laboratory, Navrangpura, Ahmadabad 380009, India

E-mail: atreya@iopb.res.in

Abstract. We have investigated the possibility of generating baryon inhomogeneities in the quark-gluon plasma phase in the early universe due to CP violating scattering of quarks and anti-quarks from moving $Z(3)$ domain walls. CP violation, here, is spontaneous in nature, and arises from the non-trivial profile of the background gauge field between different $Z(3)$ vacua. We have calculated the spatial variation of the background gauge field across the $Z(3)$ interface from the profile of the Polyakov loop $L(x)$ for the $Z(3)$ interface and determined the reflection of quarks and anti-quarks using the Dirac equation. Our results have shown that the reflection coefficients of quarks and anti-quarks can differ by a large amount. We have discussed the implications of this for baryon inhomogeneity generation from collapsing large $Z(3)$ walls.

1. Introduction

Our Universe has gone various phase transitions throughout its expansion history. One such transition is confinement-deconfinement phase transition that is supposed to have occurred when universe had cooled to around $T = 200\ MeV$. The order parameter for confinement-deconfinement phase transition is Polyakov loop, and defined as:

$$L(x) = \frac{1}{N} Tr\left[\mathbf{P}\ exp\left(ig\int_0^\beta A_0(\vec{x},\tau)d\tau\right)\right]. \quad (1)$$

In confining phase, $\langle L(x)\rangle = 0$, while $\langle L(x)\rangle \neq 0$ in deconfined phase. Under $Z(N)$, which is centre of $SU(N)$, $L(\vec{x}) \longrightarrow e^{i\phi} \times L(\vec{x})$, where $\phi = 2\pi m/N$, and $m = 0, 1, \cdots, (N-1)$. This leads to N-fold (for QCD, $N = 3$) degeneracy of ground states in deconfined state. As a result, domains with different $L(\vec{x})$ values will form and interfaces will exist between different domains. Note that in the presence of quarks, Polyakov loop is not an exact order parameter. So, we have worked with only pure QCD. Also, from now on, we have used $L(\vec{x})$ in place of $\langle L(x)\rangle$ for brevity.

An effective potential for Polyakov loop, as given by Pisarski [1] is:

$$V(L) = \left(-\frac{b_2}{2}|L|^2 - \frac{b_3}{6}\left(L^3 + (L^*)^3\right) + \frac{1}{4}(|L|^2)^2\right)b_4 T^4. \quad (2)$$

For $T > T_c$, second term leads to the three degenerate vacua. Parameters b_2, b_3 and b_4 are fixed using lattice results [2, 3, 4].

2. CP violating quark scattering from $Z(3)$ walls

The scattering of quarks from these $Z(3)$ walls has been studied in [5]. There, the profile of $L(\vec{x})$ was calculated (Figure 1 (a)) by energy minimization and quark scattering along with it's cosmological implications was discussed. However, in [5], no CP violation is present as $L(\vec{x})$ couples with $q, \bar{q}$ in an identical manner.

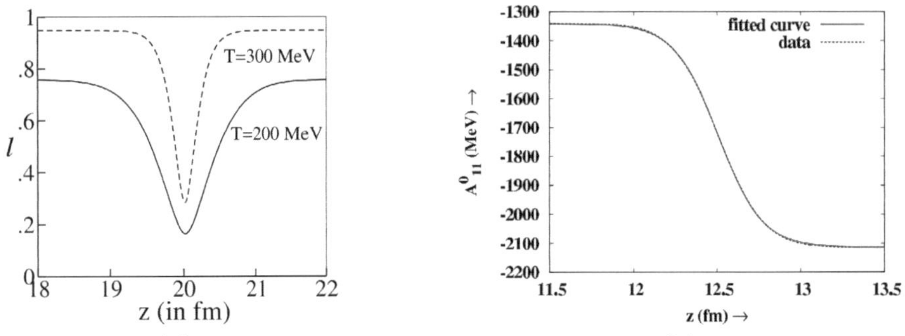

(a) (b)

Figure 1. (a) $L(\vec{x})$ profile. (b) A_0 profile. Only $(1, 1)$ component is shown.

2.1. Background A_0 profile

In [6], exact background profile was calculated by making the gauge choice given as:

$$A_0 = \frac{2\pi T}{g}\big(a\lambda_3 + b\lambda_8\big), \tag{3}$$

where a and b are constants, λ_3 and λ_8 are Gell-Mann matrices. Eq. (3) is then inserted in Eq. (1), and solved numerically for the profile given by the Figure 1 (a) to get the background A_0 (Figure 1 (b)). Reflection coefficients were calculated and it was also shown that CP violation is stronger for heavier quarks. The origin of CP violation is spontaneous in nature. (See [6] for detailed discussion on this aspect.)

3. Generation and evolution of baryon inhomogeneites

3.1. Physical picture

After inflation, the universe starts reheating and eventually the temperature is higher than critical temperature for confinement-deconfined transition. At this stage, the universe is in deconfined state and as a result, a network of $Z(N)$ interfaces is formed. The exact details of the formation of these networks itself makes an engrossing study, and is not very clear. However, one may expect it to depend on the details of reheating itself. For example, whether universe slowly reheats above the T_c or whether it quenches to a temperature above T_c may be one of the important factors in determining the network of these interfaces.

After reheating, the formation of $Z(3)$ domains is by Kibble mechanism. For $T \gg \Lambda_{QCD}$, the energy scale for these walls is set by the temperature of the universe. However, in the presence of quarks, there is a pressure difference between the true vaccum and the meta-stable vacua [7, 8]. This leads to a preferential shrinking of meta-stable vacua. As the collapse of these regions can be very fast (simulations indicate $v_w \sim 1$ [9, 10]), these can survive upto QCD phase transition only in low energy inflation models. However, there is a possibility that when effects of friction experienced by domain wall are taken into account their collapse may be slower. In that case, one may do away with the requirement of low energy inflation. (For a detailed discussion of these aspects see [5].) In such a scenario, due to CP violating effects, q and $\bar{q}$ scatter differently, and this results in the segregation of baryon number.

3.2. Evolution of inhomogeneites

While studying the evolution of these inhomogeneites, we have assumed constant temperature. This is possible if we ignore the effects coming from the reheating due to decreasing surface area as the wall collapses. This also allows us to ignore the expansion of the universe. This is consistent with the very fast moving domain walls as they collapse in time smaller than the Hubble time.

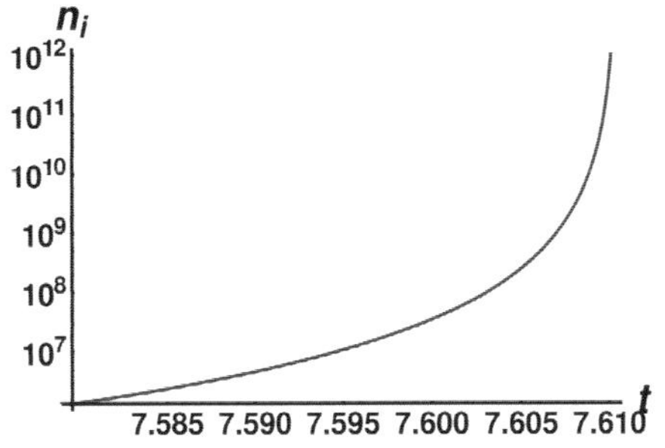

(a)

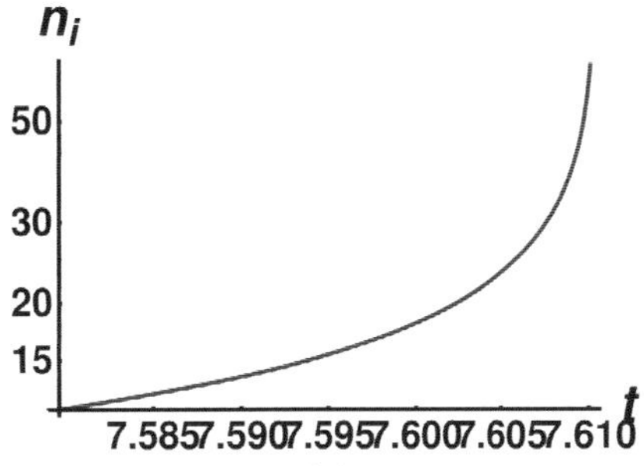

(b)

Figure 2. Number density evolution: (a) For charm-quark, and (b) For anti-charm.

The equations for studying quark number density concentration inside and outside the domain wall are [5] given by:

$$\dot{n_i} = \left(-\frac{2}{3}v_w T_w n_i + \frac{n_o T_- - n_i T_+}{6}\right)\frac{S}{V_i} - n_i \frac{\dot{V_i}}{V_i}, \quad and \tag{4}$$

$$\dot{n_0} = \left(-\frac{2}{3}v_w T_w n_i - \frac{n_o T_- - n_i T_+}{6}\right)\frac{S}{V_i} + n_o \frac{\dot{V_i}}{V_o}. \tag{5}$$

The above equations are simultaneously solved to get the number densities. Figures 2 (a) and 2 (b) show the evolution of number densities for charm-quark and anti-quark inside the collapsing domain wall at $T = 400\ MeV$. The domain wall is approximated by the step potential.

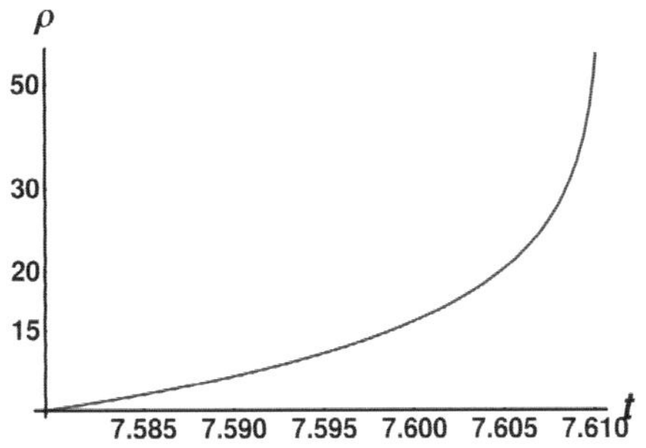

Figure 3. Evolution of baryon density profile.

It is clear that the number of quarks contained in the domain wall is several orders of magnitude higher than the number of anti-quarks. As the wall collapses, it leaves a profile of baryon density behind it. For a collapsing spherical wall, the baryon density at position R from the centre of the wall is given by [5]:

$$\rho(R) = \frac{\dot{N}_i}{4\pi v_w R^2}. \tag{6}$$

Figure 3 shows the density profile of charm anti-quark. As the quarks are completely reflected, they do not leave any density profile behind. It is important to note that Eqs. (4) and (5) assume that baryons homogenize in the two regions, while Eq. (6) does not take into account the diffusion of baryons through the wall.

4. Implications

It has been argued that the baryon inhomogeneities of sufficient initial magnitude will survive until nucleosynthesis [11]. This in turn can effect the nuclear abundances. Moreover, as these inhomogeneities are produced above quark hadron phase transition, they may alter the dynamics of phase transition [12].

These shrinking domain walls have a net baryon number concentrated in them, that can lead to the formation of stable baryonic lumps called *quark nuggets* [13]. Also, the inhomogeneties that are produced above electro-weak phase transition will change the standard baryogenesis scenario.

Acknowledgements

We thank Sanatan Digal, Partha Bagchi, Ranjita Mohapatra and Saumia P S for very valuable comments and useful discussions.

References

[1] Pisarski R D 2000 *Phys. Rev. D* **62** 111501 [hep-ph/0006205]
[2] Scavenius O, Dumitru A and Lenaghan J T 2002 *Phys. Rev. C* **66** 034903 [hep-ph/0201079]
[3] Dumitru A and Pisarski R D 2001 *Phys. Lett. B* **504** 282 [hep-ph/0010083]
[4] Dumitru A and Pisarski R D 2002 *Phys. Rev. D* **66** 096003 [hep-ph/0204223]
[5] Layek B, Mishra A P, Srivastava A M and Tiwari V K 2006 *Phys. Rev. D* **73** 103514 [hep-ph/0512367]
[6] Atreya A, Srivastava A M and Sarkar A 2012 *Phys. Rev. D* **85** 014009 [1111.3027]
[7] Dixit V and Ogilvie M C 1991 *Phys. Lett. B* **269** 353
[8] Korthals Altes C P, Lee K M and Pisarski R D 1994 *Phys. Rev. Lett.* **73** 1754 [hep-ph/9406264]
[9] Gupta U S, Mohapatra R K, Srivastava A M and Tiwari V K 2010 *Phys. Rev. D* **82** 074020 [1007.5001]
[10] Gupta U S, Mohapatra R K, Srivastava A M and Tiwari V K 2011 [1111.5402]
[11] Jedamzik K and Fuller G M 1994 *Ap. J.* **423** 33 [astro-ph/9312063]
[12] Sanyal S 2003 *Phys. Rev. D* **67** 074009 [hep-ph/0211208]
[13] Witten E 1984 *Phys. Rev. D* **30** 272

Journal of Physics: Conference Series **484** (2014) 012054 doi:10.1088/1742-6596/484/1/012054

Detailed black hole state counting in loop quantum gravity

G J F Barbero

Instituto de Estructura de la Materia, CSIC, Serrano 123, 28006 Madrid, Spain

E-mail: fbarbero@iem.cfmac.csic.es

Abstract. The combinatorial problems associated with the counting of black hole states in loop quantum gravity can be analyzed by using suitable generating functions. These can be used to perform a statistical analysis of the black hole degeneracy spectrum, study the interesting sub-structure found in the entropy of microscopic black holes, and its asymptotic behaviour for large horizon areas. They are also relevant for the discussion of the thermodynamic limit for black holes, and the understanding of sub-leading corrections to the Bekenstein-Hawking law.

1. Introduction

The two main goals of this contribution are to briefly describe the mathematical methods employed in the computation of black hole entropy in loop quantum gravity (LQG), and to discuss some applications of the resulting formalism. In the LQG framework, black holes are modelled with the help of isolated horizons [1, 2]. The quantization is carried out by combining the mathematical framework of LQG with the treatment of Chern-Simons models. By doing this, the Hilbert space appropriate for the model can be written as $\mathcal{H} = \mathcal{H}_{\text{bulk}} \otimes \mathcal{H}_{\text{surface}}$ and the entropy obtained from a maximally mixed density matrix by tracing over bulk states. In all the different proposals that have appeared in the literature [3, 4, 5] to obtain the black hole entropy (see [6, 7] for discussions on their relative merits), the actual computations can be phrased as well defined combinatorial problems involving the spin labels associated with the edges of spin networks piercing the BH horizon. These problems can be exactly solved by using number theoretic ideas. These methods are necessary for several reasons:

(i) Interesting (quasi) periodic behaviour of the entropy, that should be explained, has been found by Corichi, Borja and Díaz Polo [8].

(ii) It is important to have a detailed understanding of the entropy asymptotics.

(iii) It is important to undertake precision studies of the entropy in the LQG framework. This type of study is, in fact, the object of a lot of activity in string inspired models at the present moment [9].

(iv) They are useful to understand the thermodynamic limit, because they can be efficiently used to obtain partition functions in the area of canonical ensemble.

2. An assortment of combinatorial problems

In the following, we have enunciated the combinatorial problems that tell us how to obtain the black hole entropy for the three main proposals that have appeared in the literature:

- The Domagała-Lewandowski (DL) prescription: *The entropy* $S^{\rm DL}_{\rm micro}(a)$ *of a quantum horizon of the classical area* a*, according to quantum geometry and the Ashtekar-Baez-Corichi-Krasnov framework is*:
$$S^{\rm DL}_{\rm micro}(a) = \log \Omega^{\rm DL}(a),$$
where $\Omega^{\rm DL}(a)$ *is one plus the number of all the finite sequences* $(m_1, \ldots, m_N)$ *of non-zero elements of* $\frac{1}{2}\mathbb{Z}$*, such that the following inequality and equality are satisfied:*
$$\sum_{I=1}^{N} \sqrt{|m_I|(|m_I|+1)} \leq \frac{a}{8\pi\gamma\ell_P^2}, \quad and \quad \sum_{I=1}^{N} m_I = 0,$$
where γ *is the Immirzi parameter of quantum geometry.*
- The Ghosh-Mitra (GM) prescription: *The entropy* $S^{\rm GM}_{\leq}(a)$ *of a quantum horizon of the classical area* a*, according to the GM prescription, is defined by computing:*
$$S^{\rm GM}_{\leq}(a) = \log \Omega^{\rm GM}(a)\,,$$
where $\Omega^{\rm GM}(a)$ *is one plus the number of all the finite, arbitrarily long, sequences* $((j_1, m_1), \ldots, (j_N, m_N))$ *of ordered pairs of non-zero half integers* j_I *and spin components* $m_I \in \{-j_I, -j_I+1, \ldots, j_I\}$*, satisfying:*
$$\sum_{I=1}^{N} \sqrt{j_I(j_I+1)} \leq \frac{a}{8\pi\gamma\ell_P^2} \quad and \quad \sum_{I} m_I = 0.$$
- The Engle-Noui-Perez (ENP) prescription: *The entropy* $S^{\rm ENP}_{\rm micro}(a)$ *of a quantum horizon of the classical area* a*, is given by:*
$$S^{\rm ENP}_{\rm micro}(a) = \log \Omega^{\rm ENP}(a)\,,$$
where $\Omega^{\rm ENP}(a)$ *is one plus the number of all the finite, arbitrarily long, sequences* $(j_1, \ldots, j_N)$ *of non-zero positive half integers* j_I *satisfying:*
$$\sum_{I=1}^{N} \sqrt{j_I(j_I+1)} \leq \frac{a}{8\pi\gamma\ell_P^2},$$
and counted with a multiplicity given by the dimension of the invariant sub-space $\mathrm{Inv}(\otimes_I [j_I])$.

The combinatorial problems associated with these three prescriptions can be solved by very similar methods that can be summarized in four steps [6]:

(i) Find the possible values of $|m_I|$ (or j_I) and their multiplicity for a given value of the area a (the so called configurations). Here, the inequalities that appear in the combinatorial problems are substituted by equalities.

(ii) Compute, for each configuration, the number of possible re-orderings of the labels.

(iii) Compute, for each re-ordering, the number of configurations compatible with the additional conditions specific of each proposal: Signs for DL, spin components for GM, or the ENP condition involving the dimension of the invariant sub-space.

(iv) Add up the contributions to the black hole degeneracy for all the areas up to a, to take into account the inequalities that appear in the definitions of the entropy.

Journal of Physics: Conference Series **484** (2014) 012054 doi:10.1088/1742-6596/484/1/012054

The solution for each of these problems can be conveniently encoded by the following generating functions (especially, useful to study the asymptotic limit of the entropy for large areas):

$$G^{\mathrm{DL}}(z;x_1,x_2,\cdots)=\left(1-\sum_{i=1}^{\infty}\sum_{\alpha=1}^{\infty}\left(z^{k_\alpha^i}+z^{-k_\alpha^i}\right)x_i^{y_\alpha^i}\right)^{-1},$$

$$G^{\mathrm{GM}}(z;x_1,x_2,\cdots)=\left(1-\sum_{i=1}^{\infty}\sum_{m=1}^{\infty}\left(\frac{z^{k_m^i+1}-z^{-k_m^i-1}}{z-z^{-1}}\right)x_i^{y_m^i}\right)^{-1}, \quad and$$

$$G^{\mathrm{ENP}}(z;x_1,x_2,\cdots)=-\frac{(z-z^{-1})^2}{2}\left(1-\sum_{i=1}^{\infty}\sum_{\alpha=1}^{\infty}\left(\frac{z^{k_\alpha^i+1}-z^{-k_\alpha^i-1}}{z-z^{-1}}\right)x_i^{y_\alpha^i}\right)^{-1}.$$

Here,

- The variables x_i are associated to square free integers p_i.
- The numbers $(k_\alpha^i,\ y_\alpha^i)$, $\alpha\in\mathbb{N}$ are solutions to the Pell equation associated to the square free p_i: $(k+1)^2-p_iy^2=1$.
- The coefficient of the term $z^0x_1^{q_1}\cdots x_i^{q_i}\cdots$ gives the number of sequences of non-zero half-integers, such that $a=\sum_i q_i\sqrt{p_i}$, and satisfying the additional constraints defining the DL, GM or ENP countings.
- For a given value of a, only finite number of variables and terms are needed.

The inequality involving the area a that appears in the definition of the entropy can be conveniently taken care of by using Laplace transforms given as:

$$\mathcal{L}\left[\sum_{n\in\mathbb{N}}\beta_n\theta(a-a_n);s\right]=\frac{1}{s}\sum_{n\in\mathbb{N}}\beta_ne^{-a_ns}.$$

This is so because, if the positions of the jumps (the area eigenvalues a_n) and their magnitudes (the black hole degeneracies β_n) can be encoded in a function that can be expanded as $\sum_{n\in\mathbb{N}}\beta_ne^{-a_ns}$, then we can get an integral representation for the BH entropy as an inverse Laplace transform. This can be done by using the generating functions given above, by substituting the x_i in $G(z;x_1,x_2,\ldots)$ by $x_i=e^{-s\sqrt{p_i}/2}$ (and also $z=e^{i\omega}$), because $x_1^{q_1}\cdots x_r^{q_r}\mapsto e^{-\frac{s}{2}(q_1\sqrt{p_1}+\cdots+q_r\sqrt{p_r})}=e^{-as}$, when $a=q_1\sqrt{p_1}+\cdots+q_r\sqrt{p_r}$ (an eigenvalue of the area operator). This way, we obtain:

$$\exp S^{DL}(a)=\frac{1}{(2\pi)^2i}\int_0^{2\pi}\int_{x_0-i\infty}^{x_0+i\infty}\frac{e^{as}}{s}\left(1-2\sum_{k=1}^{\infty}e^{-s\sqrt{k(k+2)}}\cos\omega k\right)^{-1}ds\,d\omega, \tag{1}$$

$$\exp S^{GM}(a)=\frac{1}{(2\pi)^2i}\int_0^{2\pi}\int_{x_0-i\infty}^{x_0+i\infty}\frac{e^{as}}{s}\left(1-\sum_{k=1}^{\infty}\frac{\sin(k+1)\omega}{\sin\omega}e^{-s\sqrt{k(k+2)}}\right)^{-1}ds\,d\omega, \tag{2}$$

and

$$\exp S^{ENP}(a)=\frac{2}{(2\pi)^2i}\int_0^{2\pi}\int_{x_0-i\infty}^{x_0+i\infty}(\sin^2\omega)\,\frac{e^{as}}{s}\left(1-\sum_{k=1}^{\infty}\frac{\sin((k+1)\omega)}{\sin\omega}\,e^{-s\sqrt{k(k+2)}}\right)^{-1}ds\,d\omega. \tag{3}$$

The asymptotic behaviour of the entropy as a function of the area can be extracted by looking at the pole structure of the integrands in the above formulas. By doing this, one easily finds the

Bekenstein-Hawking law $S(a) \propto a$. The 1/4 pre-factor can be obtained by fixing the value of the Immirzi parameter γ. It is important to point out that the extra integration in ω gives rise to logarithmic corrections to the asymptotic behaviour for large areas. The results obtained from an asymptotic analysis along the lines just described are summarized in the following table:

Approach	γ	**Logarithmic correction**
DL	$\gamma_{\mathrm{DL}} = 0.237\cdots$	$-1/2\log(a/\ell_P^2)$
GM	$\gamma_{\mathrm{GM}} = 0.274\cdots$	$-1/2\log(a/\ell_P^2)$
ENP	$\gamma_{\mathrm{ENP}} = \gamma_{\mathrm{GM}}$	$-3/2\log(a/\ell_P^2)$

3. Conclusion

The methods that we have described above can be used to obtain important results. It is shown in a plausible way that the staircase structure of the entropy that is found for microscopic black holes is not present in the large area limit [10]. This result can be derived by introducing new generating functions that isolate the configurations contributing to each step in the entropy. In any case, it is important to realize that the counting (or statistical) entropy considered up to this point is not strictly equivalent to the entropy of the system unless one goes to a suitable thermodynamic limit. In this limit, classic theorems [11] show that the entropy has some nice smoothness and concavity properties, necessary to ensure that the standard formalism of thermodynamics can be used, and to guarantee the stability of the system respectively.

The analysis of the thermodynamic limit in the LQG black holes description [12] can be carried out, because the partition function in the so called *canonical area ensemble* can be directly read off from Eqs. (1), (2) and (3). The main conclusions derived from this analysis are that the staircase behaviour of the entropy is absent in this limit (as is to be expected from general convexity results) and also that the sub-dominant corrections to the large area behaviour of the entropy differ from the ones obtained from the counting entropy. This means that one should exercise some care when deriving physical conclusions from this type of sub-dominant corrections, in particular, when several approaches to the computation of the black hole entropy are compared.

Acknowledgements

The author wants to thank the organizers of the 7th International Conference on Gravitation and Cosmology (ICGC-2011) in Goa for their kind invitation to participate in such an interesting meeting. He also wants to thank Eduardo J S Villaseñor, for his very interesting and timely comments on the subject of this contribution. This work was supported by the Spanish MICINN Project FIS2009-11893, and the Consolider-Ingenio 2010 Programme CPAN (CSD2007-00042).

References

[1] Ashtekar A, Baez J C, Corichi A and Krasnov K V 1998 *Phys. Rev. Lett.* **80** 904
[2] Ashtekar A, Baez J C and Krasnov K V 2000 *Adv. Theor. Math. Phys.* **4** 1
[3] Domagala M and Lewandowski J 2004 *Class. Quant. Grav.* **21** 5233
[4] Ghosh A and Mitra P 2006 *Phys. Rev. D* **74** 064026
[5] Engle J, Noui K and Perez A 2010 *Phys. Rev. D* **82** 044050
[6] Agullo I, Barbero G J F, Borja E F, Diaz-Polo J and Villaseñor E J S 2010 *Phys. Rev. D* **82** 084029
[7] Barbero G J F, Lewandowski J and Villaseñor E J S 2012 [arXiv:1203.0174 [gr-qc]]
[8] Corichi A, Borja E F and Diaz-Polo J 2007 *Phys. Rev. Lett.* **98** 181301
[9] Mandal I and Sen A 2010 *Class. Quant. Grav.* **27** 214003
[10] Barbero G J F and Villaseñor E J S 2011 *Phys. Rev. D* **83** 104013
[11] Griffiths R B 1965 *J. Math. Phys.* **6** 1447
[12] Barbero G J F and Villaseñor E J S 2011 *Class. Quant. Grav.* **28** 215014

Effects of radiation on primordial non-Gaussianity

S Das

Tata Institute of Fundamental Research, Homi Bhabha Road, Colaba, Mumbai 400005, India

E-mail: suratna@tifr.res.in, suranta.das.2007@gmail.com

Abstract. We have studied the non-Gaussian features in single-field slow-roll inflationary scenario, where inflation is preceded by a radiation era. In such a scenario, both bispectrum and trispectrum non-Gaussianities are enhanced. Interestingly, the trispectrum in this scenario does not depend up on the slow-roll parameters, and thus, τ_{NL} is larger than f_{NL}, which can be a signature of such a pre-inflationary radiation era.

1. Introduction

Study of non-Gaussian features in primordial perturbations generated during inflation has become a subject of great importance, as the precise determination of these primordial non-Gaussianities can quantify the dynamics of the early universe [1]. In generic single-field slow-roll inflationary scenario, the preferred initial vacuum chosen for the inflaton perturbations is that of Bunch-Davies. It is shown in [2] that if inflation is preceded by a radiation era then the inflaton fluctuations will have an initial thermal distribution where the initial vacuum will depart from the standard Bunch-Davies one. The presence of pre-inflationary radiation era enhances the power spectrum of scalar modes by an extra temperature dependant factor $\coth(k/2T)$. The enhanced power spectrum is in accordance with the observations if the comoving temperature T of the primordial perturbations is less than 10^{-3} Mpc^{-1} [2]. In this paper, we have shown that presence of pre-inflationary radiation era not only enhances the power spectrum, but also generates large bispectrum and trispectrum, and these non-Gaussianities will carry signatures of such pre-inflationary radiation era [3], which have been discussed in detail.

2. Bispectrum and trispectrum in single-field slow-roll inflation

The derivations shown in this paper are done in *spatially flat gauge*, which is preferred, as in this gauge, the comoving curvature perturbation $\mathcal{R}(t,\mathbf{x})$ is proportional to the inflaton fluctuations $\delta\phi(t,\mathbf{x})$ as:

$$\mathcal{R}(t,\mathbf{x}) = \frac{H}{\dot{\phi}}\delta\phi(t,\mathbf{x}), \tag{1}$$

where H is the Hubble parameter and the overdot represents derivative with respect to cosmic time t. Thus, in this gauge, the comoving curvature power spectrum, i.e., the 2-point correlation function of $\mathcal{R}(t,\mathbf{k})$ in Fourier space, is directly related to inflaton's power spectrum as:

$$\mathcal{P}_{\mathcal{R}}(k) = \frac{k^3}{2\pi^2}\langle\mathcal{R}(k)\mathcal{R}(k)\rangle \longleftrightarrow \left(\frac{H}{\dot{\phi}}\right)^2 \langle\delta\phi(k)\delta\phi(k)\rangle. \tag{2}$$

The comoving curvature power spectrum is measured through observations of the TT anisotropy spectrum of CMBR, which is nearly scale-invariant.

Bispectrum, the non-vanishing 3-point correlation function of primordial fluctuations, is the lowest order departure from Gaussianity of those primordial perturbations. The non-Gaussianity arising from bispectrum is quantified by a non-linear parameter f_{NL}, which is constrained by several experiments as: (i) the WMAP-5 data yields $-151 < f_{NL}^{eq} < 253$ (95% CL) [4], (ii) the PLANCK mission is sensitive to probe bispectrum upto $f_{NL} \sim 5$ [5], and (iii) in future experiments, if the primordial non-Gaussianities imprinted in 21-cm background is measured, then $f_{NL} < 0.1$ can be probed [6, 7]. In a free theory, as the primordial perturbations are Gaussian in nature, the 3-point correlation function vanishes, yielding no non-Gaussianity. It is shown in [8] that self-interactions of inflaton field of the kind $V(\phi) = \lambda\phi^3$ generates non-vanishing bispectrum proportional to λ/H, but the non-Gaussianity is too small ($\sim \mathcal{O}(10^{-7})$) to be probed by any existing or future experiments. On the other hand, in a generic single-field slow-roll inflationary model, non-linearities in the evolution of primordial perturbations $\mathcal{R}(t, \mathbf{k})$ can also generate primordial non-Gaussianities in CMBR. In the non-linear limit, one can write:

$$\mathcal{R}_{NL}(t, \mathbf{x}) = \frac{H}{\dot{\phi}}\delta\phi_L(t, \mathbf{x}) + \frac{1}{2}\frac{\partial}{\partial\phi}\left(\frac{H}{\dot{\phi}}\right)\delta\phi_L^2(t, \mathbf{x}) + \mathcal{O}(\delta\phi_L^3), \tag{3}$$

which yields a non-vanishing 3-point correlation function of $\mathcal{R}_{NL}$ in terms of 4-point correlation function of inflaton perturbations $\delta\phi_L$ as:

$$\langle \mathcal{R}_{NL}\mathcal{R}_{NL}\mathcal{R}_{NL}\rangle \simeq \left(\frac{H}{\dot{\phi}}\right)^2 \frac{1}{2}\frac{\partial}{\partial\phi}\left(\frac{H}{\dot{\phi}}\right)\langle \delta\phi_L\delta\phi_L\delta\phi_L^2\rangle, \tag{4}$$

even when the initial perturbations $\delta\phi_L$ are Gaussian in nature. The 4-point correlation function on the right hand side can be written in terms of product of two 2-point correlation functions, and defining f_{NL} as:

$$\langle \mathcal{R}(\mathbf{k}_1)\mathcal{R}(\mathbf{k}_2)\mathcal{R}(\mathbf{k}_3)\rangle = (2\pi)^{-\frac{3}{2}}\delta^3(\mathbf{k}_1 + \mathbf{k}_2 + \mathbf{k}_3)\frac{6}{5}f_{NL}\left(\frac{P_{\mathcal{R}}(k_1)}{k_1^3}\frac{P_{\mathcal{R}}(k_2)}{k_2^3} + 2 \text{ perms.}\right), \tag{5}$$

one can show that the non-linear parameter is of the order of slow-roll parameters $f_{NL} = \frac{5}{6}(\delta - \epsilon)$ [11], which is also too small $(\mathcal{O}(10^{-2}))$ to be detected by any present or forthcoming experiments. The delta-function in the above equation ensures that the three momenta form a triangle and f_{NL} is determined in several such triangle configurations, and some of them which we have considered in this paper are: (i) *Squeezed configuration* ($|\mathbf{k}_1| \approx |\mathbf{k}_2| \approx k \gg |\mathbf{k}_3|$), (ii) *Equilateral configuration* ($|\mathbf{k}_1| = |\mathbf{k}_2| = |\mathbf{k}_3| = k$), and (iii) *Folded configuration* ($|\mathbf{k}_1| = |\mathbf{k}_3| = \frac{1}{2}|\mathbf{k}_2| = k$).

The connected part of 4-point correlation function of primordial fluctuations is called the *trispectrum*, given as:

$$\begin{aligned}\langle \mathcal{R}(\mathbf{k}_1)\mathcal{R}(\mathbf{k}_2)\mathcal{R}(\mathbf{k}_3)\mathcal{R}(\mathbf{k}_4)\rangle_c \equiv & \langle \mathcal{R}(\mathbf{k}_1)\mathcal{R}(\mathbf{k}_2)\mathcal{R}(\mathbf{k}_3)\mathcal{R}(\mathbf{k}_4)\rangle - (\langle \mathcal{R}_L(\mathbf{k}_1)\mathcal{R}_L(\mathbf{k}_2)\rangle\langle \mathcal{R}_L(\mathbf{k}_3)\mathcal{R}_L(\mathbf{k}_4)\rangle \\ & +2 \text{ perm }).\end{aligned} \tag{6}$$

The non-linear parameter τ_{NL} quantifies the non-Gaussianity arising from trispectrum and it is constrained by observations as: (i) WMAP constraints trispectrum as $|\tau_{NL}| < 10^8$ [9], (ii) PLANCK is expected to reach the sensitivity upto $|\tau_{NL}| \sim 560$ [10], and (iii) future 21-cm experiments can probe trispectrum up to the level $\tau_{NL} \sim 10$ [7]. A free-scalar theory yields vanishing trispectrum like vanishing bispectrum. But non-linear evolution of $\mathcal{R}$ as given in Eq. (3) yields a trispectrum, where $\tau_{NL} = \left(\frac{6}{5}f_{NL}\right)^2$ [11], which being proportional to the square of slow-roll parameters is too small to be detected by any present or future experiments. It is to be noted that the generic single-field slow-roll inflation predicts a trispectrum, which is smaller than the bispectrum by orders of magnitude.

3. Inflation with prior radiation era and enhanced non-Gaussianity

If inflation is preceded by a radiation era, then the inflaton field will have an initial thermal distribution, where the thermal vacuum $|\Omega\rangle$ will have finite occupation as $N_k|\Omega\rangle = n_k|\Omega\rangle$ (the number operator $N_k \equiv a^\dagger_{\mathbf{k}} a_{\mathbf{k}}$). Also, there will be a probability of the system to be in an energy state $\varepsilon_k \equiv n_k k$ as:

$$p(\varepsilon_k) \equiv \frac{e^{-\beta n_k k}}{\sum_{n_k} e^{-\beta n_k k}} = \frac{e^{-\beta n_k k}}{z}. \tag{7}$$

Due to this probability distribution, the correlation functions have to be thermal averaged. Taking into account the initial thermal distributions of the primordial fluctuations and the probability distribution due to pre-inflationary radiation era, the thermal averaged inflaton's power spectrum will have an enhancement factor $1 + 2f_B(k)$, where $f_B(k)$ is the distribution function of primordial perturbations. For inflaton (scalar) perturbations, it will be Bose-Einstein distribution function $\left(f_B(k) \equiv \frac{1}{e^{\beta k}-1}\right)$, and thus, the enhancement factor will be $1 + 2f_B(k) = \coth(\beta k/2)$, where $\beta \equiv \frac{1}{T}$ [2, 3]. This enhanced power spectrum is in accordance with the observations when $T < 10^{-3}\ \mathrm{Mpc}^{-1}$ [2].

As the 2-point correlation function is thermal averaged due to the effects of pre-inflationary radiation era, the other higher-point correlation functions have also to be thermal averaged in a similar way. For 3-point correlation function, this will be:

$$\langle \mathcal{R}_{NL}\mathcal{R}_{NL}\mathcal{R}_{NL}\rangle_\beta \simeq \left(\frac{H}{\dot\phi}\right)^2 \frac{1}{2}\frac{\partial}{\partial\phi}\left(\frac{H}{\dot\phi}\right)\langle\delta\phi\delta\phi\delta\phi^2\rangle_\beta, \tag{8}$$

but now the probability of occupancy of a state with four energies ϵ_r will be:

$$p(k_1, k_2, k_3, k_4) \equiv \frac{\prod_r e^{-\beta n_{k_r} k_r}}{\prod_r \sum_{n_k} e^{-\beta n_{k_r} k_r}}. \tag{9}$$

Due to thermal averaging, f_{NL} is enhanced. We have computed the non-Gaussianity in this scenario arising from bispectrum in different triangle configurations: (i) *Squeezed configuration*: The enhanced non-Gaussianity is $f^{\mathrm{th}}_{NL} = f_{NL} \times 2\left(1 + 3.72\coth\left(\frac{\beta k}{2}\right)\right)$, where f_{NL} is enhanced by a factor of 64.82, (ii) *Equilateral configuration*: The enhanced non-Gaussianity is $f^{\mathrm{th}}_{NL} = f_{NL}\times \left(3 + \frac{5}{4\sinh^2\left(\frac{\beta k}{2}\right)}\right)$, where f_{NL} is enhanced by a factor of 90.85, and (iii) *Folded configuration*: The enhanced non-Gaussianity is $f^{\mathrm{th}}_{NL} = f_{NL} \times \left(3 + \frac{1}{\sinh^2\left(\frac{\beta k}{2}\right)}\right)$, where f_{NL} is enhanced by a factor of 73.28. It can be seen that f_{NL}, which is enhanced due to effects of pre-inflationary radiation era, is within the sensitivity of future 21-cm experiments, where primordial non-Gaussianities can be detected [6, 7]. It is also to be noted that the maximum non-Gaussianity can arise in the equilateral configuration when inflation is preceded by a radiation era. In a later work [12], similar analysis is done when perturbations are already present in the initial vacuum. It is found in [12] too that initial presence of quanta in the vacuum can significantly enhance non-Gaussianity arising from bispectrum, which is in agreement with the results presented here and in [3].

It is very interesting to note at this point that due to thermal averaging, the 4-point correlation function is not equal to the product of two 2-point correlation functions, which can yield a non-vanishing connected part as:

$$\begin{aligned}\langle\mathcal{R}(\mathbf{k}_1)\mathcal{R}(\mathbf{k}_2)\mathcal{R}(\mathbf{k}_3)\mathcal{R}(\mathbf{k}_4)\rangle_c \neq\ & \langle\mathcal{R}(\mathbf{k}_1)\mathcal{R}(\mathbf{k}_2)\mathcal{R}(\mathbf{k}_3)\mathcal{R}(\mathbf{k}_4)\rangle_\beta \\ -\ & \Big(\langle\mathcal{R}_L(\mathbf{k}_1)\mathcal{R}_L(\mathbf{k}_2)\rangle_\beta\langle\mathcal{R}_L(\mathbf{k}_3)\mathcal{R}_L(\mathbf{k}_4)\rangle_\beta + 2\mathrm{perm}\Big). \end{aligned} \tag{10}$$

Thus, defining the trispectrum in such a situation as:

$$\langle \mathcal{R}(\mathbf{k}_1)\mathcal{R}(\mathbf{k}_2)\mathcal{R}(\mathbf{k}_3)\mathcal{R}(\mathbf{k}_4)\rangle_c = \tau_{NL}\left[\frac{P_{\mathcal{R}}(k_1)}{k_1^3}\frac{P_{\mathcal{R}}(k_2)}{k_2^3}\delta^3(\mathbf{k}_1+\mathbf{k}_3)\delta^3(\mathbf{k}_2+\mathbf{k}_4) + 2\text{ perm.}\right], \quad (11)$$

and then, we have seen that as the linear perturbations can generate non-vanishing connected part due to thermal averaging, the non-linear parameter $|\tau_{NL}|$ will not depend up on slow-roll parameters and can be as large as 42.58 [3], which is within the detection range of future 21-cm background anisotropy experiments [7]. Hence, we have seen that the presence of pre-inflationary radiation era yields larger trispectrum non-Gaussianity than bispectrum.

4. Conclusion

In a generic single-field slow-roll model of super-cool inflation, non-linear evolution of primordial fluctuations generate bispectrum non-Gaussianity, which is proportional to the slow-roll parameters [11], and thus, too small to be detected by any present or future experiments. Non-linear evolution of primordial fluctuations also generates trispectrum non-Gaussianity, where τ_{NL} is proportional to the square of slow-roll parameters [11]. Thus, this generic inflationary scenario predicts trispectrum non-Gaussianity, which is much smaller than the bispectrum non-Gaussianity.

We have shown that if such a generic inflationary scenario is preceded by a radiation era, it can yield large bispectrum and trispectrum non-Gaussianities [3], which are within the range of detection of future 21-cm background anisotropy experiments [6, 7]. Due to the presence of pre-inflationary radiation era, the initial vacuum will contain thermal fluctuations, and also the energy states will have a probability distribution. Accordingly, the thermal averaged 3-point correlation function generates large non-Gaussianity, where the enhancement of f_{NL} over the generic scenario is largest in the equilateral configuration. An interesting situation arises in the case of trispectrum as the thermal averaged 4-point correlation function is not equal to the product of two thermal-averaged 2-point correlation functions. Thus, the linear primordial perturbations can generate a non-vanishing connected part of 4-point correlation function due to thermal averaging, and τ_{NL} in such a case will not depend up on the slow-roll parameters. We have computed that in such a scenario, $|\tau_{NL}|$ can be as large as 43 [3]. Thus, a significant signature of such pre-inflationary radiation era is that it yields larger trispectrum than bispectrum. This signature can distinguish between an inflationary scenario preceded by a radiation era and the generic scenario of single-field slow-roll super-cool inflation.

Acknowledgements

The author thanks the collaborator, Subhendra Mohanty.

References

[1] Bartolo N, Matarrese S and Riotto A 2005 *Nucl. Phys. Proc. Supp.* **148** 56
[2] Bhattacharya K, Mohanty S and Rangarajan R 2006 *Phys. Rev. Lett.* **96** 121302 [arXiv:hep-ph/0508070]
[3] Das S and Mohanty S 2009 *Phys. Rev. D* **80** 123537 [arXiv:0908.2305[astro-ph.CO]]
[4] Komatsu E *et al* [WMAP Collaboration] 2009 *Ap. J. Supp. Ser.* **180** 330 [arXiv:0803.0547[astro-ph]]
[5] Komatsu E and Spergel D N 2001 *Phys. Rev. D* **63** 063002 [arXiv:astro-ph/0005036]
[6] Cooray A 2006 *Phys. Rev. Lett.* **97** 261301 [arXiv:astro-ph/0610257]
[7] Cooray A, Li C and Melchiorri A 2008 *Phys. Rev. D* **77** 103506 [arXiv:0801.3463 [astro-ph]]
[8] Falk T, Rangarajan R and Srednicki M 1993 *Ap. J.* **403** L1 [arXiv:astro-ph/9208001]
[9] Boubekeur L and Lyth D H 2006 *Phys. Rev. D* **73** 021301 [arXiv:astro-ph/0504046]
[10] Kogo N and Komatsu E 2006 *Phys. Rev. D* **73** 083007 [arXiv:astro-ph/0602099]
[11] Byrnes C T, Sasaki M and Wands D 2006 *Phys. Rev. D* **74** 123519 [arXiv:astro-ph/0611075]
[12] Agullo I and Parker L 2011 *Phys. Rev. D* **83** 063526 [arXiv:1010.5766 [astro-ph.CO]]

Singularity resolution in cosmologies using AdS/CFT

A Ghosh

Department of Physics and Astronomy, University of Kentucky, Lexington, KY 40506, USA

E-mail: `archisman.ghosh@uky.edu`

Abstract. Singularities in general relativity are expected to be cured by a quantum theory of gravity. String theory is a prospective candidate for quantum gravity and holographic correspondences like AdS/CFT, that arise from string theory, might shed some light into the nature of singularities. While the theory of gravity might not be tractable near the singularity, the corresponding field theory might be well-behaved. We have come up with a model of cosmology, where curvatures become large at some time, thus, resembling a big crunch, but whose field theory dual remains well-behaved. Our analysis in the field theory indicates that we have a bouncing cosmology with a thermal distribution of radiation, and a possible formation of a small black hole after the bounce.

1. Motivation and overview

The theory of general relativity has a unique property of predicting its own failure: It breaks down at places called *singularities*. Our usual notions of spacetime, break down near these singularities. At spacelike or null singularities, like those in the interiors of neutral black holes, "time" begins or ends, the meaning of which is not clear. Particularly, disturbing are the cosmological singularities like the big bang or the big crunch, which no observer can bypass.

Near these singularities, where the curvatures have become large compared to some scale, a quantum theory of gravity is expected to take over. Holographic correspondences like AdS/CFT [1] arising from string theory provide a non-perturbative formulation of quantum gravity. Here, gravity in the bulk is an effective description of a non-gravitational field theory in a lower number of dimensions. We ask the question: Whether the dual description following these correspondences can give us some insight as to what happens near or past these singularities? In these models, the spectrum at late times can be interpreted as the primordial fluctuations that lead to structure formation in the early universe. An incomplete, but long list of attempts made in this direction has been given in [2]. (For reviews, refer [3].) The bulk of this article is based on the work done in [4].

2. The AdS/CFT correspondence

The AdS/CFT correspondence relates a theory of gravity to a gauge theory with no dynamical gravity. In the simplest example of the correspondence relevant to our discussion, the theory of gravity is a type IIB String Theory on the $AdS_5 \times S^5$ background given by the metric:

$$ds^2 = -(1+\frac{r^2}{R_{AdS}^2})dt^2 + \frac{dr^2}{1+\frac{r^2}{R_{AdS}^2}} + r^2\, d\Omega_3^2 + R_{AdS}^2\, d\Omega_5^2 \,, \qquad (1)$$

and the dual gauge theory is the $\mathcal{N} = 4$ supersymmetric $SU(N)$ Yang-Mills theory on $S^3 \times \mathbb{R}$. Here, R_{AdS} is the length scale associated with the geometry, which we have eventually set to unity. The Yang-Mills coupling g_{YM} and the N of the gauge group in the gauge theory are related to the parameters of the string theory, the string length and coupling $l_s \equiv \sqrt{\alpha'}$ and $g_s = e^{\Phi}$ respectively[1], via the relations:

$$g_{YM}^2 = 4\pi g_s \quad and \quad \lambda \equiv g_{YM}^2 N = \frac{R_{AdS}^4}{l_s^4}. \tag{2}$$

Here, λ is the 't Hooft coupling, the relevant coupling of gauge theories in the large N limit. In this limit ($N \to \infty$, $g_{YM} \to 0$, $\lambda \equiv g_{YM}^N =$ finite), the dual description is a *classical* string theory with the quantum corrections suppressed by $1/N$. Moreover, the above relations tell us that the large 't Hooft coupling limit ($\lambda \to \infty$), corresponds to the low-energy super-gravity truncation of the string theory, where the stringy modes are suppressed by α'. This is because, the massive string modes, which have masses of $O(1/\sqrt{\alpha'})$, are heavy compared to the energy scale of the theory $1/R_{AdS}$, and are not excited. So we are left only with the massless super-gravity modes.

3. Our setup in super-gravity

We have considered a Yang-Mills theory with a time-dependent 't Hooft coupling $\lambda(t)$. In the dual bulk (via AdS/CFT), this corresponds to a non-normalizable dilaton mode, whose boundary value $\Phi_0(t)$ is being dialled. The dilaton evolves following the super-gravity equations of motion and produces a non-trivial metric by back-reaction. The equations to solve are, thus, the Einstein's dilaton equations:

$$R_{ab} = \Lambda g_{ab} + \frac{1}{2}\partial_a\Phi\partial_b\Phi, \ \text{(with } \Lambda = -4), \quad and \quad \nabla^2\Phi = 0\,. \tag{3}$$

We have chosen a profile such that the 't Hooft coupling λ is large at early and late times. The dual theory, thus, has a good gravity description at these times. We have chosen the initial state to be the vacuum of the gauge theory, which in the bulk corresponds to pure $AdS_5 \times S^5$. At intermediate times, λ is allowed to become $O(1)$, as the curvatures in the bulk reach the string scale at this stage, and super-gravity description breaks down. We have fixed N to remain large, so that the curvatures do not reach the Planck scale, and the system remains classical. Thus, the singularities that we have described are rather restrictive.

Furthermore, we have varied the 't Hooft coupling, or equivalently the boundary value of the dilaton *slowly*. The curvature scale of the AdS or equivalently the radius of S^3 on which the field theory is defined sets a scale in the theory. We can quantify this as:

$$\Phi_0 = f(\epsilon t),\ \epsilon \ll 1, \quad and \quad \dot{\Phi}_0 = \epsilon f'(\epsilon t)\,, \tag{4}$$

where each time derivative comes with an additional power of ϵ.

Although, the change is slow, the total change in the value of the dilaton accrues over time, and we can eventually have a large change in the value of the dilaton. Our solution, therefore, includes the case where the value of the dilaton becomes small, the bulk curvatures become large compared to the string scale, and super-gravity approximation in the bulk breaks down.

3.1. Our solution

We have expanded the dilaton and the bulk metric order by order in ϵ as:

$$\Phi(r,t) = \Phi_0(t) + \epsilon\Phi_1(r,t) + \epsilon^2\Phi_2(r,t) + \dots\,, \quad and \quad g_{ab} = g_{ab}^{(0)} + \epsilon g_{ab}^{(1)} + \epsilon^2 g_{ab}^{(2)} + \dots\,, \tag{5}$$

[1] In String Theory, the coupling g_s is not an independent parameter, but is dynamically determined by the value of the scalar field Φ, the *dilaton*.

and plug them in the Einstein's equations. Solving to the lowest non-trivial order in ϵ, we get:

$$\Phi_2(r,t) = \Phi_0(t) + \tfrac{1}{4}\ddot{\Phi}_0(t)\left[\tfrac{1}{r^2}\log(1+r^2) - \tfrac{1}{2}\Big(\log(1+r^2)\Big)^2 - \mathrm{dilog}(1+r^2) - \tfrac{\pi^2}{6}\right],$$
$$g_{tt} = 1 + r^2 - \tfrac{1}{12}\dot{\Phi}_0^2\Big[3 - \tfrac{1}{r^2}\ln(1+r^2)\Big] \quad \textit{and} \quad g^{rr} = 1 + r^2 - \tfrac{1}{12}\dot{\Phi}_0^2\Big[1 - \tfrac{1}{r^2}\ln(1+r^2)\Big]. \quad (6)$$

The solution is smooth everywhere: There are no horizons, and no black holes are formed. This is a consequence of the slow variation. For a variation that is fast enough, we can form black holes even within the super-gravity approximation [5].

We have now considered the situation where the value of the dilaton has become small at intermediate times and super-gravity approximation has broken down. We have turned to the gauge theory to see if it can give us any clue as to what happens in the stringy regime and what happens afterwards when super-gravity is restored.

4. Field theory analysis

The boundary gauge theory is a quantum mechanical system with a time dependent parameter - the coupling - that is *slowly varying*. For the field theory on S^3, the spectrum has a gap of order $1/R_{AdS}$ for all values of 't Hooft coupling. So we can use the *adiabatic approximation* in quantum mechanics to study the system. The standard adiabatic approximation of quantum mechanics will turn out to be valid only if $N\epsilon \ll 1$, which is rather stringent. We have formulated an alternate *adiabatic approximation in terms of coherent states.*

4.1. Standard adiabatic approximation

Consider a quantum mechanical system with Hamiltonian $H(\zeta(t))$ that depends on a slowly-varying parameter $\zeta(t)$. In our case, $\zeta(t)$ is the time dependent 't Hooft coupling $\lambda(t)$ or equivalently, the boundary value of the dilation field $\Phi_0(t)$. The instantaneous eigenstates of the Hamiltonian are $H(\zeta)|\phi_m(\zeta)\rangle = E_m(\zeta)|\phi_m(\zeta)\rangle$. The evolution of the ground state $|\phi_0\rangle$ of $H(\zeta_0)$ in the adiabatic approximation is simply given by: $|\psi_0(t)\rangle \simeq |\phi_0(\zeta)\rangle e^{-i\int_{-\infty}^{t} E_0(\zeta)dt}$.

The approximation is good if the first corrections are small: $\left|\left\langle \phi_n|\frac{\partial H}{\partial \zeta}|\phi_0\right\rangle\dot{\zeta}\right| \ll (\Delta E)^2$, where ΔE is the energy difference between the ground and the first excited states. In our case, $\dot{\zeta}(t) = \dot{\Phi}_0(t) \sim O(\epsilon)$ and using AdS/CFT [2], $\left\langle \phi_n|\frac{\partial H}{\partial \Phi_0}|\phi_0\right\rangle = -\left\langle \phi_n|\hat{\mathcal{O}}_{l=0}|\phi_0\right\rangle \sim N$.

Thus, the adiabatic approximation is valid if $N\epsilon \ll 1$. This condition is, however, more stringent than the condition $\epsilon \ll 1$, which we had in the super-gravity analysis.

4.2. Adiabatic approximation with coherent states

A general coherent state has the form $|\Psi(t)\rangle = \exp\left[\sum_n \lambda_n(t)a_n^\dagger\right]|0\rangle$, where $a_n^\dagger$ are the creation operators, and $\lambda_n(t)$ are the coherent state parameters. For large values of λ_n, the system goes over to a classical configuration. In our problem, the creation and annihilation operators need to be replaced by the creation and annihilation parts $\hat{\mathcal{O}}_{(\pm)}$ of $\hat{\mathcal{O}}$. In general, the operators $\hat{\mathcal{O}}_{(\pm)}$ have a non-trivial operator algebra, which mixes all of them and makes the evolution of $\chi(t)$ and $\lambda_n(t)$, and quite complicated. However, the situation drastically simplifies for a slow variation. The expectation value of the 3-point function taken in a coherent state can be shown to be $\left\langle \psi|\hat{\mathcal{O}}_1\hat{\mathcal{O}}_2\hat{\mathcal{O}}_3|\psi\right\rangle \sim \epsilon$. For $\epsilon \ll 1$, the three and higher point functions are, thus, parametrically suppressed, and all we are left with the 2-point functions. The operators $\hat{\mathcal{O}}_{(\pm)}$ *decouple*, and their algebra reduces to that of a free harmonic oscillator.

[2] Turning on a field Φ with a boundary value Φ_0 corresponds to the deformation of the gauge theory action by a gauge invariant operator $\hat{\mathcal{O}} = \frac{\delta S_{\rm Cl}}{\delta \Phi_0}$, where $S_{\rm Cl}$ is the classical super-gravity action.

After plugging in the mode expansion in the Schrödinger equation, subsequent integration by parts leads to a perturbative expansion for λ_n, we get: $\lambda_n(t) = -\frac{iN\dot{\Phi}_0(t)}{(2n)^2} + \frac{N\ddot{\Phi}_0(t)}{(2n)^3} + \cdots$. The above expansion is valid if the second and subsequent terms are small compared to the first term, i.e., $|\frac{\ddot{\Phi}_0}{2n\dot{\Phi}_0}| \ll 1 \; \forall \; n$. This is identical to the condition $\epsilon \ll 1$. Moreover, the state represents a "classical" deformation only when the coherent state parameter λ is large. That holds when $N\epsilon \gg 1$.

4.3. Results

For small 't Hooft coupling, the coupling between the oscillators is still individually suppressed by ϵ and most of the above analysis goes through. However, now we have $O(N^2)$ *stringy modes*, whose frequencies are comparable to those of the super-gravity modes. *So there is a possibility of thermalization.* The energy that has been pumped in might get distributed between the various modes. Then at late times, one could be left with $O(N^2\epsilon^2)$ thermalized energy in the system. It is difficult to say whether thermalization would indeed occur, since the time scale of variation is same as the thermalization time scale.

However, we can assume that all the energy left behind in the system has been thermalized ($E_{\text{Thermal}} \sim N^2\epsilon^2$), and from entropic considerations in super-gravity figure out the worst case scenarios for various values of ϵ. For $\epsilon \ll \lambda^{5/4}N^{-1}$, a gas of *super-gravity modes* is favoured. For $\epsilon \gtrsim \lambda^{5/4}N^{-1}$, we can have a gas of *massive string modes*. For $\epsilon \gtrsim \lambda^{-7/8}$, a *small black hole* can form. In order to form a *large black hole*, one would require $\epsilon \gtrsim 1$. So a large black hole is never formed in our slowly varying solution.

5. Conclusions and outlook

We have an example of a situation in which the gauge theory has a smooth time-evolution across a region whose dual bulk region has a singularity. An explicit calculation of the final state is possible in principle, but made difficult in practice, because we do not have a complete AdS/CFT dictionary in the stringy regime. Of the possible outcomes, a thermal distribution at late times would imply a Gaussian spectrum while formation of a black hole would source a non-Gaussianity.

Acknowledgements

The author thanks his collaborators Adel Awad, Sumit Das, Jae-Hyuk Oh and Sandip Trivedi, and also Robert Brandenberger, Shiraz Minwalla and K Narayan, with whom he has discussed in various stages. He also thanks the organizers of ICGC-2011 for putting together a very successful conference, and the Huffaker Foundation for a Huffaker Travel Support Grant that made my participation in the conference possible. This work is partially supported by the NSF-PHY-0855614.

References

[1] Maldacena J M 1998 *Adv. Theor. Math. Phys.* **2** 231, 1999 *Int. J. Theor. Phys.* **38** 1113 [arXiv:hep-th/9711200], Witten E 1998 *Adv. Theor. Math. Phys.* **2** 253 [arXiv:hep-th/9802150], Gubser S S, Klebanov I R and Polyakov A M 1998 *Phys. Lett. B* **428** 105 [arXiv:hep-th/9802109], For a standard review, Aharony O, Gubser S S, Maldacena J M, Ooguri H and Oz Y 2000 *Phys. Report* **323** 183 [arXiv:hep-th/9905111]

[2] Hertog T and Horowitz G T 2004 *JHEP* **0407** 073 [hep-th/0406134], Craps B, Sethi S and Verlinde E P 2005 *JHEP* **0510** 005 [hep-th/0506180], Das S R and Michelson J 2005 *Phys. Rev. D* **72** 086005 [hep-th/0508068], Das S R and Michelson J 2006 *Phys. Rev. D* **73** 126006 [hep-th/0602099], Das S R, Michelson J, Narayan K and Trivedi S P 2006 *Phys. Rev. D* **74** 026002 [hep-th/0602107], Das S R, Michelson J, Narayan K and Trivedi S P 2007 *Phys. Rev. D* **75** 026002 [hep-th/0610053], Awad A, Das S R, Narayan K and Trivedi S P 2008 *Phys. Rev. D* **77** 046008 [arXiv:0711.2994[hep-th]], Awad A,

Das S R, Nampuri S, Narayan K and Trivedi S P 2009 *Phys. Rev. D* **79** 046004 [arXiv:0807.1517[hep-th]], Turok N, Craps B and Hertog T [arXiv:0711.1824[hep-th]], Craps B, Hertog T and Turok N [arXiv:0712.4180[hep-th]]

[3] Das S R *String Cosmology* ed Erdmenger J 231, Das S R 2009 *PoS ISFTG* 006

[4] Awad A, Das S R, Ghosh A, Oh J H and Trivedi S P 2009 *Phys. Rev. D* **80** 126011 [arXiv:0906.3275[hep-th]]

[5] Bhattacharyya S and Minwalla S 2009 *JHEP* **0909** 034 [arXiv:0904.0464[hep-th]]

Multi-field inflation: Formulation, effective theory and phenomenology

J-O Gong

Theory Division, CERN, CH-1211 Genève 23, Switzerland

E-mail: jinn-ouk.gong@cern.ch

Abstract. We have described how to obtain the non-perturbative low energy effective field theory of single field inflation from a generic multi-field model by integrating out heavy fields. The features of heavy physics is described by the effective speed of sound, which leaves distinctive observational signatures in the correlation functions of the curvature perturbation.

1. Introduction

The absence of the relevant inflaton in the standard model of particle physics demands that inflation be described in the context of the speculative high energy theories, where plenty of scalar fields which can contribute to the inflationary dynamics exist. Further, in multi-field system, we can obtain interesting observational signatures, which deviate from the predictions of the single field models of inflation and can be detected in near future. Thus, we have both theoretical and phenomenological motivations to study multi-field inflation.

However, the very model of inflation relevant for the observed universe is still veiled yet, based on the unverified high energy theories. To cope with our ignorance, we may take the effective field theory approach to integrate out heavy fields from the parent theory, which contains multiple fields. Then, we are left with the light "inflaton", giving rise to effectively single field inflation.

Here, we have presented the systematics of effective single field description of multi-field inflation. The effects of the heavy fields are left as a non-trivial speed of sound c_s. This leaves non-trivial features in the correlation functions of the curvature perturbation $\mathcal{R}$. We explicitly compute the 2- and 3-point functions in a simple but illustrative case. This article is based on [1, 2, 3, 4].

2. Multi-field inflation with a single light field

For clarity, we have considered a 2-field model. But our approach can immediately be generalized to more complicated models. We begin by considering the action:

$$S = \int d^4x\sqrt{-g}\left[\frac{m_{\rm Pl}^2}{2}R - \frac{1}{2}g^{\mu\nu}\partial_\mu\phi^a\partial_\nu\phi_a - V(\phi^a)\right] , \tag{1}$$

where the field indices is raised and lowered by the field space metric. Now, assume that we have a set of background solutions $\phi_0^a(t)$ and $a(t)$ with given boundary conditions. Then, because of the invariance of the action under finite time translations $t \to \hat{t} = t + \xi^0$, we can generate a family of non-trivial solutions of the form:

$$\phi^a_\Delta(t) = \phi^a_0(t + \Delta\mathcal{T}) \,, \quad and \tag{2}$$

$$a_\Delta(t) = a(t + \Delta\mathcal{T}) e^{\Delta\mathcal{R}} \,, \tag{3}$$

where $\Delta\mathcal{T}$ and $\Delta\mathcal{R}$ are arbitrary constants of our choice.

We have now considered perturbations around the background solution. We have begun by giving some definitions as: The fields $\pi = \pi(t, \boldsymbol{x})$, and $\mathcal{F} = \mathcal{F}(t, \boldsymbol{x})$ to represent departures from the homogeneous and isotropic background solution $\phi^a_0(t)$ as:

$$\phi^a(t, \boldsymbol{x}) = \phi^a_0(t + \pi) + N^a(t + \pi)\mathcal{F} \,, \tag{4}$$

where $N^a(t+\pi)$ stands for the vector N^a evaluated at $t+\pi$. Notice that π represents deviations from $\phi^a_0(t)$ exactly along the path defined by the background solution, whereas $\mathcal{F}$ parametrizes deviations off the trajectory, but evaluated at $t + \pi$. In fact, the field $\mathcal{F}$ lives in the tangent space spanned by $N^a(t + \pi)$. For gravitational sector, we have written the spatial metric h_{ij} as:

$$h_{ij} = a(t + \pi) e^{2\mathcal{R}} \delta_{ij} \,. \tag{5}$$

Then, it immediately follows that there must exist a non-trivial solution for π, $\mathcal{R}$ and $\mathcal{F}$ such that,

$$\pi = \text{constant} \,, \quad \mathcal{R} = \text{constant} \,, \quad \mathcal{F} = 0 \,. \tag{6}$$

Furthermore, provided that $\mathcal{F}$ is sufficiently massive, integrating it out will necessarily result in a theory where π =constant and $\mathcal{R}$ =constant are preserved as non-trivial solutions to all orders in perturbation theory. The structure of the perturbation action must be in the form which has these non-trivial solutions.

Given that we have been eventually interested in integrating out $\mathcal{F}$, it is convenient to work in the comoving gauge, where $\pi = 0$. A special advantage of the comoving gauge is that the gauge invariant comoving curvature perturbation exactly coincides with $\mathcal{R}$, thus, any property which our original $\mathcal{R}$ possesses is inherent in the comoving gauge. This is not the case, for example, in the flat gauge. Then, we can systematically expand the action in powers of $\mathcal{R}$ and $\mathcal{F}$. For definiteness, let us explicitly compute up to cubic action. Then, to integrate out $\mathcal{F}$, we have solved for it to *linear* order and plugged the formal solution back into the action wherever $\mathcal{F}$ appears[1], and expanded in powers of M^2_{eff}, the effective mass of $\mathcal{F}$, which to first order means dis-regarding spacetime derivatives acting on $\mathcal{F}$. This gives:

$$\mathcal{F} = -\frac{2\dot{\phi}_0 \dot{\theta}}{H M^2_{\text{eff}}} \dot{\mathcal{R}} \,, \tag{7}$$

where $\dot{\phi}^2_0 = \dot{\phi}^a \dot{\phi}_a$ and $\dot{\theta} = N^a V_a / \dot{\phi}_0$ is the angular velocity for the trajectory. Plugging $\mathcal{F}$ back into the action gives us the desired effective single field theory for $\mathcal{R}$, which we have written as: $S_{\text{eff}} = S_2 + S_3$ with:

$$S_2 = \int d^4x \frac{a^3 \epsilon m^2_{\text{Pl}}}{c^2_s} \left[\dot{\mathcal{R}}^2 - c^2_s \frac{(\nabla\mathcal{R})^2}{a^2} \right] \,, \quad and \tag{8}$$

$$S_3 = \int d^4x a^3 \left[-\epsilon m^2_{\text{Pl}} \mathcal{R} \frac{(\nabla\mathcal{R})^2}{a^2} + 3 \frac{\epsilon m^2_{\text{Pl}}}{c^2_s} \dot{\mathcal{R}}^2 \mathcal{R} + \epsilon m^2_{\text{Pl}} \frac{\left(1 - c^2_s\right)^2 - 2}{2c^4_s} \frac{\dot{\mathcal{R}}^3}{H} \right.$$
$$\left. + \frac{m^2_{\text{Pl}}}{2a^4} \left\{ \left(3\mathcal{R} - \frac{\dot{\mathcal{R}}}{H} \right) \left[\psi^{,ij} \psi_{,ij} - (\Delta\psi)^2 \right] - 4\mathcal{R}^{,i} \psi_{,i} \Delta\psi \right\} \right] \,, \tag{9}$$

where $\psi = -\mathcal{R}/H + \epsilon \Delta^{-1} \dot{\mathcal{R}}/c^2_s$, and $c^{-2}_s = 1 + \dot{\theta}^2 / M^2_{\text{eff}}$.

[1] This prescription exactly reproduces the tree level effective action, while loop contributions are not captured.

3. Features from heavy physics

Given the quadratic action by Eq. (8), we may insert the canonical quadratic action as:

$$S_{2,\mathrm{canonical}} = \int d^4x a^3 \epsilon m_{\mathrm{Pl}}^2 \left[\dot{\mathcal{R}}^2 - \frac{(\nabla\mathcal{R})^2}{a^2}\right] , \tag{10}$$

where $c_s = 1$, by writing the whole effective action as:

$$S_{\mathrm{eff}} = S_{2,\mathrm{canonical}} + \underbrace{\int d^4x a^3 \epsilon m_{\mathrm{Pl}}^2 \left(\frac{1}{c_s^2} - 1\right) \dot{\mathcal{R}}^2}_{\equiv S_{2,\mathrm{int}}} + S_3 . \tag{11}$$

That is, instead of absorbing the second term into the "free" part of the action, and hence, being able to recast the quadratic effective action for the curvature mode as one with a modifed speed of sound c_s^2 , we can also consider it as a perturbation. A particular advantage of working in this manner is that the Green's functions for the free part of our theory are simple, corresponding to those of a light scalar field in quasi de Sitter spacetime, with $d\tau = dt/a$ being the conformal time,

$$\mathcal{R}_k(\tau) = \frac{iH}{\sqrt{4\epsilon k^3} m_{\mathrm{Pl}}} (1 + ik\tau) e^{-ik\tau} . \tag{12}$$

This is in contrast to the Green's functions for the theory, where we have included the interaction term in the free part, which exhibits a modified speed of sound. The Green's function for this theory would only be available if c_s^2 were a *constant*, a situation we are not particularly interested in restricting ourselves to. The two approaches would, of course, give us the same answers if the latter were implementable.

Given an interaction Lagrangian, it is convenient to adopt the in-in formalism. The expectation value evaluated at a time t of a time dependent operator $\widehat{\mathcal{O}}(t)$ is written as:

$$\left\langle \widehat{\mathcal{O}}(t) \right\rangle = \sum_{n=1}^{\infty} i^n \int_{t_{\mathrm{in}}}^{t} dt_n \int_{t_{\mathrm{in}}}^{t_n} dt_{n-1} \cdots \int_{t_{\mathrm{in}}}^{t_2} dt_1 \left\langle 0 \middle| \left[H_{\mathrm{int}}(t_1), \left[H_{\mathrm{int}}(t_2), \cdots \left[H_{\mathrm{int}}(t_n), \widehat{\mathcal{O}}(t)\right] \cdots \right]\right] \middle| 0 \right\rangle , \tag{13}$$

where t_{in} is some early "in" time when the interaction is turned on. It is described by the interaction Hamiltonian $H_{\mathrm{int}}(t) = -L_{\mathrm{int}}(t)$ for cubic or higher order. We can then straightforwardly compute the correction to the featureless power spectrum, which is easily evaluated via Eq. (13). We have found that the contributions to the featureless power spectrum $\mathcal{P}_{\mathcal{R}}$ arising from the interaction $S_{2,\mathrm{int}}$ is given by:

$$\frac{\Delta\mathcal{P}_{\mathcal{R}}}{\mathcal{P}_{\mathcal{R}}} = \kappa \int_0^{\infty} dt \left(1 - \frac{1}{c_s^2}\right) \sin(2\kappa t) , \tag{14}$$

where $\kappa \equiv k/k_\star$ with $k_\star$ being a fiducial reference, and the leading bispectrum by:

$$\begin{aligned} B_{\mathcal{R}}(\boldsymbol{k}_1, \boldsymbol{k}_2, \boldsymbol{k}_3) =& 2\Re\Bigg\{ - 2i\widehat{\mathcal{R}}_{k_1}(0)\widehat{\mathcal{R}}_{k_2}(0)\widehat{\mathcal{R}}_{k_3}(0) \Bigg[\frac{3}{2}\epsilon\frac{m_{\mathrm{Pl}}^2}{H^2} \int_{-\infty}^{0} d\tau \left(1 - \frac{1}{c_s^2}\right)^2 \tau^{-1} \frac{d\widehat{\mathcal{R}}_{k_1}^*(\tau)}{d\tau}\frac{d\widehat{\mathcal{R}}_{k_2}^*(\tau)}{d\tau}\frac{d\widehat{\mathcal{R}}_{k_3}^*(\tau)}{d\tau} \\ & - \epsilon\frac{m_{\mathrm{Pl}}^2}{H^2} \int_{-\infty}^{0} d\tau \frac{\epsilon - 3 + 3c_s^2}{c_s^4} \tau^{-2} \frac{d\widehat{\mathcal{R}}_{k_1}^*(\tau)}{d\tau}\frac{d\widehat{\mathcal{R}}_{k_2}^*(\tau)}{d\tau} \widehat{\mathcal{R}}_{k_3}^*(\tau) + 2 \text{ perm} \\ & + \epsilon\frac{m_{\mathrm{Pl}}^2}{H^2} (\boldsymbol{k}_1 \cdot \boldsymbol{k}_2 + 2 \text{ perm}) \int_{-\infty}^{0} d\tau \frac{\epsilon - 2s + 1 - c_s^2}{c_s^2} \tau^{-2} \widehat{\mathcal{R}}_{k_1}^*(\tau)\widehat{\mathcal{R}}_{k_2}^*(\tau)\widehat{\mathcal{R}}_{k_3}^*(\tau) \\ & + 2\epsilon(\epsilon - \eta_{\|})\frac{m_{\mathrm{Pl}}^2}{H^2} \int_{-\infty}^{0} d\tau \frac{s}{c_s^4} \tau^{-3} \frac{d\widehat{\mathcal{R}}_{k_1}^*(\tau)}{d\tau} \widehat{\mathcal{R}}_{k_2}^*(\tau)\widehat{\mathcal{R}}_{k_3}^*(\tau) + 2 \text{ perm} \Bigg]\Bigg\} , \end{aligned} \tag{15}$$

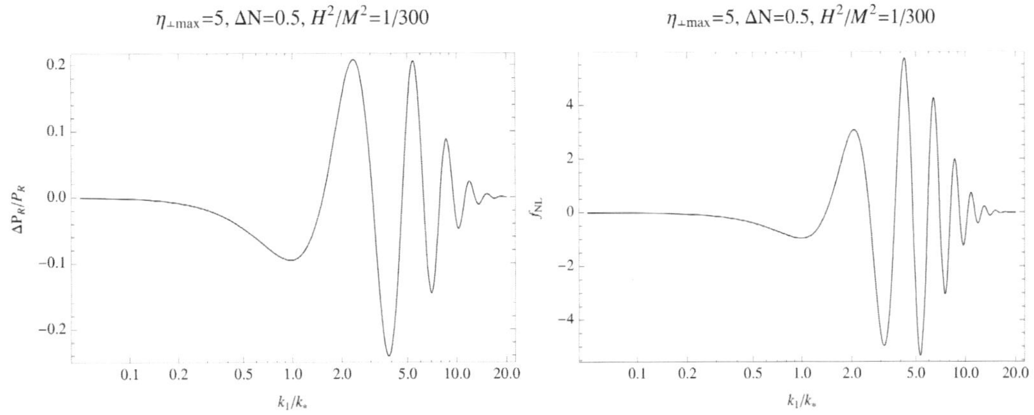

Figure 1. Features of the (left) power spectrum and (right) $f_{\rm NL}$ from the curvilinear trajectory given by Eq. (16). As the turn is experienced ($k = k_\star$) oscillatory features arise, and vanish as we probe smaller scales.

with $s = \dot{c}_s/(Hc_s)$ and $\eta_\| = -\ddot{\phi}_0/(H\dot{\phi}_0)$.

As an illustrative example, we have considered the case that there is a smooth single turn in the otherwise straight trajectory. We have parametrized the turn using the hyperbolic cosine function as:

$$\eta_\perp = \frac{\eta_{\perp\max}}{\cosh^2\left[2\log(\tau/\tau_\star)/\Delta N\right]}\,, \tag{16}$$

where $\eta_\perp = \dot{\theta}/H$ and $\star$ denotes the moment when the bending is at its maximum. Then, the resulting features in the power spectrum and the non-linear parameter $f_{\rm NL}$ are shown in Figure 1 [4]. We have seen that the features are correlated, which could be a distinctive observational signature for future experiments.

4. Conclusions

We have presented the effective single field theory of multi-field inflation. When the mass hierarchy is large, we can integrate out the heavy fields, and have single field desciption of inflation. The footprints of the heavy fields are given as a non-trivial speed of sound c_s. This leaves features in the correlation functions of the curvature perturbation. Importantly, these features are correlated with the degree of non-Gaussianity enhanced, which may be detected in near future observations.

Acknowledgements

The author thanks Ana Achucarro, Gonzalo Palma and Subodh Patil for fruitful collaborations, and has been partly supported by Korean-CERN fellowship.

References

[1] Achucarro A, Gong J O, Hardeman S, Palma G A and Patil S P 2011 *Phys. Rev. D* **84** 043502 [1005.3848 [hep-th]]
[2] Achucarro A, Gong J O, Hardeman S, Palma G A and Patil S P 2011 *JCAP* **01** 030 [1010.3693 [hep-ph]]
[3] Achucarro A, Gong J O, Hardeman S, Palma G A and Patil S P 2011 [1201.6342 [hep-th]]
[4] Achucarro A, Gong J O, Palma G A and Patil S P *To appear*

Journal of Physics: Conference Series **484** (2014) 012058 doi:10.1088/1742-6596/484/1/012058

Primordial power spectrum features and consequences

G Goswami

IUCAA, Post Bag 4, Pune University Campus, Ganeshkhind, Pune 411007, India

E-mail: gaurav@iucaa.ernet.in

Abstract. The present Cosmic Microwave Background (CMB) temperature and polarization anisotropy data is consistent with not only a power law scalar primordial power spectrum (PPS) with a small running but also with the scalar PPS having very sharp features. This has motivated inflationary models with such sharp features. Recently, even the possibility of having nulls in the power spectrum (at certain scales) has been considered. The existence of these nulls has been shown in linear perturbation theory. What shall be the effect of higher order corrections on such nulls? Inspired by this question, we have attempted to calculate quantum radiative corrections to the Fourier transform of the 2-point function in a toy field theory and address the issue of how these corrections to the power spectrum behave in models in which the tree-level power spectrum has a sharp dip (but not a null). In particular, we have considered the possibility of the relative enhancement of radiative corrections in a model in which the tree-level spectrum goes through a dip in power at a certain scale. The mode functions of the field (whose power spectrum is to be evaluated) are chosen such that they undergo the kind of dynamics that leads to a sharp dip in the tree level power spectrum. Next, we have considered the situation in which this field has quartic self interactions, and found one loop correction in a suitably chosen renormalization scheme. Thus, we have attempted to answer the following key question in the context of this toy model (which is as important in the realistic case): In the chosen renormalization scheme, can quantum radiative corrections be enhanced relative to tree-level power spectrum at scales, at which sharp dips appear in the tree-level spectrum?

1. Introduction

It is a well known fact that the correlations of CMB anisotropies depend not only on the values of various (late time) background cosmological parameters (e.g., Ω_b, Ω_c, etc.) but also on the assumed form of the PPS. Cosmological inflation has been a very actively investigated paradigm for explaining the origin of primordial metric perturbations that lead to anisotropies in CMB sky as well as to the large scale structure of the universe. There exist models of inflation that give a smooth, nearly scale invariant scalar PPS (with possibly small running of the spectral index). But, there also exist models [1, 2], in which the PPS is not so smooth, so has very sharp features. Theoretically, at this stage, there is no way to favour models that give smooth (featureless) PPS with those that do not. Interestingly, at this stage, even observationally, it is not possible to favour one of these. In many such theoretical scenarios, the scalar PPS has cuspy dips that sometimes correspond to a null in the PPS, i.e., precisely zero scalar power at some wave number. Also, for a range of modes near such a feature, the tensor power overtakes scalar power. Such cusp-like dips in scalar PPS were reported in the literature but their origin was not satisfactorily understood until recently (see [3] and references therein). An exact null in scalar PPS can have interesting consequences, such as on processed non-linear matter power spectrum. This leads to an interesting possibility: Since the power spectrum calculation is done perturbatively, and since the leading order answer is too small (or zero) at some scale, could the higher order corrections be important at this scale? If there is no power at some scale, can higher order corrections become so important that they become dominant at this scale? We have attempted to answer such questions in this paper in the context of a toy field theory.

2. Two-point functions

Recall that in cosmological perturbation theory, one is concerned with the theory of fluctuations around a background (inflationary) solution. One quantizes these fluctuations and calculates their correlation functions (on a constant time hyper-surface) using the (well known) in-in formalism. The leading higher order corrections to the 2-point function involve the one loop corrections. In the most extreme case of our interest, the tree-level contribution to the 2-point function is zero and the power spectrum at this scale gets generated purely by radiative corrections. We are interested in the 2-point function given by:

$$\langle\phi(\eta,\vec{x}_1)\phi(\eta,\vec{x}_2)\rangle = \int \frac{d^3\vec{k}_1}{(2\pi)^3}\frac{d^3\vec{k}_2}{(2\pi)^3} e^{i(\vec{k}_1\cdot\vec{x}_1+\vec{k}_2\cdot\vec{x}_2)} \times (2\pi)^3\delta^3(\vec{k}_1+\vec{k}_2)\mathcal{G}(k_1,\eta), \qquad (1)$$

where the delta function on the right hand side is due to spatial homogeneity of the background. The connected part of the quantity $\mathcal{G}$, introduced in Eq. (1), can be evaluated perturbatively as:

$$\mathcal{G}(p,\eta) = C_0(p,\eta) + C_1(p,\eta)\lambda + C_2(p,\eta)\lambda^2 + \cdots . \qquad (2)$$

Since the limit $\lambda \to 0$ recovers the free theory, $C_0(p,\eta) = |\phi_p(\eta)|^2$ where $\phi_p(\eta)$ is the mode function which determines the tree level PPS. In this article, we have attempted to calculate $C_1(p,\eta)$ perturbatively.

3. Correlation functions

To illustrate the effect we are interested in, we have tried to calculate correlation functions in a field theory, which mimics the dynamics of comoving curvature perturbation $\mathcal{R}(\eta,\vec{x})$. We have solved for the correlations of a fictitous field σ with the following (classical) action:

$$S = \int d\eta\, d^3\vec{x}\, z^2\left(\frac{(\sigma')^2}{2} - \frac{(\vec{\nabla}\sigma)^2}{2} - z^2\frac{\lambda\sigma^4}{4!}\right), \qquad (3)$$

where we have replaced all occurances of a in the action of a test scalar field (on an expanding unvierse) by $z = -a\sqrt{2\epsilon}$ (here, a is the scale factor and ϵ is the slow-roll parameter) and the function $z(\eta)$ shall be chosen such that the tree-level power spectrum has sharp features. The realistic case is quite difficult to deal with, so we have made a few simplifying assumptions. To begin with, let us assume quartic self interactions of the fluctuation field (whose power spectrum (PS) has a sharp dip at tree level). This is because the one-loop one-vertex diagram will go as $\mathcal{R}^4$, while the one-loop two-vertex diagram will go as $\mathcal{R}^6$, $\mathcal{R}$ being the comoving curvature perturbation. It is important to realize that the action in Eq. (3) is not as arbitrary as it may look naively. When one writes down the fourth-order action for fluctuations (e.g., comoving curvature perturbation), one gets too many terms (see e.g., [4], where this is done without assuming slow-roll approximation for $\delta\phi$ in spatially flat gauge, in this gauge $\mathcal{R} = \frac{H}{\dot{\phi}}\delta\phi$, so we can express the action in terms of $\mathcal{R}$). The term with quartic self-interactions is one among many of these terms (corresponding to the first term in Eq. (36) of [4], notice that the coefficient of quartic term is such that we will get the factor of z^4 in the coefficient as is assumed above), and we shall be finding out what will happen to the loop corrections to the power spectrum due to this term alone. As a second simplifying assumption, let us attempt the case in which though the tree level PS has a sharp dip, this dip does not lead to a null in the PS. This will ensure that the one-loop correction is not dominant over the tree level answer, only gets enhanced (thus, perturbation theory will still be valid). The quadratic part of the above action determines the tree level power spectrum and we have chosen the dynamics of z such that the tree level power spectrum has a sharp dip, but not a null (we have chosen z to take the form it takes in Starobinsky model [2]).

3.1. Correlation functions in canonical formalism
One can canonically quantize such a theory in the usual way. We have calculated the correlation functions in canonical formalism in an interaction picture, in which the time-dependence of the field operators shall be governed by the part of the Hamiltonian quadratic in the field, which satisfies a linear differential equation. Given the mode functions of these free fields, any correlation function can be evaluated.

While evaluating correlation functions at one loop, one encounters (just like in flat spacetime field theory) divergences. Apart from the usual UV divergences and IR divergences (the comoving curvature perturbation behaves very much like a massless test field), one also encounters (extra) late time divergences. This is because when we evaluate the 2-point function in the in-in formalism (for an interaction with an even power of field) we get:

$$\langle\Omega|\phi_H(\eta,\vec{x}_1)\phi_H(\eta,\vec{x}_2)|\Omega\rangle = \langle 0|\phi_I(\eta,\vec{x}_1)\phi_I(\eta,\vec{x}_2)|0\rangle$$
$$+i\int_{-\infty}^{\eta} d\eta' a(\eta')\langle 0|[H_I(\eta'),\phi_I(\eta,\vec{x}_1)\phi_I(\eta,\vec{x}_2),]|0\rangle + \cdots \,, \tag{4}$$

and the contribution when the external time η is set to zero, diverges.

Due to the presence of the above stated late time divergences, it is natural that the loop corrections to the correlations cannot be kept small. There is a vast literature on this subject, attempting to understand the origin and resolution of these divergences. It is important to understand that this is not the kind of enhancement we are interested in: We are talking about a situation in which the loop correction gets enhanced relative to tree level result, because the tree level contribution itself has a sharp dip with very low power. To get rid of the late time divergences, we shall cut the (one-loop) time integral. The prescription (which we have used) to do this is that, just like the tree level PS, even the loop correction to the power spectrum for the field on de Sitter background is scale invariant.

It is important to note that if one takes into account all the self interactions of the metric fluctuations (e.g., the comoving curvature perturbation) dictated by the symmetries of the problem and evaluates the one loop corrections in such a realistic case carefully, one gets no late time divergences [5] and the loop integral receives most contributions near the times, at which the mode is crossing the Hubble radius. So, by putting a cut-off in time integral (a little after Hubble crossing) in the toy field theory with quartic interactions we are attempting to simulate this.

In renormalized pertrubation theory, the bare Lagrangian (consisting of bare field, bare mass, bare self coupling and bare (non-minimal) coupling to gravity ($\xi R\sigma^2$), all of which depend in the UV regulator) is to be partitioned into a renormalized Lagrangian and a counter-term Lagrangian, the way this partition is done determines the choice of renormalization scheme. For scalar fields, in four dimensions, at one-loop (and polynomial interactions), there is no field strength renormalization. Since, we wish to keep the field massless and minimally coupled, the physical mass and physical value of ξ should both vanish. Unless we are using the so called 'On-Shell' (OS) renormalization scheme, the value of the parameters in the renormalized Lagrangian

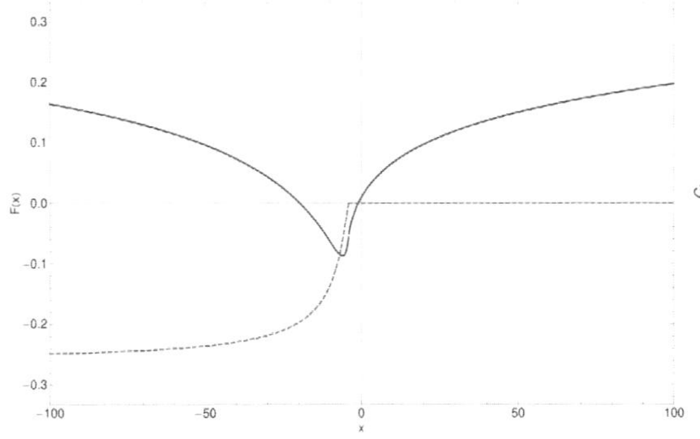

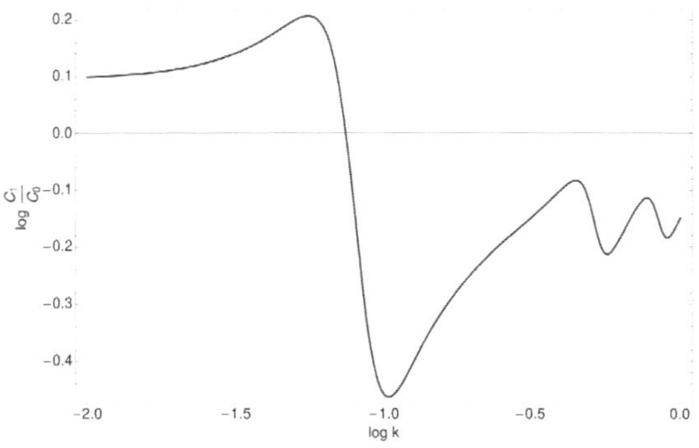

Figure 1. *Left*: For ϕ^3 theory in 6 spacetime (flat) dimensions: The variation of real part of $F(x)$ (blue, continuous) and imaginary part of $F(x)$ (red, dashed) against $x = k^2/m^2$. This function has modest values even when x becomes large. *Right*: For σ theory (Eq. (3)): Logarithm of the ratio C_1/C_0 as a function of $\log k$ for the case in which A_-/A_+ is 0.2 in MS renormalization scheme. Notice that C_1 increases with respect to C_0 and takes up its largest value at a scale just larger than the scale at which the dip in the tree-level power spectrum arises. For this value of A_-/A_+, the enhancement is very small, C_1/C_0 increases from (roughly) 26 to 58 per cent. At other values of A_-/A_+, this enhancement will be larger.

need not be the physical values of those parameters. One can implement a form of minimal subtraction renormalization scheme in a fairly straighforward way. One can easily regularize the UV divergences by putting a cut-off in three dimensional momentum space (and tune the corrections such that the field stays massless).

All this can be easily compared to what happens in the usual flat spacetime field theory. Consider ϕ^3 theory in $d = 6$ spacetime dimensions (the theory does not have a true ground state but perturbation theory does not "know" about it). Suppose we evaluate the one-loop corrections to the (four dimensional Fourier transform of) two-point function in this theory. We know the tree level answer, which is (we are using the signature $(-,+,+,+)$ and $x = k^2/m^2$) $\Delta(k^2) = \frac{1}{k^2+m^2} = \frac{1}{m^2(1+x)}$. Loop corrections are of the order of g^2 and if we call $g^2/(4\pi)^3$ to be α, then the 2-point function (one-loop) is:

$$\Delta(x) = \frac{1}{m^2(1+x)}\left[1 + \alpha F(x)\right]. \tag{5}$$

Since we are working with a weakly coupled theory, we could be working with (say) $g = 0.1$ and then $\alpha = \mathcal{O}(10^{-8})$, and the next order correction is supposed to be of the order of 10^{-16}, even if $F(x)$ becomes fairly large, perturbative calculation shall remain valid. Now, we have used dimensional regularization and OS renormalization scheme to get the function $F(x)$ as shown in the left part of Figure 1. What we are calculating is the equivalent of this for the σ field. There exist a small range of modes, near the dip in the tree-level power spectrum, where the tree-level contribution is at its lowest value while the one-loop contribution is not so small, this causes a relative enhancement in the value of C_1 with respect to C_0 (see Figure 1). The result is shown for the case in which the value of the parameter A_-/A_+ of [2] is 0.2. As one changes this parameter, the depth of the dip in tree level spectrum can be increased, thus, increasing the relative size of loop corrections. We have avoided the cases in which the depth in the dip is too deep because in those cases, the loop correction shall become very large signalling break down of perturbation theory.

Journal of Physics: Conference Series **484** (2014) 012058 doi:10.1088/1742-6596/484/1/012058

4. Conclusion

We have attempted to study how the loop corrections to the correlation functions of cosmological perturbations get affected by the actual dynamics of the mode functions. We have chosen the mode functions of the field such that the tree level power spectrum has a sharp dip, and we have realized that the power spectrum at one loop, renormalized in MS scheme can get enhanced (at scale near the tree level dip) relative to the tree level power spectrum.

References

[1] Jain R K *et al* 2009 *JCAP* **0901** 009 [arXiv:astro-ph/0809.3915], Jain R K *et al* 2010 *Phys. Rev. D* **82** 023509 [arXiv:astro-ph/0904.2518]
[2] Starobinsky A A 1992 *Pis'ma Zh. Éksp. Teor. Fiz.* **55** 477, 1992 *JETP Lett.* **55** 489
[3] Goswami G and Souradeep T 2011 *Phys. Rev. D* **83** 023526
[4] Seery D, Lidsey J E and Sloth M S 2007 *JCAP* **0701** 027 [astro-ph/0610210]
[5] Senatore L and Zaldarriaga M 2010 *JHEP* **1012** 008 [hep-th/0912.2734]

Study on caustic formation in Dirac-Born-Infeld type scalar field systems

U D Goswami[1*], H Nandan[2] and M Sami[3]

[1]Department of Physics, Dibrugarh University, Dibrugarh 786004, Assam, India
[2]Department of Physics, Gurukula Kangri Vishwavidayalaya, Haridwar 249404, Uttarakhand, India
[3]Center for Theoretical Physics, Jamia Millia Islamia, Jamia Nagar, New Delhi 110025, India

E-mail: *umananda2@gmail.com

Abstract. Formation of caustics in the Dirac-Born-Infeld type scalar field systems has been investigated for two generic of potentials, viz., exponentially decreasing and inverse power law potentials. The study reveals that in the case of exponentially decreasing rolling massive potential, there are multi-valued regions and regions of likely to be caustics in the field configuration. The formation of caustics is inevitable for the inverse power law potentials under consideration in Minkowski spacetime, whereas caustics do not form in this case in the Friedmann-Robertson-Walker Universe.

1. Introduction

The widely accepted idea of the cause of accelerated expansion of the present Universe [1] is the existence of a mysterious form of energy, dubbed as *dark energy*. The dark energy exerts a large negative pressure for the present accelerated expansion and is the dominant fraction of the energy content of our Universe. This idea believes on the validity of the general theory of relativity (GTR) over all energy scales, however, there are many alternative ideas as well, which see the phenomenon of accelerated expansion from different perspectives (e.g., see [2]). As a viable physical explanation of the observed late time cosmic acceleration, a variety of scalar field models have been proposed as a source of dark energy over the years, which are basically motivated by the fundamental theories of high energy physics [2]. The Dirac-Born-Infeld (DBI) scalar field model, referred as the tachyon field, is the string theoretical model, in which, at early time when the field remain sub-dominant and frozen, the expansion dynamics is governed by the background fluid, and as the background energy density becomes comparable to the field energy density with the evolution of time, the field begins to evolve and subsequently overtakes the background to become the dominant component of the Universe [3]. Although the models, based upon the DBI scalar, have exciting features of the cosmological dynamics, they may lead to the formation of caustics, where the second, and the higher order derivatives of the field become singular. Caustic formation is an undesirable consequence in the field theoretical models in cosmology, which can be used to investigate the fundamental shortcoming of the field theory for a specific potential.

In this paper, we have investigated the issue of caustic formation in the DBI scalar field systems for inverse power law potentials, and also for the exponentially decreasing potential, by extending the analysis given in [4], where it is shown that caustics inevitably form in tachyon system with potentials decaying faster than ϕ^{-2} at infinity [5]. In Section 2, we have discussed briefly about the DBI type general scalar field equation. Numerical results on the formation of caustics are presented in the Section 3. We have summarized our main results in the last Section.

2. The Dirac-Born-Infeld type scalar field system

The DBI type action for a scalar field (tachyon field) ϕ is given by [3, 4]:

$$S = -\int d^4x\sqrt{-g}\,V(\phi)\sqrt{1+\partial_\mu\phi\partial^\mu\phi}\,, \tag{1}$$

where $V(\phi)$ is the potential of the field ϕ. The field equation derived from the action in Eq. (1) is:

$$\nabla_\mu\nabla^\mu\phi - \frac{\nabla_\mu\nabla_\mu\phi}{1+\nabla_\alpha\phi\nabla^\alpha\phi}\nabla^\mu\phi\nabla^\nu\phi - \frac{V_{,\phi}(\phi)}{V(\phi)} = 0, \tag{2}$$

where the covariant derivative of the field ϕ with respect to the metric $g_{\mu\nu}$ is denoted by ∇_μ. In the 1+1-dimensional Minkowski space, the Eq. (2) takes the form:

$$\ddot{\phi} = \left(1+\phi'^2\right)^{-1}\left[\left(1-\dot{\phi}^2+\phi'^2\right)\left(\phi''-\frac{V_{,\phi}(\phi)}{V(\phi)}\right)+2\phi'\dot{\phi}\dot{\phi}'-\phi'^2\phi''\right], \tag{3}$$

where the dot and prime over ϕ indicate the time and space derivatives of the field ϕ respectively.

Since, in this paper, we are interested only on the formation of caustics in the tachyon field, we have only considered the inhomogeneous tachyon field where ϕ', $\phi'' > 0$, however, they should be sufficiently less then unity for a realistic scalar field in cosmology [4]. Under this consideration, the Eq. (3) can be written as:

$$\ddot{\phi} = \left(1-\dot{\phi}^2+\phi'^2\right)\left(\phi''-\frac{V_{,\phi}(\phi)}{V(\phi)}\right) = P(\phi)Q(\phi), \tag{4}$$

where

$$P(\phi) = 1-\dot{\phi}^2+\phi'^2, \quad Q(\phi) = \phi''-\frac{V_{,\phi}(\phi)}{V(\phi)}. \tag{5}$$

The equation of $P(\phi)$ contains both the time and space derivatives of the field, and therefore, this parameter can be used as an indicator of the pattern of evolution of the field governed by a particular field potential. For example, if the field $\phi(x,t)$ rapidly approaches the configuration $P(\phi)=0$, it then indicates that the time and space evolution of the field is such that it rapidly approaches $\dot{\phi}^2-\phi'^2=1$. Similarly, if $P(\phi)\approx 1$, it implies that, $\dot{\phi}^2\approx\phi'^2$, i.e., the type of evolution of the field $\phi(x,t)$ with respect to space and time are nearly equal. The extensive numerical solutions of the Eq. (4) for the scalar field with the two different classes of field potentials of our interest clearly showed the above behaviour of $P(\phi)$ [5]. Thus, the Eq. (4) has two attractors, one for $P(\phi)=0$ and the other for $P(\phi)=1$, depending on the field potential. Corresponding to the above two attractors, we have obtained the parametric solutions for the field $\phi(x,t)$ [5]. This analysis can be easily enhanced to the more realistic case of expanding, spatially flat, Friedmann-Robertson-Walker (FRW) universe, assuming a small inhomogeneous field perturbation for a specific field potential [5].

3. Caustics in DBI system

We have studied the possibility of the formation of caustics in the DBI techyon system with the aforesaid two generic classes of potentials, and the numerical results are presented.

The exponentially decreasing rolling massive scalar field potential is given by:

$$V(\phi) = V_0 e^{-\frac{1}{2}M^2\phi^2}, \tag{6}$$

where V_0 and M are constants. The inhomogeneous scalar field given in Eq. (4) for this potential can be written as:

$$\ddot{\phi} = \left(1 - \dot{\phi}^2 + \phi'^2\right)\left(\phi'' + M^2\phi\right). \tag{7}$$

We have solved the Eq. (7) for different initial values of the field ϕ and for different values of the parameters M, ϕ' and ϕ''. We have taken the optimum values of all these parameters from our extensive numerical simulation to use the solution of the Eq. (7) for $\phi(x,t)$, to study the patterns of variations of $P(\phi)$ and $Q(\phi)$ with respect to time [5]. From this study, we have considered the solution of $P(\phi)$ with $M = 1$ and $\phi_0 = 0$, and the initial field configuration as $\phi_i(q) = exp(-q^2)$ [4], to obtain the characteristic curves, which are shown in the left panel of the Figure 1. From these characteristic curves, it is observed that there are likely to be caustics as well as multi-valued regions in the field profile, which are independent of the initial conditions of the field [5].

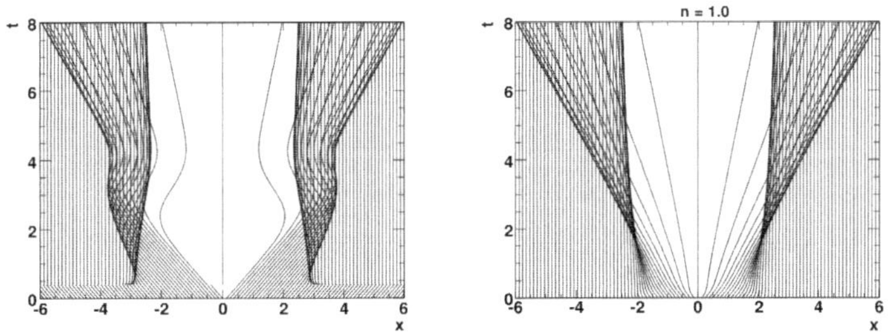

Figure 1. The particle trajectories in 1 + 1 Minkowski space obtained for potentials given in Eqs. (6) and (8) respectively.

Next, we have considered the inverse power law potential of the form, given by:

$$V(\phi) = V_0\phi^{-n}, \quad 0 < n < 2. \tag{8}$$

The late time accelerated expansion of the universe corresponds to $0 \leq n \leq 2$, and accordingly we have restricted the value of the exponent n within this range for our work. The inhomogeneous scalar field given in Eq. (4), in this case takes the form:

$$\ddot{\phi} = \left(1 - \dot{\phi}^2 + \phi'^2\right)\left(\phi'' + \frac{n}{\phi}\right). \tag{9}$$

We have followed the same numerical procedure as in the previous case for this potential also. The plots of the characteristic curves for this potential are shown in the right panel of the Figure 1, for $n = 1.0$ with $\phi_0 = 1.0$ [5]. This panel of the figure clearly shows the formation of caustics and multi-valued regions for the inverse power law potential.

As expansion works against caustics formation, we have expected not to encounter caustics in the case of exponentially decreasing potential, but even caustics are more distinctly form for

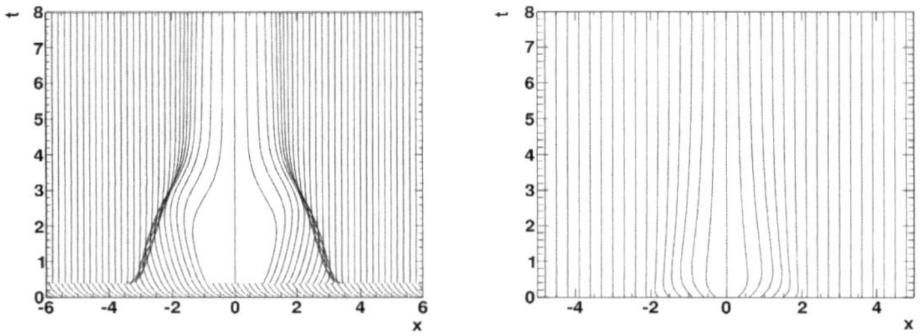

Figure 2. The particle trajectories in FRW expanding universe. The left panel is for the potential in Eq. (6), and the right is for the potential in Eq. (8). There are caustics and multivalued regions in the field configuration for the potential in Eq. (6).

the expanding universe in this case, which is clear from the left panel of the Figure 2. In the case of the inverse power law potential, there is no caustics in the field configuration as seen from in the right panel of the Figure 2. Since, in the case of this potential, the dark energy behaves as a late time attractor, and the characteristic curves, obtained by taking the form of $a(t) \sim t^{2/3(1+w)}$, incorporating all time dark energy effect, are very similar to the characteristic curves of 1 + 1-dimension at late time [5].

4. Conclusion

The caustic formation in mass free space is an important issue for the viability of the various scalar field models in the context of the dark energy paradigm. We have examined the formation of caustic in the tachyon system for two generic classes of potential, viz, massive rolling scalar potentials and inverse power law potentials. Our study is supported by analytical estimation as well as the extensive numerical simulations [5]. We have observed that caustics are formed in the tachyon field with exponentially decreasing massive scalar potential for both 1 + 1-dimensional Minkowski space, and flat FRW expanding universe. The caustic formation is more definite in the case of the expanding universe than the case of 1 + 1-dimensional Minkowski space. However, the expanding universe dilutes the effect of caustics as clear from the figures, but cannot render the situation caustic free, which is generally true for a tachyon potential that decays faster than $1/\Phi^2$ at infinity. For inverse power law potentials, caustics are formed with the multi-valued regions in the Minkowski spacetime. It is interesting to observe that no caustics are formed in the field of the flat FRW expanding universe for the inverse power law potentials as the effect of expansion can compete with the tendency of caustic formation. Dark energy as a late time attractor of the dynamics in this case gives rise to cosmic repulsion, allowing one to avoid caustics and multi-valued regions in the field profile. This behaviour of the field with inverse power law potentials agrees very well with the result obtained by assuming dark energy dominance at all times. Hence, the inverse power law is found suitable to explain the late time cosmic acceleration [5].

References

[1] Riess A G *et al* 1998 *Astron. J.* **116** 1009, Perlmutter S *et al* 1999 *Ap. J.* **517** 565
[2] Copeland E J, Sami M and Tsujikawa S 2006 *Int. J. Mod. Phys. D* **15** 1753, Padmanabhan T 2002 *Phys. Rev. D* **66** 021301, Bagla J S, Jassal H K and Padmanabhan T 2003 *Phys. Rev. D* **67** 063
[3] Sen A 1999 *JHEP* **10** 008
[4] Felder G N, Kofman L and Starobinsky A 2002 *JHEP* **0209** 026
[5] Goswami U D, Nandan H and Sami M 2010 *Phys. Rev. D* **82** 103530

Journal of Physics: Conference Series **484** (2014) 012060 doi:10.1088/1742-6596/484/1/012060

Loop quantum cosmology in the cosmic microwave background

J Grain

Institut d'Astrophysique Spatiale, CNRS/Université Paris XI
Centre Universitaire d'Orsay, Bât. 120-121, 91405 Orsay CEDEX, France

E-mail: julien.grain@ias.u-psud.fr

Abstract. The primordial universe can be used as a laboratory to set constraints on quantum gravity. In the framework of Loop Quantum Cosmology, we show that such a proposal for quantum gravity, not only solves for the big bang singularity issue, but also naturally generates inflation. Thanks to a quantitative computation of the amount of gravity waves produced in the loopy early universe, we show that future cosmological data on the polarized anisotropies of the Cosmic Microwave Background can be used to probe LQC model of the universe.

1. Introduction

Building a quantum theory of gravity is probably the most outstanding problem of modern physics, but remains also one of the most difficult tasks. Facing such a great challenge, and retrospectively, looking at the many attempts developed since the last decades, one could be tempted to consider that searching for quantum gravity is a never ending quest and that the wise attitude would be to withdraw such a challenge. However, quantum gravity is not *optional*! Several heuristic arguments pin down the necessity of a successful building of quantum gravity [1]. More importantly, the singularity theorems derived by Penrose and Hawking [2] show that the general relativity ontologically fails: A way out reconciling quantum mechanics and general relativity is mandatory!

A new physical paradigm is needed and to be successful, and such a potentially new paradigm requires a well-established theoretical framework. On the one hand, this assumes a well-posed theoretical problem[1], namely, how to quantize general relativity or alternatively, how to 'gravitize' quantum field theory? Many tentative theories have been proposed such as string theory [3] and Loop Quantum Gravity (LQG) [4] to mention the most popular ones, and though some of them have made great progresses, none of them is fully set. On the other hand, this also implies a new kind of paradigmatic experiments or observations. This is a mandatory step to discriminate theoretical proposals, which are deeply on the wrong track from those which are not, as well as to guide theoretical developments. This well-posed phenomenological and experimental problem is the second main difficulty of quantum gravity as quantum gravitational effects should pop up at an energy scale of $M_{\rm Pl} \sim 10^{19}$ GeV, far beyond any high energy physics experiments. Fortunately, the failure of general relativity precisely points towards those phenomena potentially probing quantum gravity: Spacetime singularities provide the territory

[1] By a well-posed theoretical problem, we mean an *ill-posed* enigma with respect to establised theories.

for quantum gravity and one, therefore, has to search for those phenomena built up of singular spacetimes. This qualifies the primordial universe, being the neighbourhood of the big bang singularity, as an ideal territory for probing quantum gravity!

Observing quantum gravity is, therefore, split into two questions: (i) How a given proposal for quantum gravity affects the physics of the primordial Universe?, and (ii) How can we probe the physics of the primordial universe? Apparently, the second question points towards observation. But, as a theoretical question, this should not be overlooked, since it strongly determined the precise implementation of the first question.[2] The main requirement for a quantum gravity proposal is to solve for the pathological big bang singularities, and this can be checked by computing the global evolution of the universe in a quantum cosmological setting. However, it should be stressed out that such a global evolution cannot be probed directly, as one cannot extract, from the universe, to 'see' how it evolves in its primary ages. We are stuck inside the Universe and from which, we are able to probe the physics of the early universe via the Cosmic Microwave Background (CMB) anisotropies. The origin of such anisotropies are *cosmological perturbations* produced during a phase of accelerated expansion dubbed *cosmic inflation*. Inflation is a key ingredient of the standard cosmological paradigm as it solves for the horizon and flatness problems and provides a mechanism for generating perturbations, the primordial seeds for galaxies and large scale structures formation. Moreover, inflation could be a high energetic phenomenon (its energy scale could be as high as $\sim 10^{16}$ GeV) and is, therefore, potentially affected by quantum gravity. Nevertheless, the inflationnary scenario is not free of any problem as it is very difficult to generate such a phase without invoking speculative physics or fine tuning.

Bridging a link between Loop Quantum Cosmology (LQC), the cosmological implementation of LQG [5], and inflation in the early universe allows us to solve for the big bang singularity *and* naturally trigger a phase of inflation. In addition, a quantitative computation of cosmological perturbations produced in such a *loopy* universe makes the early universe a potential laboratory for testing quantum gravity. Our methodology is the following: The evolution of the universe is derived in the framework of *effective* LQC, which includes first order quantum corrections coming from holonomies. On top of modified cosmological background, we will quantitatively compute the amount of cosmological perturbations of tensor type (*i.e.*, primordial gravity waves), and investigate how it impacts on the B-mode angular power spectrum of CMB anisotropies. Such an angular power spectrum is mainly observable, and can finally be used to forecast how future CMB observations could be used to probe LQC. We consider the early Universe to be filled with a massive scalar field Φ as matter sources, and our time variable is cosmic time t.

2. Background evolution

With holonomy corrections at first order, the modifed Friedman and Klein-Gordon equations are (denoting $H = \dot{a}/a$, the Hubble parameter and m_Φ, the mass of the scalar field):

$$H^2 = \frac{8\pi G}{3}\rho\left(1 - \frac{\rho}{\rho_c}\right) \quad \text{and} \quad \ddot{\Phi} + 3H\dot{\Phi} + m_\Phi^2\Phi = 0, \tag{1}$$

where $\rho_c \sim 0.8 \times \rho_{\rm Pl}$ is a critical energy density, which cannot be overcome and encoded LQC corrections. Because of $(1 - \rho/\rho_c)$, the universe is not singular, as the big bang is now replaced by a big bounce. The basic history of the universe is, thus, a contracting phase, followed by an expanding phase. The *regular* transition from the contracting phase to the expanding one is

[2] As further discussed in the core of this note, this *observational* question requires the computation of cosmological perturbations. In the case of, *e.g.*, Loop Quantum Cosmology, implementing those perturbations on top of a quantum modified background spacetime leads to theoretical developments to cancel anomalies in the algebra [11], and points towards a change of signature, that is, an *Euclideanization* of the background, during the cosmic history [12].

ensured by the quantum corrections and the bounce occurs at $\rho = \rho_c$. More interestingly, this peculiar evolution of the universe naturally leads to inflation! Indeed, a phase of accelerated expansion can start right after the bounce if the scalar field is in the appropriate energy state for the slow-roll conditions to be fulfilled. For a massive field, this is translated into $\Phi(t_i) \sim 3.1\ M_{\rm Pl}$ with t_i denoting some time *after* the bounce. This condition is not easily met in the standard cosmological scenario. But in the LQC bouncing universe, the Hubble parameter H is negative valued during contraction. It, therefore, acts as an *anti-friction* term making Φ to climb up its potential. Right after the bounce, it appears that such a field is precisely in the appropriate energy setting for a sufficiently long phase of inflation to start [6]. The key point about this scenario has been raised by Ashtekar and Sloan, who showed that generating a phase of inflation in the LQC universe with the mandatory amount of at least 60 e-folds is very close to unity, making inflation rather natural in this framework [7].

3. CMB power spectra

To derive the amount of gravity waves produced at the end of inflation, one needs to solve the following equation, describing tensor perturbations evolving on top of the modified FLRW background:

$$\ddot{h}^i_a + 3H\dot{h}^i_a - \frac{1}{a^2}\nabla^2 h^i_a + 12\pi G\frac{\rho}{\rho_c}\left(\frac{1}{3}\rho - V(\Phi)\right)h^i_a = 0, \tag{2}$$

where the last term, acting as a time-dependant mass term, encodes the 'quantum deformation' of the background. This equation of motion [8] has been derived from an algebra, which is anomaly-free at all orders and can be safely used throughout the entire history of the bouncing universe[3]. The main characteristics of a 'bouncy' power spectrum for tensor modes are the following: The IR (large scales) part is k^2 suppressed. This is due to the freezing of very large-scale modes in the Minkowski vacuum. Those modes indeed exit the horizon long *before* the bounce and naturally exhibit a quadratic spectrum. The UV part is identical to the standard prediction. Small scales, indeed, experience a history basically similar to that of the big bang scenario. They exit the horizon during inflation and re-enter later, leading to the standard nearly scale-invariant spectrum. Intermediate scales, around $k \approx k_\star$, exhibit both a bump of amplitude R and damped oscillations[4]. This is mostly due to the fact that all modes are inevitably in causal contact at the bounce (the Hubble parameter vanishes, therefore, leading to an infinite Hubble radius). Those characteristics have been fully determined by numerically solving the equations of motion of tensor perturbations propagating in the LQC-corrected, {bouncing+inflationary} universe [6].

Using a Boltzmann code, such a primordial tensor power spectrum serves as an input for computing the footprints of primordial gravity waves on the CMB B-mode angular power spectrum (see left panel of Figure 1). As compared to the standard prediction, the distortion of LQC B-mode power spectrum depends on the value of $k_\star$ [9]. (Roughly speaking, the suppression of power at large scales can be 'seen' in the CMB only for values of $k_\star$ corresponding to scales *smaller* than the Hubble horizon today.) If $k_\star < k_{\rm Hubble}$ corresponding to a length scale greater than the Hubble scale today, the tail of the bump for $k > k_\star$ translates into a slight bump at large angular scales (*i.e.*, small values of the multipole ℓ). However, for $k_\star > k_{\rm Hubble}$, the B-mode angular spectrum shows, as compared to the standard prediction, a suppression at large angular scales due to the suppression at large length scales in the primordial spectrum, and a

[3] The issue of a closed algebra, in effective LQC for both scalar and vector perturbations has been recently solved in [11].

[4] The critical wavenumber (locating the bump and determining the typical scales, for which the power spectrum changes from its k^2 shape to its nearly scale-invariant shape) is a phenomenological parameter related to m_Φ and to the fraction of potential energy at the time of the bounce. Similarly, R is a phenomenological parameter related to m_Φ (see [9] for further details).

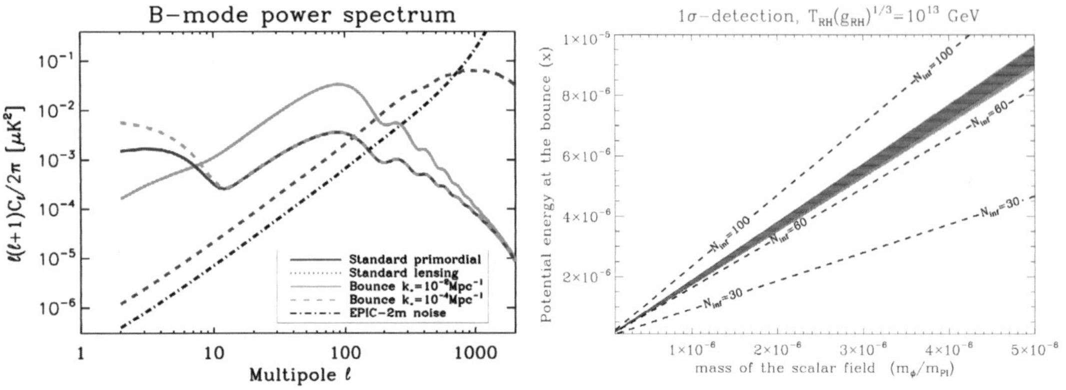

Figure 1. B-mode angular power spectrum (left panel) and detectable region of the parameters (m_Φ, x) describing a loopy universe (right panel). Figure is adapted from [9].

bump at intermediate angular scales $\ell \sim 100$. Finally, if $k_\star$ is greater than $\sim 10^{-1}$ Mpc^{-1}, the suppression is effective up to ℓ-values of a few hundreds, and the primordial part of the B-mode is, therefore, systematically pushed below the lensing-induced B-mode.

Using those angular power spectra as an observational probe, and considering both the instrumental noise as expected for a future dedicated, B-mode experiment[5] and the lensing-induced B-mode as a 'foreground' masking the detection of primordial B-mode, one can forecast some constraints to be set on the fundamental parameters describing the LQC model of the universe [9]. The detectable values at 1-σ of the mass of the scalar field m_Φ and its fraction of potential energy at the bounce, denoted x, are depicted by the blue band in the right panel of Figure 1. This roughly corresponds to a detectable range of $k_\star$ from 2×10^{-4} Mpc^{-1} to 3×10^{-1} Mpc^{-1}, considering *degeneracies* with other cosmological parameters. The upper part is not detectable as it corresponds to $k_\star \ll k_{\text{Hubble}}$, making the B-mode power spectrum *undistorted* as compared to the standard general relativistic prediction. The lower part is not detectable as it corresponds to $k_\star \gg 10^{-1}$, making the primordial B-mode systematically smaller than the lensing-induced part. Though measurements of the LQC parameters are not possible in this second case, a discrimination with pure general relativity is still possible as the suppression induced by the bounce is 'seen' via the masking of the primordial B-mode[6]; that is via its *non-detection.*

4. Conclusion

LQG offers an appealing approach to build a quantum theory of gravity. Its implementation to the symmetry reduced case of FLRW metric describing our whole Universe shows that the big bang singularity is cured and that inflation is *naturally* triggered. Moreover, thanks to a quantitative computation of gravity waves produced in the early Universe, subsequently impacting on the CMB polarized anisotropies of B-type, a possible window to constraint those loopy models of the universe using cosmological data is now opened. As stressed out in [10], the case of LQC illustrates a possible future way to test for quantum gravity using cosmological/astronomical observations.

[5] The noise level is set to 2.2 μK-arcmin., the beam width to 8 arcmin. and the sky fraction to 70%.

[6] This 'suppression' could also be interpreted as a very low value of the tensor-to-scalar ratio T/S. However, in the precise case of LQC, the three parameters T/S, $k_\star$ and R are fully determined by the *two* parameters m_Φ and x, thus, breaking the degeneracy between $k_\star$ and T/S (see [9] for details).

Acknowledgements
The author warmly thanks his collaborators Aurélien Barrau, Thomas Cailleteau and Jakub Mielczarek.

References
[1] Oriti D (Ed) 2009 *Quantum Gravity* (Cambridge University Press), Birrell N and Davies P 1982 *Quantum Field in Curved Space* (Cambrdige University Press)
[2] Hawking S and Ellis G 1973 *The Large Scale Structure of Spacetime* (Cambridge University Press)
[3] Green M, Schwarz J and Witten E 1987 *Superstring Theory* (Cambridge University Press)
[4] Rovelli C 2007 *Quantum Gravity* (Cambridge University Press), Rovelli C 1998 *Living Rev. Rel.* **1** 1, Smolin L 2004 [hep-th/0408048v3], Thiemann T 2003 *Lect. Notes Phys.* **631** 41, Perez A 2004 [gr-qc/0409061v3], Dona P and Speziale S 201 [1007.0402V1]
[5] Bojowald M 2008 *Living Rev. Rel.* **11** 4, Ashtekar A 2009 *Gen. Rel. Grav.* **41** 707
[6] Mielczarek J, Cailleteau T, Grain J and Barrau A 2010 *Phys. Rev. D* **81** 104049
[7] Ashtekar A and Sloan D 2010 *Phys. Lett. B* **694** 108, Ashtekar A and Sloan D 2010 *Gen. Rel. Grav.* **43** 3619
[8] Bojowald M and Hossain G 2008 *Phys. Rev. D* **77** 023508
[9] Grain J, Barrau A, Cailleteau T and Mielczarek J 2010 *Phys. Rev. D* **82** 123520
[10] Barrau A and Grain J [1206.1192v1]
[11] Mielczarek J, Cailleteau T, Barrau A and Grain J *To be published in Class. Quant. Grav.* [1106.3744v1], Cailleteau T, Mielczarek J, Barrau A and Grain J *To be published in Class. Quant. Grav.* [1111.3535v2]
[12] Bojowald M and Paily G M [1112.1899v1]

An underlying theory for gravity

Y K Ha

Department of Physics, Temple University, Philadelphia, Pennsylvania 19122, USA

E-mail: yuanha@temple.edu

Abstract. A new direction to understand gravity has recently been explored by considering classical gravity to be a derived interaction from an underlying theory. This underlying theory would involve new degrees of freedom at a deeper level, and it would be structurally different from classical gravitation. It may conceivably be a quantum theory or a non-quantum theory. The relation between this underlying theory and Einstein's gravity is similar to the connection between statistical mechanics and thermodynamics. We discuss the apparent lack of evidence of any quantum nature of spacetime and the meaning of quantum gravity in this context.

1. The meaning of quantum gravity

Quantum gravity has been conjectured for almost 80 years since the introduction of the graviton. It is commonly believed that gravity is a fundamental interaction and as such, it would obey quantization similar to electrodynamics. However, it is significant to point out that there is not a single observational evidence so far showing the need of a quantum theory of gravity. On the theoretical side, despite enormous efforts in the past decades, there is still no consistent quantum gravity theory with any predictive power. The only great result from quantum gravity efforts over the last 50 years is the renormalization of Yang-Mills theory by 't Hooft [1] and Veltman [2], using the techniques developed by Feynman [3] and Dewitt [4] for perturbative calculations in general relativity.

It may be helpful to look at the problem of quantum gravity in a different perspective by asking: What is not quantum gravity? At once, we see it is not about motion in spacetime, for motion with a precise trajectory is a classical concept. It is not about translation in spacetime. It is not about Lorentz transformation. It is not about elementary particles, or necessarily about unification of forces. Thus, quantizing general relativity with special relativity as a limit is a contradiction. We recognize that the quantum gravity domain is naturally of Planck size 10^{-33} cm. Furthermore, quantum gravity domain must be physical and contain real degrees of freedom, and those degrees of freedom should also carry energy, momentum, spin and other attributes like any other excellent degrees of freedom in physics.

We should also point out that the term quantum gravity has come to acquire different meaning to different researcher. The proper case is that quantum gravity should start only with Einstein's equation and not some modification of it by adding extra terms to its Lagrangian as in modified gravity theories. We can illustrate this situation with electrodynamics. In quantum electrodynamics, the equations are exactly the same as Maxwell's equations as in classical electrodynamics, and not some modification of Maxwell's equations. It is well known that a remarkable modification of Maxwell's equations has become the Yang-Mills theory. Although, the modified equations contain the original equations in form, the theory is completely different, and it is no longer electrodynamics. Quantizing a modified theory does not lead to a successful quantum theory of electrodynamics and its spectacular experimental confirmation. Therefore, current modified gravity theories are not what they intend to accomplish, namely, a quantum theory of gravitation based on general relativity. They may end up having nothing to do with ordinary gravity in the $\hbar \to 0$ limit.

2. Quantum nature of gravity

It is also common to find that in many quantum gravity theories, there is little discussion about what constitute the quantum nature of gravity. A list of conceivable quantum effects of gravity as logical outcomes of quantum idea applied to gravity is the following:

1. Existence of graviton,
2. Spacetime fluctuations and spacetime foam,
3. Discreteness of spacetime,
4. Wave-particle duality for gravity,
5. Linear superposition of spacetimes,
6. Interference of spacetimes,
7. Uncertainty relations for spacetime metric,
8. Tunnelling in gravity, and
9. Entanglement in gravity.

So far none of these phenomena has been observed and the existence of any one of the above effects would be a strong indication of the validity of quantum gravity. Some of these phenomena, such as superposition and interference of spacetimes are most difficult to perceive. Even the familiar graviton is only a hypothetical particle. All quantum field theories of gravity require the graviton as a mediating particle. The graviton corresponds to the weak field perturbation on a flat background spacetime. However, the graviton propagating in a flat spacetime contradicts the background independence of general relativity. Graviton theories are known to be divergent at high energies and they are not renormalizable. On the experimental side, it would be impossible to detect a single graviton due to its very low frequency and energy. Analysis shows that a detector must be enormous in size in order to be comparable to the graviton's wavelength. For practical observation, such a detector would become physically too massive and collapse to form a black hole, in which case the detection outcome would be lost [5]. On the other hand, an indirect evidence for the graviton would be the observation of gravitational black body radiation with a spectrum like Planck's radiation curve. Such an experimental feat would be equally daunting.

3. Astrophysical observations

There are at least a few efforts to search for the quantum nature of gravity in astrophysical settings. The idea of spacetime fluctuations and spacetime foam is a cornerstone prediction of quantum gravity. The observation of gamma ray bursts from a distant source could indicate a possible granular structure of spacetime at the smallest scale. A significant observation of the highest energy gamma rays has been conducted with the gamma ray burst GRB 090510 by the Fermi Gamma-Ray Space Telescope [6]. Gamma rays of energies up to 31 GeV from the burst are all seen to arrive within 1 second after travelling a distance of 7.3 billion light years. If the spread in time between the highest and the lowest-energy gamma rays is all attributed to quantum effects rather than production time difference, then the result rules out those quantum gravity theories in which the speed of light c varies linearly with photon energy E. A variation of photon speed is an indication that Lorentz invariance is violated. A thorough analysis shows that there are no quantum effects until the distance is down to 0.8 L_{Pl}, below the Planck length. The result indicates that spacetime is classical above the Planck length and special relativity is right.

A recent investigation also set to probe the quantum nature of spacetime is carried out on observation of the farthest quasars [7]. Light from the farthest quasars with redshift $4 < z < 5.4$ are observed by the Hubble Space Telescope. The aim is to set new limits on the fluctuations of spacetime foam and quantum gravity models. An observable effect would be the degradation of the diffraction images of distant sources. Nevertheless, a careful analysis of the diffraction images of the most distant quasars shows no blurring effect that could be caused by the interaction of photons with quantum gravity fluctuations. Photons and their quantum states can travel undisturbed for billions of light years to reach the Hubble Space Telescope. The analysis takes into account the possibility of spacetime fluctuations accumulated over cosmological distances and partly compensated by the effects of phase variations. The results exclude a number of spacetime foam models based on the holographic principle. Without phase variation corrections, all major models of quantum gravity and spacetime foam are ruled out. The results also show the absence of any directional dependence of quantum gravity effects, and therefore, the validity of the cosmological principle by an independent method.

It is puzzling to find that these observations do not reveal any quantum nature of spacetime above the Planck length. Future detections using gamma ray bursts with even higher energy photons and observation of quasars at higher red shift will provide a more stringent limit on the validity of quantum gravity. If persistent observation cannot yield any signature of quantum spacetime, then ultimately one has to come to the conclusion that spacetime fluctuations and spacetime foam are only preconceived notions. We have seen that parity conservation law in weak interactions was a preconceived notion without any support. The deeper question is: Why spacetime is so smooth and classical all the way down to the Planck length level?

4. Higher-dimensional theories

A number of quantum gravity theories involve higher spacetime dimensions. Although, these theories are highly appealing, they suffer in general from instability and causality problems. In Kaluza-Klein type theories of pure gravity in higher dimensions, the difficulty already exists at the classical level [8]. For point-like masses, these theories cannot produce the gravitational field, which corresponds to known classical gravitational tests. The classical tests of general relativity are the following:

1. Gravitational frequency shift,
2. Perihelion shift of Mercury,
3. Deflection of light,
4. Radar echo delay, and
5. Parameterized post-Newtonian parameters.

A non-relativistic weak field approximation is sufficient for the treatment of motion in the Solar system. When the equations of motion are applied to the classical tests in general relativity, only the ordinary $3 + 1$ dimensional spacetime case is found to agree with observations. The result is independent of the size of the extra dimensions as long as they are compact and have the geometry of tori. All multi-dimensional Kaluza-Klein theories face severe challenge when confronted with motions in the Solar system.

5. Is gravity non-quantum?

At present, there is a total lack of evidence of any quantum nature of gravity, including spacetime foam, discrete spacetime, breaking of Lorentz invariance, and extra dimensions. Is it possible that quantum gravity is not necessary? What if gravity is intrinsically classical and it becomes a derived interaction from an underlying theory at a deeper level [9]? We presume that the underlying theory would be a completely different theory from gravitation. It may be a quantum theory or a non-quantum theory, and the important point is that classical gravity is not the $\hbar \to 0$ limit of this underlying theory. The relation between classical gravitation and the underlying theory is more like that between thermodynamics and statistical mechanics. In classical thermodynamics, the basic variables are the pressure P, volume V, temperature T, and entropy S. They are macroscopic variables chosen for performing heat experiments. The underlying degrees of freedom of the system are the microscopic atoms and molecules. They obey the laws of quantum mechanics or classical mechanics. As we now know, it would be impossible to quantize thermodynamics to achieve the smallest units of pressure, volume, temperature and entropy in order to arrive at statistical mechanics. Classical thermodynamics is not the $\hbar \to 0$ limit of quantum statistical mechanics. In fact, quantum mechanics is only used to count the number of states available at the microscopic level.

The real connection between thermodynamics and statistical mechanics is through the partition function. In statistical mechanics, the partition function of a system is defined as:

$$Z = \sum_j e^{-\beta E_j} \ . \tag{1}$$

Here, $\beta = 1/(kT)$, k is the Boltzmann constant, and E_j are the energy levels. The thermodynamic variables are then defined as:

Internal energy

$$U = -\frac{\partial \ln Z}{\partial \beta} \ , \tag{2}$$

Pressure

$$P = \frac{1}{\beta}\frac{\partial \ln Z}{\partial V}, \quad \textit{and} \tag{3}$$

Entropy

$$S = k \ln Z + \frac{U}{T} \ . \tag{4}$$

We believe that gravity is in a similar situation. It would be impossible to quantize general relativity to achieve the smallest units of space and time in order to arrive at an underlying theory of gravity. A new structure with new degrees of freedom is needed. It is the statistical mechanics of spacetime.

References

[1] 't Hooft G 2000 *Rev. Mod. Phys.* **72** 333
[2] Veltman M 2000 *Rev. Mod. Phys.* **72** 341
[3] Feynman R 1963 *Acta Phys. Polonica* **24** 697
[4] DeWitt B S 1964 *Phys. Rev. Lett.* **12** 742
[5] Rothman T and Boughn S 2006 *Found. Phys.* **36** 1801
[6] Abdo A A *et al* 2009 *Nature* **462** 331
[7] Tamburini F *et al* 2011 *A. & A.* **A71** 553
[8] Eingorn M and Zhuk A 2010 *Class. Quant. Grav.* **27** 205014
[9] Ha Y K 2012 *Int. J. Mod. Phys.: Conf. Ser.* **7** 219

Generation of helical magnetic fields from inflation

R K Jain, R Durrer and L Hollenstein

Département de Physique Théorique and Center for Astroparticle Physics,
Université de Genève, 24 Quai Ernest Ansermet, CH–1211 Genève 4, Switzerland

E-mail: Rajeev.Jain@unige.ch

Abstract. The generation of helical magnetic fields during single field inflation due to an axial coupling of the electromagnetic field to the inflaton has been discussed. We have found that such a coupling always leads to a blue spectrum of magnetic fields during slow roll inflation. Though the helical magnetic fields further evolve during the inverse cascade in the radiation era after inflation, we have concluded that the magnetic fields generated by such an axial coupling can not lead to observed field strength on cosmologically relevant scales.

1. Introduction

There exists persuasive observational evidence for large scale magnetic fields with typical strength of μ Gauss in various cosmic structures from stars to galaxies and galaxy clusters [1, 2, 3]. Recently, using Fermi and HESS datasets, an interesting lower bound of $B \geq 10^{-16}$ Gauss has been obtained on the strength of magnetic fields in the intergalactic medium [4].

All such observations naturally lead to the question of the origin of these magnetic fields. Various mechanisms of magnetic field generation can be broadly classified into two categories: Early universe or primordial, and late universe or astrophysical. Among the former ones, magnetic fields produced during inflation are most interesting as an arbitrary spectrum for the magnetic fields can be produced, and therefore, can lead to sufficient amplitude on large scales to provide seeds for the observed fields in galaxies and clusters [5, 6].

In what follows, we have briefly discussed the magnetic field generation during inflation due to an axial coupling and shown that contrary to a non-helical coupling, a helical coupling always leads to a magnetic spectral index $n = 1$ during slow roll [7]. We have further shown that the backreaction turns out to be negligible as the magnetic energy density at the end of inflation is too small. Even though, the inverse cascade in the radiation era after inflation transfers power from small to large scales, the resulting field strength on cosmologically interesting scales is still insufficient to provide seeds for the observed magnetic fields in cosmic structures [7].

2. Axial coupling and the Fourier mode equation

Since the standard electromagnetic (EM) action is conformally invariant, the EM field fluctuations are not amplified in the conformally flat expanding background of inflation. In order to generate magnetic fields, one needs to break the conformal invariance of the EM field, e.g., by coupling the EM field to a scalar or a pseudo-scalar field or to a curvature invariant. Here, we have investigated the first possibility and studied a helical coupling given by:

$$\mathcal{L}_I(\phi, A_\mu) = \frac{1}{4} f(\phi) F_{\alpha\beta} \widetilde{F}^{\alpha\beta} . \tag{1}$$

It describes a coupling of the scalar field ϕ or the inflaton to the parity-violating term $F\widetilde{F}$, where $\widetilde{F}$ is the dual of the EM field tensor, and is defined as $\widetilde{F}^{\mu\nu} \equiv \frac{1}{2}\eta^{\mu\nu\alpha\beta}F_{\alpha\beta}$, where $F_{\mu\nu} \equiv \partial_\mu A_\nu - \partial_\nu A_\mu$, and $\eta^{\mu\nu\alpha\beta}$ is the totally anti-symmetric tensor with $\eta^{0123} \equiv (-g)^{-1/2}$.

Adopting the Coulomb gauge, where $A^\mu = (0, A^i)$ with $\partial_i A^i = 0$, we have found that the inhomogeneous Maxwell's equation with the axial current becomes:

$$\ddot{A}_i - \nabla^2 A_i = -f'(\phi)\dot{\phi}\,\epsilon_{ijk}\partial_j A_k\,, \tag{2}$$

where $f'(\phi) \equiv df/d\phi$, and the overdots denote derivative with respect to the conformal time t. Note that, for a constant axial coupling, $f'(\phi) = 0$, the sourced Maxwell's equation reduces to the standard free wave equation and no EM fluctuations are amplified during inflation. After quantization of the vector potential, we have found that the Fourier modes $\mathcal{A}_h(t,k)$ corresponding to the helicity states $h = \pm$ satisfy the wave equation [7]:

$$\ddot{\mathcal{A}}_h + \left[k^2 + hkf'(\phi)\dot{\phi}\right]\mathcal{A}_h = 0\,. \tag{3}$$

The power spectrum of a primordial stochastic magnetic field, statistically homogeneous and isotropic, is given by two scalar functions and can be written as:

$$\langle \widetilde{B}_i(t,\boldsymbol{k})\widetilde{B}^*_j(t,\boldsymbol{q})\rangle = \frac{(2\pi)^3}{2}\delta(\boldsymbol{k}-\boldsymbol{q})\Big\{(\delta_{ij} - \hat{k}_i\hat{k}_j)P_S(t,k) - i\epsilon_{ijn}\hat{k}_n P_A(t,k)\Big\}, \tag{4}$$

where $\widetilde{B} \equiv a^2 B$ is the re-scaled magnetic field by the expansion factor a, and $P_{S/A}$ are the symmetric/anti-symmetric parts of the power spectrum respectively. With respect to the helicity basis, the spectra can directly be written as:

$$P_{S/A}(t,k) = k^2\left(|\mathcal{A}_+(t,k)|^2 \pm |\mathcal{A}_-(t,k)|^2\right), \tag{5}$$

and the re-scaled magnetic energy density per logarithmic wave number is given by:

$$\frac{d\widetilde{\rho}_B}{d\ln k}(t,k) = \frac{k^3}{(2\pi)^2}P_S(t,k)\,. \tag{6}$$

3. The magnetic power spectra in slow roll inflation

In order to treat the theory perturbatively, the strength of the interaction between the scalar field and the EM field should be small at all times, which then leads to $|f_N| < 1$, where f_N is the dimensionless part of the coupling term, defined as: $f_N \equiv df/d\ln a = f'(\phi)\dot{\phi}/\mathcal{H}$. In the slow roll regime, where f_N is approximately constant, the mode Eq. (3) can be solved analytically in terms of Coulomb wave functions. By requiring the solution to approach the free wave solution as initial condition, the full normalized solution can be written as:

$$\mathcal{A}_h(k,t) = \frac{1}{\sqrt{2k}}\Big[G_0(-hf_N/2, -kt) + iF_0(-hf_N/2, -kt)\Big]. \tag{7}$$

Using the asymptotic behaviour of this solution at late times, we have found that the power spectra are given by [7]:

$$P_S(k) = k\,\frac{\sinh(\pi f_N)}{\pi f_N} \quad \text{and} \quad P_A(k) = k\,\frac{\cosh(\pi f_N) - 1}{\pi f_N}\,. \tag{8}$$

Note that both the spectra are proportional to k, and only their amplitude changes with the coupling strength f_N, which tend to $(2\pi f_N)^{-1}\exp(\pi f_N)$ for large values of f_N. Using the symmetric power spectrum, the magnetic energy density per logarithmic wave number at the end of inflation can directly be computed to be:

$$\frac{d\rho_B}{d\ln k}(t_{\rm end}, k) = \frac{k^4}{a_{\rm end}^4}\frac{\sinh(\pi f_N)}{4\pi^3 f_N} \equiv \frac{k^4}{a_{\rm end}^4}\mathcal{S}^2(f_N), \tag{9}$$

which then leads to:

$$\rho_B(t_{\rm end}) \simeq \frac{1}{4}H_{\rm end}^4\,\mathcal{S}^2(f_N), \tag{10}$$

where $H_{\rm end}$ is the Hubble parameter at the end of inflation. Now, using the Friedmann equation, we can write:

$$\Omega_B(t_{\rm end}) \equiv \frac{\rho_B}{\rho_{\rm tot}} \simeq \frac{\mathcal{S}^2(f_N)}{12}\left(\frac{H_{\rm end}}{m_P}\right)^2. \tag{11}$$

The generic condition for back-reaction to be negligible is $\Omega_B < 1$ which then leads to:

$$\mathcal{S}(f_N) \lesssim \frac{m_P}{H_{\rm end}} \simeq \left(\frac{m_P}{T_*}\right)^2 \simeq 1.5\times 10^{10}\left(\frac{10^{14}\,{\rm GeV}}{T_*}\right)^2, \tag{12}$$

where T_* is the reheating temperature, given by $T_*^4 \simeq m_P^2 H_{\rm end}^2$. In terms of f_N, the back-reaction bound becomes:

$$f_N \lesssim 17\left(\frac{10^{14}\,{\rm GeV}}{T_*}\right). \tag{13}$$

Therefore, no back-reaction on the background evolution is expected if inflation ends well below the Planck scale [7]. Typically, the effective coupling term can be large in a situation wherein the EM field is coupled to a large number of pseudo-scalar fields [8].

4. Final strength of the magnetic fields

After inflation, due to helicity conservation, the magnetic field evolves by inverse cascade, which can move power from small scales to large scales without affecting the spectral shape of the spectrum on large scales [9]. On the other hand, on small scales, the magnetic fields are exponentially damped due to viscosity of the plasma. A sketch of the process of inverse cascade is shown in Figure 1. Taking into account the maximal amplification of the magnetic field due to inverse cascade for our case, we have found that the final strength of the magnetic field is given by:

$$\widetilde{B}(k) \simeq 3\times 10^{-19}\,{\rm Gauss}\,\mathcal{S}(f_N)\left(\frac{k}{10^{12}/{\rm Mpc}}\right)^2\left(\frac{T_*}{10^{14}\,{\rm GeV}}\right)^{9/11}. \tag{14}$$

At a typical scale of $\lambda = 0.1\,{\rm Mpc}$ or $k = 10/{\rm Mpc}$, the field strength becomes:

$$\widetilde{B}(k=10/{\rm Mpc}) \simeq 3\times 10^{-41}\,{\rm Gauss}\,\mathcal{S}(f_N)\left(\frac{T_*}{10^{14}\,{\rm GeV}}\right)^{9/11}. \tag{15}$$

which is by far insufficient for subsequent dynamo amplification, which requires seed fields of the order of at least 10^{-20} Gauss [10]. Furthermore, to estimate how large this field can become for our best case, we require $\mathcal{S}(f_N) < (m_P/T_*)^2$ and therefore,

$$\widetilde{B}(k=10/{\rm Mpc}) \leq 10^{-32}\,{\rm Gauss}\left(\frac{10^{14}\,{\rm GeV}}{T_*}\right)^{13/11}. \tag{16}$$

In order to obtain the minimal necessary field for dynamo amplification of about 10^{-20} Gauss, we would require $T_* \leq 10^4\,{\rm GeV}$, which corresponds to a rather low inflation scale, but not completely excluded [7].

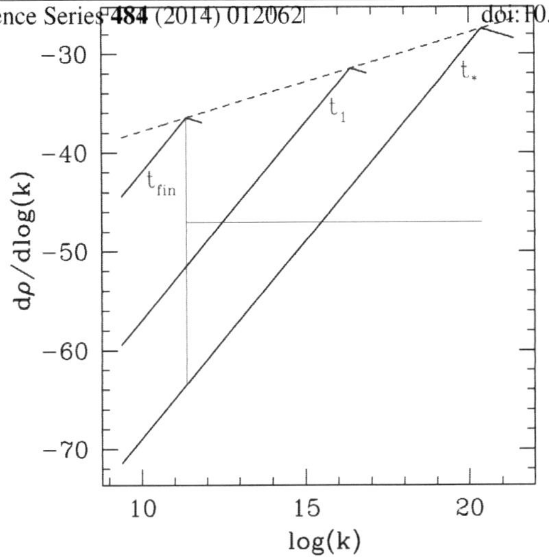

Figure 1. The process of inverse cascade is depicted. The horizontal line indicates the total increase in the correlation scale during the inverse cascade, while the vertical line represents the total amplification factor. On large scales, the shape of the spectrum remains unchanged while the amplitude increases, because the correlation scale grows and the total helicity is conserved.

5. Conclusions

We have discussed the generation of helical magnetic fields during inflation induced by an axial coupling term. During slow roll inflation, we have shown that the magnetic field power spectrum is always blue with spectral index $n = 1$. Since the helical coupling is active only around horizon crossing, and is nearly constant during slow roll, all modes are coherently amplified, which leaves the shape of the spectrum invariant with respect to the initial one. Although, after inflation and reheating, an inverse cascade can transfer power from small to large scales, it turns out that for typical reheating temperatures, the magnetic fields are of $\mathcal{O}(10^{-40})$ Gauss, which are largely insufficient for dynamo amplification [10]. Recently, we have shown that even the resonant amplification of magnetic fields due to an oscillating coupling fails to provide sufficient amplitude for the observed field strength due to extremely large back-reaction [11].

Acknowledgement

We acknowledge the financial support from the Swiss National Science Foundation.

References

[1] Kronberg P P 1994 *Report Prog. Phys.* **57** 325
[2] Pentericci L, Van Reeven W, Carilli C L, Rottgering H A J and Miley G K 2000 *A. & A. Supp. Ser.* **145** 121 [astro-ph/0005524]
[3] Battaglia N, Pfrommer C, Sievers J L, Bond J R and Ensslin T A 2009 *MNRAS* **393** 1073 [arXiv:0806.3272[astro-ph]]
[4] Neronov A and Vovk I 2010 *Science* **328** 73 [arXiv:1006.3504[astro-ph.HE]]
[5] Martin J and Yokoyama J 2008 *JCAP* **0801** 025 [arXiv:0711.4307[astro-ph]]
[6] Subramanian K 2010 *Astron. Nachr.* **331** 110 [arXiv:0911.4771[astro-ph.CO]]
[7] Durrer R, Hollenstein L and Jain R K 2011 *JCAP* **1103** 037 [arXiv:1005.5322[astro-ph.CO]]
[8] Anber M M and Sorbo L 2006 *JCAP* **0610** 018 [astro-ph/0606534]
[9] Campanelli L 2007 *Phys. Rev. Lett.* **98** 251302 [arXiv:0705.2308[astro-ph]]
[10] Brandenburg A and Subramanian K 2005 *Phys. Report* **417** 1 [astro-ph/0405052]
[11] Byrnes C T, Hollenstein L, Jain R K and Urban F R 2012 *JCAP* **1203** 009 [arXiv:1111.2030[astro-ph.CO]]

Tunnelling in a time dependent quartic potential: Possible implications for cosmology

R Kabir and A Mukherjee

Department of Physics and Astrophysics, University of Delhi, Delhi 110007, India

E-mail: kabirakesh@gmail.com, amimukh@gmail.com

Abstract. The theory of a real scalar field with an arbitrary potential plays an important role in cosmology, particularly in the context of inflationary scenarios. However, in most applications, the potential is treated as independent of time, whereas in an evolving universe, for example, before the onset of inflation, the potential is actually likely to be changing with time. As pointed out by Berry in the context of single-particle quantum mechanics, the existence of multiple time scales can lead to results that are qualitatively different from those obtained with a static potential. The present paper reports on numerical investigations in a scalar field theory with a double-well potential that depends explicitly on time. The transition rate per unit volume for the decay of the false vacuum is found to depend strongly on time. Possible implications for old inflation are discussed.

1. Introduction

The quantum tunnelling in a time-dependent asymmetric double-well potential appears in a variety of physical systems. A typical asymmetric double-well potential will have local (energetically dis-favoured) minima in addition to the global minimum which defines the true vacuum. The purpose of this paper is to investigate how the tunnelling rate behaves depending on whether the relative depths of two minima of the potential change *slowly* or *rapidly*. Berry had studied quantum mechanics of a single particle in a time-varying potential. He had used the quantum mechanical wave function in Schrödinger's time dependent equation and concluded that results are significantly different from the static potential [1]. Berry's work was the main motivation for our work, and we have aimed to generalize his results for the field theoretic context. We have used the instanton method, which is by now a well-known tool for dealing with quantum tunnelling phenomena [2]. In a study of the decay of metastable states by quantum tunnelling, Coleman had introduced the concept of the bounce, and had shown that the bounce dominates the quantum tunnelling [3].

In the original version of cosmological inflation [4], the phase transition which ends the inflationary epoch is of first order, and proceeds through the quantum tunnelling of scalar fields from a false vacuum state to a true vacuum state. However, this model of inflation is plagued by the graceful exit problem [5]. We have suggested a possibility for solving the problem, using the time-dependent tunnelling rate.

2. Basic formulation

In this section, we have reviewed the formalism used for calculating tunnelling probabilities. This discussion will be applicable to the single scalar field ϕ with static as well as the time dependent asymmetric double-well potentials, although with slight modifications in the latter case.

Let us consider a scalar field theory with a Lagrangian density, given by:

$$\pounds(x,t) = \frac{1}{2}\partial^{\mu}\partial_{\mu}\phi - V(\phi), \tag{1}$$

where the potential $V(\phi)$ has two minima at ϕ_f (false vacuum) and ϕ_t (true vacuum). A typical asymmetric double-well potential may look like:

$$V(\phi) = \frac{1}{2}A\phi^2 - \frac{1}{3}B\phi^3 + \frac{1}{4}C\phi^4, \tag{2}$$

where the coefficients A, B and C are constants and satisfy the inequality, $B^2 > 9/2AC$ [6].

Figure 1. The asymmetrical double-well potential for two different values of A. Note that as time evolves the depth of the true vacuum increases whereas the false vacuum remains fixed at the origin.

If the system is initially in the false vacuum phase, barrier tunnelling leads to the appearance of bubbles of a new phase with $\phi = \phi_t$. In the zero temperature limit, the rate, i.e., decay probability per unit time per unit volume, for this process is (upto leading order in $\hbar$):

$$\Gamma = K\ e^{-S_E[\phi]}, \tag{3}$$

where $S_E[\phi]$ is the action corresponding to the O(4)-symmetric solution of Euclidean equation of motion:

$$\frac{d^2\phi}{d\rho^2} + \frac{3}{\rho}\frac{d\phi}{d\rho} = \frac{dV(\phi)}{d\phi}, \tag{4}$$

where ρ is the four-dimensional radial coordinate, with boundary conditions:

$$\lim_{\rho\to\infty}\phi = \phi_f = constant, \quad and \quad \lim_{\rho\to 0}\frac{d\phi}{d\rho} = 0. \tag{5}$$

K is the determinantal factor of the order η^4, where η is the energy scale of phase transition. Our focus has been on the calculation of the action $S_E[\phi]$, which in the O(4)-symmetric case is given by:

$$S_E = 2\pi^2\int_0^{\infty} d\rho\rho^3\left[\frac{1}{2}\left(\frac{d\phi}{d\rho}\right)^2 + V(\phi)\right]. \tag{6}$$

2.1. Potentials with explicit time dependence

To get explicit time dependence, it is enough to make any one coefficient of the potential given in Eq. (2) a function of time. Although, there could be various other ways, we make the coefficient A, time dependent in the following way:

$$A(t) = a_2 + \frac{(a_1 - a_2)}{\cosh(t/\lambda)}, \tag{7}$$

where λ can be considered to be the time scale for the potential change. In this model, $A(t)$ starts from a_1 and goes to a_2 as $t \to \infty$. To solve the time-dependent case, we have taken $a_1 = 1.9$ (nearly degenerate) and $a_2 = 0.1$ (highly asymmetric) (see Figure 1). Further, we have fixed $B = 3.0$ and $C = 1.0$ for all calculations.

3. Numerical simulations

The computation of the Euclidean action for explicit time dependent potential requires us to modify Eq. (6). We have used the following for $S_E[\phi]$ [7] as:

$$S_E = \int d\tau S_3, \quad where \quad S_3 = 4\pi \int_0^\infty dr r^2 \left[\frac{1}{2}\left(\frac{d\phi}{dr}\right)^2 + V(\phi, \tau) \right]. \tag{8}$$

The procedure to calculate S_E in the quasi-static approximation is as follows: For a fixed λ (say 8.0), we have divided the allowed time period into N (say 100) steps. At the midpoint of each step, we have calculated the value of A using Eq. (7), and subsequently solve the static equation of motion using O(3)-symmetry (see Figure 2). Then, we have calculated the action S_3 corresponding to this instantaneous static solution. We have repeated this for the midpoints of subsequent steps. Finally, we have summed (integrated) up the contributions, thus, obtained to get the so called quasi-static action S_E for a single λ. Although, this procedure is not strictly justifiable, it is a good approximation for large enough λ (in practice $\lambda > 5$). Table 1 shows actions S_E for different values of λ.

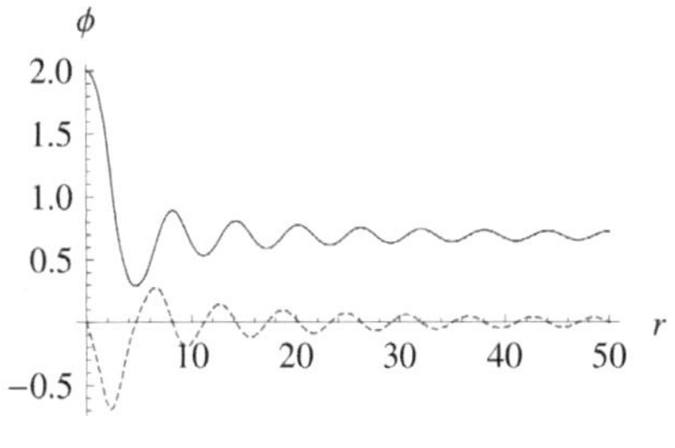

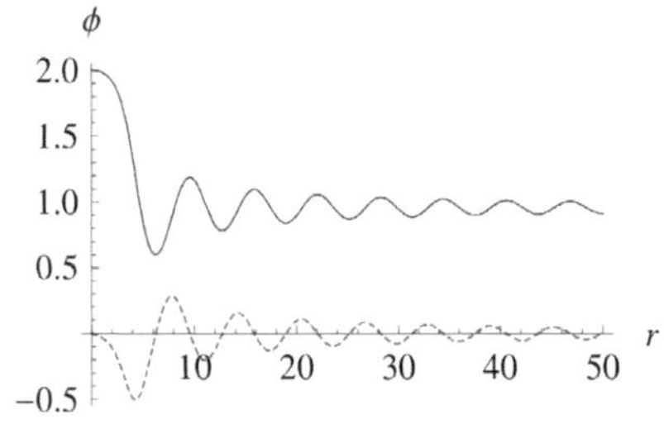

Figure 2. Numerically obtained bounce solutions as functions of r for two different time scales. The solid line represents ϕ, while dashed one represents $d\phi/dr$.

3.1. Inferences from the numerical study

The observation, that as we vary the parameter (here time scale), the tunnelling rate (negative exponential of the action) changes by orders of magnitude, leads to the following conclusion: If the potential is explicitly time dependent, then in particular, the tunnelling probability is quite different from the case of a static potential; and in general, physics can be quite different [1, 8]. The key result of this paper can be stated qualitatively as: *A slowly varying potential renders*

Table 1. Action as a function of time scale.

Time Scale (λ)	Action (S_E)
6.0	1504.87
7.0	1561.08
8.0	2009.40
10.0	2513.93
12.0	3018.46
14.0	3523.00

the tunnelling rate small, a rapidly varying potential increases it. Here, 'rapidly varying' is a relative term, since all values of λ are chosen to lie within the domain of validity of the quasi-static approximation. Apart from cosmology, this result may be relevant to any study dealing with tunnelling in time-dependent potentials, for example, Bose-Einstein condensates [9], and condensed matter physics [10, 11].

4. Implications for the old inflationary models

The quantity of interest for cosmology is $\epsilon = \Gamma/H^4$, where ϵ is the number of bubbles nucleated per Hubble volume per Hubble time. In theory, inflation ends when true vacuum bubbles nucleate, expand and collide. However, in Guth's model, percolation never occurs as bubbles inflate away faster than they form and grow. As bubbles do not merge, no reheating takes place. In terms of ϵ, the horizon and flatness problems are solved if $\epsilon <\ 0.001$, whereas percolation occurs only if $\epsilon\ \approx\ 0.1$ [12].

In Guth's model, both Γ and H are constant, and the requirements of sufficient inflation and percolation are incompatible (graceful exit problem). Obviously, if ϵ is allowed to vary with time, the graceful exit problem can be solved. As we have observed in the numerical study, the tunnelling rate Γ is strongly dependent on the time scales of the time dependent potentials. Hence, if somehow a time dependence can be introduced that causes the potential to change slowly in the beginning and rapidly later, ϵ will be small initially, allowing sufficient inflation to occur, and will become large later on for the percolation to occur; this might solve incompatibility problem.

Although, we have used a toy model for the time dependence, in a cosmological setting the effective potential automatically acquires time dependence. Perhaps there is a case for revisiting old inflationary models with explicitly time dependent potentials. Of course, one needs to carry out a self-consistent calculation. Work on this is in progress.

Acknowledgements

We appreciate the discussions with David Wands and Raghavan Rangarajan. One of us (RK) wishes to thank CSIR, Government of India, for the award of JRF (No. 09/045(0930)/2010-EMR-I).

References

[1] Berry M V 1984 *J. Phys. A: Math. Gen.* **17** 1225
[2] Vainshtein A I *et al* 1982 *Sov. Phys. Usp.* **195**
[3] Coleman S 1977 *Phys. Rev. D* **15** 2929, Coleman S 1985 *Aspects of Symmetry* (Cambridge University Press)
[4] Guth A H 1981 *Phys. Rev. D* **23** 347
[5] Guth A H and Weinberg E J *Nucl. Phys. B* **212** 321
[6] Zhou B, *et al* 2000 *Phys. Lett. A* **271** 26
[7] Widyan H, *et al* 1999 *Phys. Rev. D* **59** 045003

[8] Keski-Vakkuri E and Kraus P 1991 *Phys. Rev. D* **54** 7404
[9] Kramer T and Moshinsky M 2005 *J. Phys. A: Math. Gen.* **38** 5993
[10] Stone A D, Azbel M Y and Lee P A 1985 *Phys. Rev. B* **31** 1707
[11] Grifoni M and Hanggi P 1998 *Phys. Report* **304** 229
[12] Widrow L M 1991 *Phys. Rev. D* **44** 2306

Proceedings of the 7th International Conference on Gravitation and Cosmology (ICGC2011) IOP Publishing
Journal of Physics: Conference Series **484** (2014) 012064 doi:10.1088/1742-6596/484/1/012064

Action for gravity: An emergent perspective

S Kolekar
IUCAA, Post Bag 4, Pune University Campus, Ganeshkhind, Pune 411007, India
E-mail: sanved@iucaa.ernet.in

Abstract. There is increasing recognition in recent years that the field equations of gravity may have the same conceptual status as the equations of fluid mechanics or elasticity, and hence, gravity could be thought of as an emergent phenomenon. We have shown that the action functional leading to the field equations of gravity too, can be given a thermodynamic interpretation, which supports the emergent paradigm of gravity. We have explored the action functional in three different ways, the key features of which we have summarized in this article.

1. Introduction

In recent years, it is recognized that the field equations of gravity may have the same conceptual status as the equations of fluid mechanics or elasticity, and hence, gravity could be thought of as an emergent phenomenon just like, say, fluid mechanics (for a review see [1]). This approach has a long history, originating from the work of Sakharov [2] and interpreted in many ways by different authors [3]. We have used the term 'emergent' in the specific and well-defined sense in terms of the equations of motion, rather than in more speculative vein, like considering the space and time themselves to be emergent, etc. The evidence for such a *specific interpretation* comes from different facts, like the possibility of interpreting the field equations in a wide class of theories as thermodynamic relations, the possibility of obtaining the field equations from a thermodynamic extremum principle, application of equi-partition ideas to obtain the density of microscopic degrees of freedom, etc. [1]. In such an approach, geometrical variables like the metric, etc. are derived concepts (e.g., pressure, density, etc. of a gas), and the dynamical equations governing them can be derived from the thermodynamic limit of an underlying micro-structure, say, by extremizing a suitably defined entropy functional. But on the other hand, we know from standard textbook description that one can obtain the field equations for gravity from an action functional in which the metric is varied. This raises the question:

If gravity is, indeed, an emergent phenomenon, should not the action functionals of gravity contain some signature of this fact?

We have shown that this is, indeed true, and the action functional does suggest that gravity is emergent. We have reviewed below three results based on earlier works in [4, 5, 6]

2. Holography in action

It is well-known that the Einstein-Hilbert action can be separated into a bulk term and a surface term. It is *only* the variations of the bulk $\Gamma\Gamma$ term that contribute to the field equations. That is, the field equations (and their solutions) do *not* depend in any way on the surface term. It is, therefore, a mystery - in the conventional approach - that the surface term, which is ignored while obtaining the field equations can be used to determine the entropy of the horizons that arise in the theory! The solution to this mystery was first pointed out in [7], where it was emphasized that the bulk and surface terms in Einstein-Hilbert action are connected by a peculiar relation:

$$L_{\rm sur} = -\partial_i \left(g_{ab} \frac{\partial L_{\rm bulk}}{\partial(\partial_i g_{ab})} \right), \tag{1}$$

which allows the information about either one to be extracted from the other. The relationship between the bulk and boundary terms in the action was termed 'holographic', because the information about the bulk action functional (from varying of which, we can obtain the dynamical equations) is encoded in the boundary action functional. We continue to use this terminology 'holographic action' with this understanding. The holographic nature of the action fits very well with the thermodynamic approach to gravity, and can be thought of as the hidden signal in the action functionals indicating that the description of gravity is an emergent one. In fact, one can provide very general arguments to suggest that the action functional describing *any* theory of gravity that obeys the principle of equivalence and principle of general covariance will have a bulk and boundary term related holographically. If this is the case, one would like to explore the connection further and see what insights it can provide. We decompose the Einstein-Hilbert action into a particular surface and bulk term for any static spacetime using the time-time component of the Einstein tensor as:

$$L_{EH} = -2G^0_0 + 2R^0_0, \tag{2}$$

where $2R^0_0$ can be shown to be total divergence in this case. In any static spacetime, one can show that $L_{bulk} = -2G^0_0$ is related to the ADM Hamiltonian in GR, while $L_{sur} = -2R^0_0$ is related to the Noether charge, associated with the diffeomorphism symmetry of the theory. One can show that the action functional then has direct thermodynamic interpretation in the Euclidean sector as:

$$A_E \equiv -\beta E + S. \tag{3}$$

The Euclidean-Einstein action has the interpretation as giving the (negative of) free energy of spacetime. One can then interpret extremizing the integral over $-2G^0_0$, while keeping the surface term fixed at the boundary, as extremizing the bulk energy of the static spacetime while keeping the entropy fixed. From the previous work, we know that the resulting field equations can be interpreted as the thermodynamic identity $dE = TdS - PdV$ on the horizon. In the usual thermodynamic systems, if we know the entropy functional $S(E,V)$, we can obtain the other thermodynamic variables like (T,P), etc. of the system. Alternatively, one can invert the form of $S(E,V)$ to get energy $E(S,V)$ in terms of entropy. Similarly, in the case of Einstein-Hilbert action, we can consider the extremization of $-2G^0_0$, as equivalent to obtaining the thermodynamic relation $dE = TdS - PdV$ (which is the same as field equations of the theory) from an energy functional. Further, if gravity is truly emergent, thermodynamic, phenomenon of an underlying microscopic theory, then one should be able to invert the energy functional $E(S,V)$ of gravity to obtain the $S(E,V)$ functional. This motivates us to ask: Can we obtain the surface term $2R^0_0$ directly from the bulk term $-2G^0_0$? Holography answers this question. We can show by direct computation that $L_{\rm sur} = 2\sqrt{-g}R^0_0$, and $L_{\rm bulk} = -2\sqrt{-g}G^0_0$ are related by the holographic relation [4] as:

$$L_{\rm sur} = -\frac{1}{[(D/2)-1]}\partial_i \left(g_{ab} \frac{\delta L_{bulk}}{\delta(\partial_i g_{ab})} + \partial_j g_{ab} \frac{\partial L_{bulk}}{\partial(\partial_i \partial_j g_{ab})} \right). \tag{4}$$

We have also found that a similar interpretation continues to hold even for higher curvature theories such as Lanczos Lovelock.

3. The partition function for Lanczos-Lovelock gravity

Consider further, that the static spacetime is also spherically symmetric in D dimensions described by the metric of the form:

$$ds^2 = -f(r)dt^2 + f(r)^{-1}dr^2 + r^2 d\Omega^2, \tag{5}$$

such that $f(r)$ vanishes at some surface $r = a$, say, with $f'(a)$ $(= B$, say) remaining finite. The class of metrics in Eq. (5) with the behaviour $[f(a) = 0, f'(a) = B]$ constitute a canonical ensemble at constant temperature, since they all have the same temperature $T = |B|/4\pi$. Let us consider the partition function for a general Lanczos-Lovelock theory, of order m for this ensemble, which is given by the path integral sum:

$$Z(\beta) \;=\; \sum_{g\epsilon\mathcal{S}} \exp(-A^{(e)}_{\rm LL}(g)). \tag{6}$$

The sum in Eq. (6) is restricted to the set $\mathcal{S}$ of all metrics of the form in Eq. (5) with the behaviour $[f(a) = 0, f'(a) = B]$, and the Euclidean-Lagrangian is a functional of $f(r)$. The spatial integration will be restricted to a region bounded by $(D-2)$-spheres at $r = a$ and at the asymptotic boundary $r = b$. For the metric of the form in Eq.(5), the bulk term $2\sqrt{-g}G^0_0$ too, becomes a total derivative and the contribution to the Euclidean action then comes only from the boundary values. However, we have ignored the contribution from the asymptotic boundary, since it will only lead to a factor Z_Q, in deciding about the dependence of $Z(\beta)$ on the form of the metric near $r = a$. The final result can be written in a very suggestive form [5]:

$$\begin{aligned} Z(\beta) &= Z_Q \exp\left[\frac{(D-2)!}{(D-2m)!}\frac{m\Omega}{4}a^{D-2m} - \beta\left(\frac{(D-2)!}{(D-2m-1)!}\frac{\Omega}{16\pi}a^{D-2m-1}\right)\right], \\ &\propto \exp\left[S^{(e)}(a) - \beta E(a)\right], \end{aligned} \tag{7}$$

where Ω is the area of a unit $(D-2)$-sphere. Here, $S^{(e)}(a)$ and $E(a)$ turn out exactly the same as the expression for Wald entropy, and energy of the horizon of the corresponding Lanczos-Lovelock theory obtained through other methods. Another feature is that the partition function also highlights the connection between the ADM Hamiltonian, and the quasi-local energy of a horizon in a spherically symmetric static spacetime.

4. Action principle for the fluid-gravity correspondence

It is known that the black hole horizon can be interpreted as a dissipative membrane with Einstein's equations projected on to it, taking a form very similar (but not identical) to the Navier-Stokes equation in fluid mechanics. This work formed the basis for the development of membrane paradigm [8]. Recently [1], it was shown that, when Einstein's equations are projected on to any null surface and the resulting equations are viewed in the freely falling frame, they become *identical* to Navier-Stokes equation. Figuratively speaking, this result shows that a spacetime filled with null surfaces can be equivalently thought of as hosting a fluid (with the fluid variables related to the structure of the null surface at any given event), which satisfies the Navier-Stokes equation in the local inertial frame. In the approachs taken in all the previous work in this subject, one usually obtains the field equation by extremizing the Einstein-Hilbert action with respect to the variations of the metric and then project them on to a black hole horizon or on to a generic null surface.

This is, however, conceptually not very satisfactory in the emergent paradigm for two reasons. First, it would be nice if equations of macroscopic dynamics could be obtained from a *thermodynamic* extremum principle rather than a *field theoretic* action principle. Second, given the fact that final equations are expressed in terms of variables defined using a null surface, it would be appropriate if the same variables are used in the extremum principle rather than the metric. Here, we have provided a thermodynamic extremum principle from which one directly obtains the Navier-Stokes equation rather than first obtaining or assuming Einstein's field equations and then deriving the Navier-Stokes equation by a projection to a null surface. The degrees of freedom varied in the action principle are the null vectors in the spacetime and *not* the metric tensor. The form of the action in terms of the fluid variables, which are in turn related to the null vectors defining the null surface is given in [6]. Here, we have pointed out an interesting feature related to this action functional.

Consider the spacetime to be completely foliated by null surfaces. Further, let the (non- affine) parameter along the null geodesics generating the null surface $\mathcal{S}$ be denoted by λ. Consider an infinitesimal cross sectional area δA of the 2-surface $\mathcal{S}_t$, which is the intersection of the null surface $\mathcal{S}$ with the constant time spacelike hypersurface Σ_t. Then, we have shown that the action can be written in a form:

$$\frac{dS}{dt} \propto -dE + TdS_{\mathcal{H}} + Pd\delta A, \tag{8}$$

where (i) dE denotes the loss in energy, because of viscous dissipation during evolution of the small area element δA of the null surface from λ_1 to λ_2 along the null congruence, (ii) the second term corresponds to rise in the gravitational entropy proportional to the increase of the area of the horizon, which is due to the familiar information loss processes, and (iii) the third term is the (virtual) work done by the horizon against the pressure P during its area expansion $d\delta A$. This suggests that we can interpret the rate of change of local action as an on-shell local entropy production rate of the given spacetime.

Thus, we have highlighted three features of the action leading to the field equations of gravity, which allow us to give a thermodynamic interpretation to the action, thus, supporting the paradigm that gravity is emergent.

Acknowledgements
This work was done in collaboration with T Padmanabhan and Dawood Kothawala.

References

[1] Padmanabhan T 2010 *Report Prog. Phys.* **73** 046901 [arXiv:0911.5004], Padmanabhan T 2011 *J. Phys. Conf. Ser.* **306** 012001 [arXiv:1012.4476]
[2] Sakharov A D 1968 *Sov. Phys. Dokl.* **12** 1040
[3] Hu B L 1996 *General Relativity as Geometro-Hydrodynamics* [arXiv:gr-qc/9607070], Jacobson T 1995 *Phys. Rev. Lett.* **75** 1260 [arXiv:gr-qc/9504004], Volovik G E 2003 *The Universe in a Helium Droplet* (Oxford University Press), Barcelo C, Liberati S and Visser M 2005 *Living Rev. Rel.* **8** No. 12 [arXiv:gr-qc/0505065]
[4] Sanved Kolekar and Padmanabhan T 2010 *Phys. Rev. D* **82** 024036 [arXiv:1005.0619]
[5] Sanved Kolekar, Dawood Kothawala and Padmanabhan T 2012 *Phys. Rev. D* **85** 064031 [arXiv:1111.0973]
[6] Sanved Kolekar and Padmanabhan T 2011 *Phys. Rev. D* **85** 024004 [arXiv:1109.5353]
[7] Padmanabhan T 2005 *Braz. J. Phys. (Special Issue)* **35** 362 [gr-qc/0412068]
[8] Thorne K S *et al* 1986 *Black Holes: The Membrane Paradigm* (Yale University Press)

Proceedings of the 7th International Conference on Gravitation and Cosmology (ICGC2011) IOP Publishing
Journal of Physics: Conference Series **484** (2014) 012065 doi:10.1088/1742-6596/484/1/012065

Trace Dynamics as a model for emergent spacetime

K Lochan and T P Singh

Tata Institute of Fundamental Research, Homi Bhabha Road, Colaba, Mumbai 400005, India

E-mail: kinjalk@tifr.res.in, tpsingh@tifr.res.in

Abstract. Trace Dynamics is a classical theory of non-commuting matrices, which uses cyclic permutation inside a trace to define the derivative with respect to an operator. We have used the methods of Trace Dynamics to construct a non-commutative special relativity. We have defined a line-element using the trace over spacetime coordinates, which are assumed to be operators. The line-element is shown to be invariant under generalized Lorentz transformations, and is used to construct a non-commutative relativistic dynamics. We have been motivated for such an operator structure at a more fundamental level, and attempt to obtain an emergent picture of classical spacetime.

1. Introduction

Trace Dynamics [TD] is the classical dynamics of (generic $N \times N$) matrices (equivalently, operators) $\{q_r\}$, whose elements can be either even grade [Bosonic sector] or odd grade [Fermionic sector] elements of Grassmann algebra [1]. The Lagrangian $\mathbf{L}[\{q_r\}, \{\dot{q}_r\}]$ in this programme has been defined as the trace of a polynomial function of the matrices $\{q_r\}$ and their time derivatives $\{\dot{q}_r\}$. The derivative with respect to an operator $\mathcal{O}$, of the trace $\mathbf{P}$ of a polynomial P made out of non-commuting operators is defined as follows: Infinitesimal variation in TrP is written and arranged as:

$$\delta TrP = \delta \mathbf{P} = Tr\frac{\delta TrP}{\delta \mathcal{O}}\delta \mathcal{O} = Tr\frac{\delta \mathbf{P}}{\delta \mathcal{O}}\delta \mathcal{O}, \tag{1}$$

with $\delta\mathbf{P}/\delta\mathcal{O}$ being designated as the *trace derivative*. By proceeding as in ordinary classical mechanics, one constructs an action; Lagrangian dynamics is derived from an action principle, and conjugate momenta p_r, a trace Hamiltonian $\mathbf{H}$ and Hamilton's equations of motion are constructed. Apart from the trace Hamiltonian, there are two other important conserved quantities: One is the 'trace Fermionic number' $N \equiv \sum_F q_r p_r$ obtained by summing over Fermionic variables. The other is the remarkable traceless and anti-self-adjoint Adler-Millard constant [2] given as:

$$\tilde{C} \equiv \sum_B [q_r, p_r] - \sum_F \{q_r, p_r\}, \tag{2}$$

which is a result of the invariance of the Lagrangian under global unitary transformations of the q_r and p_r. The subscripts B/F denote sum over commutators/anti-commutators of Bosonic/Fermionic matrices. It is profound that such a conserved commutator should appear in a classical theory, in which the matrices and their commutators/anti-commutators take arbitrary values. The presence of this matrix-valued Noether charge, makes TD different from ordinary classical mechanics. However, at the effective level, when we do statistical mechanics with these degrees of freedom, the statistical average of this charge is expected to throw some light on the effective commutation relations.

2. Non-commutative spacetime in Trace Dynamics approach

In TD, a background spacetime is given, and is made up of the usual spacetime coordinates $(\mathbf{x}, t)$ of a Minkowski spacetime, and dynamics obeys Poincarè invariance. Here, we ask if the techniques of TD can be used to raise the coordinates $(\mathbf{x}, t)$ to the level of non-commuting operators, and a non-commutative classical dynamics constructed, while still preserving the metric under certain spacetime diffeomorphisms (Poincarè like invariance). We have considered a set of four non-commuting finite dimensional operators $(\hat{t}, \hat{x}, \hat{y}, \hat{z})$ having arbitrary commutation relations (which are not fixed and are completely general as of now), and from which, a **real** line-element is defined as follows:

$$ds^2 = Trd\hat{s}^2 \equiv Tr[d\hat{t}^2 - d\hat{x}^2 - d\hat{y}^2 - d\hat{z}^2]. \tag{3}$$

We ask for the most general linear transformations as:

$$\begin{aligned} \hat{t}' &= A_1\hat{t} + B_1\hat{x} + C_1\hat{y} + D_1\hat{z}, \\ \hat{x}' &= A_2\hat{t} + B_2\hat{x} + C_2\hat{y} + D_2\hat{z}, \\ \hat{y}' &= A_3\hat{t} + B_3\hat{x} + C_3\hat{y} + D_3\hat{z}, \quad and \\ \hat{z}' &= A_4\hat{t} + B_4\hat{x} + C_4\hat{y} + D_4\hat{z}, \end{aligned} \tag{4}$$

which keep the line element invariant. Here, $A_i,\ B_i,\ C_i,\ and\ D_i$ belong to a graded vector space given as:

$$x = x_0 + \sum_i x_i\theta_i + \sum_{i<j} x_{ij}\theta_i\theta_j + ..., \qquad , \tag{5}$$

where $\{x_0, x_i, x_{ij}, ...\} \in \mathcal{C}$. In the above equation, θ_is are anti-commuting Grassmann numbers. Taking clue from the standard Lorentz transformation for a boost along the x-axis, and the symmetry between the y- and z- axes, we propose:

$$\begin{aligned} \hat{t}' &= A\hat{t} + B\hat{x} + \alpha\tilde{C}\hat{y} + \alpha\tilde{C}\hat{z}, \\ \hat{x}' &= A\hat{x} + B\hat{t} + \alpha\tilde{C}\hat{y} + \alpha\tilde{C}\hat{z}, \\ \hat{y}' &= \hat{y} + \tilde{B}_3\hat{x} + \tilde{C}_3\hat{t} + \tilde{D}_3\hat{z}, \quad and \\ \hat{z}' &= \hat{z} + \tilde{B}_3\hat{x} + \tilde{C}_3\hat{t} + \tilde{D}_3\hat{y}, \end{aligned} \tag{6}$$

where α is an ordinary real number. By substituting these transformations in Eq. (3), and by demanding the invariance of $ds^2 = Trd\hat{s}^2$, it can be shown that:

(i) All $A, B, \tilde{C}, \tilde{D}_i$s must belong to Grassmannn even sector.

(ii) Demanding that $Trd\hat{s}^2$, $Trd\hat{t}^2$, $Trd\hat{x}^2$, $Trd\hat{y}^2$, $Trd\hat{z}^2$ are real, and $\hat{t}$, $\hat{x}$, $\hat{y}$, $\hat{z}$ are all forced to be either self-adjoint or anti-self-adjoint. Furthermore, for them to remain real in a generic transformation, and for invariance of adjointness type, it is required of all of them to be of same adjointness type and $A, B, \tilde{C}, \tilde{D}_i$s to be Grassmann real.

(iii) A and B are Grassmann even elements, which satisfy $A^2 - B^2 = 1$. *This relation implies that generically A and B are non-Grassmann real numbers* (this assertion also requires that the transformations form a group).

(iv) $\tilde{C}_3 = -\tilde{B}_3 \equiv E$ are those Grassmann even elements, which have their squares identically zero.

(v) $\tilde{D_3}$ is another Grassmann even element proportional to E, i.e., $\tilde{D_3} = \kappa E$, where κ is an ordinary real number.

(vi) $E = \alpha(B - A)\tilde{C}$, where $\tilde{C}$ is the Grassmann element, and α is a real number.

The requirement of existence of a symmetry group of these transformations (product of two transformations with the above properties must be a transformation with the same properties) further forces the choice $\alpha = 0$. Therefore, one is very rigidly constrained to restrict the transformation to the standard Lorentz transformation, with $\beta = -B/A$, and $A^2 - B^2 = 1$, as in ordinary special relativity. We have set $c = 1$; here, it is a universal constant without any further physical interpretation. With these choices of parameters, the transformation of symmetry is devoid of any Grassmann variable and with this structure, we clearly have obtained a symmetry group (the *Lorentz group*) over the field of real numbers as:

$$\begin{pmatrix} \frac{1}{\sqrt{1-\beta^2}} & \frac{-\beta}{\sqrt{1-\beta^2}} & 0 & 0 \\ \frac{-\beta}{\sqrt{1-\beta^2}} & \frac{1}{\sqrt{1-\beta^2}} & 0 & 0 \\ 0 & 0 & 1 & 0 \\ 0 & 0 & 0 & 1 \end{pmatrix}.$$

We have shown next that it is possible to construct a Poincarè invariant dynamics. We have defined a four velocity $\hat{u}^\mu = d\hat{x}^\mu/ds$, and for a particle, a four-momentum $\hat{p}^\mu$ which satisfies $Tr[\hat{p}^\mu \hat{p}_\mu] = m^2$ (some invariant). An action operator $\hat{S}$ and the Trace Action are introduced as:

$$S = Tr\hat{S} = \int ds \; Tr\hat{\mathcal{L}}(\hat{x}, \dot{\hat{x}}), \tag{7}$$

where dot represents derivative with respect to the parameter s. Extremization of the action leads to the Lagrange equations for the Trace Lagrangian $L = Tr\hat{\mathcal{L}}$ as:

$$\frac{\partial L}{\partial \hat{x}} - \frac{d}{ds}\frac{\partial L}{\partial \dot{\hat{x}}} = 0. \tag{8}$$

Hence,

$$\frac{\delta L}{\delta \dot{\hat{x}}} = \int ds \frac{\delta L}{\delta \hat{x}} = \frac{\delta}{\delta \hat{x}} \int ds L. \tag{9}$$

If we define $\delta L/\delta \dot{\hat{x}}$ as momentum conjugate to $\hat{x}$, then we have:

$$\hat{P}_x = \frac{\delta S}{\delta \hat{x}}. \tag{10}$$

Derivatives with respect to operators are to be understood as trace derivatives. Momentum is the tangent to the trajectory in configuration space. Therefore, $\hat{P}_{x^\mu}$ is tangent to the curve drawn in the configuration space of x^μ co-ordinates. Hence,

$$\hat{P}_{x^\mu} = \hat{p}_\mu, \quad and \quad \hat{P}_\mu = \frac{\delta S}{\delta \hat{x}^\mu}.$$

Therefore, the Hamilton-Jacobi equation of motion is:

$$Tr\left[\left(\frac{\delta S}{\delta \hat{t}}\right)^2 - \left(\frac{\delta S}{\delta \hat{x}}\right)^2 - \left(\frac{\delta S}{\delta \hat{y}}\right)^2 - \left(\frac{\delta S}{\delta \hat{z}}\right)^2\right] = m^2. \tag{11}$$

The Hamiltonian analogue of the dynamics is obtained by going to phase-space, and constructing the Hamiltonian as:

$$\mathcal{H} = Tr\left(\sum_r \hat{p}_r \dot{\hat{x}}_r - \hat{L}\right),$$

and the Hamilton's equations of motion follow, by considering the variation as:

$$\delta\mathcal{H} = Tr\sum_r (\epsilon_r \dot{\hat{x}}_r \delta\hat{p}_r - \dot{\hat{p}}_r \delta\hat{x}_r), \tag{12}$$

where the Trace Hamiltonian is a trace functional of operators $\hat{x}_r$, $\hat{p}_r$. Thus,

$$\frac{\delta\mathcal{H}}{\delta\hat{x}_r} = -\dot{\hat{p}}_r\ , \quad and \quad \frac{\delta\mathcal{H}}{\delta\hat{p}_r} = \epsilon_r \dot{\hat{x}}_r. \tag{13}$$

Now, the derivative can be performed element-wise, following Adler's scheme and $\epsilon_r = \pm 1$, if the element belongs to Bosonic/Fermionic sector.

This phase-space can be equipped with a generalized Poisson bracket. For traced operators $\mathbf{A}(\hat{x}_r, \hat{p}_r, \tau)$, $\mathbf{B}(\hat{x}_r, \hat{p}_r, \tau)$, the generalized Poisson bracket is:

$$\{\mathbf{A}, \mathbf{B}\}_{GPB} = Tr\sum_r \epsilon_r \left(\frac{\delta\mathbf{A}}{\delta\hat{x}_r}\frac{\delta\mathbf{B}}{\delta\hat{p}_r} - \frac{\delta\mathbf{A}}{\delta\hat{p}_r}\frac{\delta\mathbf{B}}{\delta\hat{x}_r}\right). \tag{14}$$

With the help of Hamilton's equations of motion given in Eq. (13), one can show that:

$$\frac{d\mathbf{A}}{d\tau} = \frac{\partial\mathbf{A}}{\partial\tau} + \{\mathbf{A}, \mathbf{H}\}_{GPB}. \tag{15}$$

The generalized Poisson bracket satisfies the Jacobi-identity [1, 3], and the Lie algebras of symmetries of the theory can be represented as Lie algebras of trace functionals under generalized Poisson bracket operation.

Moreover, the line element defined above has its symmetry group as the Poincaré group, whose generators commute with the generators of global unitary transformations. Thus, Poincaré invariant theories will have, in complete analogy with the construction of Adler and Millard, a Noether charge corresponding to global unitary invariance of $\mathcal{H}$, given by:

$$Q = \sum_{r\in B}[\hat{x}_r, \hat{p}_r] - \sum_{r\in F}\{\hat{x}_r, \hat{p}_r\}. \tag{16}$$

We have, thus, demonstrated that it is possible to construct a special relativity for non-commuting coordinate operators, by defining an infinitesimal distance using the Trace function.

As a more general extension of TD, we would like to treat space and time also as operators, and then derive the classical spacetime, and a quantum theory on the classical spacetime background, as thermodynamic approximation to this generalized Trace Dynamics. Further details and the motivation for such a construction can be found in [4].

Acknowledgements

It is a pleasure to thank Angelo Bassi, Cenalo Vaz and Satyabrata Sahu for discussions.

References

[1] Adler S L 2004 *Quantum Theory as an Emergent Phenomenon* (Cambridge University Press) [arXiv:hep-th/0206120]
[2] Adler S L and Millard A C 1996 *Nucl. Phys. B* **473** 199
[3] Adler A L, Bhanot G V and Weckel J D 1994 *J. Math. Phys.* **35** 531 [arXiv:1106.0911]
[4] Lochan K and Singh T P 2011 *Phys. Lett. A* **375** 43

Journal of Physics: Conference Series **484** (2014) 012066 doi:10.1088/1742-6596/484/1/012066

Remarks on non-Gaussian fluctuations of the inflaton and constancy of ζ outside the horizon

N Mahajan and R Rangarajan*

Theoretical Physics Division, Physical Research Laboratory, Navrangpura, Ahmedabad 380009, India

E-mail: *raghavan@prl.res.in

Abstract. We have pointed out that the non-Gaussianity arising from cubic self interactions of the inflaton field is proportional to ξN_e. For scales of interest $N_e = 60$, and for models such as new inflation, natural inflation, and running mass inflation ξ is large compared to the slow roll parameter. Therefore, the contribution from self interactions should not be outrightly ignored while retaining other terms in the non-Gaussianity parameter $f_{\rm NL}$. But the N_e dependent term seems to imply the growth of non-Gaussianities outside the horizon. Therefore, we have briefly discussed the issue of the constancy of correlations of the curvature perturbation ζ outside the horizon. We have then presented our results on the 3-point function of ζ_k, and found that the N_e dependent contribution to $f_{\rm NL}$ from self interactions of the inflaton field is cancelled by contributions from other terms associated with non-linearities in cosmological perturbation theory.

1. Introduction

Non-Gaussianities in the curvature perturbation ζ in single field inflation can arise from self interactions of the inflaton and from non-linearities in cosmological perturbation thoery. The contribution arising from self interactions of the inflaton field is usually treated as negligible compared to that from non-linearities in cosmological perturbation theory [1, 2], because it is proportional to higher order slow roll parameters. For example, consider a cubic self interaction term $\mu\phi^3$, for the inflaton field ϕ. The contribution from this term to $\langle\zeta^3\rangle \sim \langle\phi^3\rangle$ is proportional to μ and thus, to the slow roll parameter $\xi \sim V'''$. This is ignored when compared to terms proportional to slow roll parameters $\epsilon \sim V'^2$ and $\eta \sim V''$. But for many models of inflation such as new inflation, small field natural inflation and running mass inflation $\xi \gg \epsilon$. Moreover, as we have pointed out below, the contribution of the self interaction term is actually proportional to ξN_e, where $N_e(t)$ is the number of e-foldings, since the mode of interest crossed the horizon, and varies from 0 to 60 by the end of inflation for our current horizon scale. The above arguments indicate that one should not outrightly ignore the contribution of the self interactions of the inflaton in generating non-Gaussianities in small field inflation models.

If the expression for the 3-point function of the curvature perturbation is proportional to $N_e(t)$, then it indicates that correlations of the curvature perturbation are growing outside the horizon. This is contrary to one's expectations. One may wish to question: Whether the time dependent result is itself correct? However, it has been obtained independently in [3, 4, 5, 6] in different contexts. So, we have now first considered the literature on the constancy of correlations of ζ outside the horizon in the context of cosmological perturbation theory, and discussed the arguments related to constancy of n-point functions of the curvature perturbation outside the horizon.

2. Discussion and results

Salopek and Bond [7] first introduced the generalization of the Bardeen-Steinhardt-Turner variable ζ [8] for the case when one wishes to consider non-Gaussianities in the curvature perturbation. The Salopek-Bond $\zeta(x)$ is constant outside the horizon. In other works [9, 10, 11, 12, 13, 14], the constancy of ζ is shown for a ζ_k mode. The above works deal with classical ζ, and the constancy of classical $\zeta(x)$ or ζ_k only implies that the quantum 2-point function $\langle\hat{\zeta}_{\mathbf{k_1}}\hat{\zeta}_{\mathbf{k_2}}\rangle = (2\pi)^3|\zeta_{k_1}|^2\delta(\mathbf{k_1}-\mathbf{k_2})$ is constant outside the horizon (at lowest order)[1]. But, it is not obvious that other higher n-point functions of $\hat{\zeta}$ are constant outside the horizon.

The n-point functions of $\hat{\zeta}$ are given by [15, 1]:

$$\begin{aligned}\langle\hat{O}(t)\rangle &= \sum_{N=0}^{\infty} i^N \int_{t_0}^{t} dt_N \int_{t_0}^{t_N} dt_{N-1}\cdots\int_{t_0}^{t_2} dt_1 \\ &\quad\times\left\langle\left[\hat{H}_I(t_1),\left[\hat{H}_I(t_2),\cdots\left[\hat{H}_I(t_N),\hat{O}_I(t)\right]\cdots\right]\right]\right\rangle, \qquad (1)\end{aligned}$$

where $\hat{O}(t)$ can be any product of $\hat{\zeta}$ operators, $\hat{O}_I(t)$ is $\hat{O}(t)$ in the interaction picture generated by the quadratic part of the Hamiltonian, and t_0 is some early time. Note that the expectation values are obtained in the in-in formalism and so the bra-s and kets refer to ${}_{in}\langle 0|$ and $|0\rangle_{in}$ respectively. $\hat{H}_I$ is the interaction Hamiltonian and includes terms that are third or higher order in $\hat{\zeta}$. For the 3-point function at lowest order, this reduces to:

$$\langle\hat{\zeta}^3(t)\rangle = i\int_{t_0}^{t} dt'\left\langle\left[\hat{H}_I(t'),\hat{\zeta}_I^3(t)\right]\right\rangle. \qquad (2)$$

If ζ_k is constant, $\hat{\zeta}\sim e^{i\mathbf{k}.\mathbf{x}}\zeta_k c_k + e^{-i\mathbf{k}.\mathbf{x}}\zeta_k^* c_k^\dagger$ is constant. But for $\langle\hat{\zeta}^3(t)\rangle$ to be constant outside the horizon, one must ensure that the contribution to the integral above from $t_{\rm ex}$ to t is suppressed, where $t_{\rm ex}$ is the time of horizon exit for the scale of interest.

The convergence of the integral for large t, and certain other conditions for the constancy of $\langle\hat{\zeta}^n\rangle$ outside the horizon have been discussed in general in [16]. But in [16] (see Eq. (29)), only Gaussian fluctuations of the inflaton are considered[2].

One may argue that after horizon exit, the curvature perturbations are classical and so the constancy of ζ should imply constancy of n-point function for $n > 2$ also. But as argued by Weinberg in [15], the perturbations are classical in the sense that commutators involving $\hat{\zeta}$ and its time derivative go to 0 for large t, i.e.,

$$[\zeta,\zeta],\ [\zeta,\dot{\zeta}]\rightarrow 0, \qquad \text{as t}\rightarrow\infty\,. \qquad (3)$$

However, this implies that quantum n-point functions of zeta can nevertheless grow outside the horizon, though only as powers of $\ln a$ and not as powers of a. It is the time dependence due to $\ln a\sim N_e$ for the 3-point function, which is the focus of our study.

In the light of the time dependent contribution from self interactions and the above discussion, we explicitly verify whether or not $\langle\hat{\zeta}^n\rangle$ is, indeed, constant outside the horizon when one includes non-Gaussian fluctuations of the inflaton.

[1] Higher order corrections from loops can give time dependent or N_e dependent contributions as mentioned in section VI of [15].

[2] Note that the authors in [7] have also considered only Gaussian fluctuations of the inflaton, since $\delta\phi$ is set equal to $H/(2\pi)$ in the evaluation of ζ in sections IIIB and IIID.

In [17], we have first calculated $\langle\delta\phi^3\rangle$ in the $\delta\phi \neq 0$ gauge using the canonical formalism for cubic inflation with $V(\phi) = V_0 - \mu\phi^3$. We have then related $\zeta_k(t)$ to $\delta\phi(k,t)$ and calculated $\langle\zeta^3(t)\rangle$, and the non-Gaussianity parameter fnl. For us, t is arbitrary, unlike in the δN formalism in which t is the time of horizon exit. This allows us to study possible time dependent growth outside the horizon.

After a detailed calculation [17], we have found that:

$$\begin{aligned}\frac{6}{5}f_{\rm NL} &= \xi\left[\frac{1}{3}+\gamma-N_e+\frac{3}{\sum_i k_i^3}\left(k_t\sum_{i<j}k_ik_j-\frac{4}{9}k_t^3\right)\right]\\ &+\frac{3}{2}\epsilon-\eta+\frac{\epsilon}{\sum_i k_i^3}\left(\frac{4}{k_t}\sum_{i<j}k_i^2k_j^2+\frac{1}{2}\sum_{i\neq j}k_ik_j^2\right),\end{aligned} \tag{4}$$

where $f_{\rm NL}$ is defined by:

$$\langle\hat{\zeta}(\vec{k}_1,t)\hat{\zeta}(\vec{k}_2,t)\hat{\zeta}(\vec{k}_3,t)\rangle \equiv (2\pi)^3\delta(\vec{k}_1+\vec{k}_2+\vec{k}_3)\frac{6}{5}f_{\rm NL}\sum_{i<j}P_\zeta(k_i)P_\zeta(k_j)\,. \tag{5}$$

Our expression for $f_{\rm NL}$ is similar to that in Eq. (38) of [6] with their t_* replaced by t, and we have replaced N_* by $-N_e = -H(t-t_{\rm ex})$. $\gamma \approx 0.577216$ is Euler's constant, and $k_t = \sum_i k_i$ and all magnitudes k_i are presumed to be approximately equal.

For our potential $V = V_0 - \mu\,\phi^3$ $(\phi > 0)$, $\xi = 0.5\,\eta^2$. Also, $\eta = 0.5\,(n_s-1)+3\,\epsilon \approx 0.5\,(n_s-1)$ [18]. If we take $n_s = 0.96$ [19], then η is $-\ 0.02$ and ξN_e is 0.012. ϵ is much smaller. Thus, we have seen that ξN_e is comparable to η and $\epsilon \ll \xi N_e$ and so the self interaction contribution proportional to ξ should not be automatically ignored[3, 4].

Now, we have considered the constancy of the 3-point function of $\hat{\zeta}$. To determine how $f_{\rm NL}$ changes during inflation after a mode crosses the horizon, we have taken the derivative of $f_{\rm NL}$ with respect to t. Now, using $d/dt = \dot{\phi}\,d/d\phi$, we get:

$$\begin{aligned}\frac{d\epsilon}{dt} &\simeq [4\epsilon^2-2\eta\epsilon]\,H,\\ \frac{d\eta}{dt} &\simeq [2\epsilon\eta-\xi]\,H, \quad and\\ \frac{d\xi}{dt} &\simeq [4\epsilon\xi-\eta\xi-\sigma]\,H,\end{aligned} \tag{6}$$

where $\sigma = V'^2V''''/V^3$. Then $df_{\rm NL}/dt \approx (5/6)\,d[-\xi N_e - \eta]/dt = 0$. Here, we have kept terms to first order in slow roll parameters as in $f_{\rm NL}$. Thus, for the terms considered above $df_{\rm NL}/dt$ is zero indicating that $f_{\rm NL}(t) = f_{\rm NL}(t_{\rm ex})$, i.e., the N_e factor in ξN_e, which corresponds to growth outside the horizon, is cancelled by changes in other terms and does not contribute to the bispectrum[5]. (The argument above for the constancy of ζ closely follows that in [6].)

[3] Interestingly, Appendix B of [20] obtains a similar result, but erroneously concludes that the N_e dependent result of [3] and the N_e independent result of [1] are the same because they are of the same order.

[4] In [6], it is argued that evaluating expectation values at the end of inflation may not be valid for large $N_e \approx 60$, because of divergences of the form $\epsilon^{m+2}N_e^m$ $(m \geq 1)$. However, for the potential, we are considering ϵ is much smaller than 1/60.

[5] Note that these arguments are for adiabatic fluctuations of the inflaton and do not apply to any non-adiabatic fluctuations or fluctuations of other scalar fields considered in [3, 4, 5].

3. Conclusion

We have argued that it is not proper to a priori, ignore the contribution of inflaton self interactions to the correlations of the curvature perturbation $\zeta_k(t)$ in the case of small field inflation. In fact, the contribution of these terms grows linearly outside the horizon. Explicitly calculating the total 3-point correlation function of ζ, we have found that the time dependent contribution from the self interaction term to the non-Gaussianity parameter $f_{\rm NL}$ is cancelled by the growth of other terms in $f_{\rm NL}$.

Reference

[1] Maldacena J M 2003 *JHEP* **0305** 013 (see Sec. 3.1) [arXiv:astro-ph/0210603]
[2] Creminelli P 2003 *JCAP* **0310** 003 [arXiv:astro-ph/0306122]
[3] Falk T, Rangarajan R and Srednicki M 1993 *Ap. J.* **403** L1 [arXiv:astro-ph/9208001]
[4] Zaldarriaga M 2004 *Phys. Rev. D* **69** 043508 [arXiv:astro-ph/0306006]
[5] Bernardeau F, Brunier T and Uzan J P 2004 *Phys. Rev. D* **69** 063520 [arXiv:astro-ph/0311422]
[6] Seery D, Malik K A and Lyth D H 2008 [arXiv:0802.0588v4 [astro-ph]] This differs in the expression for $f_{\rm NL}$ from the (earlier) published version, *JCAP* **0803** 014
[7] Salopek D S and Bond J R 1990 *Phys. Rev. D* **42** 3936
[8] Bardeen J M, Steinhardt P J and Turner M S 1983 *Phys. Rev. D* **28** 679
[9] Lyth D H and Wands D 2003 *Phys. Rev. D* **68** 103515 [arXiv:astro-ph/0306498]
[10] Rigopoulos G I and Shellard E P S 2003 *Phys. Rev. D* **68** 123518 [arXiv:astro-ph/0306620]
[11] Malik K A and Wands D 2004 *Class. Quant. Grav.* **21** L65 [arXiv:astro-ph/0307055]
[12] Lyth D H, Malik K A and Sasaki M 2005 *JCAP* **0505** 004 [arXiv:astro-ph/0411220]
[13] Langlois D and Vernizzi F 2005 *Phys. Rev. Lett.* **95** 091303 [arXiv:astro-ph/0503416]
[14] Langlois D and Vernizzi F 2005 *Phys. Rev. D* **72** 103501 [arXiv:astro-ph/0509078]
[15] Weinberg S 2006 *Phys. Rev. D* **74** 023508 [arXiv:hep-th/0605244]
[16] Weinberg S 2008 *Phys. Rev. D* **78** 123521 [arXiv:0808.2909 [hep-th]]
[17] Mahajan N and Rangarajan R 2011 *Phys. Rev. D* **83** 043510 [arXiv:1004.5492 [astro-ph.CO]]
[18] Alabidi L and Lyth D H 2006 *JCAP* **0608** 013 [arXiv:astro-ph/0603539]
[19] Komatsu E *et al* 2009 [WMAP Collaboration] *Ap. J. Supp. Ser.* **180** 330 [arXiv:0803.0547 [astro-ph]]
[20] Barnaby N and Cline J M 2007 *JCAP* **0707** 017 [arXiv:0704.3426 [hep-th]]

Classical and quantum correlation of quantum fluctuation

Y Nambu

Department of Physics, Graduate School of Science, Nagoya University, Chikusa, Nagoya 464-8602, Japan

E-mail: nambu@gravity.phys.nagoya-u.ac.jp

Abstract. We have investigated the classical and quantum correlations of a quantum field in the inflationary universe using a particle detector model. We have found that the entanglement between two comoving detectors becomes zero after their physical separation exceeds the Hubble horizon. We have also found that the quantum discord, which is defined as the quantum part of total correlation, approaches zero on the super horizon scale.

1. Introduction

According to the inflationary scenario of cosmology, all structure in the Universe can be traced back to primordial quantum fluctuations during an accelerated expanding phase of the very early universe. Short wavelength quantum fluctuations generated during inflation are considered to lose quantum nature when their wavelength exceed the Hubble horizon length. Then, the statistical property of generated fluctuations can be represented by classical distribution functions. This is the assumption of the quantum to classical transition of quantum fluctuations generated by the inflation. As the structure in the present Universe is classical objects, we must explain or understand how this transition occurred and the quantum fluctuations changed to classical fluctuations.

We have investigated this problem from the viewpoint of entanglement [1, 2]. In our previous study, we have defined spatially separated two regions in the inflationary universe and investigated the bipartite entanglement between these regions. We have found that the entanglement between these two regions becomes zero after their physical separation exceeds the Hubble horizon. This behaviour of the bipartite entanglement confirms our expectation that the long wavelength quantum fluctuations during inflation behave as classical fluctuations and can become seed fluctuations for the structure formation in the Universe. Our previous analysis concerning the entanglement of quantum fluctuations in the inflationary universe relies on the separability criterion for continuous bipartite systems [3, 4], of which dynamical variables are continuous. The applicability of this criterion is limited to systems with Gaussian states: The wave function or the density matrix of the system is represented in a form of Gaussian functions. Thus, we cannot say anything about the entanglement for the system with non-Gaussian state such as excited states and thermal states. Furthermore, from a viewpoint of observation or measurement, information on quantum fluctuations can be extracted via interaction between quantum fields and measurement devices. Hence, it is more natural to consider a setup that the entanglement of quantum field is probed using detectors.

In this paper, we have considered particle detectors [5, 6, 7] with two internal energy levels, which are assumed to interact with a scalar field. By preparing spatially separated two equivalent detectors interacting with the scalar field, we can extract the information on entanglement of the scalar field by evaluating the entanglement between these two detectors. As a pair of such detectors is two qubits system, we have the necessary and sufficient condition for entanglement of this system [8, 9].

2. Two detectors system

We have considered a system with two equivalent detectors interacting with the massless scalar field in an expanding universe [10, 11]. The detectors have two energy level state $|\uparrow\rangle, |\downarrow\rangle$ and their energy difference is given by Ω. The interaction Hamiltonian is assumed to be:

$$V = g(t) \sum_{j=A,B} \left(\sigma^+_{(j)} + \sigma^-_{(j)}\right)\phi(\boldsymbol{x}_j(t)), \quad \sigma^+ = |1\rangle\langle 0|, \quad \sigma^- = |0\rangle\langle 1|. \tag{1}$$

Two detectors are placed at $\boldsymbol{x}_{A,B}(t)$ and $(t, \boldsymbol{x}_{A,B}(t))$, to represent their world lines. We have assumed that the detectors are comoving with respect to cosmic expansion. Strength of the coupling $g(t)$ is controlled in accordance with the Gaussian window function $g(t) = g_0 \exp\left(-(t-t_0)^2/2\sigma^2\right)$.

The initial state of the total system is assumed to be $|\Psi_0\rangle = |\downarrow\downarrow\rangle|0\rangle$. Then the final state $(t \to +\infty)$ of the detector after interaction becomes:

$$\rho^{AB} = \begin{pmatrix} X_4 & 0 & 0 & X \\ 0 & E & E_{AB} & 0 \\ 0 & E_{AB} & E & 0 \\ X^* & 0 & 0 & 1-2E-X_4 \end{pmatrix}. \tag{2}$$

The matrix elements are expressed by integral of the Wightman function of the scalar field.

3. Entanglement and correlations of detectors

As a measure of the entanglement between two detectors, we have considered the negativity [12], defined via a partial transpose operation to the density matrix given in Eq. (2) with respect to detector B's degrees of freedom:

$$\mathcal{N} = |X| - E. \tag{3}$$

The negativity gives the necessary and the sufficient condition of the entanglement for two qubits systems [8, 9], and two detectors are entangled when $\mathcal{N} > 0$ and separable when $\mathcal{N} < 0$.

To define classical part of correlation between two detectors (two qubits), we have performed a local projective measurement of detectors' state. The measurement operators for each detector are:

$$M^A_\pm = \frac{I \pm \boldsymbol{n}_A \cdot \boldsymbol{\sigma}}{2}, \quad and \quad M^B_\pm = \frac{I \pm \boldsymbol{n}_B \cdot \boldsymbol{\sigma}}{2}, \qquad |\boldsymbol{n}_A| = |\boldsymbol{n}_B| = 1, \tag{4}$$

where $\pm$ denotes output of the measurement. $\boldsymbol{\sigma} = (\sigma_1, \sigma_2, \sigma_3)$ are the Pauli matrix and $\boldsymbol{n}_A, \boldsymbol{n}_B$ represent internal direction of measurement. The joint probability p_{jk} attaining measurement result j for detector A, and k for detector B $(j, k = \pm 1)$ is obtained as:

$$p_{jk} = \mathrm{Tr}\left(M^A_j \otimes M^B_k \, \rho^{AB}\right). \tag{5}$$

Using the joint probability obtained by the measurement, the classical mutual information I_C is defined by [13]:

$$I_C(p) = H(p_j) + H(p_k) - H(p_{jk}), \quad H(p_j) = -\sum_j p_j \log_2 p_j, \quad H(p_{jk}) = -\sum_{jk} p_{jk} \log_2 p_{jk}$$

where $H(p)$ is a Shannon entropy for a probability distribution p. The measure of classical correlation is defined via the maximization done over all possible projective measurements as:

$$\mathcal{C}(p) = \sup_{\{\boldsymbol{n}_A, \boldsymbol{n}_B\}} I_C(p) = \max\left[\frac{1}{\ln 2}\left(E^2 - X_4 + X_4 \ln\left(\frac{X_4}{E^2}\right)\right), \frac{2}{\ln 2}\Big(|E_{AB}| + |X|\Big)^2\right]. \tag{6}$$

On the other hand, the quantum mutual information I_Q of the bipartite state ρ^{AB} is defined independent of measurement procedure given as:

$$I_Q(\rho^{AB}) = S(\rho^A) + S(\rho^B) - S(\rho^{AB}), \qquad S(\rho) = -\mathrm{tr}(\rho \log_2 \rho), \tag{7}$$

where $S(\rho)$ is the von Newmann entropy for the state ρ. This quantity represents the total correlations of the bipartite system including both quantum and classical part of correlations. The quantum mutual information for the state in Eq. (2) is given by:

$$\begin{aligned} I_Q(\rho) = &\frac{1}{\ln 2}\Big[-2E \ln E + (E + E_{AB}) \ln(E + E_{AB}) + (E - E_{AB}) \ln(E - E_{AB})\Big] \\ &+ \frac{1}{\ln 2}\Big[E^2 - X_4 + |X|^2 - 2X_4 \ln E + (X_4 - |X|^2) \ln(X_4 - |X|^2)\Big] + O(g^6). \end{aligned} \tag{8}$$

The quantum discord [14, 15, 16] is introduced as the difference between the quantum mutual information and the classical mutual information given as:

$$\mathcal{Q}(\rho) = I_Q(\rho) - \mathcal{C}(p). \tag{9}$$

For arbitrary local projective measurements, it can be shown that $\mathcal{Q} \geq 0$. Thus, the quantum mutual information can be decomposed into a positive classical mutual information and a positive quantum discord, which represents the quantum part of the total correlation. The state with zero quantum discord can be said as a classical state. For pure state, we have $\mathcal{Q} = \mathcal{C}$ and the state with zero quantum discord has no classical correlations.

4. Behaviour of entanglement and correlations of scalar field

We have considered asymptotic estimation of the matrix elements E_{AB}, X, X_4 for $\Omega\sigma \gg 1, H\sigma \ll 1$, where H is the Hubble parameter and σ is duration time of interaction between detectors and the scalar field. For the massless minimal scalar field in de Sitter spacetime, for sufficiently large e-foldings $Ht_0 \gg 1$, the matrix elements of the density matrix are:

$$\begin{aligned} X &\approx -(2\pi g_0^2)\frac{e^{-(\Omega\sigma)^2}}{4\pi^2}(H\sigma)^2\Big[-\ln(Hr_p) + Ht_0\Big], \\ E_{AB} &\approx (2\pi g_0^2)\frac{e^{-(\Omega\sigma)^2}}{4\pi^2}(H\sigma)^2\left[\frac{1}{4\sin^2(H\Omega\sigma^2)} - \ln\sqrt{(Hr_p)^2 + 4\sin^2(H\Omega\sigma^2)} + Ht_0\right], \end{aligned}$$

and the negativity for the minimal scalar field is:

$$\mathcal{N} = \frac{g_0^2}{2\pi} e^{-\Omega^2\sigma^2}(H\sigma)^2\left[\ln\left(\frac{2\sin(H\Omega\sigma^2)}{Hr_p}\right) - \frac{1}{4\sin^2(H\Omega\sigma^2)}\right]. \tag{10}$$

The separability condition $\mathcal{N} < 0$ yields:

$$Hr_p \gtrsim 2\sin^2(H\Omega\sigma^2)\exp\left(-\frac{1}{4\sin^2 H\Omega\sigma^2}\right) \sim 1.0, \tag{11}$$

where the numerical value is obtained for $H\Omega\sigma^2 = 1$. Thus, we can find sufficiently large r_p at which detectors are separable for any value of detector's parameters Ω, σ. As r_p grows in time t_0, entangled two detectors $r_p < H^{-1}$ evolves to a separable state after their separation exceeds the Hubble horizon scale. This behaviour is consistent with our previous analysis of entanglement using the lattice model and the coarse-grained model of the scalar field [1, 2]: Bipartite entanglement of the scalar field between spatially separated regions in de Sitter spacetime disappears beyond the Hubble horizon scale.

The ratio of classical and total correlation of the minimal scalar field is:

$$\frac{\mathcal{C}}{I_Q} \approx 4E \sim g_0^2 e^{-(\Omega\sigma)^2}(H\sigma)^2(Ht_0) = g_0^2 e^{-(\Omega\sigma)^2}(H\sigma)^2 \ln\left(\frac{r_p}{r_0}\right).$$

For sufficiently large distance (large e-folding Ht_0) given by:

$$\ln\left(\frac{r_p}{r_0}\right) \sim \frac{e^{(\Omega\sigma)^2}}{g_0^2(H\sigma)^2} \gtrsim \frac{1}{g_0^2}\left(\frac{\Omega}{H}\right)^2, \tag{12}$$

$\mathcal{C} \approx I_Q$, and this means the quantum state of detectors approaches zero quantum discord state and can be regarded as "classical" in the sense that measurement procedure does not alter the state. Therefore, this behaviour of correlations supports that the long wavelength quantum fluctuations of the massless minimal scalar field in de Sitter spacetime can be treated as classical fluctuations.

5. Summary

We have investigated quantum and classical correlations of the quantum field in de Sitter spacetime using the detector model. Entanglement of the scalar field is swapped to that of two detectors interacting with the scalar field, and we can measure the entanglement of the quantum field by this setup of experiment. In de Sitter spacetime, the entanglement between detectors disappears on super horizon scale, and this behaviour is consistent with our previous analysis using the lattice model and the coarse-grained model of the scalar field [1, 2]. However, the behaviour of correlations shows different behaviour depending on type of scalar fields [7]. For the massless minimal scalar field, the ratio of classical correlation to the total correlation approaches unity for sufficiently large e-foldings. On the other hand, for the massless conformal scalar field, that ratio approaches a constant value smaller than unity, and the condition for classicality is not achieved. These results support the long wavelength quantum fluctuation of the minimal scalar field, which can be treated as classical fluctuations and can become seed fluctuations for the structure in the our Universe.

References

[1] Nambu Y 2008 *Phys. Rev. D* **78** 044023 [arXiv:0805.1471[gr-qc]]
[2] Nambu Y and Ohsumi Y 2009 *Phys. Rev. D* **80** 124031 [arXiv: 0909.2750[gr-qc]]
[3] Simon R 2000 *Phys. Rev. Lett.* **84** 2726
[4] Duan L, Giedke G, Cirac J I and Zoller P 2000 *Phys. Rev. Lett.* **84** 2722
[5] Unruh W G 1976 *Phys. Rev. D* **14** 870
[6] Birrell N D and Davis P C W 1982 *Quantum Fields in Curved Space* (Cambridge University Press)
[7] Nambu Y and Ohsumi Y 2011 *Phys. Rev. D* **84** [044028 arXiv:1105.5212[gr-qc]]
[8] Peres A 1996 *Phys. Rev. Lett.* **77** 1413
[9] Horodecki M, Horodecki P and Horodeck R 1996 *Phys. Lett. A* **223** 1
[10] Reznik B, Retzker A and Silman J 2005 *Phys. Rev. A* **71** 042104
[11] Steeg G V and Menicucci N C 2009 *Phys. Rev. D* **79** 044027
[12] Vidal G and Werner R F 2002 *Phys. Rev. A* **65** 032314
[13] Nielsen M A and Chuang I L 2000 *Quantum Computation and Quantum Information* (Cambridge University Press)

[14] Ollivier H and Zurek W H 2002 *Phys. Rev. Lett.* **88** 01790
[15] Vedral V 2003 *Phys. Rev. Lett.* **90** 050401
[16] Luo S and Zhang Q 2009 *J. Stat. Phys.* **136** 165

Journal of Physics: Conference Series **484** (2014) 012068 doi:10.1088/1742-6596/484/1/012068

Cosmological singularities, AdS/CFT and de Sitter deformations

K Narayan

Chennai Mathematical Institute, SIPCOT IT Park, Padur PO, Siruseri 603103, India

E-mail: narayan@cmi.ac.in, krishnan.narayan@gmail.com

Abstract. We have reviewed aspects of certain time-dependent deformations of AdS/CFT, containing cosmological singularities and their gauge theory duals. Towards understanding these solutions better, we have explored similar singular deformations of de Sitter space and argued that these solutions are constrained, possibly corresponding to specific initial conditions.

1. Cosmological singularities and AdS/CFT

General relativity breaks down at cosmological singularities, with curvatures and tidal forces typically diverging: notions of spacetime, thus, break down. There is a rich history of string theory explorations of such singularities [1, 2]. We have focused on describing certain time-dependent deformations of AdS/CFT [3, 4, 5, 6], where the bulk gravity theory develops a cosmological singularity, and breaks down while the holographic dual field theory, a sensible Hamiltonian quantum system typically subjected to a time-dependent gauge coupling, can potentially be addressed in the vicinity of the singularity. The bulk string theory on $AdS_5 \times S^5$ (in Poincare slicing) with constant dilaton (scalar) is deformed to:

$$ds^2 = \frac{1}{z^2}(\tilde{g}_{\mu\nu}dx^\mu dx^\nu + dz^2) + d\Omega_5^2 \,,$$

with $\tilde{g}_{\mu\nu}, \Phi$ functions of x^μ alone ($\Phi = \Phi(t)$ or $\Phi(x^+)$ then gives time-dependence). This is a solution if

$$\tilde{R}_{\mu\nu} = \frac{1}{2}\partial_\mu\Phi\partial_\nu\Phi, \qquad and \qquad \tilde{\Box}\Phi = 0,$$

where $\tilde{\Box} \equiv \frac{1}{\sqrt{-\tilde{g}}}\partial_\mu(\sqrt{-\tilde{g}}\tilde{g}^{\mu\nu}\partial_\nu)$: these include, *e.g.*, AdS-Kasner, FRW, BKL (based on the Bianchi classification), etc. In many cases, it is possible to find new coordinates such that the boundary metric $ds_4^2 = \lim_{z\to 0} z^2 ds_5^2$ is flat at least as an expansion about $z = 0$. This suggests that the dual is the $\mathcal{N}$=4 super Yang-Mills theory with the gauge coupling $g_{YM}^2 = e^\Phi$, deformed to have external time-dependence. It is useful to focus on sources approaching $e^\Phi \to 0$ at some finite point in time: For instance, a coupling of the form $g_{YM}^2 \to t^p$, $p > 0$, gives rise to $R_{tt} \sim \frac{1}{2}\dot{\Phi}^2 \sim \frac{1}{t^2}$, *i.e.*, a bulk singularity with curvatures and tidal forces diverging near $t = 0$.

Analyzing the gauge theory is possible in some cases. While at first sight, one might imagine that the dual in such cases to be weakly coupled, this is not the case and interactions are important in general [6]. For instance, the gauge kinetic terms $\frac{1}{g_{YM}^2(t)}TrF^2$ can be transformed to canonical ones for redefined gauge fields (as in standard perturbation theory), but this gives rise to new (tachyonic, divergent) mass-terms stemming from the time derivatives of the coupling, which ensure that the field variables get driven to large values as $t \to 0$. With gauge kinetic terms $\frac{1}{g_{YM}^2(t)} TrF^2$, it turns out that the gauge theory Schrödinger wavefunctional near singularity ($t \to 0$) has a "wildly oscillating" phase for $p > 1$. Furthermore, the energy expectation value generically diverges as $\langle H \rangle \sim \frac{1}{g_{YM}^2(t)} \langle V \rangle$. This suggests that if the coupling vanishes strictly, $g_{YM}^2 = e^{\Phi} \to 0$, the gauge theory response is singular. Deforming the gauge coupling so that $g_{YM}^2 = e^{\Phi}$ is small but non-zero near $t = 0$ leads to finite, but large phase oscillation and energy production: $\dot{\Phi} \sim \frac{\dot{g}_{YM}}{g_{YM}}$ is now finite, so the bulk is also non-singular (but stringy). The eventual gauge theory endpoint depends on the details of energy production, but one might expect thermalization on long time scales if the sources turn off.

With the gauge coupling $g_{YM}^2 = e^{\Phi(x^+)}$ being a function of lightcone time x^+, the physics is quite different [3, 4]: In this case, $\tilde{g}_{\mu\nu} = e^{f(x^+)}\eta_{\mu\nu}$ in the bulk, which is engineered to acquire a null singularity at, say, $x^+ = 0$ (with $e^{\Phi} \sim (x^+)^p$). Redefining A_μ to absorb the coupling gives canonical kinetic terms, and here, the potentially problematic mass-terms, in fact, vanish due to the lightlike coupling. With these new gauge theory variables, the interaction terms become unimportant as $e^{\Phi} \to 0$. The near singularity lightcone Schrödinger wavefunctional then appears regular, suggesting weakly coupled Yang-Mills theory. These variables appear to be dual to stringy objects in the bulk (see also related work on worldsheet string descriptions of certain null Kasner-like singularities [7]). We have noted the potential questions about renormalization effects, however, *e.g.*, introducing a "short-time" cutoff near singularity, and studying contributions to the gauge theory effective action from sufficiently high frequency modes (relative to $\dot{\Phi}$).

An interesting point in this discussion has to do with the initial conditions for the time evolution and eventual cosmological singularity. While in many cases, the sources can be seen to turn off in the far past, thus, suggesting the initial state is the vacuum (for the $\mathcal{N}$=4 SYM theory), this is subtle. The fact that there is a bulk curvature singularity in the deep interior ($z \to \infty$) [5] and the observation that the deformations are constrained (*e.g.*, $\tilde{R}_{\mu\nu} = \frac{1}{2}\partial_\mu\Phi\partial_\nu\Phi$, $\tilde{\Box}\Phi = 0$), suggest that, in fact, the initial conditions are constrained or fine-tuned. Exploring this further turns out to be more fruitful in a related context, that of similar deformations in de Sitter space.

2. de Sitter boundary deformations and dS/CFT

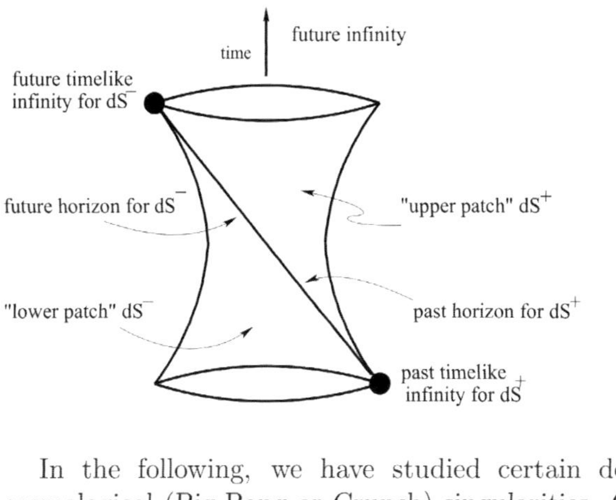

Figure 1. de Sitter space dS_{d+1} in the planar coordinates foliation has the metric $ds^2 = \frac{1}{\tau^2}\left[-d\tau^2 + \delta_{ij}dx^i dx^j\right]$: This covers half the space ($dS^{\pm}$), and $\tau \to \pm\infty$ with x^i fixed corresponds to past/future timelike infinity, while light rays $x^i \sim \tau$ give the horizon.

In the following, we have studied certain deformations of de Sitter space, containing cosmological (Big-Bang or Crunch) singularities. Consider de Sitter space dS_{d+1} in the planar coordinates foliation (Figure 1), which is a solution to $R_{MN} = dg_{MN}$ (with positive cosmological constant), τ being conformal time. We have introduced possible deformations of the d-dim Euclidean boundary:

$$ds^2 = \frac{1}{\tau^2}\left[-d\tau^2 + \tilde{g}_{ij}dx^i dx^j\right], \tag{1}$$

where the spatial metric $\tilde{g}_{ij}$ is a function solely of the spatial coordinates x^i (*i.e.*, not involving time τ). We have seen that these are in a sense analytic continuations of the AdS-cosmologies described above, although various physical features are quite different qualitatively. These metrics have $R_{\tau\tau} = -\frac{d}{\tau^2}$, $R_{ij} = \tilde{R}_{ij} + \frac{d}{\tau^2}\tilde{g}_{ij}$, and are, thus, solutions to Einstein's equations $R_{MN} = dg_{MN}$ (with no additional matter) if $\tilde{R}_{ij} = 0$, *i.e.*, the d-dim Euclidean metric is Ricci-flat. In general, we have considered regular $\tilde{g}_{ij}$. We have then seen that the generic spatial metric $\tilde{g}_{ij}$ (even if regular) gives rise to singularities at $|\tau| \to \infty$ due to diverging $R^{ABCD}R_{ABCD} \sim \tau^4 \tilde{R}^{ijab}\tilde{R}_{ijab} + \ldots$ ($|\tau| \to \infty$ is the analogue of $z \to \infty$ in the AdS interior).

Now, consider dS_{d+1}-deformations sourced by a background scalar field ϕ. If the scalar has purely spatial dependence $\phi(x^i)$, this is a solution if the d-dim part is an Einstein scalar system:

$$\tilde{R}_{ij} = \frac{1}{2}\partial_i\phi\partial_j\phi\,, \qquad and \qquad \tilde{\Box}\phi = 0\,, \tag{2}$$

with $\tilde{\Box} = \frac{1}{\sqrt{\tilde{g}}}\partial_i(\tilde{g}^{ij}\sqrt{\tilde{g}}\partial_j)$. This scalar is non-dynamical and could represent non-trivial initial or final conditions sourcing the deformation of the spacetime. Using Eq. (2), these solutions have the invariants:

$$R = d(d+1) + \frac{\tau^2}{2}(\tilde{\partial}\phi)^2, \qquad and \qquad R_{AB}R^{AB} = d^2(d+1) + \frac{3\tau^2}{2}(\tilde{\partial}\phi)^2 + \frac{\tau^4}{4}\Big((\tilde{\partial}\phi)^2\Big)^2,$$

where $(\tilde{\partial}\phi)^2 = \tilde{g}^{ij}\partial_i\phi\partial_j\phi$. Thus, these invariants diverge at early/late times $|\tau| \to \infty$, sourced by the scalar (even if the scalar energy density itself is finite).

We have noticed now that purely gravitational dS_4-deformations do not exist: $\tilde{R}_{ij} = 0$ has only trivial solutions, pure 3-dim gravity being trivial. Non-trivial dS_4-deformations require a non-trivial source, *e.g.*, the scalar field $\phi(x^i)$ above[1]: *e.g.*, consider the dS_4-deformation:

$$ds^2 = \frac{1}{\tau^2}\Big[-d\tau^2 + dr^2 + (r+C)d\varphi^2 + 2(r+C)dz^2\Big], \qquad and \qquad e^{\phi} = \frac{1}{r+C}, \tag{3}$$

with $R = 12 + \frac{\tau^2}{2(r+C)^2}$, $R_{\mu\nu}R^{\mu\nu} = 36 + \frac{3\tau^2}{(r+C)^2} + \frac{\tau^4}{(r+C)^4}$, and $R_{\mu\nu\rho\sigma}R^{\mu\nu\rho\sigma}$ has similar structure. Then for finite r, these invariants diverge as $|\tau| \to \infty$, signalling a singularity at past/future timelike infinity. Light ray trajectories include geodesics of the form $\tau = r$, reaching the past horizon as $|\tau| \to \infty$, and here, these invariants are finite. If higher order invariants are also finite, this would give a spacelike singularity at $|\tau| \to \infty$.

It is interesting to compare these solutions with conventional investigations of cosmological perturbations about de Sitter space, *e.g.*, [8]. In an initial value formulation, the metric family $ds^2 = -N^2dt^2 + h_{ij}dx^idx^j$, with $N(t)$ the lapse function and h_{ij}, the spatial metric, has the action:

$$S = \int d^4x\sqrt{h}N\Big[K_{ij}K^{ij} - K^2 + \tilde{R}^{(3)} - 2\Lambda + \frac{1}{2}N^{-2}\dot{\phi}^2 - \frac{1}{2}h^{ij}\partial_i\phi\partial_j\phi\Big], \tag{4}$$

[1] For dS_5-deformations (and higher dimensions), there are non-trivial solutions to $\tilde{R}_{ij} = 0$, *e.g.*, 4-dim Ricci-flat spaces, which include ALE spaces (non-compact $\mathbb{C}^2/\mathbb{Z}_N$ singularities and their complete resolutions which are smooth) and (compact) K3-surfaces. The dS_4-deformation above has been found starting with the spatial metric $d\sigma^2 = dr^2 + (f(r))^2d\varphi^2 + h(r)dz^2$, $\phi = \phi(r)$. $C \neq 0$ ensures regularity at $r = 0$; also ϕ is regular for large r.

with $R^{(3)}$ is the 3-curvature, the extrinsic curvature being $K_{ij} = -\frac{1}{2}(\nabla_i n_j + \nabla_j n_i) = -\frac{1}{2N}\partial_\tau h_{ij}$ ($K = h^{ij}K_{ij}$, and n is a unit normal to $t = constant$ surfaces). The action for small gravitational fluctuations $h_{ij} = a^2(t)(\delta_{ij} + \gamma_{ij})$ becomes:

$$S \sim \int d^4x\Big(Na^d\frac{1}{4N^2}(\dot\gamma_{ij})^2 - Na^d\frac{1}{a^2}(\partial_k\gamma_{ij})^2\Big),$$

for dS_4, using $N = a = \frac{1}{\tau}$ leads to the familiar Bunch-Davies vacua (after imposing appropriate regularity conditions). On the other hand, the solutions in Eq. (1) arise from the above action in the limit, where the γ_{ij} become essentially time-independent. The Hamiltonian constraint:

$$K_{ij}K^{ij} - K^2 = \tilde{R}^{(3)} - 2\Lambda - \frac{1}{2}N^{-2}\dot\phi^2 - \frac{1}{2}h^{ij}\partial_i\phi\partial_j\phi\ , \tag{5}$$

can be seen to be satisfied for Eq. (1) with $\tilde{R}_{ij} = \frac{1}{2}\partial_i\phi\partial_j\phi,\ \Lambda = \frac{d(d-1)}{2}$. The on-shell action for these solutions contains no divergence arising from the near singularity region $|\tau| \to \infty$. Similarly, the boundary term $S_B \sim \int d^3x\sqrt{g}K$ is also regular. This apparent non-singularity of the wavefunction of the universe $\Psi = e^{iS}$ could, of course, easily be invalidated by possibly singular higher derivative terms: one might expect stringy effects are important.

From the point of view of dS/CFT [9], using the discussion in [8], we expect that the wavefunction of the universe (related to the partition function for the dual field theory) is effectively an analytic continuation of that for similar solutions in Euclidean AdS, as they are correlation functions (while observables such as n-point functions of bulk fluctuations are not). The precise analytic continuation from Euclidean AdS_{d+1} to dS_{d+1} (in planar coordinates) is obtained by $z \to -i\tau,\ \ R_{AdS} \to -iR_{dS}$. For the deformations we are considering, in the limit of small deformations, we have seen that sources are turned on for the operators dual to bulk graviton and scalar ϕ modes. This leads us to guess that the dual CFT lives on the space $\tilde{g}_{ij}$ and has been deformed by a source for the operator $\mathcal{O}^\phi$ (although for our solutions, the deformations are not small). In this perspective, these solutions are dual to corresponding deformations of the Euclidean CFT dual to de Sitter space.

Correlation functions for operators dual to bulk scalar modes can be calculated by differentiating with respect to the boundary sources. For instance, by using a Fourier decomposition in terms of eigen-mode functions on $\tilde{g}_{ij}$ and evaluating the action, it can be seen that the momentum space 2-point function in these dS_4-deformed theories is just as in dS_4. The holographic stress tensor can be calculated using the usual counter term prescriptions for these theories using [10, 5], giving (on one of the dS-patches, Figure 1):

$$T^{ij} = \frac{1}{8\pi G_{d+1}}\left[K^{ij} - Kh^{ij} + (d-1)h^{ij} + \frac{1}{2}G^{ij} - \frac{1}{4}\partial^i\phi\partial^j\phi + \frac{1}{8}h^{ij}(\partial\phi)^2\right]\ . \tag{6}$$

For the present dS_{d+1}-deformations, with boundary metric $h_{ij} = \frac{1}{\tau^2}\tilde{g}_{ij}$, the extrinsic curvature is $K_{ij} = h_{ij}$, so that the stress tensor vanishes identically, using $\tilde{R}^{ij} = \frac{1}{2}\partial^i\phi\partial^j\phi$.

Recall that if a source for a bulk field is turned on, we generically expect a non-zero 1-point function for the dual operator as a response to the source. Here, with the metric source $\tilde{g}_{ij}$, we expect $\langle T_{ij}\rangle \neq 0$: This is at variance with the vanishing stress tensor above. In fact, requiring that the holographic stress tensor vanishes (relatedly, trace anomaly encoding conformality of the Euclidean CFT on the curved space) gives the conditions $\tilde{R}^{ij} = \frac{1}{2}\partial^i\phi\partial^j\phi$.

Now consider the ($\tau \sim 0$) Fefferman-Graham expansion for an asymptotically locally de Sitter spacetime:

$$ds^2 = -\frac{d\tau^2}{\tau^2} + \frac{1}{\tau^2}\Big[g^0_{ij}(x^i) + \tau^2 g^2_{ij}(x^i) + \ldots\Big]dx^i dx^j.$$

With the leading source g^0_{ij}, the sub-leading coefficients g^2_{ij}, $g^4_{ij}, \ldots$ are generically non-zero [11], and encode information about the state of the dual CFT. Consider then the deformations above as applied to the lower patch dS^-: We have specified initial conditions for the bulk metric and equivalently, the initial conditions/state for the dual Euclidean CFT. In the present case, we have seen that, in fact, only $g^0_{ij} \equiv \tilde{g}_{ij} \neq 0$ with $g^n_{ij} = 0,\ n > 0$, requiring that the higher order coefficients vanish, which leads to our solutions above. Using the small-τ Fefferman-Graham expansion for both metric and scalar $\phi = \tau^{(d-\Delta)/2}(\phi^0 + \tau^2\phi^2 + \ldots)$, we need to solve $R_{MN} = dg_{MN} + \frac{1}{2}\partial_M\phi\partial_N\phi$ iteratively, and this gives:

$$g^2_{ij} \sim R^0_{ij} - \frac{1}{2}\partial_i\phi\partial_j\phi - \frac{1}{2(d-1)}\Big(R - \frac{1}{2}(\partial\phi)^2\Big)g^0_{ij} \quad and \quad g^2_{ij} = 0 \quad \Rightarrow \quad R^0_{ij} = \frac{1}{2}\partial_i\phi\partial_j\phi\,, \qquad (7)$$

(for a massless scalar $\Delta = d$) also implying the higher order coefficients vanish. Likewise, with $\Box^0$ being the Laplacian with respect to g^0_{ij}, we have also obtained $\phi^{(2)} \sim \Box^0\phi^0$: thus $\phi^{(2)} = 0$ implies $\Box^0\phi^0 = 0$. The vanishing of the stress tensor is also related to these sub-leading coefficients.

Our solutions have constrained these sub-leading pieces of the metric and scalar in the small τ expansion to vanish. These conditions on the g^n_{ij}, ϕ^n, $n > 0$, are highly non-generic and appear to be non-trivial constraints on the CFT state.

3. Discussion

We have argued that these deformations of AdS and dS are constrained from a Fefferman-Graham perspective, leading to certain singular structures. This suggests to turning on the sub-leading coefficients in appropriate fashion towards de-singularizing them, and we hope to explore this further. We believe these arguments also apply to the solutions in [12], which although, quite different in interpretation, are related to the AdS null solutions (section 1 above) by coordinate transformations [5]. Here, diverging tidal forces (with finite curvature invariants) lead to the large-z singularity [13].

Acknowledgements

We thank the ICGC-2011 organizers for a stimulating conference. Section 2 on dS-deformations is based on work in progress with Sumit Das. Also we thank S Minwalla, M Rangamani and S Trivedi for useful conversations. This work was partially supported by a Ramanujan Fellowship, DST, Government of India.

References

[1] Craps B and Evnin O 2011 *Class. Quant. Grav.* **28** 204006 [arXiv:1103.5911[hep-th]]
[2] Burgess C P and McAllister L 2011 *Class. Quant. Grav.* **28** 204002 [arXiv:1108.2660[hep-th]]
[3] Das S R, Michelson J, Narayan K and Trivedi S P 2006 *Phys. Rev. D* **74** 026002 [hep-th/0602107]
[4] Das S R, Michelson J, Narayan K and Trivedi S P 2007 *Phys. Rev. D* **75** 026002 [hep-th/0610053]
[5] Awad A, Das S R, Narayan K and Trivedi S P 2008 *Phys. Rev. D* **77** 046008 [arXiv:0711.2994[hep-th]]
[6] Awad A, Das S, Nampuri S, Narayan K, Trivedi S 2009 *Phys. Rev. D* **79** 046004 [arXiv:0807.1517[hep-th]]
[7] Madhu K, Narayan K 2009 *Phys. Rev. D* **79** 126009 [arXiv:0904.4532[hep-th]], Narayan K 2010 *Phys. Rev. D* **81** 066005 [arXiv:0909.4731[hep-th]], 2011 *JHEP* **1101** 145 [arXiv:1012.0113[hep-th]]
[8] Maldacena J M 2003 *JHEP* **0305** 013 [astro-ph/0210603]
[9] Strominger A 2001 *JHEP* **0110** 034 [hep-th/0106113] (See also Witten E [hep-th/0106109])
[10] Balasubramanian V, de Boer J and Minic D 2002 *Phys. Rev. D* **65** 123508 [hep-th/0110108]
[11] de Haro S, Solodukhin S and Skenderis K [hep-th/0002230], Skenderis K [hep-th/0209067]
[12] Balasubramanian K and Narayan K 2010 *JHEP* **1008** 014 [arXiv:1005.3291[hep-th]]
[13] Horowitz G T and Way B 2012 *Phys. Rev. D* **85** 046008 [arXiv:1111.1243[hep-th]].

Perturbations in loop quantum cosmology

W Nelson, I Agullo and A Ashtekar
Institute for Gravitation and the Cosmos, The Pennsylvanian State University, USA
E-mail: nelson@gravity.psu.edu

Abstract. The era of precision cosmology has allowed us to accurately determine many important cosmological parameters, in particular via the CMB. Confronting Loop Quantum Cosmology with these observations provides us with a powerful test of the theory. For this to be possible, we need a detailed understanding of the generation and evolution of inhomogeneous perturbations during the early, quantum gravity phase of the universe. Here, we have described how Loop Quantum Cosmology provides a completion of the inflationary paradigm, that is consistent with the observed power spectra of the CMB.

1. Open issues with inflation

The standard picture we have of early universe cosmology is that of inflation. This beautiful idea contains a very few basic assumptions, yet it produces, with spectacular agreement, the power spectra of density fluctuations observed in the Cosmic Microwave Background (CMB). Nevertheless, inflation remains a paradigm in search of a model, and there are several key questions that remain unanswered. Broadly speaking, these can be split into two categories: (i) Difficulties facing the particle physics interpretation of inflation, and (ii) those we expect to be related to quantum gravity. Examples of each type are given in Table 1.

Inflation is a period of quasi-de Sitter expansion, and typically the assumption is that slow-roll inflation began with the quantum state describing perturbations in its 'natural' vacuum state: the Bunch-Davies vacuum. A priori, this is an odd assumption, since it says that the quantum state is tuned to the subsequent (quasi-de Sitter) evolution of the geometry. Essentially, this implies that the pre-inflationary dynamics of the universe and the 'true' initial state conspired in such a way as to ensure that we arrived at the onset of slow-roll inflation with no particles present (relative to the Bunch-Davies vacuum). The intuition behind this assumption was that even if there were particles present at the onset of inflation, the exponential expansion would rapidly dilute them and hence, their consequences can safely be ignored. This intuition misses the important fact that quantum fields in a dynamical spacetime experience *both* spontaneous and simulated creation of particles [2]. The latter effect actually compensates for the exponential growth of the volume in such a way that the particle number density remains approximately constant.

In classical general relativity, inflation is inevitably preceded by the Big Bang and the only natural place to give initial conditions is at this singularity. An important open question then is: Can one find a quantum gravity completion to the inflationary paradigm? To be viable, such a completion should be non-singular and agree with current observations. It would also open up the exciting possibility that we can directly observe the pre-inflationary universe. If it were

Particle physics issues	Quantum gravity issues
What is the inflaton? Why is the potential flat? How does it couple to the standard model? What interactions are present?	What are the initial conditions? Is there a singularity? Why is the perturbation theory valid? What happens when the frequencies become trans-Planckian?

Table 1. Some examples of the unresolved issues facing inflation.

possible to see observational consequences of the quantum state at the onset of slow-roll, we would be able to probe the dynamics of the quantum gravity era of the universe.

2. Loop Quantum Cosmology

Loop Quantum Gravity is particularly a well developed approach to quantizing gravity [3]. It is a Hamiltonian quantization that maintains the fundamental relationship between geometry and gravity and is fully non-perturbative. However, a rigorous understanding of the dynamics of the theory are still lacking. One useful way to make progress is to consider simplified (truncated) systems of the full theory, which on the one hand, can be completely understood, and on the other are physically interesting. This approach has been applied with great success to the study of black holes (see Fernando Barbero's paper in this session), and graviton propagators within Loop Quantum Gravity. One can also consider the truncation of full general relativity to homogeneous systems, and then use Loop Quantum Gravity techniques to quantize these. This leads to Loop Quantum Cosmology (LQC) [4], which has turned out to be very successful at answering many of the difficulties facing classical cosmology. In particular, it has been shown in detail how the classical singularity is replaced by a 'Big Bounce' and how the late time, low energy, limit reproduces the standard expectations of general relativity. It has also been shown that the probability of having a sufficiently long phase of (single scalar field driven) slow-roll inflation, within LQC, is very close to one [5].

However, in order to make predictions about LQC effects on the CMB, first, one has to extend the underlying approach to include (perturbative) inhomogeneities. There have been several promising attempts to include such inhomogeneities based on consistent alterations of the homogeneous formulation of LQC. Here, we have described a systematic approach to extending the underlying formulation.

3. LQC and perturbations

The first step is to find a suitable truncation of the phase-space of classical general relativity that allows for cosmologies with perturbative inhomogeneities. Since we work in the Hamiltonian theory, we restrict our attention to cosmologies whose spatial slices are (flat) tori. One can then show that the full phase-space decomposes into homogeneous and *purely* inhomogeneous parts:

$$\Gamma_{\text{Full}} = \Gamma_{\text{H}} \times \Gamma_{\text{IH}} \ . \qquad (1)$$

The canonical variables of the homogeneous phase-space (Γ_{H}) are $(a, \pi_a, \phi, \pi_\phi)$, i.e., the scale factor a and the scalar field ϕ, and their conjugate momenta. The canonical variables of the inhomogeneous phase-space (Γ_{IH}) are the corresponding perturbations, $\left(h_{ab}(x), \pi^{ab}(x), \varphi(x), \pi_\varphi(x)\right)$, all of which are purely inhomogeneous.

All the canonical structures of the phase-space decompose in this way, in particular, the symplectic structure and the Possion brackets factor. One can now consider the inhomogeneous fields as perturbations, expand the constraints and find gauge invariant degrees of freedom [6]. In particular, the gauge invariant scalar modes of the perturbations satisfy the constraint:

$$\mathcal{C} = \frac{\pi_\phi^2}{2l^3} + \frac{m^2V^2}{2l^3} - \frac{3}{8\pi G l^3}b^2V^2 + \int \mathrm{d}^3k \left(\frac{1}{2}P_k^2 + f\left(V, b, \phi, \pi_\phi; k\right) Q_k^2\right) \approx 0, \tag{2}$$

where $V \sim a^3$ and $b \sim \pi_a/a^2$ are canonically conjugate background variables, l is the coordinate size of the 3-torus and (P_k, Q_k) are the gauge invariant, scalar degree of freedom (related to the Mukhanov variable). The important point is that gauge invariant scalar (and tensor) perturbations behave exactly as test scalar fields with a time dependent mass $f\left(V, b, \phi, \pi_\phi; k\right)$.

In [7], it was shown how to quantize such a system, and we schematically sketch this procedure here. Writing the background and perturbation parts of Eq. (2) as:

$$\mathcal{C} = \frac{\pi_\phi^2}{2l^3} + H_0^2 + \int \mathrm{d}^3k H_{\tau,k}\ , \tag{3}$$

where H_0 is the Hamiltonian for the background geometry (and the potential term) and $H_{\tau,k}$ is the (time dependent) Hamiltonian for the perturbations. These are then promoted to operators and we have de-parameterized with respect to the scalar field ϕ, to find,

$$-i\hbar\partial_\phi\Psi = \left(\widehat{H_0^2} + \int \mathrm{d}^3k\widehat{H}_{\tau,k}\right)^{1/2}\Psi \approx \left[\widehat{H}_0 + \widehat{H}_0^{-1/2}\left(\int \mathrm{d}^3k\widehat{H}_{\tau,k}\right)\widehat{H}_0^{-1/2}\right]\Psi, \tag{4}$$

where $\widehat{H_0^2} = \hbar^2\widehat{\Theta}_0$ is the operator that governs the LQC evolution of the background [4], and the right hand side has been approximated via a series expansion.

This is the only approximation that is made in the procedure, and it can be viewed as a test-field approximation. Essentially, ones attention is to those states in which the $\widehat{H}_0$ is dominant, and $\widehat{H}_{\tau,k}$ does not alter the background dynamics[1]. The second term on the right hand side of Eq. (4) is precisely the Hamiltonian for the perturbations associated to the relational time defined by ϕ.

Finally, one decomposes the wave-function as a tensor product of background and perturbation pieces as:

$$\Psi\left(V, Q_k, \phi\right) = \Psi_0\left(V, \phi\right) \otimes \psi\left(Q_k, \phi\right), \tag{5}$$

and considers those states for which Ψ_0 is sharply peaked with respect to some particular classical background geometry at late times, i.e., a semi-classical background geometry. Taking expectation values of Eq. (4) with respect to Ψ_0, one arrives at the standard Hamiltonian for test quantum fields in a curved spacetime, with the background scale factor $a\left(\phi\right)$ replaced by $\left.\langle\Psi_0|\hat{a}|\Psi_0\rangle\right|_\phi$.

4. Results and conclusions

In the previous Section, we have briefly sketched how one can extend the formulation of LQC to include (perturbative) inhomogeneities. With this in hand, one can now look for consequences of the pre-inflationary era and in particular, we can look for deviations from the standard initial conditions for inflation. The standard approach in inflation is to assume the existence of a scalar field with a suitable potential, which gives oneself some initial conditions for the quantum state of the perturbations and hence, calculate the late time power spectra. Here, we have taken exactly the same approach: We have specified initial conditions at the Big Bounce, and calculated the

[1] There are many important issues to do with ensuring that the integral over k results in a well defined operator that are not being described here (See [1] for details.

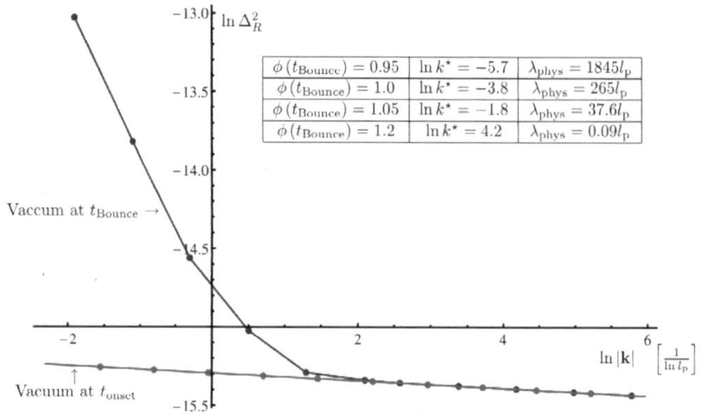

Figure 1. The power spectra for tensor perturbations at the end of slow-roll inflation, given an initial vacuum at the bounce. We have set $a\left(t_{\text{bounce}}\right) = 1$ and hence, the physical scale at (for example) decoupling depends on the amount of inflation that has occurred. Note, in particular, the agreement with the predictions of standard inflation provided $\phi\left(t_{\text{bounce}}\right) > 1.2$.

resulting power spectra (here, we have concentrated only on tensor perturbations). Since, at the bounce, the wavelength of all the observable modes is much smaller than the curvature scale, we have taken the initial state (for these modes) to be the Minkowski vacuum (see [1] for a discussion on this point).

Figure 1 shows the resulting late time (dimensionless) power spectra $\Delta^2_{\mathcal{R}}$, defined as:

$$\delta\left(\mathbf{k} + \mathbf{k}'\right) \frac{\Delta^2_{\mathcal{R}}(k)}{4\pi|\mathbf{k}|^3} = \left\langle 0|\hat{R}_{\mathbf{k}}\hat{R}_{\mathbf{k}'}|0\right\rangle, \tag{6}$$

where the variable $R_{\mathbf{k}}$ is related to the Mukhanov variable $Q_{\mathbf{k}}$ by $R_{\mathbf{k}} = \frac{\dot{\phi}}{H}Q_{\mathbf{k}}$, and represents the gravitational potential on constant ϕ hyper-surfaces.

The important result, here, is that provided the background scalar field ϕ, at the bounce is large enough, i.e., $\phi\left(t_{\text{bounce}}\right) \geq 1.2$, the power spectra is entirely consistent with that of standard inflation. Hence, we have a completion of the inflationary paradigm, including the quantum gravity era, which agrees with the observations. Note that there remains a (small but important) window $(0.95 < \phi\left(t_{\text{bounce}}\right) < 1.2)$ in which, we may hope to see corrections to the largest scales (smallest k) of the observed power spectra, and hence, directly observe pre-inflationary (i.e., quantum gravity) physics.

Acknowledgements
This work was supported in part by the NSF grants PHY0854743 and PHY1068743.

References

[1] Agullo I, Ashtekar A and Nelson W *In preparation*
[2] Agullo I and Parker L 2011 *Phys. Rev. D* **83** 063526
[3] Ashtekar A, Bojowald M and Lewandowski J 2003 *Adv. Theor. Math. Phys.* **7** 233, Ashtekar A and Lewandowski J 2004 *Class. Quant. Grav.* **21** R53
[4] Ashtekar A and Singh P 2011 *Class. Quant. Grav.* **28** 213001, Bojowald M 2008 *Living Rev. Rel.* **11**
[5] Germani C, Nelson W and Sakellariadou M 2007 *Phys. Rev. D* **76** 043529, Ashtekar A and Sloan D 2011 *Gen. Rel. Grav.* **43** 3619, Corichi A and Karami A 2011 *Phys. Rev. D* **83** 104006
[6] David Langlois 1994 *Class. Quant. Grav.* **11** 387
[7] Ashtekar A, Kaminski W and Lewandowski J 2009 *Phys. Rev. D* **79** 064030

Journal of Physics: Conference Series **484** (2014) 012070 doi:10.1088/1742-6596/484/1/012070

Information and cosmological physics

R R Parwani

Department of Physics, National University of Singapore, Kent Ridge, Singapore

E-mail: parwani@nus.edu.sg

Abstract. We have reviewed an information-theoretic approach to quantum cosmology, summarizing the key results obtained to date, including a suggestion that an accelerating universe will eventually turn around.

1. Preamble

There are several intriguing approaches to quantum spacetime, and quantum cosmology, in particular. Here, we have described the work of a number of people involved in one endeavour: An information-theoretic approach to quantum cosmology. Spacetime limit has made this review brief and qualitative.

2. A method of inference

The maximum entropy principle (MEP) [1] originated in statistical mechanics as an inference principle, which allowed one to obtain relevant probability distributions in a conceptually appealing manner. For example, consider a statistical system with unknown probability distribution $p(x)$, but with specified mean energy $E = \int \epsilon(x)p(x)dx$. In the MEP approach, one maximizes the Gibbs entropy, given by:

$$I_{GS} = -\int p(x)\ ln\ p(x)dx\ , \tag{1}$$

under the given constraint to determine the form for the probability distribution: Introducing the Lagrange multiplier β and maximizing $I_{GS} - \beta E$ with respect to variations in $p(x)$ gives the well known canonical probability distribution $p(x) \propto exp(-\beta\epsilon(x))$.

Although, in physics, the quantity in Eq. (1) is usually associated with problems in statistical mechanics, an identical expression was derived independently by Claude Shannon in his search for a measure that could be used to quantify the information content, or uncertainty, in a system. Indeed, the original MEP of physics is actually an example of a more general maximum uncertainty principle (MUP), a method of inference used in diverse fields of study [2]. The basic idea is that one should provide the most unbiased description of the state of the system, since maximizing the uncertainty measure acknowledges our ignorance of a more detailed structure.

In some cases, one may already have some *a priori* information about the system, encoded in the form of a reference probability distribution $r(x)$. Then the relevant measure, that is maximized in a relative uncertainty measure, called the Kullback-Leibler (KL) information, is given by:

$$I_{KL}(p,r) = -\int p(x)\ ln\frac{p(x)}{r(x)}dx. \tag{2}$$

If there is no useful *a priori* information, then $r(x)$ can be taken to be a uniform distribution and the KL measure then reduces to the Gibbs-Shannon entropy.

Although the MUP is elegant, it does require some specific input: The precise form of the information measure is to be used. As Shannon has shown, the measure in Eq. (1) is the simplest measure that satisfies certain axioms appropriate for the context of some problems. Different situations may require one to relax those assumptions and thus, lead to measures such as in Eq. (2) or others like the Fisher measure to be discussed below.

3. Quantum physics

It is useful to re-examine the Schrödinger equation, which describes the evolution of probability amplitudes, from the perspective of the MUP [3]. In one dimension, the transformation $\psi = \sqrt{p}\, e^{iS/\hbar}$ can be used to re-write the Schrödinger equation in terms of two real functions, p and S as:

$$\frac{\partial S}{\partial t} + \frac{1}{2m}\left(\frac{\partial S}{\partial x}\right)^2 + V + Q = 0, \quad \textit{and} \tag{3}$$

$$\frac{\partial p}{\partial t} + \frac{1}{m}\frac{\partial}{\partial x}\left(p\frac{\partial S}{\partial x}\right) = 0, \tag{4}$$

with $Q = -\frac{\hbar^2}{2m}\frac{1}{\sqrt{p}}\frac{\partial^2\sqrt{p}}{\partial x^2}$, the "quantum potential". For $Q = 0$, the Eq. (3) is just the Hamilton-Jacobi equation for a classical *ensemble* of particles described by a probability distribution $p(x,t)$, and with the function $S(x,t)$ related to the velocity of a particle given by $v = \frac{1}{m}\frac{\partial S}{\partial x}$. The Eq. (4) is just the expression for conservation of probability .

It was noted [3] that the classical $Q = 0$ limit of Eqs. (3) and (4) may be obtained by minimizing the action:

$$\Phi_A \equiv \int p\left(\frac{\partial S}{\partial t} + \frac{1}{2m}\left(\frac{\partial S}{\partial x}\right)^2 + V\right) dx\, dt,$$

through a variation of both p and S. The full quantum equations follow if simultaneously the Fisher information (an inverse uncertainty measure) $I_F = \int \frac{1}{p}\left(\frac{\partial p}{\partial x}\right)^2 dxdt$ is also minimized: That is, $\Phi_A + \xi I_F$ is minimized with respect to both p and S. If the Lagrange multiplier ξ is set to $\hbar/8m$, then one obtains the Eqs. (3) and (4), which are equivalent to the time-dependent Schrödinger equation.

Thus, the principle of minimum Fisher information is an example of the MUP, and the result above may be viewed as an extension of the method of inference to quantum mechanics. More specifically, one may argue that the method of inference allows one to supplement classical ensemble dynamics with additional fluctuations to arrive at quantum mechanics. The method is easily extended to many particles in higher dimensions.

4. Non-linear Schrödinger equation

The pioneering approach in [3] did not provide a derivation of the Fisher measure in the same way that the Gibbs-Shannon measure was derived: From axioms appropriate for the information-theoretic context. Such a derivation was provided in [4], and it has clarified that the Fisher measure was the simplest to satisfy suitable assumptions, such as locality and separability. One of the assumptions [4] was that the number of derivatives in the information measure be at a minimum. Clearly then, a generalization would lead to higher derivative information measures and a generalized Schrödinger equation.

However, a direct expansion in derivatives would likely lead to highly singular expressions in the generalized Schrödinger equation, and so a different approach would be to construct a suitable information measure, which may be viewed as a "sum of higher derivative" terms. It turns out that there already exists such a measure! If in Eq. (2) one chooses the reference distribution $r(x)$ to be the same as $p(x)$, but with infinitesimally shifted arguments, that is, $r(x) = p(x + \Delta x)$, then to lowest order:

$$I_{KL}(p(x), p(x + \Delta x)) = -\frac{(\Delta x)^2}{2} I_F(p(x)) + O(\Delta x)^3 \ . \tag{5}$$

So to the lowest order, minimizing the Fisher information is the same as maximizing the relative entropy for two probability distributions that are close to each other, $r(x) = p(x + \Delta x)$.

An interpretation of Δx is readily available. For example, many heuristic arguments combining quantum theory and gravity suggest the existence of a minimal position uncertainty, thus, Δx in Eq. (5) may be taken to be the scale at which the coordinates become distinguishable. Beyond leading order, the expansion in Eq. (5) involves higher derivatives. These too have a natural interpretation as describing situations, where fluctuations at increasingly shorther scales become important, as one might expect when quantum physics affects the spacetime.

This motivates one to study a generalized Schrödinger equation that results from using the KL measure, instead of the Fisher measure in the derivation of the last Section. It leads to a non-linear Schrödinger equation [5], whose consequences have been studied in a number of papers. Though non-linear, the equation still shares some important properties of the linear Schrödinger equation: The continuity Eq. (4) is unchanged, so that $p = \psi^* \psi$ is still conserved, and has a sensible interpretation as probability density; also, the equation does not depend on the normalization of the wave function.

A conservative attitude would be to view the information-theoretic non-linear Schrödinger equation as an effective equation, which models, perhaps in an approximate manner, the unknown. In passing, we have noted that non-linear generalizations of the Dirac equation using information-theoretic measures have been discussed in [6].

5. Quantum cosmology of de Sitter space

In quantum cosmology, one studies the universe as a single quantum entity, which in the Wheeler-DeWitt (WDW) approach is described by a wave function of the universe. Usually, situations of restricted symmetry are studied in the mini-superspace scheme, which reduces the functional differential equation to a manageable Schrödinger-like equation with a few degrees of freedom.

In the spirit of the MUP, it was suggested in [7] that an information-theoretic extension of the usual WDW equation be used to model the unknown structure of quantum spacetime. This was motivated in part by the fact that the usual linear WDW equation could not always resolve cosmological singularities.

The non-linear WDW equation for the de Sitter space studied in [7] was just the non-linear Schrödinger equation of [5], but using the potential appropriate for quantum cosmology. As the equation was complicated, being a non-linear difference-differnetial equation, the first attempt involved studying a truncated and linearized version, with an effective potential, which represented approximately the new ingredients. Encouraging results have been obtained. It was shown that the original Big Bang of the linear WDW equation was replaced by creation of the universe through tunnelling.

Recently [8], we have managed to transform the full non-linear difference-differential equation into a purely difference equation for the probability density by solving first, the current conservation constraint. The difference equation could then be studied easily, though still mostly numerically. The full treatment has shown some new features not seen in the previous approximate study, such as the existence of a minimum and amximum allowable size to the quantum de Sitter universe.

At the quantum mechanical level, a_{min} and a_{max} corresponded to the location where the probability density p vanished. Unlike the case for normal quantum mechanical systems, the vanishing of p in the non-linear WDW equation with a de Sitter potential was shown to imply a termination of the evolution of the difference equation.

Since, for the de Sitter case, there is no other variable in the WDW equation to play the role of an internal clock, a semi-classical analysis was performed to interpret a_{min} and a_{max}. It was shown that in the effective classical dynamics, the location of nodes in p corresponded to the position of barriers, which caused bounces at short and large distances.

The de Sitter model is reasonable at early times of our Universe. It will also be a reasonable model at late times if the current acceleration continues. However, our results, taken at face values, suggest that eventually the acceleration will stop and the universe collapse. The above results hold even when the cosmological constant varies slowly [9].

6. The quantum FRW - ϕ universe

A massless scalare field may be used as an internal clock. In our first approximate study [7], using the truncated version of the non-linear WDW equation, we have found that the zero size singularity of the classical FRW - ϕ model was resolved by a bounce in the effective classical dynamics.

Recently, we [9] have extended the study of the FRW - ϕ model by using the full non-linear difference-differential equation in a similar manner mentioned in [8]. By using a suitable ansatz, wave packets with appropriate physical properties were constructed. It was observed that the quantum evolution of the wave packets displayed bounces at short and large distances leading to cyclic evolution, though the cycles observed so far were neither periodic nor everlasting.

7. Outlook

The results from our studies of toy quantum universe, within the information-theoretic approach, have been encouraging. Some natural extensions would involve adding other matter fields, studying less symmetrical situations, and examining the robustness of the results to deformations of the information measure used.

Acknowledgements

The quantum cosmology resluts reported, here, were obtained in collaboration with L H Nguyen, S N Tarih and M Ashwinkumar. R P thanks the organisers of the ICGC-2011 for the opportunity to present this summary.

References

[1] Jaynes E T 1957 *Phys. Rev.* **106** 620
[2] Jaynes E T 2004 *Probability Theory, The Logic of Science* (Cambridge University Press), Buck B and Macaulay V A 1991 *Maximum Entropy in Action* (New York: Oxford University Press)
[3] Prieden B R 1989 *Am. J. Phys.* **57** 1004, Reginatto M 1998 *Phys. Rev. A* **58** 1775
[4] Parwani R 2005 *J. Phys. A: Math. Gen.* **38** 6231
[5] Parwani R 2005 *Ann. Phys.* **315** 419
[6] Ng W K and Parwani R 2011 *Mod. Phys. Lett. A* **26** 681
[7] Nguyen L H and Parwani R 2009 *Gen. Rel. Grav.* **41** 2543, Nguyen L H and Parwani R 2009 *AIP Conf. Proc.* **1115** 180
[8] Tarih S N and Parwani R [gr-qc/1107:3347v2]
[9] Ashwinkumar M and Parwani R *In preparation*

Journal of Physics: Conference Series **484** (2014) 012071 doi:10.1088/1742-6596/484/1/012071

Upper bound of the time-space non-commutative parameter from gravitational quantum well experiment

A Saha

Department of Physics, West Bengal State University, Barasat 700126, India

E-mail: anirban@iucaa.ernet.in

Abstract. Starting from a field theoretic description of the gravitational well problem in a canonical non-commutative spacetime, we have studied the effect of time-space non-commutativity in the gravitational well scenario. The corresponding first quantized theory reveals a spectrum with leading order perturbation term of non-commutative origin. GRANIT experimental data are used to estimate an upper bound on the time-space non-commutative parameter.

1. Introduction

Non-commutative (NC) spacetime [1], where the coordinates x^μ satisfy the non-commutative algebra,

$$[x^\mu, x^\nu] = i\Theta^{\mu\nu}, \tag{1}$$

has regained prominence in the recent past, and field theories defined over the NC spacetime are currently a subject of very intense research [2, 3]. Various gauge theories including gravity are being studied in a NC perspective formally [4, 5, 6, 7, 8, 9, 10] as well as phenomenologically [11, 12, 13, 14]. A part of the endeavour is to find the order of the NC parameter and exploring its connection with observations [15, 16, 17, 18, 19, 20, 21].

In particular, GRANIT experimental data [22], which shows the quantum states of the neutrons trapped in earth's gravitational field, have been used to set an upper bound on the momentum space NC parameters [23, 24] by analyzing the gravitational well problem using NC quantum mechanics, where non-commutativity is introduced among the phase-space variables at the Hamiltonian level. Therefore, non-commutativity in the time-space sector, i.e., $\theta^{0i} \neq 0$ is not accounted for. Time-space non-commutativity poses certain difficulties regarding unitarity and causality [25, 26, 27], which could be avoided by a perturbative approach [28, 29, 30, 31]. Therefore, a search for any possible upper bound on the time-space NC parameter is very much desirable. In this paper, we have studied the effect of time-space NC (if any) on the spectrum of a cold neutron trapped in a gravitational quantum well, starting from a NC Schrödinger field theory.

2. The NC Schrödinger action

To model a non-relativistic particle with a constant background gravitational field in NC spacetime, we start with the NC Schrödinger action in the deformed phase-space, where the ordinary product is replaced by the star product [31, 32, 33]. Here, the fields are defined as functions of the phase-space variables and the redefined product of two fields $\hat{\phi}(x)$ and $\hat{\psi}(x)$ is given by:

$$\hat{\phi}(x) \star \hat{\psi}(x) = \left(\hat{\phi} \star \hat{\psi}\right)(x) = e^{\frac{i}{2}\theta^{\alpha\beta}\partial_\alpha \partial'_\beta} \hat{\phi}(x)\hat{\psi}(x')|_{x'=x} \, . \tag{2}$$

The action for the system in vertical $x-y$ $(i=1,2)$ plane and gravitational background in the x-direction reads:

$$\hat{S} = \int dx\, dy\, dt\, \hat{\psi}^\dagger \star \left(i\hbar\partial_0 + \frac{\hbar^2}{2m}\partial_i\partial_i - mg\hat{x}\right) \star \hat{\psi}. \tag{3}$$

Under $\star$ composition, the Moyal bracket between the coordinates is $[\hat{x}^\mu, \hat{x}^\nu]_\star = i\Theta^{\mu\nu}$, where the non-trivial components are: $\Theta^{12} = -\Theta^{21} = \theta$, $\Theta^{10} = -\Theta^{01} = \eta$ and $\Theta^{20} = -\Theta^{02} = \eta'$. Since the effect of non-commutativity is expected to be small, we have expanded the star product and considered only the first order correction terms. A physically irrelevant re-scaling[1] to the field variable, re-definition of the observable mass, and the partial derivative (∂_y) are given by:

$$\psi \mapsto \tilde{\psi} = \sqrt{\left(1 - \frac{\eta}{2\hbar}mg\right)}\, \psi, \qquad \tilde{m} = \left(1 - \frac{\eta}{2\hbar}mg\right)m, \quad \textit{and} \quad \partial_y = \left(\partial_y - \frac{i\theta}{2\hbar^2}\tilde{m}^2 g\right), \tag{4}$$

give the final effective NC Schrödinger action as:

$$\hat{S} = \int dx\, dy\, dt\, \tilde{\psi}^\dagger \left[i\hbar\partial_t + \frac{\hbar^2}{2\tilde{m}}\left({\partial_x}^2 + {\partial_y}^2\right) - \tilde{m}gx - \eta\left(\frac{\tilde{m}^2 g^2}{\hbar}\right)x\right]\tilde{\psi}, \tag{5}$$

which gives the equation of motion for the field $\tilde{\psi}(x)$.

3. Reduction to first quantized theory

In a field theoretic setting, we have imposed non-commutativity and found the only non-trivial change in the Schrödinger equation is, indeed, originating from time-space non-commutativity. Specifically, it shows up only in the direction of the external gravitational field $\mathbf{g} = -g\mathbf{e_x}$. Since the first and second quantized formalisms are equivalent as far as Galilean systems are concerned, we can, hereafter, reinterprete $\tilde{\psi}$, the basic field, as a wave function and carry out an equivalent NC quantum mechanical analysis.

From Eq. (5), we easily read off the Hamiltonian as:

$$H = H_0 + H_1 = \frac{({p_x}^2 + {p_y}^2)}{2\tilde{m}} + \tilde{m}gx + \eta\frac{\tilde{m}^2 g^2}{\hbar}x. \tag{6}$$

The last term in Eq. (6) represents a perturbation H_1 in the gravitational quantum well problem described by H_0, which we now briefly review.

3.1. Ordinary gravitational quantum well

The first two terms in Eq. (6) describe the quantum states of a particle with mass $\tilde{m}$ trapped in a gravitational well. Since the particle is free to move in y-direction, its energy spectrum is continuous along y and the corresponding wave function will be a collection of plane waves $\tilde{\psi}(y) = \int_{-\infty}^{+\infty} g(k)e^{iky}dk$, where $g(k)$ determines the group's shape in phase-space. The analytical

[1] Such re-scalings are only viable in a region of spacetime, where variation of the external field is negligible. Since the results we have derived are to be compared with the outcome of a laboratory-based experiment, we can safely assume a constant external gravitational field throughout.

solution of the Schrödinger equation in x-direction $H_0\tilde{\psi}_n = E_n\tilde{\psi}_n$, is well known [34]. The eigenfunctions can be expressed in terms of the Airy function $\phi(z)$ as $\psi_n(x) = A_n\phi(z)$, with eigenvalues given by the roots of the Airy function α_n, with $n = 1, 2\ldots$ as:

$$E_n = -\left(\tilde{m}g^2\hbar^2/2\right)^{1/3}\alpha_n\,. \tag{7}$$

The dimensionless variable z is related to the height x by $z = \left(2m^2g/\hbar^2\right)^{1/3}(x - E_n/\tilde{m}g)$. The normalization factor for the n-th eigenstate is given by:

$$A_n = \left[\left(\hbar^2/2m^2g\right)^{\frac{1}{3}}\int_{\alpha_n}^{+\infty} dz\phi^2(z)\right]^{-\frac{1}{2}}. \tag{8}$$

The wave function for a particle with energy E_n oscillates below the classically allowed height $x_n = \frac{E_n}{\tilde{m}g}$, and above x_n it decays exponentially. This was realized experimentally by Nesvizhevsky *et al* [22] by letting cold neutrons flow with horizontal velocity 6.5 ms^{-1} through a horizontal slit formed between a mirror below and an absorber above. The number of transmitted neutrons as a function of absorber height is recorded, and the classical dependence is observed to change into a stepwise quantum-mechanical dependence at a small absorber height. The experimentally found value of the classical height for the first quantum state is $x_1^{exp} = 12.2 \pm 1.8$ (syst.) $\pm\, 0.7$ (stat.) μm, and the corresponding theoretical value can be determined from Eq. (7) for $\alpha_1 = -2.338$, yielding $x_1 = 13.7$ μm. This value is contained in the error bars and allow for maximum absolute shift of the first energy level with respect to the predicted values:

$$\Delta E_1^{exp} = 6.55\times 10^{-32}\ \mathrm{J} = 0.41\ \mathrm{peV}. \tag{9}$$

The values of the constants taken in this calculations are $\hbar = 10.59\times 10^{-35}$ Js, $g = 9.81$ ms^{-2}, *and* $\tilde{m} = 167.32\times 10^{-29}$ kg.

3.2. Analysis of the perturbed energy spectrum

Going back to the effective NCQM theory, we have now analyzed the perturbed system in Eq. (6). The perturbative potential given by $H_1 = \eta\,(\tilde{m}^2g^2/\hbar)\,x$ is a direct manifestation of time–space non-commutativity. So, we have worked out an upper bound for the time-space NC parameter following [23] by demanding that the correction in the energy spectrum should be smaller or equal to the maximum energy shift allowed by the experiment [22]. We have worked out the theoretical value of the leading order energy shift of the first quantum state numerically. It is just the expectation value of the perturbation potential, given by:

$$\Delta E_1 = \eta\frac{\tilde{m}^2g^2}{\hbar}\int_0^{+\infty} dx\ \tilde{\psi}_1^*(x)\ x\ \tilde{\psi}_1(x) = \eta\frac{\tilde{m}^2g^2}{\hbar}\left[\left(\frac{2\tilde{m}^2g}{\hbar^2}\right)^{-\frac{2}{3}}A_1^2I_1 + \frac{E_1}{\tilde{m}g}\right], \tag{10}$$

where $I_1 \equiv \int_{\alpha_1}^{+\infty} dz\phi(z)z\phi(z)$. The values of the first unperturbed energy level E_1 is determined from Eq. (7) with $\alpha_1 = -2.338$ as:

$$E_1 = 2.259\times 10^{-31}\ (\mathrm{J}) = 1.407\ (\mathrm{peV}). \tag{11}$$

The normalization factor A_1 is calculated from Eq. (8). The integrals in A_n and I_1 are numerically determined for the first energy level $A_1 = 588.109$, *and* $I_1 = -0.383213$. The first order correction in the energy level is given by $\Delta E_1 = 2.316\times 10^{-23}\eta$ J. Comparing with the experimentally determined value of the energy level from Eq. (9), we have found the bound on the time-space NC parameter is:

$$|\eta| \sim 2.83 \times 10^{-9}\ \mathrm{m}^2. \qquad (12)$$

Interestingly, the value of the upper bound derived, here, can be shown [18] to be consistent with the results of [23, 24].

4. Conclusions

In this paper, we have obtained an effective NC quantum mechanics (NCQM) for the gravitational well problem starting from a NC Schrödinger action coupled to external gravitational field. We have re-interpreted this one particle field theory as a first quantized theory and obtained an effective NCQM. The outcome of our calculation shows that the time-space sector of the NC algebra introduces non-trivial NC effects in the energy spectrum of the system. We have demanded that the calculated perturbation in the energy level should be less than or equal to the maximum energy shift allowed by the GRANIT experiment [22]. This comparison leads to an upper bound on the time-space NC parameter. However, one should keep in mind that this value is only in the sense of an upper bound, and not the value of the parameter itself.

References

[1] Snyder H S 1947 *Phys. Rev.* **71** 38
[2] Seiberg N and Witten E 1999 *JHEP* **09** 032
[3] See for example, Szabo R J 2003 *Phys. Report* **378** 207 *and the referances therein*
[4] Saha A [arXiv:0909.4057[hep-th]] *and the references therein*
[5] Chamseddine A H 2001 *Phys. Lett.* B **504** 33
[6] Aschieri P *et al* 2005 *Class. Quant. Grav.* **22** 3511
[7] Calmet X and Kobakhidze A 2005 *Phys. Rev. D* **72** 045010
[8] Alvarez-Gaume L *et al* 2006 *Nucl. Phys. B* **753** 92
[9] Mukherjee P and Saha A 2006 *Phys. Rev. D* **74** 027702
[10] Mukherjee P and Saha A 2008 *Phys. Rev. D* **77** 064014
[11] Sheikh-Jabbari M M 1999 *Phys. Lett. B* **455** 129, Sheikh-Jabbari M M 2000 *Phys. Rev. Lett.* **84** 5625
[12] Riad I F and Sheikh-Jabbari M M 2000 *JHEP* **0008** 045
[13] Chair N and Sheikh-Jabbari M M 2001 *Phys. Lett. B* **504** 141
[14] Jaeckel J *et al* 2006 *JHEP* **0602** 028, Abel S A *et al* 2006 *JHEP* **0601** 105
[15] Chaichian M *et al* 2001 *Phys. Rev. Lett.* **86** 2716
[16] Mocioiu I *et al* 2000 *Phys. Lett. B* **489** 390
[17] Carroll S M *et al* 2001 *Phys. Rev. Lett.* **87** 141601
[18] Saha A 2007 *Euro. Phys. J. C* **51** 199, Saha A 2010 *Phys. Rev. D* **81** 125002
[19] Stern A 2008 *Phys. Rev. Lett.* **100** 061601
[20] Saha A and Gangopadhyay S 2009 *Phys. Lett. B* **681** 96
[21] Saha A *et al* 2011 *Phys. Rev. D* **83** 025004
[22] Nesvizhevsky V V *et al* 2002 *Nature* **415** 297, Nesvizhevsky V V *et al* 2003 *Phys. Rev. D* **67** 102002, Nesvizhevsky V V *et al* 2005 *Euro. Phys. J. C* **40** 479
[23] Bertolami O *et al* 2005 *Phys. Rev. D* **72** 025010
[24] Banerjee R *et al* 2006 *Phys. Rev. D* **74** 045015
[25] Gomis J and Mehen T 2000 *Nucl. Phys. B* **591** 265
[26] Alvarez-Gaume L *et al* 2001 *JHEP* **05** 057
[27] Seiberg N *et al* 2000 *JHEP* **06** 044
[28] Chu C S *et al* 2002 *Nucl. Phys. B* **632** 219
[29] Eliezer D A and Woodard R P 1989 *Nucl. Phys. B* **325** 389
[30] Chen T C *et al* 2002 *Nucl. Phys. B* **625** 151
[31] Saha A *et al* 2006 *Phys. Lett. B* **638** 292, Saha A *et al* 2008 *Mod. Phys. Lett. A* **23** 2947
[32] Chakraborty B *et al* 2004 *Phys. Rev. D* **70** 107707
[33] Mukherjee P and Saha A 2006 *Mod. Phys. Lett. A* **21** 821, Mukherjee P and Saha A 2007 *Mod. Phys. Lett. A* **22** 1113

[34] Landau L D and Lifshitz E M 1965 *Quantum Mechanics. Nonrelativistic Theory* (London, Pergamon)

Proceedings of the 7th International Conference on Gravitation and Cosmology (ICGC2011) IOP Publishing
Journal of Physics: Conference Series **484** (2014) 012072 doi:10.1088/1742-6596/484/1/012072

Universal property of quantum gravity implied by uniqueness theorem of Bekenstein-Hawking entropy

H Saida

Department of Physics, Daido University, Nagoya 457-8530, Japan

E-mail: saida@daido-it.ac.jp

Abstract. We have searched for a universal property of quantum gravity, where "universal" means to be independent of any existing model of quantum gravity (such as the super string theory, loop quantum gravity, causal dynamical triangulation, and so on). To do so, we have investigated carefully the basis of black hole thermodynamics. The uniqueness of Bekenstein-Hawking entropy lets us extract a reasonable universal property of quantum gravity from the conditions justifying Boltzmann formula. The universal property indicates a repulsive gravity at Planck length scale, otherwise stationary black holes can not be regarded as thermal equilibrium states of gravity.

1. Introduction

As mentioned in the abstract, we have investigated the basis of black hole thermodynamics in order to search for universal properties of quantum gravity, where "universal" means to be independent of any existing model of quantum gravity. The basic concept of black hole thermodynamics is the following consideration: *Because a stationary black hole emits the thermal radiation (Hawking radiation), the black hole is in a thermal equilibrium state of gravity. And, usually, the microscopic origin of black hole is expected to be the quantum states of gravity.* This basic concept depends on the thermal radiation theory, which indicates that the source of "Planck spectrum" is a thermal equilibrium object.

By accepting the above concept of black hole thermodynamics, it is reasonable to require that the quantum gravity satisfies the pre-supposition of thermal radiation theory, otherwise we can not regard black holes as thermal equilibrium states by observing the Hawking radiation. Here, the pre-supposition of thermal radiation theory is that the ordinary thermodynamics and quantum statistical mechanics work well on any thermal equilibrium object, at least in laboratory systems. Therefore, it is natural to require that *the basic idea of statistical mechanics is applicable to thermal equilibrium states of not only laboratory systems but also quantum gravity (black holes).* Here, the basic idea of statistical mechanics is, for example, the Boltzmann formula, $S_{BH} = \ln \Omega_g$, where S_{BH} is the Bekenstein-Hawking entropy, and Ω_g is the number of states of black hole counted by the underlying quantum gravity.

From the above, we adopt the following two suppositions for our purpose:

Supposition 1: A stationary black hole is a thermal equilibrium state composed of micro-states of underlying quantum gravity. This is based on the Hawking radiation.

Supposition 2: Basic formalism of statistical mechanics (e.g., Boltzmann formula, grand partition function, and so on) is applicable to underlying quantum gravity so as to describe the thermal equilibrium states (black holes). This is based on the pre-supposition of thermal radiation theory.

The issues for each supposition are as follows:

Issue on supposition 1 (Uniqueness of BH entropy): Note that, if the Boltzmann formula $S_{BH} = \ln \Omega_g$ holds for black holes, then it defines the Bekenstein-Hawking entropy S_{BH} *uniquely* in microscopic point of view. This means that the Boltzmann formula can never work well unless the uniqueness of entropy is proved in macroscopic point of view of thermodynamics (without using any microscopic theory) [1, 2, 3]. In ordinary thermodynamics, the uniqueness of entropy is proven by some basic requirements of thermodynamics [1, 2]. However, as summarized in Section 2, some basic requirements of ordinary thermodynamics do not hold in black hole thermodynamics [4, 5, 6]. Then, it is not necessarily manifested whether the uniqueness of S_{BH} holds in black hole thermodynamics. Hence, we have to prove the uniqueness of S_{BH} before the discussion of Boltzmann formula.

Issue on supposition 2 (Justification of Boltzmann formula): Once the uniqueness of Bekenstein-Hawking entropy S_{BH} is proved in macroscopic point of view, we can proceed to the investigation of Boltzmann formula. Note that the validity of Boltzmann formula in ordinary quantum statistical mechanics is justified by some intrinsic properties of Hilbert space of ordinary quantum mechanics. Those intrinsic properties appear in Ruelle-Tasaki theorem and Dobrushin theorem [3, 7, 8] as summarized in Section 3. *Then, if S_{BH} is given by the Boltzmann formula, we can expect that those theorems of ordinary quantum mechanics are related with some property of underlying quantum gravity. Some universal property of quantum gravity may be extracted from those theorems.*

In Section 4 we have shown a conclusion about the universal property of quantum gravity.

2. Uniqueness theorem of Bekenstein-Hawking entropy

In the rigorous axiomatic formulation of ordinary thermodynamics [1, 2], the basic axioms consist of not only the "four laws of thermodynamics" but also some more requirements. One of the axioms of ordinary thermodynamics modified in black hole thermodynamics is the classification requirement of state variables, all thermodynamic state variables are classified into two categories: extensive variables (e.g., energy, entropy, and so on) and intensive variables (e.g., temperature, pressure, and so on). Here, the extensivity and intensivity are defined by the scaling behaviour as follows: Under the scaling of "system's size" $V \to \alpha V$, and $N \to \alpha N$ (where V is volume, and N is mol-number for a gas in a container), the extensive variable X is scaled as $X \to \alpha X$, and the intensive variable Y is invariant $Y \to Y$. This classification is used in showing many theorems of ordinary thermodynamics including the uniqueness theorem of entropy.

On the other hand, in the framework of black hole thermodynamics, e.g., for Schwarzschild black hole of radius R_{BH}, some representative state variables are as follows: Bekenstein-Hawking entropy $S_{BH} = \pi R_{BH}^2$, Hawking temperature $T_{BH} = (4\pi R_{BH})^{-1}$ and the internal energy $E_{BH} = R_{BH}/2$ (in units $c = 1$, $G = 1$, $\hbar = 1$). For the Schwarzschild black hole, the basic scaling is given by the scaling of length size $R_{BH} \to \lambda R_{BH}$, and we find the scaling behaviours as [5]. :

$$S_{BH} \to \lambda^2 S_{BH}, \quad T_{BH} \to \lambda^{-1} T_{BH}, \quad \textit{and} \quad E_{BH} \to \lambda E_{BH} \ .$$

This implies that the state variables in black hole thermodynamics should be classified not into two categories, but into three categories: The extensive variables (e.g., S_{BH}), intensive variables (e.g., T_{BH}), and energy variables (e.g., E_{BH}). In this classification, the energy variables, which are extensive in ordinary thermodynamics, form one independent category in black hole thermodynamics, since the scaling behaviour of energy is different from that of extensive variables. Furthermore, the scaling behaviour of extensive and intensive variables are different from that of ordinary thermodynamics. But, the scaling behaviour of [extensive variable]×[intensive variable] (e.g., $T_{BH}S_{BH}$) is the same with that of energy variables [4, 6].

Because the classification of state variables of black holes are changed, the uniqueness of S_{BH} is not manifested. However, as shown in [4], the uniqueness of S_{BH} can be proved with modifying the proof of uniqueness in ordinary thermodynamics. The statement of theorem is:

Uniqueness theorem of entropy: Let K be an extensive variable, which increases along irreversible adiabatic processes. Then, there exist two constants a (> 0) and b such that $K = aS_{BH} + b$, where S_{BH} is the Bekenstein-Hawking entropy. (See [4] for the proof.)

Once this theorem is proved, we can proceed to the investigation of Boltzmann formula. In this sense, we can regard the conclusion in Section 4 as a result implied by the uniqueness of S_{BH}.

3. Conditions justifying Boltzmann formula

Consider the ordinary quantum mechanics. For simplicity, consider the system of N identical particles in a region of volume V with no external field, and:

- Interaction potential: $\Phi(\vec{x}_1, \cdots, \vec{x}_N) = \sum_{j=1}^{\infty} \phi^{(j)}(\vec{x}_{i_1}, \cdots, \vec{x}_{i_j})$, where $i_j \in (1, \cdots, N)$.
- Energy eigenvalue: $E_k(V, N)$, where $k = 0, 1, 2, \cdots$ and $E_k \le E_{k+1}$ ("=" is for degeneracy).
- Number of states: $\Omega(V, N; U) :=$ "Number of eigen states satisfying $E_k \le U$" $= \max_{E_k \le U} k$.

First, let us show the Ruelle-Tasaki theorem, which clarifies the sufficient conditions for the validity of Boltzmann formula [4, 7, 3]:

Ruelle-Tasaki theorem: Suppose the following two conditions of $\Phi(\vec{x}_1, \cdots, \vec{x}_N)$:

Condition A: Arbitrary j-particle interaction $\phi^{(j)}$, becomes negative for sufficiently large distribution of j particles. That is, there exists a constant r_A (> 0), such that $\phi^{(j)}(\vec{x}_{i_1}, \cdots, \vec{x}_{i_j}) \le 0$, for $r_A \le \min_{k,l=1,\cdots,j} |\vec{x}_{i_k} - \vec{x}_{i_l}|$.

Condition B: Interaction potential Φ is bounded below. That is, there exists a constant ϕ_B (> 0), such that $\Phi(\vec{x}_1, \cdots, \vec{x}_N) \ge -N\,\phi_B$.

Then, the limit $\sigma(\varepsilon, \rho) := \lim_{t.l.} \dfrac{\ln \Omega(V, N; U)}{V}$ exists uniquely, where $\lim_{t.l.}$ is thermodynamic limit defined by $V \to \infty$ with fixing $\rho := N/V$, and $\varepsilon := U/V$ at constant values.

By this theorem, the thermodynamics limit of $\ln \Omega(V, N; U)$ is defined well. Furthermore, it can be also proved that $\sigma(\varepsilon, \rho)$ is concave about its arguments (ε, ρ) and monotonically increasing about ε, and remains constant along *reversible* adiabatic processes [3]. These behaviours are some characteristic properties of entropy already known in thermodynamics. Then, in statistical mechanics, it is assumed that $\ln \Omega(V, N; U)$ is equal to the entropy (the Boltzmann formula).

Obviously, the above conditions A and B are the sufficient conditions for the validity of Boltzmann formula. Furthermore, a system, which holds Boltzmann formula but violates conditions A or B, has not been found experimentally. *Hence, at least in laboratory systems, it is reasonable to require that any physical system satisfies the conditions A and B.*

Next, let us show the Dobrushin theorem, which clarifies the necessary condition for the existence of thermal equilibrium states [4, 7, 8]:

Dobrushin theorem: Consider the case satisfying the following two pre-suppositions:

Pre-supposition C: The $j(\neq 2)$-particle interactions disappear, and the total interaction potential is a sum of two-particle interactions $\Phi(\vec{x}_1, \cdots, \vec{x}_N) = \sum_{1 \le i < j \le N} \phi^{(2)}(\vec{x}_i, \vec{x}_j)$.

Pre-supposition D: $\phi^{(2)}(\vec{x}, \vec{y}) \propto |\vec{x} - \vec{y}|^{-\alpha}$, ($\alpha > 0$) for sufficiently large $|\vec{x} - \vec{y}|$.

Under these pre-suppositions, if the ground partition function can be defined uniquely (i.e., if thermal equilibrium states exist), then the following inequality holds:

$$I_V^{(2)} := \frac{1}{V^2} \int\int_V d^3x_1\, d^3x_2\, \phi^{(2)}(\vec{x}_1, \vec{x}_2) \geq 0 \,. \tag{1}$$

Obviously, $I_V^{(2)} \geq 0$ is the necessary condition for the existence of thermal equilibrium states. Note that the pre-suppositions C and D are natural when we consider the gravity.

4. Conclusion

Our supposition 2 in Section 1 implies that the ordinary quantum mechanics and quantum gravity shares the same properties, which justify the validity of Boltzmann formula and the existence of thermal equilibrium states. The sufficient conditions A, B and the necessary condition $I_V^{(2)} \geq 0$ are expected to be such shared properties [1]. Note that these three conditions (A, B and $I_V^{(2)} \geq 0$) are for the interaction potential.

On the other hand, it is not clear whether the full quantum gravity is expressed by using the interaction potential Φ or Hamiltonian. However, the semi-classical expression of quantum gravity should be expressed by an effective Lagrangian using an interaction potential.

Hence, our suggestion is:

Implication from the above: The interaction potential in the effective Lagrangian of quantum gravity should satisfy the conditions A, B and $I_V^{(2)} \geq 0$, in order to regard a stationary black hole as a thermal equilibrium state of quantum gravity, whose entropy is given by the Boltzmann formula. The typical form of the effective potential is shown in Figure 1. This means that **the semi-classical correction to Einstein-Hilbert action expresses a repulsive gravity at a short length scale (which may be Planck scale).**

In this paper, any existing model of quantum gravity is not used. (See [4] for detail.) Therefore, this implication can be regarded as a universal property of quantum gravity.

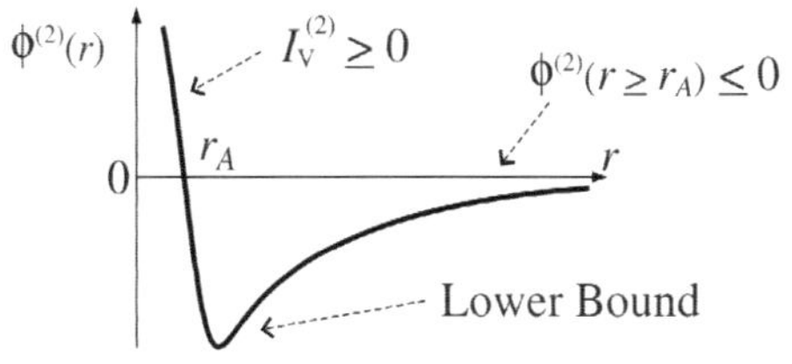

Figure 1. Typical form of the effective potential of quantum gravity at semi-classical level. Thermal equilibrium states of gravity (black holes) may not exist unless a repulsive region appears, where r_A may be Planck scale.

References

[1] Lieb E H and Yngvason J 1999 *Phys. Report* **310** 1
[2] Tasaki H 2000 *Thermodynamics* (Netus-Rikigaku) Baifu-kan Publ. (Published in Japanese)
[3] Tasaki H 2008 *Statistical Mechanics* (Tokei-Rikigaku) Baifu-kan Publ. (Published in Japanese)
[4] Saida H 2011 *Entropy* **13** 1611 (ed J Bekenstein)
[5] Jr.York J W 1986 *Phys. Rev. D* **33** 2092
[6] Saida H 2010 *Prog. Theor. Phys.* **122** 1515
[7] Ruelle D 1999 *Statistical Mechanics - Rigorous Results* (Imperial College Press and World Scientific Publ.)
[8] Dobrushin R L 1964 *Teorija Verojatn. i ee Prim.* **9** 626

[1] So far, as a fact, no counter-example to conditions A and B is found experimentally. Then, it may be reasonable to extend this fact to quantum gravitating systems, so that quantum gravity also satisfies those conditions.

Proceedings of the 7th International Conference on Gravitation and Cosmology (ICGC2011) IOP Publishing
Journal of Physics: Conference Series **484** (2014) 012073 doi:10.1088/1742-6596/484/1/012073

Highlights of non-commutative spectral geometry

M Sakellariadou

Theoretical Particle Physics and Cosmology Group, Department of Physics, King's College, University of London, Strand WC2R 2LS, London, UK

E-mail: `Mairi.Sakellariadou@kcl.ac.uk`

Abstract. A summary of non-commutative spectral geometry as an approach to unification has been presented. The role of the doubling of the algebra, the seeds of quantization and some cosmological implications have been briefly discussed.

1. Introduction

In the various attempts to quantize gravity, it is either assumed a purely gravitational theory without any matter fields, or it is considered that the interaction between gravity and matter is the most important element of the dynamics. We have adopted here the second approach.

Much below the Planck energy scale, gravity can be considered as a classical theory, and the laws of physics can be described to a good approximation by an effective action and continuum fields. As the energies, however, approach the Planck scale, the quantum nature of spacetime becomes apparent, and the simple prescription, dictating that physics can be described by the sum of the Einstein-Hilbert, and the Standard Model (SM) action ceases to be valid. In the framework of non-commutative spectral geometry (NCSG) [1], gravity and the SM fields were put together into matter and geometry on a non-commutative space made from the product of a four-dimensional standard commutative manifold by a non-commutative internal space.

In what follows, we have briefly presented the elements of NCSG as an approach to unification [2], and highlight the relation between the doubling of the algebra and the gauge fields [3], an essential element to make the link with the SM. We have then discussed how the doubling of the algebra is related to dissipation, which incorporates the seeds of quantization [3], and finally discussed briefly some cosmological consequences [4, 5, 6, 7, 8], since the model lives by construction in high energy scales, offering a natural framework to address early universe cosmology.

2. Elements of non-commutative spectral geometry

NCSG [9, 10] is based on a two-sheeted space, made from the product of a four-dimensional smooth compact Riemannian manifold $\mathcal{M}$ (a continuous geometry for spacetime), by a discrete non-commutative space $\mathcal{F}$ (an internal geometry for the SM) composed by only two points. The non-commutative nature of the discrete space $\mathcal{F}$ is given by a spectral triple $(\mathcal{A}, \mathcal{H}, \mathcal{D})$, where $\mathcal{A}$ is an involution of operators on the finite-dimensional Hilbert space, $\mathcal{H}$ of Euclidean Fermions, and $\mathcal{D}$ is a self-adjoint unbounded operator in $\mathcal{H}$. All information about space is encoded in the algebra of coordinates $\mathcal{A}$, which is related to the gauge group of local gauge transformations.

Assuming the algebra $\mathcal{A}$ to be symplectic-unitary, then $\mathcal{A} = M_a(\mathbb{H}) \oplus M_k(\mathbb{C})$, with $k = 2a$ and $\mathbb{H}$ denoting the algebra of quaternions [11]. The choice $k = 4$ is the first value that produces the correct number ($k^2 = 16$) of Fermions in each of the three generations, with the number of generations being a physical input. While the choice of algebra $\mathcal{A}$ constitutes the main input of the theory, the choice of Hilbert space $\mathcal{H}$ is irrelevant. The operator $\mathcal{D}$ corresponds to the inverse of the Euclidean propagator of Fermions, and is given by the Yukawa coupling matrix, which encodes the masses of the elementary Fermions and the Kobayashi–Maskawa mixing parameters. The SM Fermions provide the Hilbert space $\mathcal{H}$ of a spectral triple for the algebra $\mathcal{A}$, while the Bosons are obtained through inner fluctuations of the Dirac operator of the product geometry.

One applies the spectral action principle [12], stating that the bare Bosonic-Euclidean action is the trace of the heat kernel associated with the square of the Dirac operator, and is of the form $\mathrm{Tr}(f(\mathcal{D}/\Lambda))$, where f is a cut-off function, and Λ fixes the energy scale. This action can be seen *à la* Wilson as the bare action at scale Λ. The Fermionic term can be included by adding $(1/2)\langle J\psi, D\psi\rangle$, where J is the real structure on the spectral triple, and ψ is a spinor in the Hilbert space of the quarks and leptons. For the four-dimensional Riemannian geometry, the trace is expressed perturbatively in terms of the geometrical Seeley-deWitt coefficients a_n [13] as:

$$\mathrm{Tr}(f(D/\Lambda)) \sim 2\Lambda^4 f_4 a_0 + 2\Lambda^2 f_2 a_2 + f_0 a_4 + \cdots + \Lambda^{-2k} f_{-2k} a_{4+2k} + \cdots \ . \tag{1}$$

Since, its Taylor expansion at zero vanishes, it reduces to:

$$\mathrm{Tr}(f(D/\Lambda)) \sim 2\Lambda^4 f_4 a_0 + 2\Lambda^2 f_2 a_2 + f_0 a_4 \ , \tag{2}$$

where f plays a rôle only through its momenta f_0, f_2, f_4, which are three real parameters, related to the coupling constants at unification, the gravitational constant, and the cosmological constant respectively. The computation of this asymptotic expression results to the full Lagrangian for the SM minimally coupled to gravity, with neutrino mixing and Majorana mass terms.

This purely geometric approach to the SM leads to the correct representations of the Fermions with respect to the gauge group of the SM, the Higgs doublet appears as part of the inner fluctuations of the metric, and spontaneous symmetry breaking mechanism arises naturally with the negative mass term without any tuning [1]. The see-saw mechanism is obtained, the 16 fundamental Fermions are recovered, and a top quark mass of $\sim$ 179 GeV is predicted [1]. The model also predicts the correct order of magnitude for the Higgs mass. Strictly speaking, the predicted Higgs mass of approximately 170 GeV is ruled out from the experimental data, nevertheless it is rather remarkable that the order of magnitude is correct, given that the NCGS approach based on the particular choice for the algebra $\mathcal{A}$ must be seen as an effective theory.

3. Dissipation and the origin of quantization

The central ingredient in the NCSG, namely the doubling of the algebra $\mathcal{A} = \mathcal{A}_1 \otimes \mathcal{A}_2$ acting on the space $\mathcal{H} = \mathcal{H}_1 \otimes \mathcal{H}_2$ is related to dissipation and to the gauge field structure [3].

To highlight the justification of this claim, let us consider the equation of the classical one-dimensional damped harmonic oscillator $m\ddot{x} + \gamma\dot{x} + kx = 0$, with time independent m, γ and k, which is a simple prototype of open systems. In the canonical formalism for open systems, the doubling of the degrees of freedom is required in such a way as to complement the given open system with its time-reversed image, thus, obtaining a globally closed system for which the Lagrangian formalism is well defined. Considering the oscillator in the doubled y-coordinate, $m\ddot{y} - \gamma\dot{y} + ky = 0$, and then using the coordinates $x_1(t) = \Big(x(t) + y(t)\Big)/\sqrt{2}$ and $x_2(t) = \Big(x(t) - y(t)\Big)/\sqrt{2}$, the Lagrangian of this closed system takes the form:

$$L = \frac{1}{2m}\Big(m\dot{x}_1 + \frac{e_1}{c}A_1\Big)^2 - \frac{1}{2m}\Big(m\dot{x}_2 + \frac{e_2}{c}A_2\Big)^2 - \frac{e^2}{2mc^2}\Big({A_1}^2 + {A_2}^2\Big) - e\Phi, \tag{3}$$

where we have introduced the vector potential as: $A_i = (B/2)\epsilon_{ij}x_j$ for $i, j = 1, 2$ with $B \equiv \gamma\, c/e$ and $\epsilon_{ii} = 0$, $\epsilon_{12} = -\epsilon_{21} = 1$. It describes two particles with opposite charges $e_1 = -e_2 = e$ in the (oscillator) potential $\Phi \equiv (k/2e)({x_1}^2 - {x_2}^2) \equiv \Phi_1 - \Phi_2$, with $\Phi_i \equiv (k/2/e){x_i}^2$ and in the constant magnetic field $\mathbf{B}$ defined as $\mathbf{B} = \nabla \times \mathbf{A}$.

The doubled coordinate, e.g., x_2 acts as the gauge field component A_1 to which the x_1 coordinate is coupled, and *vice versa*. The energy dissipated by one of the two systems is gained by the other. The gauge field acts as the bath or reservoir in which the system is embedded [3].

The NCSG classical construction carries implicit in the doubling of the algebra seeds of quantization [3]. 't Hooft has conjectured that, provided some specific energy conditions are met and some constraints are imposed, loss of information might lead to a quantum evolution [14]. By considering the classical damped harmonic oscillator and its time–reversed image, we have shown [3] that the obtained Hamiltonian belongs to the class of Hamiltonians considered by 't Hooft. We have shown [3] that the dissipation term in the Hamiltonian is responsible for the zero point contribution to the energy, which is the signature of quantization.

4. Cosmological consequences

In the low-energy limit, the corrections to the background Einstein's equations do not occur at the level of a Friedmann-Lemaître-Robertson-Walker (FLRW) background [4]. One may have naively claimed that this was expected, arguing that in a spatially homogeneous spacetime, the spatial points are equivalent and any non-commutative effects are then expected to vanish. However, this argument does not apply here; the non-commutativity is incorporated in the internal manifold $\mathcal{F}$ and the spacetime is a commutative four-dimensional manifold. The coupling between the Higgs field and the background geometry can no longer be neglected once the energies reach the Higgs scale, in which case, the non-minimal coupling of Higgs field to curvature leads to corrections to Einstein's equations even for homogeneous and isotropic cosmological models [4]. The effect of the non-minimal coupling of the Higgs field can be seen in two ways: It leads to an effective gravitational constant, or it increases the Higgs mass [4].

The non-minimal coupling between the Higgs field and the Ricci curvature may turn out to be crucial in early universe cosmology [5, 6]. Such a coupling has been introduced *ad hoc* in the literature, in an attempt to drive inflation through the Higgs field. However, the value of the coupling constant between the scalar field and the background geometry should be dictated by the underlying theory. Actually, even if classically, the coupling between the Higgs field and the Ricci curvature could be set equal to zero, a non-minimal coupling would be induced once quantum corrections in the classical field theory were considered.

We have thoroughly investigated [6] where the Higgs field could play the role of the inflaton leading to a sufficient period of inflation with induced temperature anisotropies, which are in agreement with the current measurements. The Higgs potential is $V(H) = \lambda_0 H^4 - \mu_0^2 H^2$, with μ_0 and λ_0 subject to radiative corrections as functions of energy. For large enough values of the Higgs field, the normalized value of μ_0 and λ_0 must be calculated. We have shown [6] that for each value of the top quark mass, there is a value of the Higgs mass, where the effective potential is about to develop a metastable minimum at large values of the Higgs field, and the Higgs potential is locally flattened. Calculating [6] the renormalization of the Higgs self-coupling up to two-loops, we have constructed an effective potential, which fits the renormalization group improved potential around the flat region. There is a very good analytic fit to the Higgs potential around the minimum of the potential [6], given as:

$$V^{\text{eff}} = \lambda_0^{\text{eff}}(H)H^4 = [a \ln^2(b\kappa H) + c]H^4, \tag{4}$$

where the parameters a, b are related to the low energy values of top quark mass m_t as:

$$a(m_t) = 4.04704 \times 10^{-3} - 4.41909 \times 10^{-5} \left(\frac{m_t}{\text{GeV}}\right) + 1.24732 \times 10^{-7} \left(\frac{m_t}{\text{GeV}}\right)^2, \quad \textit{and}$$

$$b(m_t) = \exp\left[-0.979261 \left(\frac{m_t}{\text{GeV}} - 172.051\right)\right]. \tag{5}$$

The parameter c encodes the appearance of an extremum and depends on the values for top quark mass and Higgs mass.

A search in the parameter space using a Monte Carlo chain has shown [6] that even though slow-roll inflation can be realized - a result which does not hold for minimally coupled Higgs field - the resulting ratio of perturbation amplitudes is too large for any experimentally allowed values for the masses of the top quark and the Higgs Boson.

Finally, by considering the energy lost to gravitational radiation by orbiting binaries and requiring the magnitude of deviations from general relativity (GR) obtained within the NCSG context, to be less than the allowed uncertainty in the data, we have imposed [7] an upper limit to the moment f_0, which is used to specify the initial conditions on the gauge couplings. In particular, by setting $\beta^2 = (5\pi)/(48Gf_0)$, we have imposed the lower limit as: $\beta > 7.55 \times 10^{-13}$ m^{-1}. This observational constraint may seem weak, however, it is comparable to the existing constraints on similar, *ad hoc*, additions to GR. Moreover, since the strongest constraint comes from systems with high orbital frequencies, future observations of rapidly orbiting binaries, relatively close to the Earth, could improve it by many orders of magnitude. Thus, by purely astrophysical observations, we were able to constrain the natural length, defined through the f_0 momentum of the cut-off function f, at which the non-commutative effects become dominant.

5. Conclusions

NCSG offers an elegant and purely geometric interpretation of the SM of electroweak and strong interactions. The doubling of the algebra is an essential element in order to get the gauge fields of the SM. Moreover, the NCSG classical construction carries implicit in its feature of the doubling of the algebra seeds of quantization. The NCSG model lives at unification scale; thus, it provides an excellent framework to address early universe cosmological questions, while to study astrophysical consequences one will have to go beyond the perturbative approach.

References

[1] Chamseddine A H, Connes, A and Marcolli M 2007 *Adv. Theor. Math. Phys.* **11** 991
[2] Sakellariadou M 2011 *Int. J. Mod. Phys. D* **20** 785
[3] Sakellariadou M, Stabile A and Vitiello G 2011 *Phys. Rev. D* **84** 045026
[4] Nelson N and Sakellariadou M 2010 *Phys. Rev. D* **81** 085038
[5] Nelson W and Sakellariadou M 2009 *Phys. Lett. B* **680** 263
[6] Buck M, Fairbairn M and Sakellariadou M 2010 *Phys. Rev. D* **82** 043509
[7] Nelson W, Ochoa J and Sakellariadou M 2010 *Phys. Rev. Lett.* **105** 101602
[8] Nelson W, Ochoa J and Sakellariadou M 2010 *Phys. Rev. D* **82** 085021
[9] Connes A 1994 *Noncommutative Geometry* (New York: Academic Press)
[10] Connes A and Marcolli M 2009 *Noncommutative Geometry, Quantum Fields and Motives* (India: Hindustan Book Agency)
[11] Chamseddine A H and Connes A 2007 *Phys. Rev. Lett.* **99** 191601
[12] Chamseddine A H and Connes 1997 *Commun. Math. Phys.* **186** 731
[13] Chamseddine A H and Connes A 2006 *J. Math. Phys.* **47** 063504
[14] 't Hooft G 1999 *Class. Quant. Grav.* **16** 3263

Proceedings of the 7th International Conference on Gravitation and Cosmology (ICGC2011) IOP Publishing
Journal of Physics: Conference Series **484** (2014) 012074 doi:10.1088/1742-6596/484/1/012074

Super-horizon evolution and the fate of f_{NL}

N Sivanandam

African Institute for Mathematical Sciences, 6 Melrose Road, Muizenberg, Cape Town 7945, South Africa

E-mail: navin.sivanandam@gmail.com

Abstract. Local form primordial non-Gaussianity is a powerful probe of the number of effective degrees of freedom during the inflationary epoch. In this paper, we have described the recent work, which attempts to further narrow down the set of multi-field models that can result in large local non-Gaussianities. This is done by imposing the constraint that any iso-curvature produced by multi-field inflation be eliminated before the end of the inflationary epoch. In the restricted class of models studied, where non-Gaussianity is produced by a sharp turn in field space and eliminated by a short phase of single field inflation, it is found that adiabaticity and a large local f_{NL} are incompatible.

1. Introduction

The inflationary paradigm has proved successful in providing a theory of the very early universe that reproduces the primordial spectrum observed through CMB and other measurements. However, the space of inflationary models is large, and exploring it is a challenging task. One valuable tool for such an exploration is primordial non-Gaussianity, the deviation from a purely Gaussian spectrum that can arise in some inflationary models.

Non-Gaussianity comes in many guises, but perhaps the easiest way to study it is through the 3-point curvature correlation function, and (as observations become more precise) through other higher order correlation functions. The power spectrum, bispectrum (3-point function) and non-Gaussianity parameter f_{NL} are defined as follows:

$$\begin{aligned}
\langle\zeta_{\boldsymbol{k}_1}\zeta_{\boldsymbol{k}_2}\rangle &\equiv (2\pi)^3\delta^{(3)}(\boldsymbol{k}_1+\boldsymbol{k}_2)\frac{2\pi^2}{k_1^3}\mathcal{P}_\zeta(k_1)\,,\\
\left\langle\delta\phi^I_{\boldsymbol{k}_1}\delta\phi^J_{\boldsymbol{k}_2}\right\rangle &\equiv (2\pi)^3\delta^{IJ}\delta^{(3)}(\boldsymbol{k}_1+\boldsymbol{k}_2)\frac{2\pi^2}{k_1^3}\mathcal{P}_*(k_1)\,,\\
P_*(k) &\equiv \frac{H_*^2}{4\pi^2}\,,\\
\langle\zeta_{\boldsymbol{k}_1}\zeta_{\boldsymbol{k}_2}\zeta_{\boldsymbol{k}_3}\rangle &\equiv (2\pi)^3\delta^{(3)}\left(\sum_i \boldsymbol{k}_i\right)\mathcal{B}_\zeta(k_1,k_2,k_3)\,, \qquad \textit{and}\\
\frac{6}{5}f_{NL} &\equiv \frac{\prod_i k_i^3}{\sum_i k_i^3}\frac{B_\zeta}{4\pi^4\sum_{i\neq j}\mathcal{P}_{\zeta_{\mathbf{k_i}}\zeta_{\mathbf{k_j}}}}\,.
\end{aligned} \tag{1}$$

In standard single field inflation, with a canonical kinetic term and a Bunch-Davies vacuum, Maldacena [1] has shown that the parameter f_{NL} must be small. As a result, a non-zero measurement of f_{NL} can be a valuable probe into the nature of the inflation epoch, by indicating a departure from the simplest scenario.

The momentum-conserving delta function in the definition of the bispectrum means that the domain of f_{NL} can be described by the set of triangles in momentum space. Moreover, different triangles are indicative of different particular deviations from canonical, single field inflation [2]. In particular, an f_{NL} signal that peaks for squeezed triangles ($k_1 \sim k_2 \gg k_3$) is indicative of multi-field inflation [3]. Such signals are denoted by $f_{NL}^{\rm local}$.

Current experimental bounds on f_{NL} are given by $-5 < f_{NL}^{\rm local} < 59$ (WMAP-7+SDSS), $-214 < f_{NL}^{\rm equil} < 266$ (WMAP-7) and $-410 < f_{NL}^{\rm orthog} < 6$ (WMAP-7). The equilateral shape has $k_1 \sim k_2 \sim k_3$ and can indicate a non-canonical kinetic term during inflation. The orthogonal shape is constructed to be orthogonal to the other two shapes.

As mentioned above, the only known models that produce a large local non-Gaussianity are multi-field models. In addition, in all of these models, such non-Gaussianity is produced by super-horizon evolution of the curvature perturbation. Super-horizon evolution occurs for modes for which $k/aH \ll 1$ and can only happen in multi-field models. A number of scenarios exist with a large $f_{NL}^{\rm local}$ arising from super-horizon evolution, including the curvature scenario, modulated reheating and multi-field models with sharp turns in field space [4, 5, 6, 7, 8, 9, 10, 11, 12, 13, 14, 15, 16, 17, 18].

As well as the possibility of a measurable $f_{NL}^{\rm local}$, multi-field models of inflation can also give rise to iso-curvature perturbations. Since (a) primordial iso-curvature is unobserved, and (b) the super-horizon evolution of the curvature perturbation will continue until the iso-curvature is driven to zero, it is important that the evolution of the perturbations be tracked until a purely adiabatic spectrum is obtained. From this point, the primordial perturbations will be time-independent, until they are once again inside the horizon.

The work described herein (carried out in detail in [19] and [20]) has studied a class of two field models, where a large local non-Gaussianity is produced as the inflaton undergoes a sharp turn in field space. To match the observed universe and preserve the super-horizon constancy of the curvature perturbation, the residual iso-curvature is eliminated by a short period (a few e-foldings) of effectively single field inflation, which takes place immediately before the end of the inflationary era. In these models (with some assumptions about the form of the potential), it is found that $f_{NL}^{\rm local}$, and all other local form non-Gaussianities are driven to zero as the iso-curvature is eliminated. This suggests that local non-Gaussianity and an adiabatic spectrum are difficult to achieve simultaneously in this class of models.

2. The model and results

The action in the models studied is:

$$S = \int d^4x\sqrt{-g}\left[\frac{1}{2}m_p^2 R + \frac{1}{2}g^{\mu\nu}\delta_{ab}\partial_\mu\phi^a\partial_\nu\phi^b - W(\vec{\phi})\right]. \tag{2}$$

Using ϕ and χ for the two scalar fields, the slow roll equations of motion are (assuming $|\dot{H}| \ll H^2$ and $|\ddot{\phi}_a| \ll H|\dot{\phi}_a|$):

$$\begin{aligned} 3H\dot{\phi} &\simeq -\partial_\phi W\,, \\ 3H\dot{\chi} &\simeq -\partial_\chi W\,, \quad \textit{and} \\ H^2 &\simeq \frac{1}{3m_p^2}W\,. \end{aligned} \tag{3}$$

The potential W can be any function of the form $F(U(\phi)+V(\chi))$, but for clarity, the treatment here is restricted to $W = (U(\phi) + V(\chi))^\gamma$.

The evolution of the perturbations at scales larger than the Hubble radius has been studied using the δN formalism [21, 22, 23, 24]. This gives the local contribution to f_{NL} as:

$$\frac{6}{5}f_{NL}^{(4)} \equiv \frac{\sum_{IJ} N_{,I}N_{,J}N_{,IJ}}{\left(\sum_K N_{,K}^2\right)^2}. \tag{4}$$

I and J denote derivatives with respect to the fields, and N is the background number of e-foldings, given by:

$$N = \int_*^c H\mathrm{d}t. \tag{5}$$

In terms of the slow roll parameters, the values of the potential at horizon exit, and at the end of inflaton, one obtains the following result for f_{NL}:

$$\frac{6}{5}f_{NL}^{(4)} = \frac{2}{\gamma}\left(\frac{\frac{x_h^2}{\epsilon_*^\phi}\left[1-\left(\frac{\gamma\eta_*^\phi}{2\epsilon_*^\phi}-\gamma+1\right)x_h\right]+\frac{y_h^2}{\epsilon_*^\chi}\left[1-\left(\frac{\gamma\eta_*^\chi}{2\epsilon_*^\chi}-\gamma+1\right)y_h\right]}{\left(\frac{x_h^2}{\epsilon_*^\phi}+\frac{y_h^2}{\epsilon_*^\chi}\right)^2}\right)$$
$$+\frac{2}{\gamma}\left(\frac{\frac{(U_c+V_c)^2}{(U_*+V_*)^2}\left(\frac{x_h}{\epsilon_*^\phi}-\frac{y_h}{\epsilon_*^\chi}\right)^2\frac{\epsilon_c^\phi\epsilon_c^\chi}{\epsilon_c}\left(\frac{\gamma\eta_c^{ss}}{\epsilon_c}-1\right)}{\left(\frac{x_h^2}{\epsilon_*^\phi}+\frac{y_h^2}{\epsilon_*^\chi}\right)^2}\right). \tag{6}$$

This can be usefully expressed in terms of the leading contributions as:

$$\frac{6}{5}f_{NL}^{(4)} \sim \mathcal{O}(\varepsilon_*)+\mathcal{O}(1)\times\frac{\epsilon_c^\phi\epsilon_c^\chi}{\epsilon_c^2}\eta_c^{ss}. \tag{7}$$

It can be shown that to eliminate iso-curvature prior to the end of inflation, the parameter η_c^{ss} must become large (>1) for a few e-foldings [19], this in turn results in the exponential suppression of note only in the iso-curvature (δs), but also $f_{NL}^{(4)}$, which are given by:

$$|\delta s| \sim \exp\left[-\frac{3}{2}\int H\mathrm{d}t\right], \quad and$$
$$f_{NL}^{(4)} \sim \mathcal{O}(\varepsilon_*)+\mathcal{O}(1)\times\eta^{ss}\exp\left[-2\int C_\eta H\eta^{ss}\mathrm{d}t\right]. \tag{8}$$

C_η is a number (>1), whose value depends on the particular direction of the inflaton as η^{ss} becomes larger than 1.

A similar exponential suppression is found for more general potentials [19] and higher order (but still local form) non-Gaussianity [20].

3. Conclusions

Local non-Gaussianities are a powerful probe of the inflationary model space. They are know to be small in single field models, and so their detection would be indication of multiple degrees of freedom during inflation. The work described here suggests that as well as ruling out single field models, a detection of $f_{NL}^{\rm local}$ would also be incompatible with a subset of multi-field models, where the non-Gaussianity arises from the super-horizon evolution of curvature during inflation, and where iso-curvature is eliminated through a period of effectively single field inflation. Of course, such models only form a small subset of possible ways to produce a large local primordial non-Gaussianity, and it remains an outstanding challenge to see if the class of models for which $f_{NL}^{\rm local}$ and adiabaticity are mutually exclusive can be expanded.

Acknowledgements
This work was carried in collaboration with Joel Meyers.

References

[1] Maldacena J M 2003 *JHEP* **0305** 013 [astro-ph/0210603]
[2] Komatsu E 2010 *Class. Quant. Grav.* **27** 124010 [arXiv:1003.6097]
[3] Creminelli P and Zaldarriaga M 2004 *JCAP* **0410** 006 [astro-ph/0407059]
[4] Linde A D and Mukhanov V F 1997 *Phys. Rev. D* **56** 535 [astro-ph/9610219]
[5] Bernardeau F and Uzan J P 2002 *Phys. Rev. D* **66** 103506 [hep-ph/0207295]
[6] Bernardeau F and Uzan J P 2003 *Phys. Rev. D* **67** 121301 [astro-ph/0209330]
[7] Lythe D H, Ungarelli C and Wands D 2003 *Phys. Rev. D* **67** 023503 [astro-ph/0208055]
[8] Wands D 2010 *Class. Quant. Grav.* **27** 124002 [arXiv:1004.0818]
[9] Byrnes C T and Tasinato G 2009 *JCAP* **0908** 016 [arXiv:0906.0767]
[10] Battefeld T and Easther R 2007 *JCAP* **0703** 020 [astro-ph/0610296]
[11] Chen X and Wang Y 2010 *Phys. Rev. D* **81** 06351 [arXiv:0909.0496]
[12] Sasaki M 2008 *Prog. Theor. Phys.* **120** 159 [arXiv:0805.0974]
[13] Gao X 2008 *JCAP* **0806** 029 [arXiv:0804.1055]
[14] Battefeld D and Battefeld T 2009 *JCAP* **0911** 010 [arXiv:0908.4269]
[15] Byrnes C T, Choi K Y and Hall L M 2009 *JCAP* **0902** 017 [arXiv:0812.0807]
[16] Cai Y F and Xia H Y 2009 *Phys. Lett. B* **677** 226 [arXiv:0904.0062]
[17] Dvali G, Gruzinov A and Zaldarriaga M 2004 *Phys. Rev. D* **69** 023505 [astro-ph/0303591]
[18] Kofman L 2003 [astro-ph/0303614]
[19] Meyers J and Sivanandam N 2010 [arXiv:1011.4934]
[20] Meyers J and Sivanandam N 2011 *Phys. Rev. D* **84** 063522 [arXiv:1104.5238]
[21] Starobinsky A A 1985 *JETP Lett.* **42** 152
[22] Sasaki M and Stewart E D 1996 *Prog. Theor. Phys.* **95** 71 [astro-ph/9507001]
[23] Lyth D H, Malik K A and Sasaki M 2005 *JCAP* **0505** 004 [astro-ph/0411220]
[24] Lyth D H and Rodriguez Y 2005 *Phys. Rev. Lett.* **95** 121302 [astro-ph/0504045]

Modified constraint algebra in loop quantum gravity and spacetime interpretation

R Tibrewala

Indian Institute of Science Education and Research, CET Campus, Thiruvananthapuram 695016, India

E-mail: rtibs@iisertvm.ac.in

Abstract. Classically, the constraint algebra of general relativity, which generates gauge transformations, is equivalent to spacetime covariance. In LQG, inverse triad corrections lead to an effective Hamiltonian constraint, which can lead to a modified constraint algebra. We have shown, using example of spherically symmetric spacetimes, that a modified constraint algebra does not correspond to spacetime coordinate transformation. In such a scenario, the notion of black hole horizon, which is based on spacetime notions, also needs to be reconsidered. A possible modification to the classical trapping horizon condition leading to consistent results is suggested. In the case where the constraint algebra is not modified, a spacetime picture is valid and one finds mass threshold for black holes and small corrections to Hawking temperature.

1. Introduction

Quantum gravity continues to remain elusive despite numerous attempts taking various forms. Loop quantum gravity (LQG) - a non-perturbative, canonical approach to quantum gravity, is one such attempt [1, 2, 3]. Underlying its construction is a discrete notion of spacetime, a quantum geometry replacing smooth geometry of general relativity. In the context of symmetry reduced cosmological models, where there are only finitely many degrees of freedom (similar to quantum mechanics), the framework has been quite successful leading to a resolution of the big bang singularity [4, 5]. However, this success comes with an important question mark: Whether the singularity resolution is an artifact of classical symmetry reduction? For this, one needs to go beyond the homogeneity of cosmology and probe inhomogeneous (in other words, field theoretic) models. Spherically symmetric models seem ideal for this purpose.

At present, the dynamics of the theory is not fully understood. However, certain effects do not crucially depend on the details of the dynamics and can be used to make preliminary studies of inhomogeneous systems. One such effect, which is easy to implement in the context of spherical symmetry, is the inverse triad effect, coming from the quantization of inverse powers of the triad variables. This effect corrects the Hamiltonian of the theory leading to modified dynamics. More importantly, it can modify the constraint algebra of the theory. The (classical) constraint algebra serves a dual purpose - generating gauge transformations in the phase-space and encoding spacetime covariance of the theory. A modification of the constraint algebra can break this correspondence. This in effect could mean that the classical spacetime concepts - the notion of black hole horizon for instance - no longer hold. We have provided an explicit example to show that this, indeed, is the case. We have suggested a modification of the classical horizon condition that leads to consistent results. In the case where the corrections do not modify the constraint algebra, classical spacetime notions are valid and lead to a mass threshold for black holes, and small corrections to Hawking temperature (detailed calculations can be found in [6]).

2. Classical theory and constraint algebra

We begin with classical theory for spherical symmetry using Ashtekar variables, which consist of $\mathfrak{su}(2)$-valued connection $A_a^i(x) = \Gamma_a^i(x) + \gamma K_a^i(x)$ and densitized triads $E_i^a(x)$. Here, Γ_a^i is the spin connection, K_a^i is related to the extrinsic curvature, γ is the Barbero-Immirzi parameter and E_i^a determines the three metric g_{ab}. Hamiltonian formulation leads to the Hamiltonian and the diffeomorphism constraints (and an additional Gauss constraint in $\mathfrak{su}(2)$ variables). After solving the Gauss constraint, one is left with a set of two canonical pairs (see [7, 8] for details):

$$\{K_x(x), E^x(y)\} = 2G\delta(x,y) \quad \text{and} \quad \{K_\varphi(x), E^\varphi(y)\} = G\delta(x,y)\,. \tag{1}$$

The spherically symmetric metric in terms of these variables is given by:

$$\mathrm{d}s^2 = -N^2\mathrm{d}t^2 + \frac{E^{\varphi\,2}}{|E^x|}(\mathrm{d}x + N^x\mathrm{d}t)^2 + |E^x|\mathrm{d}\Omega^2, \tag{2}$$

where $N(t,x)$ is the lapse function and $N^x(t,x)$ is the shift vector. The Hamiltonian constraint $H[N]$, and the diffeomorphism constraint $D[N^x]$ (including generic matter contributions) are:

$$\mathbf{H}[N] = -\frac{1}{2G}\int \mathrm{d}x\, N|E^x|^{-\frac{1}{2}}\left[K_\varphi^2 E^\varphi + 2K_\varphi K_x E^x + (1-\Gamma_\varphi^2)E^\varphi + 2\Gamma_\varphi' E^x - 8\pi G E^\varphi |E^x|\rho\right] \approx 0, \tag{3}$$

$$\textit{and} \quad \mathbf{D}[N^x] = \frac{1}{2G}\int \mathrm{d}x\, N^x\left[2E^\varphi K_\varphi' - K_x E^{x\prime} - 8\pi G E^\varphi\sqrt{|E^x|}J_x\right] \approx 0\,, \tag{4}$$

where $\Gamma_\phi = -E^{x\prime}/2E^\phi$. The gravitational part of these satisfies the *surface deformation algebra*:

$$\{H_{\rm grav}[N], D_{\rm grav}[N^x]\} = -H_{\rm grav}[N^x N'], \quad \textit{and} \tag{5}$$

$$\{H_{\rm grav}[N], H_{\rm grav}[M]\} = D_{\rm grav}\left[|E^x|(E^\varphi)^{-2}(NM' - MN')\right]. \tag{6}$$

As discussed in [9], Eqs. (5) and (6) imply that dynamics takes place on spacelike hypersurfaces.

3. Inverse triad corrections and constraint algebra

Next, we have considered LQG effect in the form of the inverse triad corrections. These arise when quantizing the inverses of triad operators appearing in the Hamiltonian. Since triads have a discrete spectrum containing zero, they do not have well defined inverses. Techniques exist [10] reproducing the inverse in the classical limit but implying corrections in the quantum domain. We have represented the effective inverse as $1/E \to \alpha(E)/E$, (see [11] for derivation of $\alpha(E)$ and Figure 1 for its plot).

The gravitational part of the Hamiltonian constraint given in Eq. (3) thus, becomes:

$$H^Q_{\rm grav}[N] = -\frac{1}{2G}\int \mathrm{d}x\, N\left[\frac{\alpha K_\varphi^2 E^\varphi}{|E^x|^{\frac{1}{2}}} + 2\bar\alpha K_\varphi K_x|E^x|^{\frac{1}{2}} + \alpha(1-\Gamma_\varphi^2)\frac{E^\varphi}{|E^x|^{\frac{1}{2}}} + 2\bar\alpha\Gamma_\varphi'|E^x|^{\frac{1}{2}}\right], \tag{7}$$

where for generality, different powers of E^x have been corrected by different functions α and $\bar\alpha$. (No corrections are included with spin connection Γ_ϕ, since it does not lead to anomaly free algebra [12]). For the quantum corrected Hamiltonian, the constraint algebra turns out to be:

$$\{H^Q_{\rm grav}[N], D_{\rm grav}[N^x]\} = -H^Q_{\rm grav}[N^x N'], \quad \textit{and} \tag{8}$$

$$\{H^Q_{\rm grav}[N], H^Q_{\rm grav}[M]\} = D_{\rm grav}\left[\bar\alpha^2|E^x|(E^\varphi)^{-2}(NM' - MN')\right]. \tag{9}$$

Journal of Physics: Conference Series **484** (2014) 012075 doi:10.1088/1742-6596/484/1/012075

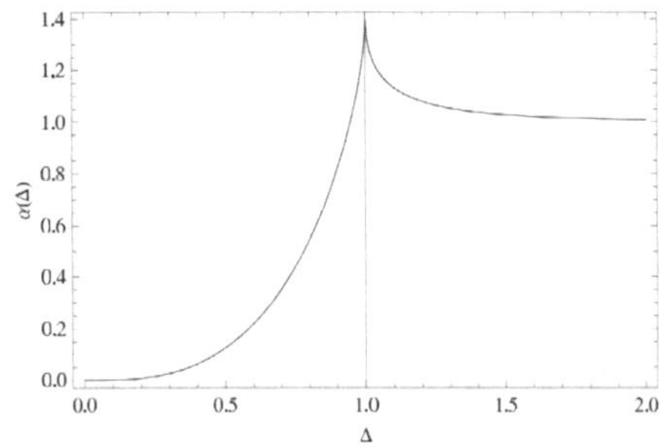

Figure 1. The correction function $\alpha(\Delta)$, where Δ is taken relative to $\Delta_* := \sqrt{\gamma/2}\ell_{\rm P}$. Note that $\alpha(\Delta) \to 1$ for $\Delta \gg \Delta*$.

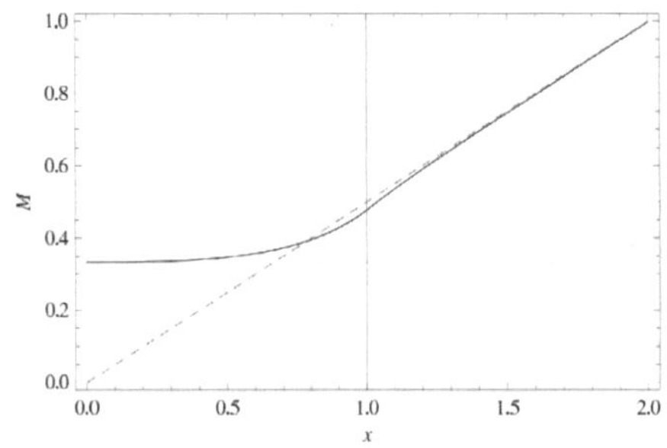

Figure 2. Horizon curve: Right hand side of $M = x/2f_\alpha(x)$ (solid) and the classical horizon curve (dashed).

If $\bar{\alpha} = 1$, the algebra is classical, implying that dynamics takes place on spacelike hypersurface. In general, $\bar{\alpha} \neq 1$ and the algebra is modified and one would like to explore its implications with regard to the spacetime covariance property of the quantum theory. We have considered two cases: (i) $\bar{\alpha} = 1$, in which, only the dynamics is modified but spacetime covariance properties are classical, and (ii) $\bar{\alpha} = \alpha$, in which, both the dynamics and spacetime properties change (situations more general can be related to these two cases [6]).

Case (i): $\bar{\alpha} = 1$

In this case, in vacuum, one can solve the constraints and the equations of motion (in the static gauge $K_x = K_\phi = N^x = 0$ and $E^x = x^2$) and obtain the quantum corrected version of the classical Schwarzschild solution (note that not all quantum gravity effects have been included) as:

$$\mathrm{d}s^2 = -g_\alpha^2 \left(1 - \frac{2Mf_\alpha}{x}\right) \mathrm{d}t^2 + \left(1 - \frac{2Mf_\alpha}{x}\right)^{-1} \mathrm{d}x^2 + x^2 \mathrm{d}\Omega^2 . \qquad (10)$$

Here, $f_\alpha = 1/g_\alpha$ are functions correcting the classical E^ϕ and N, and tend to one in the limit $x \to \infty$ giving back the classical solution. The horizon equation is given implicitly by $M = x/2f_\alpha$, and gives a mass threshold below which black hole does not form (see Figure 2), consistent with the observation of [13]. Away from deep quantum regime, one can repeat Hawking's analysis to find modified Hawking temperature $k_{\rm B}T = \hbar\alpha(x_{\rm h})g_\alpha^2(x_{\rm h})/8\pi M$, which goes to the standard expression when the horizon $x_h \gg \sqrt{\gamma/2}l_p$. More importantly, since the constraint algebra is classical, spacetime covariance properties persist. This can be verified by doing a coordinate transformation to the Painlevé-Gullstrand coordinates, in terms of which the metric becomes:

$$\mathrm{d}s^2 = -\frac{1}{f_\alpha^2}\left(1 - \frac{2Mf_\alpha}{x}\right) \mathrm{d}T^2 + f_\alpha^{-2}\mathrm{d}x^2 + 2f_\alpha^{-2}\sqrt{f_\alpha^2 - 1 + \frac{2Mf_\alpha}{x}}\mathrm{d}T\mathrm{d}x + x^2\mathrm{d}\Omega^2 , \qquad (11)$$

and checking that for this metric, the constraints are identically satisfied (see [6] for details).

Case (ii): $\bar{\alpha} = \alpha$

As already noted, in this case, both the Hamiltonian and the constraint algebra get modified. Assuming that here also the usual spacetime properties continue to hold, we solve the constraints and the equations of motion to find the 'metric' corresponding to the classical Schwarzschild solution:

$$\not{d}s^2 = -\alpha^{-2}\left(1 - \frac{2M}{x}\right) \not{d}t^2 + \left(1 - \frac{2M}{x}\right)^{-1} \not{d}x^2 + x^2 \not{d}\Omega^2 . \qquad (12)$$

Following the previous procedure, if we now go to the corresponding 'Painlevé-Gullstrand' coordinates, the 'metric' becomes (slash on ds and the quotes indicate that these quantities do not have the usual geometric meaning as made clear from statement after Eq. (13)):

$$\not{d}s^2 = -\alpha^{-2}\left(1-\frac{2M}{x}\right)\not{d}T^2 + \alpha^{-2}\,\not{d}x^2 + 2\alpha^{-2}\sqrt{\alpha^2 - 1 + \frac{2M}{x}}\,\not{d}x\,\not{d}T + x^2\,\not{d}\Omega^2. \qquad (13)$$

It turns out that Eq. (13) does not solve the constraints, thus, showing that for modified constraint algebra, coordinate transformations do not map solutions of constraints to other solutions.

4. Implications and conclusions

Having shown that the conventional spacetime properties (or even a concept like metric) do not hold for modified constraint algebra, the obvious next question is: What happens to the concept of horizon? Do the entropy calculations in LQG [14, 15, 16], based on classical concept of horizon, continue to hold? To explore such questions, one can include matter (in the form of scalar field, say) as a second order perturbation to the classical solutions and evaluate various horizon conditions like the trapping horizon [17] and the isolated horizon [18]. For classical spherically symmetric backgrounds (where classical algebra holds), these two notions of horizon are equivalent. However, for modified algebra the two notions of horizon turn out to be inequivalent (details are omitted here, for want of space, can be found in [6]).

In fact, the trapping horizon, given by $E^x/(E^\varphi)^2 - (N^x/N)^2 = 0$, becomes gauge dependent giving different results in different 'coordinates' (defined through the choice of N, since as seen above, the concept of coordinates is not meaningful in this case). The trapping horizon condition can be modifed in a simple way to $E^x/(E^\varphi)^2 - (N^x/\bar{\alpha}N)^2 = 0$, such that different coordinate systems give equivalent results (modification obtained looking at equations of motion and not based on some understanding of horizons in quantum gravity). With such a scenario, it seems that black hole entropy calculations, where the isolated horizon conditions are fixed before quantization, miss out some quantum features relevant for the horizon. This, for instance, can have bearing on the value of Barbero-Immirzi parameter. But more crucially, it is to be noted that for corrected algebra even the concept of isolated horizon is difficult to define suggesting that a reconsideration of entropy calculations might be needed. On the other hand, we also had an example where the quantum correction did not modify the algebra and classical definitions continue to give consistent results, and therefore, entropy calculations in such a situation would also be consistent. However, here, only one type of quantum gravity correction is considered. More generally, there will be holonomy and quantum back reaction effects (which so far have been difficult to implement in full generality in inhomogeneous situations) and which most likely would lead to a modified algebra. In such a case, one might need a quantum definition of the mapping from phase-space to spacetime and even a quantum notion of classical concepts of geometry.

Acknowledgements

The author would like to thank Martin Bojowald for suggesting improvements to the draft. This work was supported under Max Planck-India Partner Group in Gravity and Cosmology.

References

[1] Rovelli C 2004 *Quantum Gravity* (Cambridge University Press)
[2] Thiemann T 2007 *Introduction to Modern Canonical Quantum General Relativity* (Cambridge University Press) [gr-qc/0110034]
[3] Ashtekar A and Lewandowski J 2004 *Class. Quant. Grav.* **21** R53 [gr-qc/0404018]
[4] Bojowald M 2001 *Phys. Rev. Lett.* **86** 5227 [gr-qc/0102069]

[5] Ashtekar A, Pawlowski T and Singh P 2006 *Phys. Rev. Lett.* **96** 141301 [gr-qc/0602086]
[6] Bojowald M, Paily G, Reyes J and Tibrewala R 2011 *Class. Quant. Grav.* **28** 185006 [arXiv:1105.1340]
[7] Bojowald M 2004 *Class. Quant. Grav.* **21** 3733 [gr-qc/0407017]
[8] Bojowald M and Swiderski R 2006 *Class. Quant. Grav.* **23** 2129 [gr-qc/0511108]
[9] Hojman S, Kuchař K, and Teitelboim C 1976 *Ann. Phys.* **96** 88
[10] Thiemann T 1998 *Class. Quant. Grav.* **15** 1281 [gr-qc/9705019]
[11] Bojowald M, Harada T and Tibrewala R 2008 *Phys. Rev. D* **78** 064057 [arXiv:0806.2593]
[12] Bojowald M, Reyes J and Tibrewala R 2009 *Phys. Rev. D* **80** 084002 [arXiv:0906.4767]
[13] Bojowald M, Goswami R, Maartens R and Singh P 2005 *Phys. Rev. Lett.* **95** 091302 [gr-qc/0503041]
[14] Ashtekar A, Baez J, Corichi A and Krasnov K 1998 *Phys. Rev. Lett.* **80** 904 [gr-qc/9710007]
[15] Kaul R and Majumdar P 1998 *Phys. Lett. B* **439** 267 [gr-qc/9801080]
[16] Engle J, Noui K and Perez A 2010 *Phys. Rev. Lett.* **105** 031302 [arXiv:0905.3168]
[17] Hayward S 1994 *Phys. Rev. D* **49** 6467
[18] Ashtekar A and Krishnan B 2004 *Living Rev. Rel.* **7** 1 [gr-qc/0407042]

Quantum gravitational collapse

C Vaz

Department of Physics, University of Cincinnati, Cincinnati, OH 45221-0011, USA

E-mail: `Cenalo.Vaz@uc.edu`

Abstract. We have outlined a canonical quantization of the LeMaître-Tolman-Bondi models, which describe the collapse of inhomogeneous, non-rotating dust. Although, there are many models of gravitational collapse, this particular class of models stands out for its simplicity and the fact that both black holes and naked singularity end states may be realized on the classical level, depending on the initial conditions. We have solved the Wheeler-DeWitt equation exactly after regularization on a spatial lattice, and discussed some consequences of the quantization. The solutions describe Hawking radiation and provide an elegant microcanonical description of black hole entropy, but they also raise some very interesting questions concerning the nature of gravity's fundamental degrees of freedom.

1. Introduction

So long as no generally agreed upon theory of quantum gravity exists, it is important to examine the quantization of particular models. This contribution addresses the quantization of the LeMaître-Tolman-Bondi (LTB) solutions, which describe the classical collapse of spherically symmetric, inhomogeneous, self-gravitating dust. The classical solutions were originally introduced by G LeMaître [1] to study cosmology, where it has found interesting applications [2]. Our principal interest here will be to develop the Hamiltonian formalism for both the classical and quantum LTB models, as such we will describe a generalization of work by Kuchař [3], who has developed a midi-superspace quantization of the Schwarzschild black hole.

In any study of quantum collapse, one would like to be able to: (a) describe Hawking radiation during collapse for black holes and naked singularities, (b) recover a microscopic model of the eternal black hole that leads naturally to the Bekenstein-Hawking entropy, and (c) understand the role played by the cosmological constant (e.g., what is the origin of the Hawking-Page transition [4] for a negative cosmological constant and how do we effectively deal with two horizons when we have a positive cosmological constant.) In this paper, we have briefly addressed the first two of these, and argued that the presence of a (negative) cosmological constant requires a rethinking of the statistics obeyed by the fundamental degrees of freedom describing the black hole.

2. Classical models

The interior of a spherically symmetric, non-rotating and self-gravitating dust ball is described by the stress energy tensor $T_{\mu\nu} = \varepsilon\, u_\mu u_\nu$ and the line element:

$$ds^2 = d\tau^2 - \frac{R'^2}{1+2E(\rho)}d\rho^2 - R^2 d\Omega^2, \qquad (1)$$

where $R = R(\tau, \rho)$ is the area radius and $\varepsilon = \varepsilon(\tau, \rho)$ is the energy density, satisfying:

$$\dot{R}(\tau, \rho) = -\sqrt{2E(\rho) + \frac{F(\rho)}{R(\tau, \rho)}}, \quad and \quad \varepsilon(\tau, \rho) = \frac{F'(\rho)}{R(\tau, \rho)^2 R'(\tau, \rho)}. \tag{2}$$

The initial data are completely specified by the two functions $F(\rho)$ and $E(\rho)$ representing respectively, the weighted mass and the total energy contained within a shell of radius ρ, and generally referred to as the mass and energy functions. An analysis of the classical solutions for these models can be found in [5]. Here, we have discussed the canonical formalism for the so called "marginally bound" models, defined by $E(\rho) = 0$ [6]. A generalization to non-marginally bound models is found in [7].

The canonical dynamics of the collapsing cloud is described by embedding the spherically symmetric ADM metric (with radial function L) in the LTB spacetime in Eq. (1), and casting the action for the Einstein-dust system in canonical form. The phase-space of non-rotating dust is described by the dust proper time τ, and its conjugate momentum, whereas the gravitational phase-space consists of the configuration space variables (R, L) and their conjugate momenta. Using a version of the canonical transformations developed by Kuchař [3], the configuration variable L is replaced by the mass density $\Gamma = F'$. In terms of the new chart $(\tau, R, \Gamma, P_\tau, P_R, P_\Gamma)$, the momentum and the Hamiltonian constraints read:

$$H_r = \tau' P_\tau + R' \bar{P}_R - \Gamma P_\Gamma' \approx 0, \quad and \quad H = \left(P_\tau{}^2 + \mathcal{F} \bar{P}_R^2\right) - \frac{\Gamma^2}{4\mathcal{F}} \approx 0, \tag{3}$$

where the primes now refer to derivatives with respect to the ADM label r. The algebra of the constraints is not, however, of the standard form (given, for example, in [8]). A short calculation shows that the Poisson bracket of the Hamiltonian with itself vanishes, in contrast with the general case, in which it closes on the momentum constraint. The other brackets coincide with the general case. The transformations generated by the Hamiltonian constraint can, thus, no longer be interpreted as hypersurface deformations. They are in general not orthogonal to the hypersurfaces, but act along the flow lines of dust.

3. Quantization

Applying Dirac's quantization procedure, the Hamiltonian constraint turns into the Wheeler-DeWitt equation and the momentum constriant ensures spatial diffeomorphism invariance. One can show, without much difficulty, that the momentum constraint is satisfied by any wave functional of the general form:

$$\Psi[\tau, R, \Gamma] = U\left(-\frac{i}{2}\int dr \Gamma(r) \mathcal{W}(\tau, R, F)\right), \tag{4}$$

where U is an arbitrary, complex valued function of its argument, provided that $\mathcal{W}$ has no explicit dependence on r. The Hamiltonian constraint requires more work: As usual, it must be regularized and there are factor ordering ambiguities.

Regularization may be performed on a spatial lattice so long as $\Gamma \neq 0$. If, moreover, one demands that the wave-functional is factorizable on the lattice, then it is possible to obtain exact and unique solutions of the Wheeler-DeWitt equation together with a natural Hilbert space measure. One finds in the exterior, the positive energy solutions given by:

$$\Psi[\tau, R, \Gamma] = \prod_j \psi_j(\tau_j, R_j, \Gamma_j) = \prod_j e^{\omega_j b_j} \times \exp\left\{-\frac{i\omega_j}{\hbar}\left[\tau_j \pm 2F_j\left[z_j - \tanh^{-1}\frac{1}{z_j}\right]\right]\right\}, \quad z > 1, \tag{5}$$

where j refers to the lattice site $\omega_j = \sigma\Gamma_j$, σ is the lattice spacing $z_j = \sqrt{R_j/F_j}$ and

$$\Psi[\tau, R, \Gamma] = \prod_j \psi_j(\tau_j, R_j, \Gamma_j) = \prod_j e^{\omega_j b_j} \times \exp\left\{-\frac{i\omega_j}{\hbar}\left[\tau_j \pm 2F_j\left[z_j - \tanh^{-1} z_j\right]\right]\right\}, \quad z < 1, \quad (6)$$

i.e., in the interior. The ψ_j represent shell wave functions, *i.e.*, wave functions describing shells of collapsing dust. In the exterior, the positive sign represents ingoing waves and the negative sign represents outgoing waves. The reverse holds in the interior. For these marginal models, the measure is uniquely given by $\mu_j = z_j$ up to a constant scaling.

4. Hawking radiation

This can be demonstrated by taking an approach that closely parallel to Hawking's original work by imagining a pre-existing, eternal black hole surrounded by tenuous dust, so that the mass function has the form $F(r) = 2M\Theta(r) + f(r)$, where $\Theta(r)$ is the Heaviside function and $f(r)$ represents the dust perturbation. Then the wave functional factorizes into a piece that represents the black hole at the centre and another that represents the dust propagating in the black hole geometry. A simple calculation of the Bogoliubov coefficient then shows that [9, 10]:

$$\left\langle \text{in}|\widehat{N}_{\text{out}}|\text{in}\right\rangle = |\beta(f,\omega)|^2 \approx \prod_j \frac{2\pi M}{\Delta f_j}\left[\frac{1}{e^{8\pi M \Delta f_j} - 1}\right], \quad (7)$$

where $\Delta f_j = f_j - f_{j-1}$ is the mass energy associated with the lattice site j. This is interpreted as the eternal black hole being in equilibrium with a thermal bath at the Hawking temperature $(8\pi M)^{-1}$.

An alternative approach, better adapted to *collapse*, was considered in [11]. By matching the shell wave functions describing gravitational collapse across the apparent horizon, it was shown that an ingoing wave on one side of the apparent horizon is necessarily accompanied by an outgoing wave on the other side. Furthermore, the relative amplitude of the outgoing wave is suppressed by the square root of the Boltzmann factor at the "Hawking" temperature of the shell. Strictly speaking, the Hawking temperature $T_H = (8\pi M)^{-1}$, refers to an eternal black hole of mass M. The temperature appearing in the Boltzmann factor from matching shell wave functions across the horizon is $T_H = (4\pi F)^{-1}$, where F is the mass function and represents twice the mass contained within the shell. Diffeomorphism invariant wave functionals describing the collapse can also be matched and yields the same picture, but now the relative amplitude of the outgoing wave functional to the ingoing one is given by $e^{-S/2}$, where S is the entropy of the final state black hole. These results can be extended to non-marginal models with or without a (negative) cosmological constant and in any dimension.

5. Entropy

One can think of an eternal black hole as a single shell represented by a mass function $F(r) = 2M\Theta(r)$. Then the black hole is, in fact, described by a free Klein-Gordon like equation, which is hyperbolic in the interior and elliptic in the exterior. The stationary states (in the interior) describe a spectrum of the form [12, 13]:

$$4GM_n L_{h,n} = A_{\text{Pl}}\left(n + \frac{1}{2}\right), \quad (8)$$

where A_{Pl} is the Planck area, n is a whole number and L_h is the proper radius of the black hole horizon. This expression holds true even in the presence of a (negative) cosmological constant and in any dimension. In 3+1 dimensions, and in the absence of a cosmological constant,

$L_h = \pi M$ and hence, one finds that the *area* of the horizon is quantized in half integer units. This is no longer true in the presence of a (negative) cosmological constant.

Now, we have treated the black hole as a single shell formed as the end state of many shells that have collapsed. Regardless of their history, we have assumed that each of the shells then occupies only the levels given above, contributing some multiple of the Planck area to the total horizon area A, of the final state. Thus, if the distribution is such that N_n shells occupy level n, then the horizon area becomes:

$$\frac{A}{4} = A_{\rm Pl} \sum_n \left(n + \frac{1}{2}\right) N_n, \tag{9}$$

and the single shell solution of Eq. (8) is to be interpreted as an excitation by $N = \sum_n N_n$ shells. It is now a simple matter to count the number of states Ω, leading to a black hole of a given horizon area A, assuming that they are distinguishable. Holding A fixed and extremizing the entropy $S = k_B \ln \Omega$ with respect to the number of shells, one easily recovers the Bekenstein-Hawking area law. In addition, the area quantization in Eq. (8) ensures that the entropy is effectively quantized in units of the Planck area.

The formalism can be used to describe AdS black holes as well. In 2+1 dimensions, one obtains $L_h = \pi l/2$, where l is the AdS length, and therefore, Eq. (8) leads to mass quantization [14]. Likewise, in higher dimensions, for large black holes, with a horizon radius that is much larger than the AdS length scale, it is not the area but the mass of the black hole that is quantized, whereas area quantization holds in the opposite limit [13]. The two limits are separated by the Hawking-Page phase transition [4].

Comments

When mass and not area quantization holds, Bose and not Boltzmann statistics must be employed to recover the Bekenstein-Hawking entropy. Thus, the Hawking-Page transition describes a change in the nature of the fundamental degrees of freedom from geometric for small AdS black holes to field theoretic for large AdS black holes. This calls for a better understanding. It would be useful to examine the statistical mechanics of AdS black holes near the Hawking-Page transition point.

Hawking radiation from naked singularities, important for cosmic censorship, should be interesting to examine. Singularity avoidance could also be examined from this point of view via the construction of wave-packets, but this also remains to be done. Finally, we do not quite understand how to effectively deal with collapse in the presence of a positive cosmological constant.

References

[1] LeMaître G 1933 *Ann. Soc. Sci. Brux. A* **53** 51 (For English translation, see 1997 *Gen. Rel. Grav.* **29** 641)
[2] Krasinski A 1997 *Inhomogeneous Cosmological Models* (Cambridge University Press)
[3] Kuchař K V 1994 *Phys. Rev. D* **50** 3961
[4] Hawking S W and Page D N 1983 *Commun. Math. Phys.* **87** 577
[5] Joshi P and Singh T P 1995 *Phys. Rev.* **51** 6778
[6] Vaz C, Witten L and Singh T P 2001 *Phys. Rev. D* **63** 104020
[7] Kiefer C, M-Hill J and Vaz C 2006 *Phys. Rev. D* **73** 044025
[8] Kiefer C 2004 *Quantum Gravity* (Clarendon Press, Oxford)
[9] Vaz C, Kiefer C, Singh T P and Witten L 2003 *Phys. Rev. D* **67** 024014
[10] Kiefer C, Mueller-Hill J, Singh T P and Vaz C 2007 *Phys. Rev. D* **75** 124010
[11] Vaz C and Wijewardhana L C R 2010 *Phys. Rev. D* **82** 084018
[12] Vaz C and Witten L 2001 *Phys. Rev. D* **64** 084005
[13] Vaz C and Wijewardhana L C R 2009 *Phys. Rev. D* **79** 084014.
[14] Vaz C, Gutti S, Kiefer C, Singh T P and Wijewardhana L C R 2008 *Phys. Rev. D* **77** 064021.

AUTHOR INDEX

AUTHOR INDEX

Institute of Physics
Dirac House, Temple Back
Bristol BS1 6BE UK

CURRAN ASSOCIATES INC
proceedings.com